Jargon

an informal dictionary of computer terms

Other books by Robin Williams

The Little Mac Book

The Mac is not a typewriter

The PC is not a typewriter

PageMaker: An Easy Desk Reference (Mac and Windows versions)

Tabs and Indents on the Macintosh

Other books by Steve Cummings

DeskJet Unlimited

Understanding Quicken

Jargon

an informal dictionary of computer terms

Robin Williams

with Steve Cummings

Peachpit Press ▾ Berkeley ▾ California

Jargon, an informal dictionary of computer terms

Peachpit Press, Inc.
2414 Sixth Street
Berkeley, California 94710
510.548.4393
510.548.5991 fax

ISBN 0-938151-8-43

0 9 8 7 6 5 4 3 2 1
Printed and bound in the United States of America.

This book is dedicated,
with great love,
respect, and
admiration,

to BMUG,

for what they gave to me,
and for what they give to the world.

Acknowledgments

Shannon Williams, my little sister, for everything imaginable (if you like my books, you should thank her for keeping me together enough to write them)

Steve Cummings, for adding so much value to this book

Carole Quandt, for proofreading and editing and suggestions (but I added and edited so much after she proofread that all typos and grammatical faux pas are entirely my fault)

John Grimes, for the wonderful cartoons

Guy Kawasaki, definitions and moral support

Clay Gordon, tech edit and friendship

Bob Weibel, tech edit

Doug McCasland, editing and tech edit

Pam Mason, for her great appendix

Harrah Argentine, illustrations under font

Olav Martin Kvern, bailing me out of corrupted file trouble

Barbara Sikora, Drew Cronk, Janet Butcher, and **Pam Mason** for writing some of the definitions

Sam Hunting, Dave deBronkart, Tom Hagan, Jim Alley, and **Glenn Fleishman** for contributing to several of the definitions

Terry O'Donnell from Adobe Systems, for clarifying *font* vs. *typeface*

Andy Baird, moral support

Phillip Russell, whom I have never met, but who gives me tremendous moral support (Phillip, I do hope this book meets your expectations!)

Jon Winokur, for the perfect quote

Hassan Herz of Typecast, Inc., for imaging this book so efficiently and kindly

Cary Norsworthy at Peachpit, for pulling this all together and being so patient with me

Ted Nace, publisher of Peachpit Press (Ted, you're too good to me)

Scarlett Williams, for the font design

My kids, for being so patient with me for the past 18 months (did they have a choice?): **Ryan** (now 15 years old), **Jimmy** (11), **Scarlett** (7), and **Brian** (18, our foreign exchange student for the year). Do you think they will ever thank me for all the things they learned this year, like how to do their own laundry, cook their own meals, mend their clothes, grocery shop, clean the house, etc. etc. etc.?

The following people, none of whom I have met (isn't it an interesting phenomenon that we can make friends electronically), added and added and added words they felt needed to be defined, which is what made this book 688 pages instead of the original 120, which also made the book a year late in the production schedule, which also made the book much more useful. Thank you!

John Merideth
Herb Schulsinger
Jim DeWitt
Jehuda Saar
John Holland
Alleta Baltes
Brad Mohr
Erik C. Thauvin
Richard M. Parres
Glenn Brown
Pam Michaelson
Brian Carter
Etienne Grosjean
Ken Schneider
Lee Hinde
Jeffrey Kane
Alan Touchberry
Dan Druliner
Michael Lilly
John Thatcher
John Raymonds
Forrest F. Carhart, III
David Bass
Robert L. Hoover

Contents

JUST BEFORE WORK, ROBIN WOULD SLIP INTO THE LADIES ROOM AND ADJUST HER PERSONALITY.

Personal note

This book was not easy to write. I'm not a very technical person, and I had to struggle with many of these esoteric concepts. As I was writing this (and constantly complaining) many people would tell me (especially male technonerds) that no one needs to know this stuff anyway. But computers are becoming part of everyone's lives and we are all bumping into technical terms in magazines, catalogs, promotional literature, manuals, and so on. I have read letters from readers to magazines complaining that the articles don't explain the acronyms, but y'know what? Even if they told you what the acronym stood for, it wouldn't make any more sense—big deal that SCSI stands for Small Computer Systems Interface.

I wrote this book because I was frustrated with every computer dictionary I own (which is 12, plus 2 on disks) and all the glossaries I had collected (which is about 20). The explanations were usually so brief, they assumed I knew a host of other things, and they never told me how this information was connected to anything else in my life or why it might be important for me to know this. After reading this definition for SIMM, (which was not alphabetized under "SIMM" but was under "Single In-line Memory Module," as if I would know what SIMM stood for), I realized I had a personal obligation to write definitions that simple people like me and my mother could understand. This was the definition for SIMM:

> *A memory module that contains the chips needed to add 256K or 1M of random-access memory to your computer. A SIMM plugs into a motherboard or a logic board.*

This definition assumes I know what *memory* means, what a *memory module* is, what a *chip* is, what *256K* could be, what *1M* stands for, what *random-access* means, what a *motherboard* is, and what a *logic board* is. If I knew what all those things were, I wouldn't have to look up Stupid In-line Memory Module in the first place. (And I don't understand why computer dictionaries can't start a paragraph with a complete sentence.)

Oh, I shouldn't be so snotty and crabby—a few of the existing dictionaries are excellent, and I can recommend the ones that helped me tremendously and that you should use if you need more than what these definitions offer. And anyway, this book has a different purpose than most dictionaries. This book is not meant to be the ultimate reference—it's purpose in life is to provide you, assuming you are a beginning to average computer user, with the information necessary to just get your everyday work accomplished. And (I do hope) you may even understand *why* you are doing what you are doing.

Steve Cummings (whom I have never met!) added so much value to this book with his PC definitions, as well as many of the generic defs. He was truly wonderful to work with, and we were able to hack away at each other's writing without remorse. I would be happy to work with him again.

Note from Steve: All I want to say is that I had a blast working on this book. Robin is as funny, gracious, and generous a collaborator as anyone could ask for, and I'm lucky I got to help out. Anytime, Robin, anytime.

Technical note

This book covers terminology for all computers. But if you are strictly a PC user, you probably don't give a hoot about the Mac references, and I know (cuz I'm one of 'em) Mac users don't like to slog through PC information. So every definition has been carefully edited so you won't accidentally read something distasteful or useless.

If there is no icon, the paragraph applies to all computers.

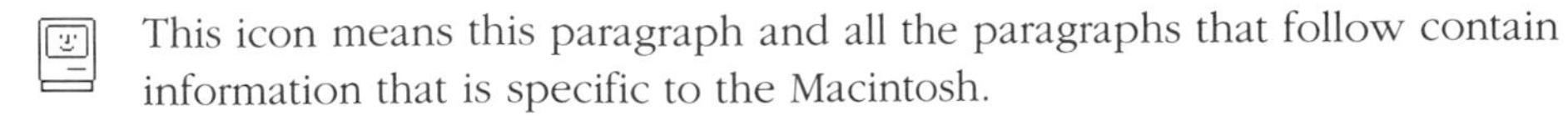

This icon means this paragraph and all the paragraphs that follow contain information that is specific to the Macintosh.

This icon means this paragraph and all the paragraphs that follow contain information that is specific to the PC ("PC" as defined below).

Throughout this book, we use "PC" as the generic term for all computers patterned after the original IBM PC, meaning that they can use the same software. Since 1981 when the original IBM PC came out, hundreds of companies have made countless computer models that work basically like the IBM PC. No matter what the official model name, all of these machines are PCs. IBM itself stopped using the name PC in 1987, but IBM's newer models still qualify as PCs because they run PC software.

Likewise, when we use the term "DOS," we mean any brand and any version of DOS for PCs. DOS, the Disk Operating System, is by far the most popular kind of master control software, or operating system, used with PCs. You might have MS-DOS—the most common brand name—or PC-DOS, IBM DOS, or DR DOS, but fundamentally, they all work the same way. And even though there are some important differences between specific versions of DOS—between, say, MS-DOS 2.11 and MS-DOS 6—the basic commands are always the same.

An *italic* word means I realize you might not know that term and it is defined as a main entry (but some words I used so often that I thought it would be irritating if I italicized it every time, words like "hard disk" or "Windows"). A word in "quotation marks" means it might be a new term to you, but it is not defined elsewhere—its definition is probably right there in that sentence.

I indexed this dictionary. I know, you might think it is a dumb idea to index a dictionary, but when I was trying to figure out what 32-bit addressing was, I found it helped me tremendously to read every definition in every book that referred to that topic. Without an index, how would you know all the possible definitions that contain information about graphic file formats or various features of hard disks or the different kinds of buses and cartridges and fonts? And how would you find all the words that don't have a separate entry but are defined as part of another definition? If you don't find what you need in the main entries, definitely try the index. Personally, I would start with the index first.

Anyway, I do hope you find this book useful. Let me know what you like or dislike and we'll make it better next time!

Have you ever wondered...

Why the ruler on your screen is not exactly one inch? See *resolution*.

Why you keep getting the message that you are "out of memory" when you know your hard disk is not full? See *RAM*.

How to install a card yourself? See *add-in board*.

If you should install Windows on your PC? See *Windows*.

When mixing red and green makes yellow? See *additive color*.

Why some fonts look smooth on your screen and some don't? See *font*.

Why you don't have an option in your Memory Control Panel to use virtual memory? See *32-bit clean*.

Where those alternate sorts of characters are that you see in other people's documents? See *Key Caps (Mac)* or *Character Map (PC)*.

Why some fonts have city-names and some don't? See *fonts (city-named)*.

What the difference is between **lines per inch** and **dots per inch**? See *lines per inch*.

Why small letters are called "lowercase"? See *lowercase*.

How to giggle online? See *baudy language*.

Why you should never move your hard disk while your computer is on? See *park*.

What this means when you see it in a menu: … ? See *ellipsis*.

Who is the best publishing house in the world? See *Peachpit Press*.

What is the important and significant difference between round radio button and square checkbox buttons? See *checkbox*.

What my older brother looks like? See *propeller head*.

Who really owns your software (it's not *you!*)? See *license agreement*.

Why you can't type a colon in a file name? See *illegal character*.

What that little piece of plastic is that came in the box with your Macintosh (don't throw it away!)? See *programmer's switch*.

If a monitor has only three colors in it (red, green, and blue) how does it show black or white? See *RGB*.

How to take a picture of the screen? See *screen capture*.

Why photographs are made of dots when they come out of your printer? See *halftone*.

What those flashing lights on your modem box mean? See *modem*.

How to recover as much as possible when you crash? See *Ctrl-Alt-Del (PC)* or *force quit (Mac)*.

How a computer gets PMS? See *Pantone Matching System*.

When a nice arpeggioed chord means Big Trouble? See *Chimes of Doom*.

How to make a movie play when you turn on your Mac? See *Startup Movie*.

Why ATM makes type so smooth on your screen and even on your non-PostScript printer? See *fonts (ATM)*

Who lives in Silicon Forest? See *Silicon Forest*.

How to use the built-in screen saver in Microsoft Word? See *screen saver*.

What to look for when shopping for a VGA monitor? See *VGA*.

If you're using System 6 or 7? See *System 7* for the instant clue.

What to do if things on your computer start acting a little weird? See *virus*.

How to use the keyboard to select and activate buttons? See *Tab key*.

Where to find men's toggle switch? See page 595.

What kind of software can create side-by-side paragraphs easily? See *table.*

What are the really important things to know? See *Three Rules of Life.*

Where all those acronyms come from? See *TLA*.

What is the difference between **margin** and **indent**? See *indent.*

What to do if you Mac is sad? See *Sad Mac*.

Why you can't connect two computers to the same external hard disk? See *SCSI address.*

If you are guilty of using the single most visible sign of unprofessional type? See *smart quotes.*

Why you shouldn't rub balloons on your head while sitting at your computer? See *antistatic device.*

How to make any Mac document a stationery pad? See *template.*

How to prevent your application from crashing so much due to lack of memory, even if you have lots of memory available? See *application heap.*

How to communicate the subtleties of your online message (wink, wink) since the other person can't read your body language? See *baudy language.*

How to get the squiggle over the n in piñata? See *tilde.*

Why we call starting the computer, "booting"? See *boot.*

How to send a file to someone else in a format that you are sure their program can open, but that will save some of the formatting, like bold and italic and font choices? See *Rich Text Format.*

What is the shortcut to selecting and activating buttons in Windows? See *buttons.*

How to get numbers from your document into the Calculator, and from the Calculator into your document? See *Calculator.*

How to make your Mac tell your lover sweet things? See *Startup Sounds.*

How Caps Lock is different from the Shift Lock on a typewriter? See *Caps Lock.*

What tips experts give to help prevent straining your wrists? See *carpal tunnel syndrome.*

What creature says, "Moof!"? See *Dogcow.*

Why you shouldn't compress your only copy of an original file? See *compress.*

If there really was a Murphy who pinpointed one of the laws of life? See *Murphy's Law.*

If there is proof of who really runs the show? See *motherboard.*

The simple trick to killing the WDEF virus? See *WDEF.*

How to speed up your computer a little bit for absolutely free? See *rebuild the desktop.*

How this book was created? See *desktop publishing.*

What makes some ROMs dirty and some clean? See *32-bit addressing.*

How a laser printer actually works? See *laser printer.*

What is the instant clue to tell whether your Macintosh fonts are downloadable or not? See *downloadable font.*

What is the problem with being too close to the front, sides, or back of a monitor? See *ELF.*

What happens late at night on computers across the country? See *electronic sex.*

In this work,
when it shall be found
that much is omitted,
let it not be forgotten
that much likewise
is performed.

Dr. Samuel Johnson
upon completion of his dictionary, 1755

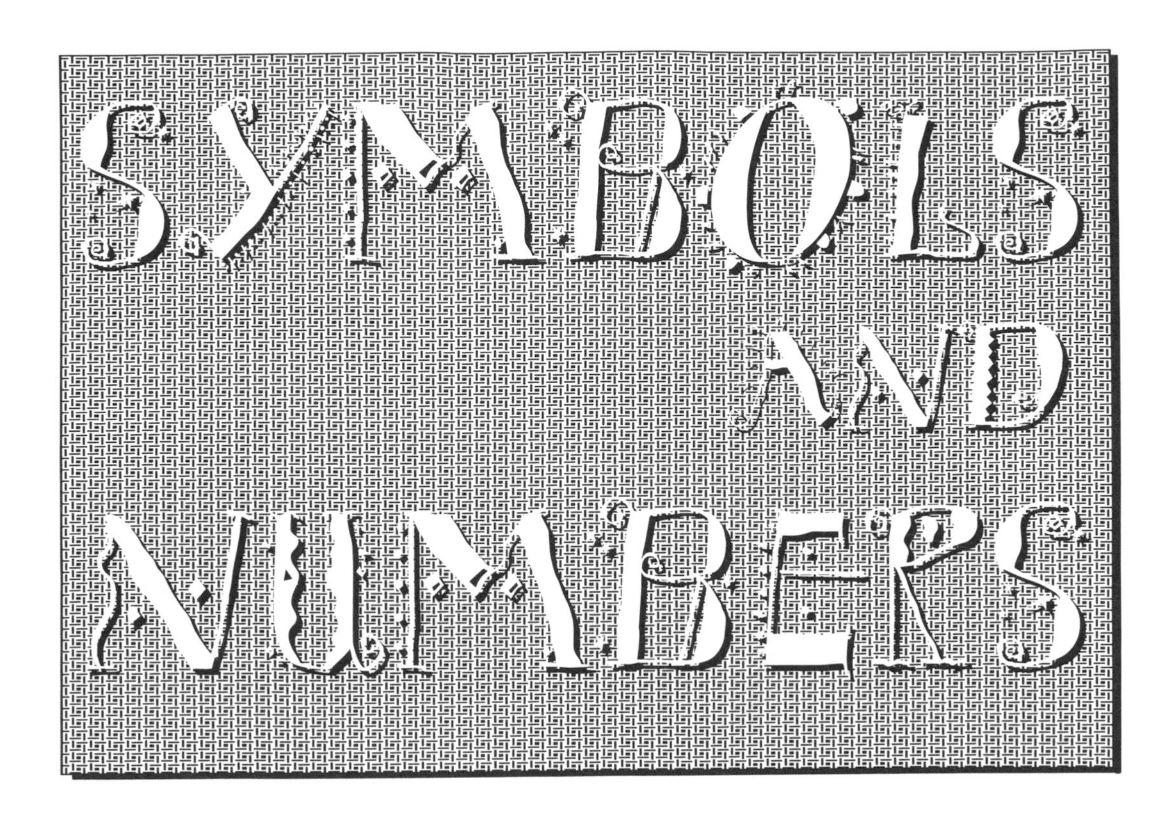

Symbols

The following symbols are all defined alphabetically within this book, unless noted.

ALT	**Alt key**
&	**ampersand**
,	**Apple key**
*	**asterisk** (also known as a **star**; sometimes called a **splat**)
\	**backslash;** see *directory*
•	**bullet**

^	**caret**
^^^	**giggle;** see *baudy language.*
:	**colon**
⌘	**Command Key**
^	**Control Key**
.	**dot** in file names (see *extension*); **point** in *version* numbers
esc	**Escape**
F1, F2, etc.	**F keys, function keys**
I	**I-beam** (pronounced "eye-beam")
⌥	**Option key**
◁	**Power On key**
#	**pound sign,** also known as the number sign or space symbol
?	**question mark cursor;** see *online help*
⇧	**Shift key**
↵	**soft Return**
*	**star;** see *asterisk*
.	**star dot star**
⇥	**Tab**
	Happy Mac
	Sad Mac
?	**?** on startup; see *Sad Mac*
X	**X** (flashing)

020, 030, 040

These numbers, pronounced "oh twenty," "oh thirty," and "oh forty," are nicknames for the *chips* in the Macintosh that are actually named 68020, 68030, and 68040. Please see their definitions later in this section.

1-bit, 2-bit, 8-bit, 16-bit, 20-bit, 24-bit, 32-bit anything

The word *"bit"* attached to a number can refer to several things, depending on the context.

> It can refer to how much information can be stored in the "registers" of the computer's *central processing unit (CPU),* which is an indication of speed and performance; see *8-bit computer.*
>
> It can refer to the size of the *bus* through which the processor sends out the information; see *8-bit computer* and *8-bit slot.*
>
> It can refer to how much information in *memory* a computer can address (use); see *24-bit* and *32-bit addressing.*
>
> It can refer to the depth of the pixels on the monitor, which creates the color and the resolution of the images; see the following definition on grayscale and color.

A *bit* is a tiny electronic signal. In any context, the bigger the bit number, the more powerful, faster, or more colorful the feature.

1-bit, 2-bit, 8-bit, 16-bit, 24-bit, 32-bit color

When referring to grayscale or color, the bit number tells you how many bits of information can be sent to the *pixels* (dots) on the screen. The more bits of info, the more colors or shades of gray you can have. And the more bits, the more memory your computer must have, which means that the number of colors on the screen are not limited by the kind of color monitor you use (except for some PC monitors), but by the amount of video *memory* available. On a color monitor, any bit depth between 1 and 32 is possible, but 8-bit and 24-bit are becoming standard.

Monochrome

If an image is **1-bit,** that means one bit of information is sent to the pixel on the screen. That bit can turn the pixel on (white) or off (black). All 1-bit images are black-and-white. On a **monochrome monitor,** the pixels can't deal with more than just that one bit of data so that's all you can ever get is black and white. (Sometimes "black" is green or amber.)

Grayscale

On a **grayscale monitor,** each pixel can accept from 1 to **8 bits** of data, which will show from 1 to 256 shades of gray.

If there are 2 bits per pixel, there are four possible combinations of on and off: on/on, off/off, on/off, and off/on. Each of these combinations displays a different shade of gray (including black and white).

If there are 4 bits per pixel (2^4), you will have 16 levels of gray

If there are 8 bits per pixel, there are 256 possible combinations (2^8). This is the maximum number of grays possible on any grayscale monitor, which is plenty because our eyes can't distinguish more than that number of grays anyway.

24-bit, 32-bit color

On a **color** monitor, each pixel has three dots arranged in a triad—one red dot, one green, and one blue. Each dot can deal with a maximum of 8 bits, which makes a total of 24 bits per pixel. With the possibility of combining the 256 levels of color in each of the three color dots, **24-bit color** gives you the awesome potential of 16.7 million colors on your screen (256 times 3). Many of these colors differ so slightly that even the most acute observer couldn't tell the difference between them. Simply stated: 16 million colors is more than enough. (How do you get black and white if there are three colored dots? If all dots are on, the pixel is white; if all dots are off, the pixel is black.)

Now, you will often hear of **32-bit color,** which there isn't, really. Those other 8 bits don't offer any extra color, but they do offer the capacity for masking and channeling (see *channels*).

4DOS

4DOS is a software product (for DOS, of course) that makes DOS itself easier to use and more capable in all kinds of little ways. For instance, 4DOS lets you copy a list of individual files with a single command; without 4DOS, you have to type the copy command again for each file. It's inexpensive, and it's *shareware,* so you can try it before you buy it. If you use standard DOS commands with any regularity, 4DOS is a great little program to have.

Technically, 4DOS is a replacement for *COMMAND.COM,* the standard "command processor" that comes with DOS (see *COMMAND.COM*).

8-bit, 16-bit, 32-bit computer

The term **32-bit computer** (or **8-bit** or **16-bit**) refers to the power of the *central processing unit (CPU)*, which is the chip that runs the computer. A CPU that can process 8 *bits* of information at a time (the minimum configuration for a computer) is called an 8-bit computer. If the CPU can process 16 bits at a time, it is (guess!) a 16-bit computer, and the same for 32-bit.

But y'know what? Even though the CPU can internally process 16 or 32 bits at a time, it doesn't mean the rest of the computer can use that big of a chunk of information. After the data gets processed, it has to get on a *bus* and be sent to the rest of the computer. Many computers with 16-bit processors (CPUs) have an external data bus that is only 8 bits wide, so essentially it doesn't matter whether the CPU is 16 bits or not—only 8 bits can get used at a time anyway. (Yes, you can think of the bus as a little thing with wheels that carts information around to where it needs to go.)

The same is true for computers advertised as 32-bit machines; on some computers the external data bus is only 16 bits wide.

Nowadays most computers are made with data buses to match the processor.

8-bit slot, 16-bit slot

PCs have several different types of *expansion slots* (connections inside the computer where you can add *cards* that expand the machine's capability). The slots in the original IBM PC could transfer only 8 *bits* of information at a time between the *motherboard* and the card in the slot; these are **8-bit slots.** The *PC/AT* was the first PC to have **16-bit slots**, and set the standard still followed by most PCs today (see *ISA* and *bus*). The 16-bit slots actually have two separate slots arranged in a line; the slot toward the back of the computer is exactly the same as an 8-bit slot. In other words, you can plug an 8-bit add-in board into a 16-bit slot (but not the other way around, of course). *EISA* slots are 32-bit slots which accept standard 8-bit and 16-bit boards, as well as special 32-bit boards. Then there are the 32-bit *Micro Channel* slots, which take only Micro Channel boards. See also *full-length slot*.

24-bit addressing, 32-bit addressing 32-bit clean, clean ROMs, dirty ROMs

Remember, your *hard disk* is where you store all your files. When you open a file to work, the computer puts a copy of that file into *memory,* sort of like when you take a document out of your filing cabinet (the hard disk) and put it on your desk (which is equivalent to the computer's memory).

You can visualize your computer's memory as billions of little cubicles, each with an *address*. The information that the computer stores in each cubicle is in the form of *bytes,* tiny electronic pulses. (Actually, each pulse is one *bit* of information; eight bits in a row is one byte.)

Now, a **24-bit address** is an address made of 24 numbers in a row. Each number is either a 1 (one) or a 0 (zero) and represents an electronic pulse of on or off. Each pulse is one *bit.* A 24-bit address might look like this: 100100111010110001010001. This long number identifies one of those little cubicles in the memory chip where information is stored.

Well, imagine if the address to your house could have 24 numbers in it. Even using only the numbers one or zero, there are 16 million different addresses available (2^{24}). If a computer's *CPU (central processing unit,* the chip that runs the computer) uses 24-bit addressing, then the computer has the potential to send data to and get data from 16 million different memory locations, or addresses. You can see that the more memory a computer can address, the bigger the projects can be and the more projects you can work on at a time.

Sixteen million bytes (since each memory location holds one byte) is 16 megabytes, so you might think you could address (use) 16 megs of memory as long as you installed the appropriate memory chips.

Well, you *can* address 16 megabytes of memory if the *processor* (the *chip*) in your computer is the *80286* used in the IBM PC/AT and compatibles. The *386* and *486* processors can take advantage of 32-bit addressing (Windows 3.0 limits the addressable memory to 16 megs, but 3.1 can use it all).

But with **Macintoshes,** there's a fly in the ointment. The early Macs that used the 68000 processor (Mac 512, Plus, and SE, as well as the original Classic) only give 4 of those megabytes to *RAM (random access memory),* which is the memory we're concerned with. The other 12 were used to address other parts of the system, like the *motherboard, monitors, ROMs,* etc. Beginning with the Mac II and the 68020 processor, the Mac only took 8 megs for the other stuff, so the machines could support another 8 megs in RAM. **With 24-bit addressing, 8MB (megabytes) of memory is all you can ever use, even if you add more memory chips.** (It's possible to extend that memory with software products like Maxima and Optima from Connectix. See below.)

32-bit addressing

If a computer uses **32-bit addressing,** that means the address can have 32 numbers. With the possibility of each number being either a one or a zero, there are 4 billion different addresses available (2^{32}). Four billion bytes (each memory location holds one byte) is 4,096 *megabytes,* which is 4 *gigabytes.* So with 32-bit addressing you can theoretically address 4 gigabytes of memory.

Alas, at this point in technology there are still some limitations in the hardware, so you can't **really** install and use (address) all 4 gigabtyes. A PC with an 80386 or 80486 microprocessor can address 4 gigabytes of memory all right, but they don't make PCs that can hold that many memory chips. The most a modular Mac (any of the Mac II family) can address is "only" 128 megabtyes, and some of the Quadras can go up to 256 megabytes.

32-bit clean, clean ROMs, dirty ROMs

The Mac, from the beginning, has always had 32-bit addressing capabilities internally. But because of other software and hardware limitations, it could only make use of 24 of those bits (because of the 24-bit address bus—it could only find 16 million different addresses instead of 4 billion). Well, when they were making Macs several years ago, this didn't seem like a big limitation. So software engineers used those other 8 bits (above the 24) that weren't being used by the computer itself. They used them for their own special purposes in the *operating system,* in the ROMs, and even in certain applications and extensions (INITs). As Sam Hunting says, it's comparable to people using the extra four numbers in a "plus-four" zip code for anything they want, such as **I❤U!**.

Then Apple started making bigger Macs (starting with the Mac II) that used a 32-bit bus. So that's great—the hardware (the bus) is now able to use all 32 bits of those addresses, but the ROMs (the software, actually called *firmware*) in the Mac II, and in the IIx, IIcx, and the SE/30 are still using those other "extra" 8 bits for other sorts of special purposes. This wasn't a big problem at that point, because the operating system couldn't figure out how to use all 32 bits anyway (the Post Office didn't need that plus-four yet).

Then Apple made two big changes: System 7 came along, with the ability to go into 32-bit addressing mode; and all the Macs made after the IIcx (like the ci, fx, si, LC, Classic II, Quadras, PowerBooks, Centris, etc.) were built with the new, "clean" ROMs ("clean" because they did not use those other 8 bits for anything). If you have one of these newer machines, you may have noticed the option in your Memory Control Panel to turn on 32-bit addressing. Because these Macs have clean ROMs, they can handle 32-bit addresses, which means they can use much more memory. (If you have installed more than 8 megabytes of memory on a Mac with clean ROMs, you **must** turn on 32-bit addressing or the computer cannot use any of the memory over 8 megs.)

But since the Mac II, IIx, IIcx, and SE/30 have "dirty" ROMs, you can't even turn on 32-bit addressing on those machines—you don't even have the option in the Memory Control Panel. The messages sent by the dirty ROMs would go to the wrong addresses because of the extra information in those last 8 bits. And the computer would get confused and freak out. It's kind of like how the Post Office has gotten dependent on the plus-four zip, so it freaks out over personal messages like **I❤U!** in the last four digits.

If you choose to use 32-bit addressing in a Mac with clean ROMs, you may still have trouble if you use an application, or more likely, an *extension* (in System 6, known as an INIT) in which the programmer took advantage of those 8 extra "bits." That information in those last 8 bits will confuse the System and it will positively crash. If the program does *not* use those extra 8 bits, then it is considered to be "32-bit clean." (If you really need to use that program, turn 32-bit addressing off through the Memory Control Panel.)

So a **32-bit clean ROM** is a *read-only memory chip* (the ones that are installed in the computer and you can't change) in which the software engineers did not monkey with the top eight bits.

A **32-bit clean application** (or extension/INIT) is an application in which the programmers did *not* use the last eight bits (the ones that used to be extra) of any address to program unusual information.

To use 32-bit addressing on a Mac II, IIx, IIcx, or SE/30, get **MODE32.** It's a Control Panel device *(cdev)* from Connectix and is distributed free. It's on every *bulletin board* and *online service* that has software for you to download, or ask your local *user group*. Or call Connectix at 800.950.5880. They have other wonderful software and the best little book in the world (free even) called *The Macintosh Memory Guide,* which describes Macintosh memory (what it is, how to buy it, etc.).

24-bit, 32-bit memory

24-bit and *32-bit memory* are the same as *24-bit* and *32-bit addressing,* since addressing refers to the space in the memory.

24-mode, 32-bit mode

24-bit mode means the computer is using *24-bit addressing,* in contrast to *32-bit mode,* which uses *32-bit addressing.*

32.bis, 42.bis

See v.32 and v.42 in the **V** section. These terms refer to *telecommunication* standards.

32-bit QuickDraw

32-bit QuickDraw, also called Color QuickDraw, gives those Macs that have it their incredible color capabilities (which in turn makes the monitors appear to have higher *resolution*). The original QuickDraw only had the capability to work with eight colors (eight colors total, not *8-bit color*). Color Quickdraw can work with 16.7 million colors. Color QuickDraw is built into the ROMs (the read only memory) of newer Macs. It is also available as a system *extension*.

286, 386, 486

These numbers, **286, 386,** and **486** are short for the *processors* (*chips* that run the computer) that are actually named *80286, 80386,* and *80486.* The definitions for these are further on in this section.

300, 1200, 2400, 9600 bps or baud

These numbers refer to the speed of a computer's *serial port* (or the serial ports of printers and other computer peripheral equipment). It may also refer to the data transmission rate of a *modem,* though many modems are more correctly rated in *bits per second,* rather than baud. Please see the definition for *baud* .

680x0

This number, **680x0** (pronounced "sixty-eight oh x oh") refers to the entire family of Motorola *microprocessors,* as noted below. It's easier to write or say 680x0 than to write or say "68000, 68020, 68030, or 68040."

68000, 68020, 68030, 68040

These numbers designate different *chips,* designed by Motorola, for Macintosh computers. Each Macintosh has one of these chips in it to run the computer. Because this chip runs the computer, it is called the *central processing unit,* also known as the *CPU* or the *microprocessor* (or even just the *processor*). Generally, the higher the number, the faster and more powerful the machine (although the chip itself is not the only thing that determines the speed, and two different computers with the same chip can run at two different speeds).

The 68000 chip is roughly comparable to the Intel 80286. The 68020, 68030, and the 68040 are roughly comparable to the Intel 80386 and the 80486 (IBM PCs and compatibles).

68000

Pronounced "sixty-eight thousand," this chip is installed in the earlier Macs and in the recent low-cost Macs, like the 128, the 512s, the Plus, the SE, the Classic, the Portable, and the PowerBook 100. It does not support *virtual memory* or *32-bit addressing*.

68020

Pronounced "sixty-eight oh twenty" (or often just "oh twenty"), this chip is installed in the Mac II and the Mac LC. It makes these Macs run about five times faster than those with the 68000. If you want to use *virtual memory* with a Mac II, you have to add a *PMMU coprocessor*. On a Mac LC there is no place to add a coprocessor, so you have to upgrade the *microprocessor* itself (the 68020 chip). The 020 is available in speeds of 16, 20, 25, and 33 *MHz*.

68030

Pronounced "sixty-eight oh thirty" (or often just "oh thirty"), this chip is installed in the SE/30, all Mac IIs (except the II, the LC, and the LCII), the Color Classic, and in most PowerBooks. It can support virtual memory without having to add anything else. If you want to have *floating point* math capabilities (important for heavy math, high-end graphics, music, and CAD applications), you need to add a special *coprocessor* (see *68881*). Various 68030 chips run at 16, 20, 25, 33, 40, or 50 *MHz*.

68040

Pronounced "sixty-eight oh forty" (or often just "oh forty"), this chip is installed in the Quadras and the Centris machines. It supports virtual memory and has built-in floating point math capabilities (well, the 68040 in the Centris 610 is a special chip, officially called the LC68040, that has no floating point math circuitry). Various 68040s run at 20, 25, and 33 *MHz*.

68881, 68882

The **68881** and the **68882** are *floating point coprocessors* (also known as math coprocessors) that are used to supplement the *68020* and the *68030* processors (the *68040* has it built in, except for the LC68040 in the Centris 610). The math coprocessor significantly speeds up the applications that use heavy math, high-end graphics, music, and CAD functions.

88000

The **88000** (pronounced "eighty-eight thousand") is a family of *RISC microprocessors,* also from Motorola. Please see *RISC* for more information about this controversial technology.

8088, 8086, 80186, 80286, 80386, 80486

Each of these numbers is the name of a *microprocessor chip* made by the Intel Corporation. All IBM-compatible PCs use a member of the "80x6" family of microprocessors as their main brain. Beginning with the 80286, it's been chic to refer to them by their last three digits, as in "the 286." Each chip in the series is compatible with the ones that came before it, and that's why any PC software that works on an older machine will work on a new one. On the other hand, each new chip generation is faster, more capable, or both, which is one big reason PCs keep getting better. However, software that takes advantage of the features of the newer chips won't run on PCs that use older chips. The latest chip in the series has a name, not a number—it's called the "Pentium."

8088

The **8088** is the chip that launched the PC revolution—IBM designed the original IBM PC around the 8088 microprocessor. The 8088 was also used in the PC/XT. Even by the standards of its day, the 8088 was slow, and it suffered from technical shortcomings that have bedeviled PC software developers ever since.

8086

The **8086** is identical to the 8088 except that it can access twice as much data in one gulp, and that makes it faster. (In technical terms, the 8086 has a 16-bit data *bus,* while the 8088 has an 8-bit bus.) IBM chose to use the 8088 instead of the faster 8086 because in those days the 8-bit support chips for the 8088 were substantially cheaper than the 16-bit support chips required for the 8086. Hence IBM could build a less expensive machine, which was crucial to their marketing plans. A few other manufacturers such as Epson did go with the 8086, making their PCs inherently a little faster.

80186

The **80186** is just an 8086 with some extra circuits added so that fewer "helper" chips are needed. Very few PCs used this chip.

80286

The **80286** was the microprocessor used in IBM's PC/AT (see *AT*). The 80286 calculates faster than the 8088. It also corrected one of the biggest problems with the 8088—its inability to access more than a small amount (1 megabyte) of *memory*. But—and this is a big catch—DOS couldn't take advantage of the extra memory. Why, you ask? Well, because DOS had to run on 8088-based computers too, so it couldn't be changed to accommodate the new chip. Windows, however, does utilize the special talents of the 80286 in "standard mode." In fact, you can't run Windows 3.1 unless your PC has an 80286 or an even newer microprocessor. However, there is a way to create software that uses the extra memory but still manages to run in DOS (see *DOS extenders*).

80386

The **80386** chip has one big advantage over the 80286: it includes fully functional, built-in circuits for *multi-tasking,* or running two or more programs at the same time. Although software like Windows, DESQview, or Unix will let you "multi-task" using lowlier chips, the 80386 does it more reliably and with less hassle. Windows, for example, running in "386 Enhanced Mode," lets you use multiple standard DOS programs at the same time, whereas you're limited to only one DOS program at a time with an 80286. Another improvement in the 80386 is its ability to access even larger amounts of memory than the 80286, and to do so with less trouble. Again, DOS is oblivious to the extra memory, but you can buy software that bypasses this limitation through a DOS extender.

The 80386 comes in two versions: the standard model, called the **DX,** and the cheaper, slightly slower **SX.** The only difference is that the DX accesses data 32 bits at a time (it has a 32-bit *bus*), while the SX can only move 16 bits at once. That makes a PC based on the 386DX somewhat faster than a PC that uses a 386SX—but nowhere near twice as fast, since both chips do everything else with equal speed.

80486

Practically speaking, the **80486**'s main advantage over the 80386 is just that the 486 is faster. Even if you compare the two chips running at the same *clock speed,* the 486 will generally finish its calculations sooner, sometimes much sooner, than the 386.

There are at least three versions of the 80486 that you may run across. The standard model, called the DX, includes a built-in *math coprocessor,* whereas the SX doesn't. PCs based on the SX are a little cheaper, but if you decide to add a math coprocessor later (it's easy to do), you'll spend more than if you bought a DX machine to start with. And there's the DX2. This model does its internal computations twice as fast as a standard DX running at the same clock speed, slowing back down when it's time to move data to or from memory. The DX2 is decidedly faster than a standard 486DX, and not much more expensive.

Abort, Retry, Ignore, Fail?

DOS displays an ominous *error message* like **Abort, Retry, Ignore, Fail?** whenever it has trouble using one of your disks or files. Why it has to be so brusque and frightening, I have no idea, but a message like this usually doesn't mean the end of the world. Here's a quick guide on what to do:

1. If you get the error message when you're trying to use a floppy disk, first be sure that the disk is all the way in the disk drive, and that the drive door (if there is one), is closed. If not, fix things, and then press **R** (for retry). If that doesn't solve the problem, go to Step 3.

2. If you tell DOS to switch to a floppy disk (by typing, say, **B:** followed by the Enter key), but there's no disk at all in the floppy drive, you'll get an error message. If the message includes the Fail option (older versions of DOS don't have the Fail choice), press **F**. That should give you a prompt something like:

   CURRENT DRIVE IS NO LONGER VALID>

Just type the name of a valid drive, like **C:** followed by the Enter key. Unless you put a disk in the drive, don't press **A** (for Abort)—if you do, DOS will just keep trying to get the floppy drive in gear, which it can't do without a disk.

3. If the error message appears when you're trying to copy or open a file, it probably means that DOS is having trouble, big trouble, reading the information on the disk. You may have lost some of the file, but maybe not all of it. In this situation, start by pressing **R** for Retry at least a few times. DOS may be able to find the correct information if you keep trying. Eventually, if the error message just keeps reappearing, you may want to press **I** for Ignore, which tells DOS to go ahead with copying the file or whatever despite the problem. The new copy will have gaps in it, but at least some of the information may be intact.

4. What you do after completing Step 3 depends on what kind of disk produced the error message.

   If it's a **floppy disk,** assume the disk is bad. Copy all the files you can to a new floppy. If important files that didn't copy correctly, you might try taking the disk to someone else's computer and copying them there. After you've made your copies, throw the bad disk away.

   If it's a **hard disk,** you're in trouble. Don't use the disk until you can get help from somebody who really knows what they're doing, or use a *utility* like Norton Disk Doctor to diagnose and fix the problem.

About

In most Macintosh and Windows programs, you can get a little background information about the software by choosing the **About** menu item. You'll see a box containing tidbits like the full name and *version number* of the program, who wrote it, and perhaps the name of the person it's *registered* to.

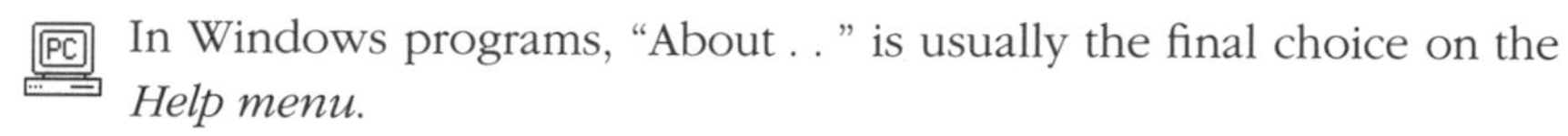

In Windows programs, "About . . " is usually the final choice on the *Help menu.*

On the Mac, "About . . ." is the first item under the *Apple menu.*

While the About window can be extremely boring, some programs use it to deliver rather creative commercials, with glitzy graphics and snazzy special effects. Sometimes if you hold down a *modifier key* (such as the Command or Control key or the Option or Alt key) as you choose About, you'll get a surprise (called an *Easter egg*).

For instance, at the *Finder* in System 7, hold down the Option key and choose "About The Finder" from under the Apple menu.

absolute cell reference

An **absolute cell reference** is a reference to one specific cell in a spreadsheet or sometimes in a database. It's easiest to explain in relation to a "relative" cell reference. In a spreadsheet, each cell has a name (like A4 or B6, kind of like a Bingo card). In formulas, you need to tell the computer which cell to act on, or which cell to refer to. A **relative** cell reference tells the spreadsheet to use the information in a cell that is in a **relative** position to this one. For instance, you may want the data in cell A5 to be divided by the data in cell A3. **You** may think you are telling A5 to divide by the cell A3, but the **computer** is hearing that you want to divide by the cell that is **two spaces above** A5. If you took the same formula from A5 and copied it to cell B5, the computer would divide by the data in cell B3, because it is still referring to **two spaces above,** relative to wherever the formula is. This is great because then you don't have to retype formulas all the time, you can just copy and paste them.

But sometimes you don't want the cell to be referenced to wherever the formula happens to be. Maybe in cell B5 you really do want the data to be divided by what is in A3, not what is in B3. Then you have to make cell A3 an **absolute** cell reference, so no matter where you copy the formula, the computer will always divide by the data in cell A3, absolutely.

The *default* (or automatic) formatting for a cell reference is usually relative, meaning if you want an absolute reference you have to tell it to be so. Different programs have different ways of making a cell reference absolute. A common method of specifying a cell to be absolute is typing a dollar sign ($) before the cell name: $A3.

accelerator, accelerator board

An **accelerator** is something you can install in your computer that makes the machine work faster. Almost always this is a **board** or *card,* a thin rectangular piece of fiberglass with a bunch of electronic parts and circuits attached to it. You install the board (or have someone else install it) into your machine—it may be designed to plug into one of the computer's *expansion slots,* or it may fit somewhere else. The accelerator board has a *microprocessor (chip)* of its own that takes over some or all of the work from your computer's original *processor.* You may actually have to remove the existing processor when you install the accelerator.

Adding an accelerator board is often an inexpensive and efficient way to upgrade your computer without having to buy a whole new system. But shop carefully—you can spend a few hundred dollars and end up barely noticing any speed boost.

access

To **access** an item (such as the information in a file) just means to use it. But the term implies that you have to do something first in order to be able to use whatever it is. You may need to switch to the folder or directory where the file you want to access is stored. You may need to enter a password before you can access it. If you can't access an item, it means you haven't told the computer where to find it, or you don't know the password, or maybe the item is *locked*.

access code

An **access code** is another term for a *password*. An access code is usually a specific combination of letters and/or numbers. You always need some sort of access code to use an *online service* or to have your computer talk to another computer.

access time

You'll usually see the term "average **access time**" in advertisements or reviews of hard disks and *CD-ROM* players. Access time refers to how fast the disk can locate and begin retrieving (accessing) a specific piece of information. Because the time needed to access any particular piece of data varies depending on where it's located on the disk, vendors use an average access time when rating the speed of their drives. Access time is measured in *milliseconds*, or *ms*, so ads say things like "average access time: 18 ms."

If you're buying a hard disk, you want one with an average access time of less than 20 milliseconds; don't settle for anything slower than 30 milliseconds.

CD-ROM players are much slower; the fastest now available have average access times of around 360 milliseconds (it takes them about a third of a second to find, say, the definition of the word "nugatory" in a CD-ROM dictionary).

Acrobat™

Acrobat is a technology developed by Adobe Systems that will allow documents created on one computer system to be read and printed on other systems, with the fonts, formatting, text attributes, and graphic elements intact. It will work in tandem with the *Multiple Masters* font technology, in which a font that exists in one computer can emulate the font in the document that was created in another computer system.

This technology will play a key role in the future of computerized communications, allowing us to send compact, well-designed, attractive, and readable documents from one computer to another, and even from one type of computer to another, without using paper. There are three key elements in this technology that make it outstanding: preserving the formatting and spacing of the original font (have you ever opened a document on your computer without the correct fonts? aaack!); the compactness of the file (generally files are quite large that can hold onto graphics and formatting); and the fact that it will work on virtually any computer system.

active

The term **active** refers to the item in use at the moment, whether you are talking about such diverse things as a hard drive, a window, or a cell in a spreadsheet.

For instance, if you have more than one window open on your screen, the window that is in front and that has lines in its *title bar* is the window that is active (see *active window*). In a *spreadsheet* or *database*, the cell or the field you selected is active. In a *dialog box,* the *edit box* that is highlighted or that contains the *insertion point* is active.

When an item is active and you do something like choose a command or type, **it will happen to the active item.** That's the whole point! For instance, if a row is active in your database and you choose the command Cut, that row will disappear. If a window is active and you create a new folder, the new folder will appear in the active window.

You may have several *hard drives* and/or several *floppy drives* (the thin slots where you insert your disk) attached to your computer, with or without disks in them. The computer has to go to one of these drives to get the information you need from the disk. The drive the computer is using at the moment is the **active drive** (compare with *boot disk*).

active matrix display

An **active matrix display** is the kind of screen used on the more expensive laptop and notebook computers, including the Macintosh PowerBook 170 and an increasing number of PC notebooks. Active matrix screens are a type of liquid crystal display, or *LCD,* like the vast majority of screens on laptop and notebook computers. But unlike the garden variety LCD screen, an active matrix screen has a separate, independent transistor for each and every dot *(pixel)* on the display. The result is a sharper image and much brighter, more lifelike color, with less of the *smearing* that makes moving images hard to see on an LCD screen. Conventional, or passive matrix, LCDs still have one big advantage: they're much cheaper than active matrix displays.

active window

Although there can be many *windows* open on the screen, only one window can be *active,* or in use, at a time. It's easy to tell which window is the **active window:** the title bar at the top of an active window looks different from the title bar of an inactive window. When you choose a command, such as "Close," the command will affect the active window. If you choose to make a new folder, the new folder will appear in the active window. To make a window active, simply click on it.

In Windows applications, the active window's title bar has a distinct color, different from the other title bars.

On the Macintosh, the active window shows horizontal lines or perhaps a pattern in its title bar, whereas the title bar of an inactive window is empty (plain white). To move a window without making it active, hold down the Command key while you *drag* its title bar.

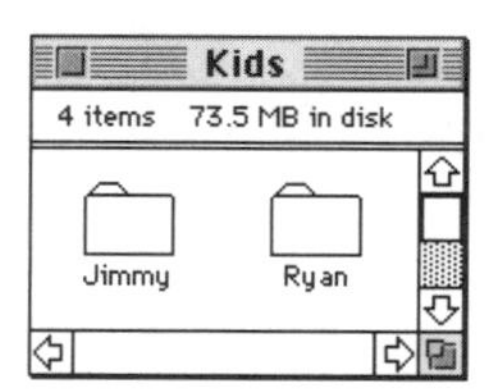

This is an active window.

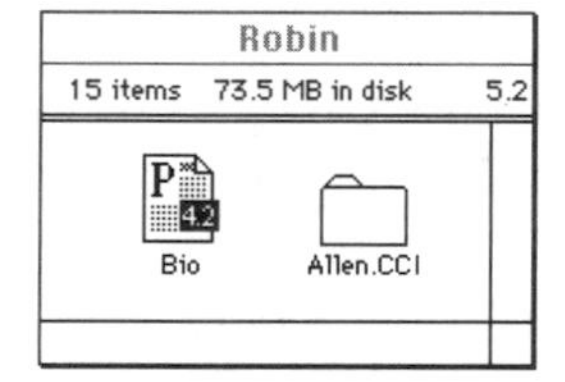

This window is not active.

A-D conversion, A/D conversion

See *A-to-D conversion.*

adapter

An **adapter** is electronic circuitry that adapts a device, such as a monitor, to your computer so that the computer can control what the device does. For example, a *VGA monitor* requires a VGA *(video graphics array)* adapter in your PC or you won't see any images on the screen.

An adapter can be an *add-in board (card)* or it may be built into the circuits on the computer's *motherboard.* A *controller* does exactly the same thing as an adapter, but the two terms tend to be used differently: the term *adapter* is usually used in reference to monitors and printers; the term *controller* is usually used in reference to disk drives and scanners.

ADB

ADB stands for **A**pple **D**esktop **B**us™, a system that uses certain hardware and software for connecting *input devices*. You use input devices (like mice, keyboards, graphic tablets) to *input*, or put in, information to the Macintosh. Older Macs, like the 512 or the Plus, don't use the ADB (they use an older and different system), which is why you can't interchange older keyboards or mice with newer ones. Also see *ADB port* and *bus*.

ADB port

This icon represents an ADB connection.

A *port* is a little socket on the back of a computer or any other *peripheral device*, such as a keyboard or an external hard disk, where you plug in a cable. An **ADB port** is the one with the diagram, or *icon*, that looks like little worms (shown to the left). *ADB* stands for **A**pple **D**esktop **B**us™.

Input devices, such as keyboards and mice, plug into ADB ports. You can plug any cord that has that little icon on its end into any ADB port. For instance, you can plug your mouse into the extra ADB port on either side of your keyboard. In fact, if you prefer a mouse and your roommate prefers a trackball, you can have the mouse connected to the ADB port on your keyboard and the trackball connected to the ADB port on the back of the computer and use them both. Older Macs, like the Plus and older, do not use the Apple Desktop Bus.

add-in board, add-in card

Add-in boards or **add-in cards** are circuit *boards* or *cards* that you insert into slots designed for them inside your computer. An add-in board consists of electronic circuits (*chips*, other parts, and the connections between them) mounted on a rectangular piece of fiberglass. The bottom of an add-in board has a row of metallic contacts on a smaller rectangular projection; you insert this part into the slot (see the illustration in the appendix).

Each board you add imparts some new capability to your computer. You can get an *accelerator board* to make your computer run faster, a "video board" to control your monitor, a "memory board" to add more memory, etc.

Opening up your computer to plug in an add-in board can be a little scary the first time you try it, but it's hard to goof up the procedure. Here are a couple of tips:

Moody River. But on other services, the addresses are gawdawful combinations of numbers and letters, like D03H1,45337. Most services offer a way to create a little electronic address book of your favorite correspondents. You can usually apply the address to your mail by clicking a button so you don't even have to type it.

Adobe Font Metrics

See *font metrics*.

Adobe Type Manager®

Adobe Type Manager (affectionately known as ATM) is a *utility* (a little software program) that uses *Type 1 PostScript fonts* (a certain variety of typefaces) to display type on the screen. No matter what type size or style you select, ATM will create the smoothest on-screen rendition that the *resolution* of your display screen allows. This makes it much easier to work with type on the screen, especially if you want to do things like *kern* your type or design pages. Also see *fonts*.

Scarlett *This is what type looks like on the screen without ATM.*

Scarlett *This is what type looks like on the screen with ATM.*

Like other *scalable fonts*, Type 1 fonts consist of mathematical formulae that describe the outline of each character. ATM takes this outline information and *scales* and *rasterizes* it, figuring out which dots your screen should display to create the smoothest and sharpest rendition of the character, at whatever size you requested (*scaling* refers to the process of creating the final character at a particular size; turning an image into dots is called *rasterizing*).

ATM can also rasterize a Type 1 font to your **printer** so that the text prints smooth and clean on a *non-PostScript printer* (such as an Apple StyleWriter, an HP LaserJet, DeskWriter, or DeskJet, or an Epson dot matrix printer). *PostScript printers* can rasterize Type 1 fonts themselves, so they don't need ATM. But no matter what type of printer you have, ATM always displays Type 1 fonts beautifully on the screen, no matter what size you've chosen.

ATM is available for both the Mac and Windows, and even for some DOS (non-Windows) programs. Now that System 7 and Windows 3.1 come with *TrueType* built in, you can get the benefits of scalable fonts on any screen or any printer without ATM. But TrueType only works with TrueType fonts—if you are using Type 1 fonts, you'll still need ATM.

AFM, AFM files

Those files on your font disks called **AFM** files hold the *Adobe Font Metrics* (on PC font disks they have the file extension .AFM). See *font metrics.*

After Dark®

After Dark is a *utility* from Berkeley Systems, called a *screen saver,* but After Dark goes far beyond most other screen savers. The idea behind a screen saver is to keep your screen dark, or keep a pattern moving, so an image doesn't get burned into the screen by staying in one place too long. Now, you **could** just dim your screen, but that would take a manual effort. You could be content with little stars moving around space, so at least you knew your computer was still on and functioning. But no—you want a little man mowing the lawn on your screen. Or fish and seahorses floating in full-color with an occasional scuba diver and flying toaster drifting through. In fact, you want an entire fleet of flying toasters because you like the gentle rhythmic sound of their little wings beating.

After Dark has an endless collection of screen-saving modules whose only purpose in life, besides protecting our screens, is to delight us. What a noble concept. Actually, once you buy the After Dark utility, you can get hundreds of modules free from any *bulletin board* or *online service,* created by other nice people who want to share the delight.

AI

See *artificial intelligence.*

alert box

An **alert box** is a message that appears on the screen to warn you of some imminent disaster, or to inform you of a situation in which your actions may have irreversible consequences. For example, if you've just chosen the command to delete a file, an alert box might warn you that the file will be gone forever, and ask you if you *really* want to delete it. You have to deal with the alert box before you can move on with your life, meaning your only option when you see something like this is to click one of the buttons it offers you. Typically the dark-bordered button (which is the *default* button) is the safest option available—and because it's the default, you can hit the Return or Enter key instead of clicking the button. If the alert is telling you the computer *crashed,* then the button may not even work, in which case you just have to turn off your machine by its switch or, if you have one, press your *reset switch* (see *programmer's switch*).

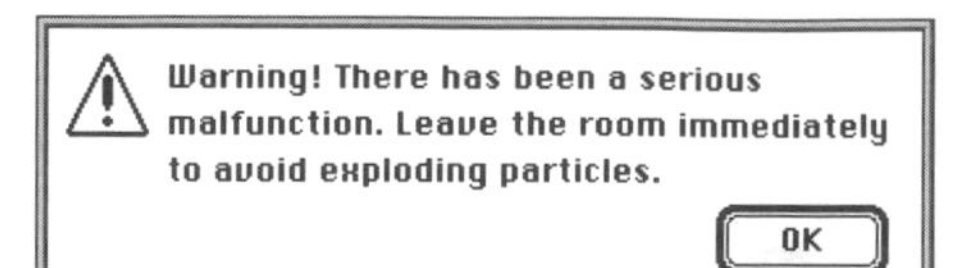

This is an example of an alert box.

algorithm

An **algorithm** is a set of rules that is supposed to give the correct answer for solving a particular problem. The rules in an algorithm must have a beginning and an end and they must be well-defined with no chance of ambiguity. A computer can't solve a problem until someone programs it with the algorithm for solving that problem.

We use algorithms in our lives all day long to get things done, from running the dishwasher to balancing a checkbook. Aldus PageMaker uses algorithms similar to those we were taught in school for hyphenating unknown words.

Unfortunately, there is no algorithm for making human relationships work; no, we tend to use the *heuristic* method, which is simply trial-and-error (supposedly **intelligent** trial-and-error).

alias

This is an example of an alias.

An **alias** is a representation of the real thing. System 7 on the Mac has a wonderful feature enabling you to create aliases of any file to aid with organization and convenience. When you create an alias of a file (any file), you are essentially creating a "gofer" (you know—go fer this, and go fer that). The alias is just an icon (using only 1 or 2 K of space) that knows where the real thing is. When you double-click or choose an alias, the alias goes and gets the **original** file and opens it. For instance, you can create an alias of MacWrite, and put the *alias* in your *Apple menu.* Then you can open MacWrite straight from the Apple menu rather than dig through folders to get to it, because when you choose the alias, that alias goes and finds the real MacWrite and opens it.

Why would you want to do this? One reason is that many applications have to be in the same folder as their accessory files, like their dictionary or preferences file, or they won't open. An alias frees you from having to open folder after folder to get to your application. Also, you can have as many aliases as you like of each file, so you can store aliases all over the place, wherever you would find it convenient. You can leave aliases of your favorite applications right out on the Desktop and use the *drag-and-drop* technique for opening documents.

To make an alias in System 7, select a file and choose "Make Alias" from the File menu. The name of an alias is always in italics (see the example, previous page), which gives you an instant visual clue that the file is an alias.

On a PC, Microsoft Windows' Program Manager lets you create icons that function similar to the aliases described above. If you drag a document icon from File Manager into a Program Manager group window, it will create a new icon in Program Manager. Double-click to launch the document's application and the document itself. You can freely rename the icon, and move or copy it to another group, but Program Manager keeps track of where the actual files are located.

Under MS DOS you might have to type some long, cryptic command at the *DOS prompt* just to copy a file from one disk to another. You can, though, create a DOS *batch file* that lets you lump a lengthy command or group of commands under a single, simpler batch command. A batch file can essentially function as an alias. For instance:

instead of typing:	COPY C:\WP51\DOCS\THISWEEK.DOC A:\BACKUP\
you could type:	THISWEEK

The MS DOS User Guide and Reference covers the use of MS DOS batch files. Certain utility programs such as *4DOS* or NDOS (which comes with Norton Utilities) or CED also help you create aliases.

aliasing

Aliasing has two definitions, depending on whether you're talking about pictures or sounds.

When a diagonal line or a curved arc drawn on the screen looks as if it was made out of bricks, when it looks like stair steps instead of a slide, the effect is technically called **aliasing.** Most of us would say it had the *jaggies*. Also see *anti-aliasing*.

Aliasing is also a type of sound distortion you can hear. It affects *digitally* reproduced sound, which is the kind of sound your computer probably makes. (The beeps you hear from a standard PC speaker aren't digital, but a sound board you plug into your PC creates sound digitally; the Mac has built-in digital sound.)

Digital sound is based on a sequence of numbers (digits) that are converted into sound waves by electronic circuits. The computer has to guess at what sound to make between each number in the sequence. If the time between each value is too long (if the "sampling rate" is low), you hear the mistaken guesses as a metallic, static distortion called aliasing. To squelch aliasing, you need a sound *card* with a sampling rate of around 40 *kilohertz* (40,000 times a second) or higher.

alignment

If you're talking about text, **alignment** refers to whether the text in a paragraph is lined up on the left or the right or centered or on both sides. For instance, in a flush left alignment the text is aligned on the left edge and is *ragged,* or unaligned, on the right. The paragraph you are reading right now has a *justified* alignment—the text is aligned on both sides.

This paragraph, however,
is centered.

And this paragraph is flush left. If you're talking about graphics, alignment means just what you'd think: the relative position of two or more graphic items, or *objects,* on the screen. Many *drawing programs* have an alignment command that automatically aligns objects so, for instance, their left sides all line up along the same imaginary vertical line, or their centers all fall on the same imaginary horizontal line.

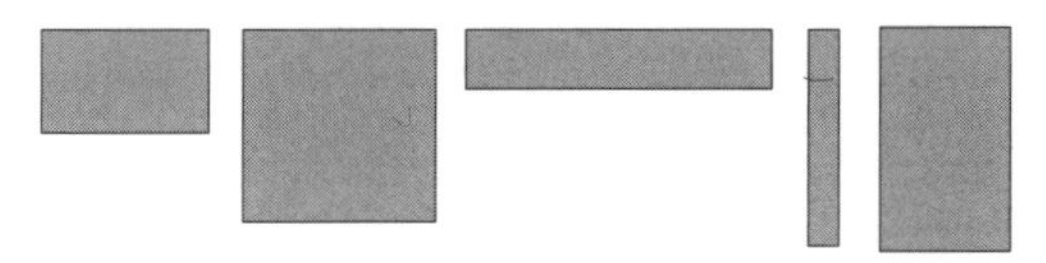

These objects are aligned across their top edges.

alphanumeric characters

The term **alphanumeric** refers to the characters you can type and print, such as the entire **alphabet** and all the **numbers**, as well as punctuation. You have keys for *Return, Tab, Alt, Option,* etc., but these are *non-printing characters* so they're not considered part of the alphanumeric group.

alpha channels

If you stumble upon a group of power users who are heavy into graphics, chances are you're gonna hear a lot about **alpha channels.** Alpha channels are pretty incredible, but it can take a long time to get the hang of them and put them to practical use.

First, consider that the Macintosh has the capability to create *24-bit* images, but does so with a technology called *32-bit QuickDraw.* So what happened to the leftover 8 bits? Those are left open for the use of alpha channels. In a 24-bit *RGB* image, the red, green, and blue channels each take up 8 bits, leaving another 8 bits open for software developers to use for special effects, masks, compositions, etc.

The most well-known application that makes use of alpha channels is Adobe Photoshop, which works even more magic by allowing you to have up to thirteen separate alpha channels (of 8 bits each!) in addition to the 24 bits used by the red, green, and blue channels. Each of these channels is a separate entity which can be modified in numerous ways without affecting the entire image. Much like we've always had in draw programs, Photoshop now gives us the ability to accurately alter many separate parts or elements of a *bitmapped* image without disturbing the nearby elements.

Think of the concept as a layering system where you can have, say, a sailboat in one layer, a lake in another layer, and a mountain range in another layer. Each of these three items can be placed in alpha channels and manipulated (colored, scaled, rotated, whatever) independently of all the others. And, if you want to manipulate the whole image, you can do so by making a composition of all three with a host of options as to how they relate to each other.

Alpha channels are a very, very cool technology, but plan on spending some time learning how to use them. If graphics are your thing, it's worth the time and effort.

alpha testing

Before being released to the public, a product goes through extensive in-house testing, often by the developers themselves. This is considered **alpha testing.** After alpha testing the product is sent to selected outside *beta testers* who find all those *bugs* and problems that the alpha testers missed.

It is sad but true that testing only uncovers the **presence** of bugs, not the absence.

alpha version

An **alpha version** of software is the first working draft. In an alpha version, all the features that are going to be in the finished product are supposedly in place and functional, but it is known that there are also *bugs*. The alpha version is tested in-house by the developers and also sent out to "alpha testers"—people who have been selected or who have shown an interest in playing with software that is almost guaranteed to *crash* their computer. After the alpha version comes one or more *beta* *versions*, and when these have been tested and had their known bugs fixed, the software is *mastered* and comes out in the "release," also called the shipping version.

Alt key

The **Alt key** on a PC is most often used in conjunction with other keys to activate commands. In programs that have menus, pressing Alt with another key is a *keyboard shortcut,* an alternative, quicker way to carry out one of the menu commands.

In some programs, including Microsoft Windows and DESQview, pressing Alt by itself does something too. In Windows, this activates the menu bar; in DESQview, it pops up the window you use to control DESQview. The Alt key's equivalent on a Macintosh is the *Option key.* Please see *modifier keys.*

America Online

America Online is an *online service.* This means it's a service business where you can do research, get the latest news, join clubs, make hotel and plane reservations, shop, meet and chat with people, attend conferences, send and receive mail, and thousands of other things, but it is all done *online* while you sit at your computer.

You need to have a *modem* and you need to have the America Online software, which is free. You access **AOL** (as it is affectionately known) by having your computer dial a local number (which to you is just a click on a button). AOL charges you $10 an hour between 8 A.M. and 5 P.M., and $5 an hour after 5. They take it right out of your checking account or add it to your credit card. It is too simple. And addictive. Everyone ends up spending hours of their life online when they first connect with AOL.

Amiga

Commodore Business Machines makes a series of personal computers called **Amiga,** ranging in use from home applications to business use and desktop publishing. Their higher-end Amiga computers are very popular for video work and animation. Standard Amigas aren't compatible with either the IBM PC or the Mac (they won't run PC or Mac software), but Commodore sells an *add-on* that makes the Amiga IBM-*compatible.*

ampersand, &

This symbol, **&**, is called the **ampersand**. The symbol stands for the word "and," and the word is a contraction of "and per se and." The ampersand has slightly different shapes in different typefaces (some of which show clearly the Latin "et" for "and"): &, &; or &, among others. Press Shift 7 to type the ampersand symbol.

In some programming languages, the ampersand is used to *concatenate,* or join, items such as words and phrases.

analog

Analog is the opposite of *digital,* and I can only explain analog in relation to digital. Analog refers to things that are in a continuous flow or that have an infinite number of values—things that are "analogous" to real life.

For instance, a watch with hands is an analog device, because as the hands sweep over the watch face, time can be shown in a continuous flow, with an infinite number of possible increments. A digital watch, on the other hand, displays time in finite chunks—seconds or tenths of seconds or perhaps even smaller chunks, but at any one moment there's one specific number showing, always countable and limited.

A photograph represents a scene or a face in analog form, in continuously varying tones of grays or colors. When you *scan* a photograph to get it onto your computer screen, the scanning process breaks up those continous tones into *digitized* (digital) *bits* of information, because your computer can only understand digital things.

Guy defines analog as "anything you can read in direct sunlight."

It is possible to make an analog computer, but digital computers are much easier to build, faster, and more reliable. Also see *digital* and *modem.*

ANSI

ANSI is the acronym for the **A**merican **N**ational **S**tandards **I**nstitute. This institute creates standards for a wide variety of industries, including computer programming languages.

As an *end user* (someone who uses the computer to get their work done, rather than programming it), you may still hear of ANSI when someone refers to the *ANSI character set.* See the character set on the following page.

ANSI character set

The **ANSI character set** is the *character set* (collection of the characters you can type in a chosen typeface) adopted by the *ANSI* organization as the standard for computers. (In real life, it isn't a standard, but it is the character set used by Microsoft Windows.) Like any character set, ANSI comprises a particular collection of letters, numerals, punctuation marks, and other symbols, each of which is assigned to its own code number. Included in the ANSI set are typographic quotation marks and dashes, business symbols like copyright marks, and lots of foreign language characters.

Character	Name	ANSI code	Macintosh keystrokes
…	ellipsis	0133	Option ;
‘	opening single quote	0145	Option]
’	closing single quote (apostrophe)	0146	Option Shift]
“	opening double quote	0147	Option [
”	closing double quote	0148	Option Shift [
•	bullet	0149	Option 8
–	en dash	0150	Option Hyphen
—	em dash	0151	Option Shift Hyphen
™	trademark symbol	0153	Option 2
©	copyright symbol	0169	Option g
®	registered trademark	0174	Option r
¼	one-quarter	0188	depends on font
½	one-half	0189	depends on font
¾	three-quarters	0190	depends on font

In Windows, to use characters in the ANSI set that aren't on your keyboard *(extended characters),* hold down the Alt key while you type the four digit code number (using the *numeric keypad*) for the character you want, as shown in the table above. This number always begins with 0 (zero). For instance, pressing Alt and typing 0233 would give you **é**. Another way to use the extended characters is to run the Windows utility *Character Map*. In the chart it displays, click on the character you want, then choose Select, followed by Copy to copy it to the Windows *Clipboard.* Then switch back to your program and choose Paste.

ANSI.SYS

ANSI.SYS (pronounced "ansee dot sis") is a *driver,* a little software module, or *controller,* that you can use with a PC running *MS-DOS*. You "install" ANSI.SYS by running it from your *CONFIG.SYS* file. With this driver installed, you can use many new commands, giving you more control over your screen and keyboard, than you have using standard MS-DOS commands.

anti-aliasing

When text or a graphic image is displayed on a *monitor,* or screen, the smoothness of the edges is limited by the resolution of the screen, which means the edges tend to be a little jagged. This jaggedness is also called *aliasing*. There are a variety of techniques used to reduce the jaggies, or the aliasing of text and graphic images, to fool our eyes into thinking the edge is smoother. For instance, in an *image editing* program you can blur the edges, or shade along the lines to make the dark-to-light transition less distinct. **Anti-aliasing,** then, means to use one of these techniques to smooth out the rough edges (the *aliasing*).

antistatic device

If there is too much static electricity hanging around, it can actually disrupt your computer, causing the screen to freeze or creating various other unpleasant disturbances that can destroy data. Static can even destroy the circuits inside your computer. Extra static can develop from the weather or from certain kinds of clothing or from activities like petting your cat while working on your computer, shuffling around on the carpet, or rubbing a balloon on your head. So a variety of **antistatic devices** have been developed to help prevent this, devices such as wrist bands, floor mats, sprays, and little metal pads that say "Touch Me."

It's particularly important to discharge any static on your own body when you open your computer to install another *card* or perhaps a *SIMM* or two. You don't need a special antistatic device to do this—just touch something metallic before you handle the cicuits. Most computers have a metal plate inside the box that's perfect for this purpose.

Any key

There really isn't a key called the **Any key.** When a direction tells you to "PRESS ANY KEY," it simply means to press whichever key you like on the keyboard.

AOL

See *America Online.*

APDA

APDA stands for **A**pple **P**rogrammers and **D**evelopers **A**ssociation. It's a support service from Apple Computer that provides documentation and technical products for people who program or develop Apple equipment.

Apollo Computer

Apollo Computer is the name of a company (now owned by Hewlett-Packard) that creates high-powered *workstations.* (A workstation is a bigger computer than we use at home and is usually specialized for scientific, engineering, or graphics applications; see *workstation.*) Apollo developed the Network Computing System which allows people to *network,* or connect, computers from different manufacturers.

append

When you **append** something, you add it on to the end of something else, such as to the end of a document or to the end of a list. This is different from "inserting" an item, which is when you add an item into a specific position in the list, not necessarily at the end.

An example of using a button or command called "append" is when you have a list of *fields* in a *database* that you want to export. When you append another field to the existing list, the one you append will be at the bottom of the list. Since the order of the exported fields is critical, be conscious of whether you need to insert or append.

Apple computer

An **Apple computer** is one of the computers made by Apple Computer, Inc. that is called "Apple," such as the Apple II, Apple IIGS, or Apple IIE. We do not call a *Macintosh* computer an Apple. An Apple and a Mac are very different machines with different *operating systems.* In fact, the Apple operating system has more in common with the operating system used by *IBM PCs* than the one used by the Mac, and Apples and Macs cannot use each other's software without special adaptations. Being able to use an Apple does not mean you automatically know how to use a Mac, and vice versa.

These days, Apples are found mostly in elementary schools and in the homes of educators since these folks were able to buy them cheap. Apple (the company) no longer sells them.

Apple Computer, Inc.

Apple Computer is a Cinderella story. Steve Wozniak and Steve Jobs founded the company in a garage on April Fool's Day in 1976, and were guided by Mike Markkula. Later that year Apple introduced the Apple I at the local *user group,* the Palo Alto Homebrew Computer Club. In 1977 they introduced the Apple II, which became the most widely used computer in the classroom and home. *VisiCalc,* the first spreadsheet on earth, was launched on the Apple II, which took the computer into the business market. The forerunner of the Macintosh computer, the Lisa, was introduced in 1983 but didn't last very long—it was slow and expensive (for a while they tried to turn it into a hybrid Mac). And of course, in 1984, the Macintosh was introduced, the machine that changed the world. By the fourth quarter of 1992, Apple was selling more Macintoshes than IBM was selling PCs.

Apple Desktop Bus™

See *ADB.*

Apple Events

Apple Events is the name of a System 7 feature that allows two-way communications between separate applications, either on a single Mac or through a *network.* For instance, while you are working in one program you can tell another program to do something. Apple Events is a component of the *Inter-Application Communication* protocol that lets applications share features with each other. Many applications are being upgraded to take advantage of this capability.

Apple File Exchange

The **Apple File Exchange** is a great little *utility,* or small program, on the Macintosh that allows you to convert a *PC* file into a Mac file. It's so cool—you just double-click on the Apple File Exchange *icon,* insert a PC disk into your Macintosh *SuperDrive* (which is only on newer Macs), select the file, and click the Translate button. Et voilà, you get a copy of the PC file that's readable by your Mac. The software comes free with your System disks. (Note that you must open the File Exchange program *first,* then insert the PC disk.)

Apple key

The **Apple key** is the key near the Spacebar that has the Apple symbol on it ( ). This key doesn't do anything all by itself, but it's used in conjunction with other keys for *keyboard shortcuts* (also see *modifier keys*).

Apple II users call this the Apple key or Open Apple. On an older Macintosh keyboard, you might not see an Apple symbol on the key; the same key is called the *Command key* on the Mac, and has the cloverleaf symbol on it. Newer keyboards have both the apple and the cloverleaf symbols on the same key.

AppleLink

AppleLink is an *electronic mail* and information service that is not open to the general public, but is reserved for Apple employees, developers, universities, user groups, dealers, and a few other privileged individuals (not including me). Like any information or *bulletin board service,* you use AppleLink through your *modem.* AppleLink provides product announcements and updates (for *third-party* products as well as for Apple products) and various technical information. And, of course, *e-mail* to other privileged individuals, which is how Guy Kawasaki wooed his wife, Beth.

Apple menu

On the Macintosh, if you position the mouse pointer over the little apple you see in the far left of your menu bar, and then press the mouse button, you get a menu which is called (guess) the **Apple menu.** This menu holds your *desk accessories,* such as the Calculator, the Scrapbook, the Chooser, and the Alarm Clock.

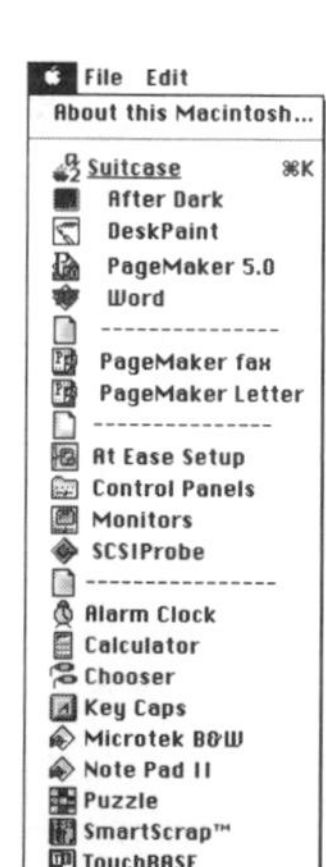

In System 7 you can add any file you want into the Apple menu simply by putting its icon or, preferably, an *alias* of its icon into the Apple Menu Items folder, located in the System Folder.

The Apple menu you see to the left is my menu that I have customized. See how it's organized into groups—applications first (aliases of applications, that is), templates for my fax sheet and letters, control panels, and then desk accessories. I was able to organize them this way by taking advantage of the fact that blank spaces get alphabetized before characters, so I added one, two, or three blank spaces in front of the names. The dividers are aliases of blank documents that have been named "--------." In fact, everything in this entire menu is an alias except the desk accessories.

AppleShare

AppleShare is the name of *file server* software from Apple. If several Macintoshes are *networked,* or connected together so they can share data and programs, there is usually one or more computers, or at least hard disks, that are dedicated to the job of just serving files, or information, to you. Working in conjunction with the server software, the AppleShare software lets you choose the server from which you want to get data. AppleShare also lets you create private servers that need a password to *access,* or to get into.

This icon represents the Appleshare software.

AppleTalk

The **AppleTalk** network is how your Mac talks to your laser printer, to other Macs, or to other sorts of machines, provided they are hooked together, or *networked,* with *cables.* (You can use other types of networks with a Macintosh, but they require special *cards,* while AppleTalk is built in.) If several computers need to share one laser printer, you can get little adapters that let more than one computer plug into the printer through AppleTalk.

Apple II, Apple IIE, Apple IIGS, etc.

These **Apple** computers belong to a family of personal computers from (take a wild guess) Apple Computer, Inc. They run under Apple's *DOS* or ProDOS operating systems, which means they function more like IBM-compatible computers than like Macintoshes. If you own a Macintosh, don't call it an Apple or people will think you have one of these Apple machines. Also see *Apple Computer.*

applets

Applets are miniature programs. The name comes from the term "applications" which is one variety of a software program (see *application*). Applets is just a cute name for what is generally called a *utility* program, but it's used almost exclusively in the Windows environment. A utility program is designed to do one little function, generally in the service of housekeeping for your computer system, rather than actually producing something. The term "applet," though, implies that these little programs are actually doing something practical, beyond what a utility would do.

application

There are several varieties of computer *programs,* but the ones most of us are familiar with are the **applications.** An application is a program you use to get some practical work done, such as word processing or accounting or illustrating.

Programs such as WordPerfect, PageMaker, Excel, and Illustrator are all examples of applications. Programs such as *Adobe Type Manager* or *Suitcase* are *utilities* because we don't create work with them—they just help us manage our computer system.

application heap

The term **application heap** refers to the amount of *memory* set aside for an *application* or software program. On a Macintosh you can choose to increase the application heap, which you may want to do if you find the program running very slow, or perhaps telling you there is not enough memory to complete an operation or even to open the application.

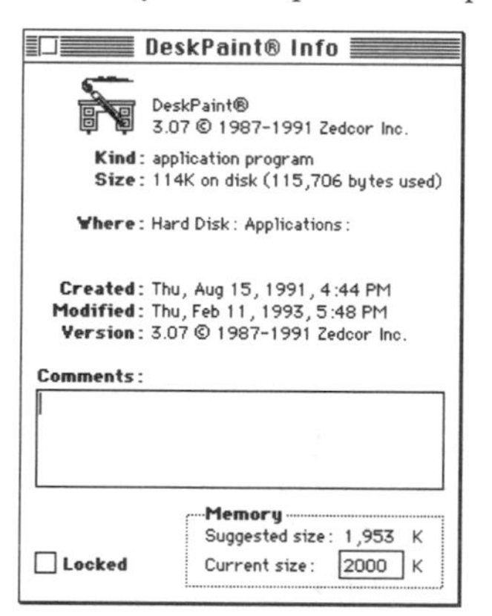

To increase the application heap, the application must be closed. Click once on the application icon, then from the File menu choose "Get Info." In the "Get Info" dialog box there are little boxes at the bottom, under "Memory." The box "Suggested size" is the minimum amount of memory needed to run the program efficiently. The box "Current size" is the amount of memory currently allocated to the application. You can change the amount shown in "Current size" by typing in it.

Notice in the example above that I increased the application heap in DeskPaint. It had previously been set to 1400 and I had trouble opening a graphic file. After giving DeskPaint a little extra memory, I had no problems.

Now, you don't want to go arbitrarily jacking up the size of this application heap. Keep in mind that any extra memory you add to *this* heap takes it away from the System heap (see *heap*), and if you don't have a lot of memory you can get into trouble. Just increase the application heap to the minimum value you need to get your work done efficiently.

application icon

An **application icon** is the icon (picture) that represents a particular *application,* or software program. The term "application icon" is used even when the program is considered a *utility* rather than an *application.* Application icons are usually fancier than the corresponding *document icons* that represent documents created with the application. (Document icons often look like a piece of paper with the corner folded down.)

This is an application icon. It's fancy looking.

This is a document icon; it's obvious which application it was created with.

In the Microsoft Windows manual, environment application icons are called "program item icons," but real people call them program icons, application icons, or just icons.

.ARC

When you see these characters, **.ARC** (pronounced "ark" or "dot ark"), at the end of a PC file name, it means that the file has been compressed to save space, or *archived,* using a program called ARC (or a compatible archiving *utility*). Archived files are usually *compressed* , so you will need to *unarchive* or *uncompress* the file before you can use it. For that, you will also need the ARC program. ARC files used to be the most popular type of archived file, but now ZIP and LZH files predominate.

architecture

Similar to the architecture of a building, the **architecture** of a computer refers to the design structure of the computer system and all its details: the *system,* the circuits, the *chips,* the *busses,* the *expansion slots,* the system *firmware, BIOS,* etc. The architecture largely determines how fast the computer is and what it can do. It also decides whether one computer is compatible with another. Can the same *boards* be used? Yes, if the architecture is compatible. Different models will have basically the same uses, but with varying degrees of performance. The architecture is what ensures *backward compatibility,* which means that your old software can run on a new computer. Also see *platform.*

If a computer has an **open architecture,** it means the company that builds the computer has published the specifications of the computer's design. Open architecture lets *third-party* developers create *add-in boards* (for customizing your computer, assuming the computer has *expansion* *slots*) and other *add-on* hardware that will work in the machine. So, open architecture means that any other company can create *memory* boards, video *adapters,* hard disks, and so on that function properly in the computer. An open architecture even allows other manufacturers to build computers that work identically. The Apple II and the IBM PC have open architectures, which is why so many *clones* (copies that work the same but usually cost less) of these machines exist.

If a computer has a **closed architecture,** the specifications for creating it are not public. That makes it difficult, if not impossible, for anyone else besides the machine's original manufacturer to create peripheral hardware to add-on to the machine. It also means no one else can recreate the same sort of computer. The Macintosh has a partially closed architecture—Apple released enough information for other manufacturers to build add-in boards and peripherals, but not enough for them to build Mac clones.

The terms "open" and "closed" architecture can also simply refer to whether or not a computer has *expansion slots* inside of it that allow you to plug in other add-in boards to expand or customize your machine. The original Macintosh had, in this sense, a closed architecture, whereas the later Macs have an open architecture.

archive

The term **archive** has the same general meanings in computer jargon as it does in English. As a verb, to archive means to store information, to back it up, generally with the idea that you are going to preserve it for a long time. As a noun, an archive is the information you store.

You archive a file to store it for safe keeping, or perhaps because you need it infrequently and so you want to keep it on a floppy disk in a filing cabinet instead of on your hard disk. Some people make a distinction between archiving files and *backing* them *up*. In the strict sense, an archive is intended as a permanent record that you may want to refer to later. For instance, architects and engineers who use computers keep archived copies of their plans in case a roof should cave in or some other catastrophe should befall them and an inspector needs to look at the plans.

Any backup file, by contrast, is intended as a safeguard in case the original file gets lost or damaged. A given backup file may serve its purpose only temporarily. If you're working on a long report, you'll create a new backup file every day or so, containing the most recent revisions. Yesterday's backup file may become more or less worthless.

In the process of archiving files, most people also *compress* them so all the backup copies don't take up so much disk space. So, you'll often hear the term *archive* used to mean the entire process of compressing and storing. (To *compress* your files, you need special software; see *compress*.) Also in the process of archiving, many people pack a set of files into a single file (with or without compressing them) to store or send them as a unit.

But be aware: *compressing an archived file may not be such a good idea. Even though you are archiving files, you never want to compress your original and only file. All it takes is to lose one bit, one electronic signal, from a compressed file and that file is destroyed. Bits get lost all the time. In an uncompressed file, it's usually not a problem, but in a compressed file, it's a catastrophe of considerable dimension.*

To **unarchive** means, logically, to go get those files that you stored away (and to uncompress them if necessary). Or perhaps you received an archive through your modem. Before you can use it you must unarchive the file.

arrow keys

Most computer keyboards have a set of four **arrow keys** that can be used to move the *cursor* or the *insertion point* around the text on your screen. Each time you press an arrow key the cursor moves one character left or right, or one line up or down. To move the cursor in larger increments, or to select text as it moves, you can often use other keys in conjunction with the arrow keys, such as the Shift, Command, or Control keys.

The *numeric keypads* on many keyboards often work as arrow keys also. You may need to press the *Num Lock key* to switch from typing numbers to moving the cursor.

artificial intelligence

Artificial intelligence is a very general term applied to computer applications that attempt to imitate human reasoning. The term is usually used to describe the capabilities of *expert systems*. Expert system programs can be taught a set of rules for handling a well-defined yet complex problem, such as choosing the right antibiotic for meningitis. Artificially intelligent software can "learn" in a rudimentary fashion, adjusting the solutions it comes up with on the basis of past experience. But no expert system can begin to imitate the flexible intelligence of human beings.

Alan Turing proposed a strict test for artificial intelligence years ago. He said, "A machine has artificial intelligence when there is no discernible difference between the conversation generated by the machine and that of an intelligent person." His test goes like this: In Room A you sit a person at a computer. In Room B you sit another person at a computer, plus you have a separate computer that claims to be intelligent, controlling itself. Using her computer, the person in Room A converses with the person and with the "intelligent" computer in Room B, one at a time, without knowing which one she is "talking" to. The person in Room A could type "How's the weather" or "What is the speed of light in a vacuum?" or "What are you going to do after breakfast?" and the person or computer in Room B would send back the answer. If the person in Room A can't tell whether she's talking to a person using a computer or to the self-controlled computer, then the computer that is controlling itself is artificially intelligent.

As Arthur Naiman says of this test, "No computer in existence today can even dream of passing the Turing test—but then again, neither can most newscasters, corporate executives, or four-star generals."

ascender

The part of a lowercase letter of the alphabet that is taller than the letter **x** is called an **ascender.** For instance, the letters d, b, l, and h each have an ascender. The parts of the letters that hang below the *baseline* (the invisible line on which the letters sit) are called descenders; the letters p, q, and j have descenders. Also see *x-height* and *baseline.*

ascending order, ascending sort

First of all, to *sort* is computer jargon for alphabetize (or, if you're talking about numbers, to organize into numeric order). So an **ascending sort** means to alphabetize starting at the bottom and going up, like starting at A and going to Z (which is the way you were taught to do it in school). If your data consists of numbers, though, rather than text, an ascending sort would arrange them with the smallest number first, the next largest number second, and so on. Or if your data is a set of calendar dates, an ascending sort would put them in chronological order from earliest to latest. The opposite of ascending sort is *descending sort,* of course. Also see *sort,* which includes the difference between sort and *find.*

ASCII

ASCII (pronounced "askee") stands for **A**merican **s**tandard **c**ode for **i**nformation **i**nterchange. It's a standardized coding system used by almost all computers and printers for letters, numerals, punctuation marks, and invisible characters such as Return, Tab, Control, etc. The fact that almost everyone agrees on ASCII makes it relatively easy to exchange information between different programs, different *operating systems,* and even different computers. It also means you can easily print basic text and numbers on any printer, with the notable exception of *PostScript printers*. If you are working in the MacWrite word processing application on the Mac and you need to send your file to someone who uses WordStar on the PC, you can save the document as an ASCII file (which is the same as *text-only*). After you transfer the file to the PC (on a disk or via a *cable* or *modem*), the other person will be able to open the file in WordStar.

In ASCII, each character has a number which the computer or printer uses to represent that character. For instance, a capital *A* is number 65 in the code. Although there are 256 possible charcters in the code, ASCII standardizes only 128 characters, and the first 32 of these are "control characters," which are supposed to be used to control the computer and don't appear on the screen. That leaves only enough code numbers for all the capital and lowercase letters, the digits, and the most common punctuation marks.

Another ASCII limitation is that the code doesn't include any information about the way the text should look (its *format*). A*SCII* only tells you which characters the text contains. If you save a formatted document as *ASCII,* you will lose all the font formatting, such as the typeface changes, the italics, the bolds, and even the special characters like ©, ™, or ®. Usually carriage returns and tabs are saved. Also see *high ASCII.*

aspect ratio

Ooops—I did not maintain the aspect ratio when I resized this graphic.

Aspect ratio is a fancy term for "proportion," or the ratio of width to height. For instance, if a direction in a software manual tells you to "hold down the Shift key while you resize a graphic in order to maintain the aspect ratio," it simply means that if you *don't* hold down the Shift key you will stretch the image out of proportion.

Some combinations of computers and printers have trouble maintaining the correct **aspect ratio** when the image goes from the screen to the printer, or when the image is transferred from one system to another, so the aspect ratio can be an important specification to consider when choosing hardware.

assembler

An **assembler** is a software program that takes what the programmer wrote in *assembly language* and translates it into a program the computer can run. (Actually, the output of the assembler must be processed by a "linker" to produce the finished program.)

assembly language

An **assembly language** is a *programming language* that allows a programmer (a human) to tell the *microprocessor* (the *chip*) in the computer exactly what to do, in terms of the specific operations the processor knows how to perform. You can think of the difference between assembly language and a *high-level language* such as BASIC, C, or Pascal this way: A program in a high-level language is like saying "point at me," while the assembly language version is like telling a person to contract the muscles that elevate the shoulder, then contract the muscles that extend the elbow, and finally contract the muscles that extend the index finger. This analogy isn't perfect, but it should give you the flavor of the difference. In assembly, each programming line corresponds directly to an instruction in the processor's *machine language*.

Besides being laborious to write, assembly language programs have another drawback: They only run on one *microprocessor family,* sometimes only on a single microprocessor. In other words, an assembly program written for the Mac won't run on a PC, and vice versa. That's because each processor knows a different set of operations.

Of course, there must be a reason people write programs in assembly language. Actually, there are three: speed, program size, and control. Assuming equal skill of the programmers, an assembly language program is almost always faster than the equivalent high-level program, and in its finished, *executable* form, it's usually much smaller (even though the assembly programmer had to write many more lines of *code*). And because you can control the microprocessor on a step-by-step basis, your program gives you exactly the results you want.

asterisk *

The **asterisk** (*) is often called a *star* (I think it's called a star because so many people can't spell it or pronounce it; it is *not* "ateriks" or "asterik," although sometimes it is a "splat.") The asterisk has several functions on a computer. One is as the symbol for multiplication in on-screen calculators, in spreadsheets and databases, in programming, etc. You can use the asterisk on your keyboard's numeric keypad to multiply, and usually the asterisk above the number 8 on the keyboard also.

Another use for this character is as a *wildcard character* in some applications and *operating systems.* For instance, if you *search* for *.* (pronounced "*star dot star*") on a *DOS* system, you can find a file with any name and extension. Or you can, in some applications, search for something like "rat*" and find rate, rationale, and ratfink, but not brat. Other applications may use a different character for a wildcard.

In *online* communications, an * means a kiss. (And this: -----/---@ is a long-stemmed rose.) See *baudy language* for a list of other important online visual communication shortcuts.

asynchronous

The word **asynchronous** is most often used when referring to transmission of data between two devices, as in "asynchronous communications." It usually applies to *serial* (as opposed to *parallel*) data transfers. Asynchronous means that the *clocks* or timers used by the two devices—such as your *modem* and the modem at the other end of the phone line—aren't running in synch with one another. For this reason, the data must include extra information to allow the two modems to agree on when each chunk (*byte*) of data starts and stops. This arrangement makes asynchronous communications less efficient than synchronous, but the latter requires sophisticated, expensive equipment. Synchronous communications are primarily used by large businesses and government agencies.

Asynchronous can also refer to events that occur at different times rather than concurrently. For instance, an exchange of e-mail involves asynchronous events: When someone sends me *e-mail,* I don't have to be *online* at that moment to receive it—I can pick up the message any time I like, which makes it an asynchronous event.

AT

Originally, **AT** was the name IBM gave to its second generation of PCs, introduced in 1984 (the full name was IBM PC-AT). The acronym stands for Advanced Technology, though a lot of people didn't think the machine lived up to its name, even when it first hit the market. As its main *microprocessor,* the AT used the *80286,* a faster, more capable chip than the one in the original IBM PC, but much less powerful than today's *80386* and *80486.* Nowadays, when someone says they have an AT, they usually just mean their PC has an 80286, whether or not the computer was made by IBM.

Despite its limitations, the AT did set standards that PCs have been following ever since. Almost all PCs sold today, even the ones with 80386 and 80486 processors, are "AT-*compatible*." This is important to you because it means these computers can run the same software and accept the same *expansion cards* (see *bus* and *ISA* for more on this last point).

AT command, AT command set

When you are using a *Hayes-compatible modem,* the collection of commands you can use to control your *modem* is called the **AT command set** (pronounced as the two letters, "a tee"). An **AT command** is one of the individual commands in the AT command set. They are called AT commands because each one must start with the letters AT, for **at**tention code.

You don't need to learn the AT command set. When you use *telecommunications software,* the program lets you control your modem easily—usually by picking from options listed in English on a menu. The software then translates your choices into the corresponding AT commands, sending them on to your modem.

A-to-D conversion

A-to-D conversion (**a**nalog-to-**d**igital conversion) is the process of converting *analog* information from life into *digital* information that a computer can understand. Most information in nature is in analog form, covering a continuous range of values (such as the passing of time, as exemplified by the hands of a regular clock). Digital information is in discrete, or separate, values, such as the countable chunks of seconds and minutes in a digital clock. You might want to use a computer to record analog information from nature by hooking it up to some kind of sensor like a light or pressure meter. But to store this information in the computer, the analog readings from these meters must be passed through an A-to-D converter to turn them into a digital form. Also see *analog* and *digital.*

attributes

Attributes means characteristics. For instance, in a database or a spreadsheet you can apply attributes to each *field* or *cell* to customize your document. As a general attribute, you can choose whether it is to be a text field or a numeric field or perhaps a computed field, whose value the application calculates for you. Then you can apply more specific attributes to the field or cell, such as making the text bold and right-aligned and perhaps in a particular typeface. If a field is numeric, you will have other attribute options available, such as how many decimal places to display, whether to use a dollar sign or a percent symbol, or whether to start a formula in the cell.

authorized Apple dealer

An **authorized Apple dealer** is, logically, a dealer who is authorized to sell Apple computer equipment. Apple only authorizes certain dealers, and they give these dealers authority to sell only specified machines. That's why you don't see Macintoshes in Home Club. You can sometimes buy Macs from unauthorized dealers, which is considered the "gray market."

Apple's *Performa* line of Macintoshes were developed to be sold in mass market stores such as Sears, Office Depot, and Circuit City, without upsetting the authorized Apple dealers.

AUTOEXEC.BAT, AUTOEXEC file

The name of this *batch file* comes from a combination of the words **auto**matically **exe**cuted **bat**ch file. Like other batch files, **AUTOEXEC.BAT** contains a series of *DOS* commands that your (*IBM-compatible) PC* runs for you, one after the other, so that you don't have to type the commands individually. What's special about AUTOEXEC.BAT is just that DOS automatically runs this particular batch file each time you turn on or restart the computer. Most people wind up calling AUTOEXEC.BAT the "AUTOEXEC" file for short.

You can customize the AUTOEXEC file yourself, filling it with exactly the commands you want to get your system up and running to your specifications and to suit the needs of your software and *peripherals*. AUTOEXEC.BAT is typically used to set the look of the DOS prompt, tell DOS which *directories* it should search when looking for programs to run, configure the *serial ports,* load the mouse *driver*, and start *memory resident* programs and utilities. If you like, you can use the AUTOEXEC file to start a particular application program (such as your word processor) or to start Windows.

Here's an excerpt from a typical AUTOEXEC.BAT file:

```
PROMPT $p$g
PATH=C:\WINDOWS;C:\DOS;C:\UTILS
SET TEMP=D:\TEMP
```

These lines set the DOS prompt to show the current directory; set the DOS *path;* and tell the system to look for temporary files in the D:\TEMP directory.

auto-repeat key

An **auto-repeat key** is a key that will automatically repeat, after a brief pause, while you keep it pressed.

Every key on a Macintosh keyboard auto-repeats unless you choose to turn it off through the Keyboard control panel. You can also choose how fast you want the keys to repeat, and how heavy or light your touch should be on the keyboard to start the repetition.

On a PC keyboard, all the keys auto-repeat, and you can control the rate they repeat—see *typematic rate.*

autosync

Autosync refers to a type of computer monitor and is synonymous with *multiscan*. Please see *multiscan* for details.

auto-save

Some programs have an **auto-save** feature that will automatically *save* your document to disk every so many minutes, just in case you forget. You can control how often you want the auto-save to interrupt your work. This is nice in theory, and it is nice for people who forget to save. Personally, I hate to be interrupted in the middle of my work as the auto-save feature stops what I am doing and does its own thing—I prefer to have control over it myself.

A/UX

A/UX (pronounced "ox") is a version of the *UNIX operating system* that can run on any of the computers in the Mac II series. If you need to run a UNIX application, as many people in government and industry do, A/UX allows you to run an A/UX-*compatible* version of the application on the Mac. Regular Macintosh applications run on A/UX as well. A/UX has enabled many government departments to requisition Macs, since the availability of UNIX for a given system is a requirement for many government branches.

The A/UX system does require generous amounts of *memory* and disk space, as well as plenty of processing power, and is not as easy to use as the standard Macintosh operating system.

background, background process

The plain ol' everday word **background** takes on a multitude of meanings in the jargon jungle. If something happens "in the background," that means it happens while you are doing something else. When you print, for instance, you might be accustomed to having to wait until the printer is finished before you can have your screen back to work on. But some systems allow you to print in the background, meaning while the printer is processing and printing your document, you can continue working on the screen (also see *spool* and *spooler.*) Some systems, including Microsoft Windows and the Macintosh, let you copy files or format disks in the background, and you can buy software that lets you *telecommunicate* or receive faxes in the background too. A **background process** is anything that's happening in the background.

In *HyperCard,* a document is called a *stack* because it is made up of a set, or stack, of "cards" on the screen. Each card has two layers—the **background**

backward compatible

Computer products—from software applications to peripherals to the computers themselves—are constantly being updated. Programs start out at *version* 1.0, then as they progress the version numbers increase. Typically once you have updated your application, any document you create in the updated version cannot be opened by an earlier version of the same application. For instance, the application PageMaker version 4.2 cannot open any publication created in PageMaker version 5.0. Some products, though, are **backward compatible** (also known as "downward compatible"), which means that a document created in an updated version of a software application can still be opened by an earlier version of the same application. For instance, documents created in Microsoft Word 5.1 (which is backward compatible) can be opened by Word 4 (which is now considered upward compatible). Of course, as you go backwards your document will lose the newer features that are part of the updated version.

Backward compatibility also applies to computers: if a newer model can run the same software as an older model, the newer model is considered to be backward compatible. Almost all Macintoshes are backward compatible with all older versions of the Mac. Machines like the IBM PC/AT are backward compatible with the IBM PC/XT.

bad sector

A disk has two sides (a top and a bottom). Each side of the disk has tracks, or concentric rings on the surface. Each ring is divided like a pie into equal wedges, or sectors, which are the smallest units of storage space on the disk. If one of these units is damaged or flawed, it is considered a **bad sector** and cannot be used. If there was already data in that sector when it got damaged, chances are slim that you can recover that data unless you have the specialized hardware and software necessary for that sort of operation. Almost all hard disks are born with bad sectors, so don't freak out if your software utility reports them. Other bad sectors should not start appearing, though, after you start using the disk.

Balloon Help

Balloon Help is a feature of System 7 on the Macintosh where little "balloons" (like cartoon-character talk) automatically pop up when you point to items on the screen. The balloons contain helpful information about the item you are pointing at. You can view the available balloons at the *Finder* (also known as the *Desktop*) and in some applications by choosing "Show Balloons" from the Help menu (the one with the question mark). Many software packages are including balloon help right in the application so you can just point (you don't have to click) to a menu item or a dialog box to see a balloon that explains how that item works.

The idea sounds good, but in reality the balloons tend to have pretty wimpy information in them. Plus they drive ya nuts the way they pop up in front of your face constantly. Some software companies have elected to ignore them altogether in their new software, while other companies say they'll be putting information that's actually useful in their balloons.

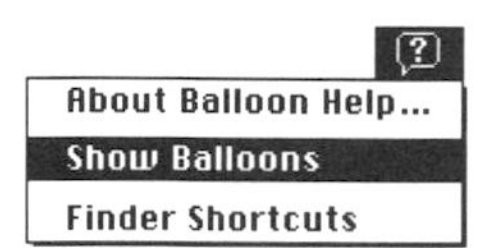

Choose "Show Balloons" from the Help menu, which is the one with the question mark in the balloon. If you don't have the question mark, then you are not on System 7 and won't have balloons.

Trash

To discard an item or eject a disk, drag it to the Trash. To get an item back, open the Trash icon and drag the item out of the Trash window.

The Trash is bulging because there is something in it.

Trash

This is an example of the balloon that appears when you point to a full trash can.

banding

Some *draw* and illustration programs let you create a fountain fill or "split fountain" effect in which the color or shade of a part of your picture changes gradually from light to dark or from one color to another. But due to the relatively low resolution of most lower-end printers, a fountain often shows the gradation from dark to light in distinct, contrasting bands rather than in one continuous flow. This effect is called **banding.** The higher the resolution of the printer, the less banding will be visible.

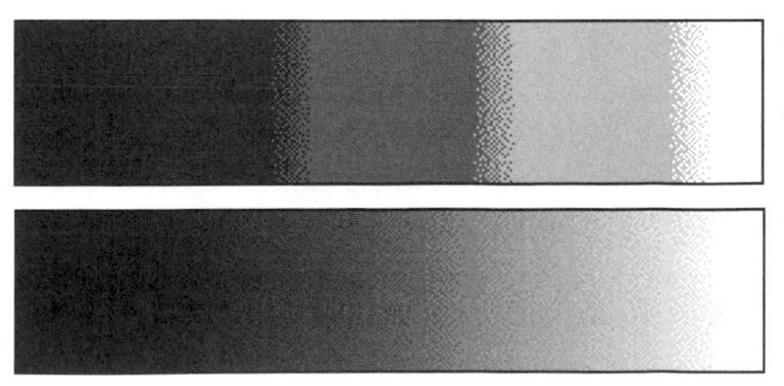

Depending on the number of colors or grays you can see on your screen (see resolution, monitor), you may see banding on the screen. That does not mean the banding will print. The rectangle on top is how it looked on a screen with only 16 colors; the rectangle on the bottom is how it printed on an imagesetter.

bandwidth

Bandwidth measures the amount of information that can flow between two points in a certain period of time. The "broader" the bandwidth, the faster the information flow. You can use the term to describe how quickly information moves from the hard disk into *memory,* or from the computer to an *add-in board* on the expansion *bus*, or from one *modem* to another across a telephone line. Depending on whether the transmission is *digital* or *analog,* the rate is measured in *bits per second (bps)* or in *hertz (cycles per second).*

The bandwidth of a particular part of your system can have a big impact on the speed of the system as a whole. Consider the situation when the computer needs to display a graphical image on the screen (*all* Macintosh and Windows screens are graphical). The computer's main processor is responsible for deciding which dots on the screen should be white, which red, and so on. But it must transfer that information through a *bus* to the video circuits that actually control your screen. No matter how fast the processor "composes" the screen, your monitor will seem sluggish if the bandwidth of the bus is narrow.

The newer expansion buses for the IBM and compatible PCs, the *Micro Channel* and the *EISA*, were developed partly to improve the bus bandwidth and so take better advantage of increased processor speed. And some new PCs have video circuits built right into the main circuit board of the computer, linked directly to the main processor via a high-speed *local* or direct *bus.*

bank

A **bank** is a collection of identical electronic devices that are connected so they can be used as a unit. For instance, a bank of memory consists of a number of separate memory chips that can be accessed as if they were one continuous section of memory.

baseline

Referring to type, the **baseline** is the invisible line that the letters sit on. Some letters, such as p, y, or g, have *descenders* that drop below the baseline. Also see *descender, ascender,* and *x-height.*

funny little piglets ←

This is the baseline, upon which all the letters sit. Notice the descenders of y, p, and g.

BASIC, BASICA

BASIC stands for **b**eginner's **a**ll-purpose **s**ymbolic **i**nstruction **c**ode, and is a computer *programming language* developed in the mid-sixties. BASIC has the advantage of English-like commands that are easier to understand and remember than those of most other languages. Even so, the latest versions of BASIC can do just about anything programming languages like *C* or *Pascal* can do.

BASIC programs have a reputation for being very slow, which they certainly were in the early days of personal computing. This sluggishness was mostly due to the fact that in those days, BASIC was an "interpreted" language; that is, every time you ran a BASIC program, you were really running an "interpreter" which executed your program code line by line, converting it *on the fly* into a form your computer could understand. That conversion process takes time. Now many good BASIC *compilers* are available. A compiler does the conversion ahead of time and only once, turning the program code into an executable program your system can run directly, at top speed. So modern, compiled BASIC is easier to use and just about as fast as C.

Interpreters do have some advantages, though. The process of writing and testing an interpreted program is actually quicker and more convenient than with a compiled program (for the explanation, see the entry for *interpreter*). A BASIC interpreter makes especially good sense for creating short, simple programs, which is all that most personal computer users would be willing to tackle. If you're interested, you can find lots of old computer books full of BASIC programs at public libraries. At any rate, some computers come with a BASIC interpreter. The best example is Microsoft's BASIC, or GW BASIC, the interpreter that comes with MS-DOS. True IBM-brand PCs had slightly different versions of the interpreter called BASIC and BASICA (Advanced BASIC) that only worked on those computers.

BAT or .BAT file

When you see the file extension **.BAT** (pronounced "dot bat") at the end of an MS-DOS file name it means the file is a batch file. Please see *batch file*.

batch file

A **batch file** is a file containing a series of commands that the operating system will carry out for you, one at a time. All you have to do is start the batch file—from that point on, the process is automatic. Batch files are great when you use a given set of commands repeatedly—instead of activating each command separately every time you want to carry out that set of commands, you can accomplish the same thing in one step.

Many computer operating systems have the built-in ability to execute batch files (or their equivalent). In *DOS*, the operating system for IBM-PCs and compatibles, you can recognize a batch file by its name, which always has the .BAT extension (as in GO.BAT, SEND.BAT, or AUTOEXEC.BAT). *DOS* includes a group of special commands that are only used in batch files.

Here's an example of a simple DOS batch file used by a writer who often sends one of his poems to someone else on floppy disk. This batch file *formats* the disk, then goes on to automatically copy the poem file there:

```
FORMAT A:
COPY C:\NEWPOEMS\%1.POM A:
```

If this batch file was named SEND.BAT, our writer would just type

```
SEND DAISY
```

to do the formatting and copy the file *DAISY.POM* to the disk. (DOS knows to substitute "whatever he types after SEND" for the "%1" in the batch file.)

Computers and operating systems with graphical user interfaces, such as Windows and the Macintosh, usually don't have batch file capability because things are so much easier to do anyway. You can buy software products that let you create batch files for these systems (Frontier for the Mac), or you can use *macros*.

batch processing

Batch processing refers to an operation in which the computer automatically performs a given command on a number of files, or performs a series of predetermined commands without further intervention on your part.

Running a *batch file* is one example of batch processing, but there are plenty of others. When you select several documents from the same application and print them all in one step (if the application allows you to do that), you are "batch printing," which is a form of batch processing. Or let's say that you want to send a whole group of files to someone else via your modem—if your *communications software* permits batch processing, you can choose all the files you want to send, and have the software send them off in a batch while you go to the kitchen for a snack. Batch processing is a good feature to have in most applications.

Battery

The **Battery** icon in your Mac's System Folder (in System 7 it is inside the Apple Menu Items folder) and the Battery item on your Apple menu are only for use with the battery-powered Macintoshes such as the PowerBooks. The Battery is a little *utility* that lets you know how much charge is left in the computer's battery. If you're using any type of Mac other than a Portable or Powerbook, you can throw away the utility (move the icon from your System Folder into the trash).

This is the icon for the Battery.

When you choose Battery from the Apple menu, you can see how much power you have left. Option-click on the picture of the battery to put the computer to sleep.

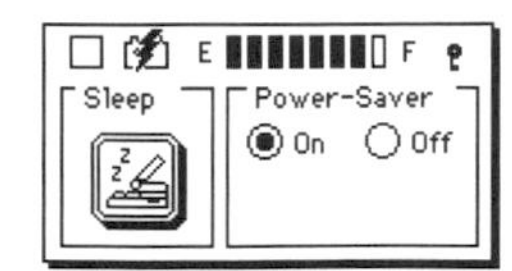

When you click the flag in the upper right corner, you'll get this tiny control panel.

baud, baud rate

When you use a modem to send information from one computer to another over the phone lines, the information moves at a certain speed. This speed is measured in *bits per second,* a *bit* being one electronic unit of information. This bits-per-second rate is also commonly called the **baud rate** (pronounced "bod"). It's a measure of how fast your modem can send and receive information. Modems most typically send at 1200, 2400, or 9600 baud.

Now, I do have to warn you that even though the masses (that's us) often use the term "baud" to mean the same thing as "bits per second," this is technically incorrect, and someday someone will surely scold you. One baud is actually one "modulation change" or one "signal event" per second on a communications channel. It's only at 300 baud that one signal event equals one bit per second. A modem operating at 1200 bits per second needs only one signal event to send 2 bits, so if you want to be persnickety you should call it a 600 baud modem. See *bps* for a discussion of common communications speeds and how they translate into speed ratings you can relate to.

The term "baud" is named after the inventor of the Baudot telegraph code, J.M.E. Baudot.

baudy language

Guy calls the collection of abbreviations and symbols used in electronic mail messages **baudy language** (see *baud*). It is an integral part of online communication to use baudy language, since many messages can be misinterpreted unless accompanied by a smile, a frown, a wink, a giggle.

The following list of symbols are to be read with your left ear on your left shoulder. As you become accustomed to using and reading them, you'll find that you will be able to understand them without bending your head. All of the symbols can have a nose added to them with a hyphen, as in :-), which is how you will commonly see them on CompuServe. People on America Online typically leave out the nose.

:) smile

;) wink

: D big smile, laughing

: > sarcastic or devilish smile

: (frown

:' (crying

: | neutral; bored

}:) horny smile

:-{) smile with a mustache, usually indicating a male

: {> devilish smile with a mustache

: O O Dear, O Woe is Me

: P tongue sticking out; raspberry; bronx cheer (really rather rude and not to be used lightly)

: X my lips are sealed

0:) angel

}: > devil

::: ::: colons surround an action that the writer is pretending to do, such as **:::going to bar and getting InfraDig a daiquiri:::** *or* **:::whacking FooFoo upside the head:::** *or* **:::getting massage oil now:::**

{{ }} indicates the person whose name is between the curly brackets is getting hugged

<--- indicates the person writing this message is the subject of the sentence, as in

ToadHall: <—- female, in answer to your question

***** kiss

P* french kiss

^ ^ ^ giggles (this solves the problem of how to spell "tee hee"); type the caret (^) with Shift 6.

-------\----@ long-stemmed rose

The following list of shortcuts are abbreviations for commonly used phrases that give the other readers important information.

!	I have a comment
?	I have a question
afk	away from keyboard
ALink	AppleLink
AOL	America Online
asap	as soon as possible
awc	after while, crocodile
bak	back at keyboard
brb	be right back
btw	by the way
byob	bring your own bottle
byom	bring your own Mac
CIS	CompuServe
CI$	CompuSpend
cul	see you later
cula	see you later, alligator
esad	eat sh*t and die
fubar	f**ked up beyond all repair
fyi	for your information
ga	go ahead
gmta	great minds think alike
ima	I might add
imho	in my humble opinion; in my honest opinion
imnsho	in my not-so-humble opinion
imo	in my opinion
lol	laughing out loud
obtw	oh, by the way
oic	oh, I see
otoh	on the other hand
rotflol	rolling on the floor, laughing out loud
rsn	real soon now
snafu	situation normal: all f**ked up
sol	sh*t out of luck
sos	help!
tanj	there ain't no justice
tanstaafl	there ain't no such thing as a free lunch
tptb	the powers that be
ttfn	ta-ta for now
wb	welcome back
wtg!	way to go!
xoxoxo	kisses and hugs

BBS

BBS stands for **b**ulletin **b**oard **s**ervice, which is a service usually set up by an organization or a club to provide or exchange information. You *access* the BBS through your *modem*. After your modem has dialed the phone number for the BBS and established a connection to the BBS's modem, you'll see on your screen the computer equivalent of a bulletin board. You can read messages, post messages of your own, ask questions, answer questions, make new friends. Also see *online service.*

BCS

The **BCS** is the **Boston Computer Sociey.** This information came directly from Pam in Boston:

The BCS is made up of over 28,000 members of *user groups* and special interest groups (*SIG*, which is a specialty subset of a user group). The user groups include: Amiga, Apple II, Atari, Commodore, ZITEL (CP/M and DOS), DEC, Hewlett-Packard, IBM PC & Compatibles, Laptops, NeXT, Macintosh, Tandy, and Texas Instruments.

The *SIG*s include: Amateur Radio, Artificial Intelligence, Beginners, Business & Management, Church & Synagogue, CASE, Construction, Consultants & Entrepreneurs, Computer-Aided Publishing, Databases, dBase, Education/Logo, E-Mail, Graphics, Hypermedia/Optical Disk, International, Investments, Legal, Lotus, Medical/Dental, Music, Networking, Programming, Real Estate, Robotics, Science & Engineering, Social Impact, Special Needs, Telecommunications, Training & Documentation, and User Interface Design."

As you can see, the BCS is hard to define because each of these user groups and SIGs is independent, having their own Directors and Activists who run each group, meetings, workshops/seminars, etc. It's pulled together by the BCS Office and a staff which handles the overall management of the groups with budgeting and finance, newsletter printing and distribution, promotion, membership, a central telephone contact, and the Resource Center, a room with about 30 different computers where members can test or try out hardware and/or hundreds of software packages.

There are 25 *bulletin board services* (BBS) set up and maintained by these user groups and SIGs, with another 12 BBSs participating in a message exchange through BCS Echomail Conference.

The BCS offers the world's largest network of information and support for personal computer users. At the BCS, there's no such thing as a too-simple or too-complex question. They offer support services for people at all levels of experience. You learn at your own pace, according to your own needs.

BCS-Mac has its own separate office and telephone number: 617.625.7080.

The **IBM-PC Group** also has its own office and telephone number, which is 617.964.2547.

The Boston Computer Society
One Kendall Square
Cambridge, MA 02139
Tel: 617.252.0600
Member relations and assistance: ext. 3314
Join or renew by phone: ext. 3316
Regular 1-year membership: $39/year
(other rates available, such as Senior Citizen, Student, etc.)

beep

Beep is the generic term for whatever sound a computer makes when it's trying to tell you something. If you press the wrong key or click on something you shouldn't, the computer will beep at you. Sometimes it beeps just to let you know that it has started or finished doing something (like copying a file). You can customize the beep sound in some computers. In particular, newer Macs and PCs with Windows 3.1 come with several sounds you can assign to various types of "events" in the system. Hundreds can be added, and you can even make your own sounds.

bells and whistles

When a product is said to have **bells and whistles,** it means the product has some fancy features that go beyond the basics. These features may allow you to work faster or more conveniently, they may entertain you, or they may be there just to look impressive.

benchmark

A **benchmark** is a test used to measure the relative performance of hardware or software products. If you want to know which of two computers is faster, you can't take the manufacturers' word. Instead, you run the same benchmark software on both computers and compare the results. Magazines generally report benchmark results in comparative reviews.

Bernoulli box

A **Bernoulli box** is a removable cartridge disk system from Iomega Corporation. Technically, it is distinctly different from other cartridge disk systems in that Bernoullis use flexible disks (sort of like floppy disks, but these can hold something like 90 megabytes of data apiece) rather than hard disks. Something about the type of disk and the fluid dynamics in the disk mechanism make Bernoullis much more stable and better able to withstand bumps and jolts, so they are considered to be the most reliable of all removable cartridge, magnetic disks. They are also the most expensive, but you have to balance the price against the cost of lost data.

beta, beta tester, beta version

Computer products go through extensive testing before they can be released to the unsuspecting and trusting public. When a product has passed the in-house testing stage (see *alpha testing*), it goes into **beta testing,** often just called **beta. Beta versions** of the product are sent out for beta testing to "normal" people who don't work for the company. (These people are then, logically, called **beta testers** or **beta sites**.) The beta testers work with the software or hardware in real-life situations and report back the things that go wrong or that need improvement.

Usually, the software company sends out several beta versions during the development process. The first one often has lots of *bugs,* and some of the features may not work at all. But at least it gets the ball rolling, so the developers can get feedback on the product design from people in the outside world. Based on this feedback, the developers may decide to revise the way the program looks and works. The next beta version should have fewer bugs, and look more like the final version. The beta testers are supposed to try every feature to find anything that doesn't work right. Based on their reports, the programmers find and fix the problems.

Ideally, this process repeats itself a time or two more, until someone decides all the bugs have been found and everyone's pretty happy about the way the program works. The sad truth: A few bugs always remain, even after beta testing. Sadder yet: Products with known bugs are sometimes released commercially, just to get the product out on the market on schedule. See *bug.*

betaware

Betaware describes a product that is in the process of going through *beta testing* before being released to the public. Also see *vaporware.*

Bézier curves

When you draw a curved line in a sophisticated *draw* or illustration application (such as Adobe Illustrator, Aldus Freehand, or Corel Draw), the application represents your line in the computer as a particular kind of mathematical formula. Curved lines defined with these formulas are called called (pronounced "bayz yay"). Bézier curves are kind of tricky to work with, but they are an efficient way to create clean, smooth graphic images. Scalable fonts, like Type 1 and TrueType fonts, are also based on Bézier curves.

You don't need to know anything about the formulas involved, but they do affect the way you work with curves on the screen. They define a Bézier curve precisely by specifying a few points: the points where the line starts and stops, the points where the curve changes direction (known variously as the anchor points, handles, or nodes) and the control points for each of the anchor points (sometimes called BCPs, Bézier control points).

These are the anchor points.

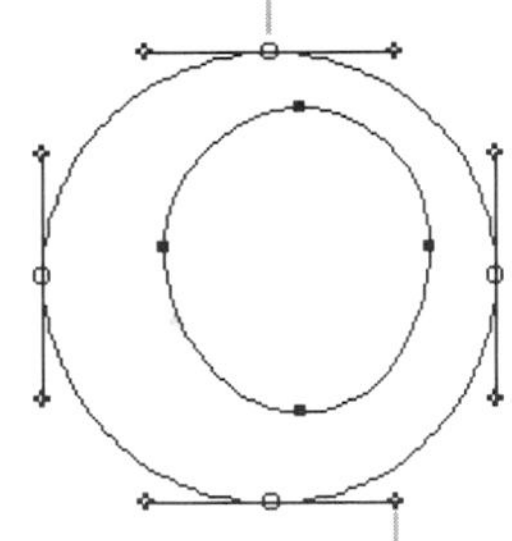

These are the control points, or BCPs.

On the screen, the anchor points (where the curve changes direction) appear as big squares or rectangles directly on the path of the curve. At each anchor point you must also know the angle of deflection, and how far the curve extends in that direction. These characteristics are represented on screen by the length and direction of dashed lines which extend out on either side of the anchor points and end in control points.

Once the anchor points and control points are defined, the program just connects the anchors with a smooth curve. Some programs let you draw freehand; after you've drawn a line, the program converts it to a Bézier, calculating where the anchor points and control points should go. In other programs, you place the anchor points by hand, one at a time. This method may sound less intuitive than the freehand approach, but it can actually be quicker and more precise. To modify the shape of the line, you can move the anchor point itself, or manipulate the control points to lengthen or shorten the deflection lines or change their angles.

All this sounds quite complicated, and it does take awhile to get the hang of these bizarre Béziers. But if you keep at it, you'll soon find that working with Bézier curves becomes a fairly intuitive process, and it definitely lets you place and structure curves exactly where you want them.

Big Blue

Big Blue is the nickname for IBM, the company. How did it get the name? Four different sources gave me four different reasons why it's called Big Blue. One said it's because the logo is blue. One said it's because all the machines were blue. Another said it's because all the typewriter and computer dust covers were blue. Another said it's because everyone was required to wear blue suits. Apparently there is some fixation on blue.

big red switch

On most PCs, the on/off switch (the power switch) is bright red and fairly large. But usually it's called the **big red switch** only in the context of something going really wrong with your system—so wrong that you can only fix it by shutting off your computer altogether, then re*booting* it. For instance, "Just as I pressed the save key, the program crashed again. I found myself reaching for the big red switch for the hundredth time that day."

bilevel bitmap

The image you see on your computer screen is made up of thousands of tiny dots, or pixels. On a monochrome monitor (black-and-white screen) each pixel on the screen can either turn on or off. Each pixel is controlled by a *bit* (a bit is the smallest unit of electronic information) that tells it whether to be black or white. Thus the **bilevel,** or two levels, of **bitmap** are black or white.

A "deep" bitmap, on the other hand, can be many more shades, or levels, of gray because it can deal with more than just one bit of data. See the entries for *bitmap* and *bit* for more information, plus the entry in the Symbols section on *2-bit, 4-bit, 8-bit (etc.) color.*

binary, binary system

The binary system is a method for working with numbers based on only two digits: 1 and 0 (binary is also known as "base two"). It's kind of a fun system to set your brain working on, if you like puzzles. In a binary number such as 10101, each digit represents a power of two; 10101 is 21 in the "base ten" system we're used to (base ten is our culture's counting system, but it isn't universal among humans). (I've always wondered how it works on powers of 2 if the only numbers are 1 and 0. But Steve reminded me that in the decimal system, which is based on powers of ten, there's no digit for "ten" either.)

On the computer, these binary numbers can easily be represented by tiny electronic signals, each of which is either "on" or "off" (1 or 0). Each one of these single signals, 1 or 0, is called a *bit.* The computer uses groups of these signals, or bits, to do all its work and to store all your information. For instance, when you press a key, a sequence of bits (such as 01000001) is sent to the computer, which in turn displays on your screen the character that corresponds to that number—in this case, the letter A. Also see the definitions for *bit, bitmapped, bilevel bitmap,* and *deep bitmap.*

For those who like number puzzles, this is from Steve: To make the conversion between binary and decimal, start at the right and multiply each digit by the next higher power of two, then add up all your answers. So, with 10101, the first digit equals 1 (1 times 1: 2 to the 0th power is 1); the next digit is 0 (2^1 equals 2, but 2 times 0 makes 0); the third digit equals 4 (1×2^2); the fourth digit is 0 again; and the 1 on the left equals 16 (1×2^4). And $16 + 0 + 4 + 0 + 1 = 21$. What fun!

BIOS, bios

Every computer is equipped with built-in instructions that tell it how to perform the really basic functions: start itself up, test itself, and communicate in a rudimentary way with the disk drives, keyboard, and screen. On an IBM-type personal computer, these built-in instructions are called the **bios** (pronounced "by ose," that's "ose" as in comatose). BIOS stands for **b**asic **i**nput-**o**utput **s**ystem. The BIOS is *firmware,* meaning it is a program built into the *read-only memory (ROM)* in your computer, rather than stored on a disk (because the BIOS is stored in ROM, it's sometimes called the ROM BIOS).

When you turn on your computer, the BIOS is responsible for checking all the hardware, including memory; it will display an error message if it finds a problem. The BIOS then loads the operating system—whether it's DOS, OS/2, Unix, or what have you—into memory from disk. Even after the operating system is running, the BIOS handles many essential chores, putting characters on the screen, getting characters from the keyboard, reading and writing sectors to the floppy or hard disk. You'll see this as your ROM BIOS chip on your computer that works with your software.

bis

The term **bis** (pronounced "biss") added to the name of a product, file, etc., indicates that the item in question (file or software) is the second version. For example, V.42*bis* is the designation of a particular standard for modem operations; two V.42*bis* modems can automatically compress data transferred between them, while plain V.42 modems can't do the compression but otherwise work the same. "Bis," which comes from Italian through Latin and Old Latin, originally meant "twice."

bisync, bisynchronous

Bisynch (pronounced "by sink") and **bisynchronous** are short for **bi**nary **synchronous,** which is a communications *protocol,* or set of rules for sending information between different machines. Bisynchronous communications are extensively used in *networks* on *mainframes*. Bisynch communications require that the *clocks* on both the computer sending the information and the computer receiving the information are synchronized before the information starts to transmit. Compare with *asynchronous* transmission.

bit

Bit is short for **binary digit**. One bit is the smallest unit of information that the computer can work with, and all the information it works with is made up of bits.

When you hit a key or click a mouse button, you send tiny electronic on/off signals to the computer. Each tiny electronic signal is **one bit.** The computer usually groups these tiny signals, or bits, into bigger chunks to work with: a series of eight bits strung together is a *byte;* a byte typically creates one character (letter, number, etc.) on your screen. A series of 1024 bytes strung together is a *kilobyte*. You've probably noticed that most computer storage devices (such as disks) and software files (your documents, for instance) are measured in kilobytes or *megabytes* (a megabyte is 1024 kilobytes). Well, those measurements are referring to how many electronic on/off signals it took to create and store the information. The more information there is, the greater the number of kilobytes or megabytes.

The fact that the computer has to work with so many bits is more than offset by another fact: it can handle bits far faster and with much cheaper circuits than it would take to work with decimal numbers directly. Also see *bitmap, binary system,* and *byte*.

By the way, you may have noticed that some of the switches that turn on your computer or other components, such as your hard drive or scanner, have a little 1 (one) and a 0 (zero) on them. Yes, they are a 1 and a 0, not an I (letter i) and an O (letter o). The 1 means on and the 0 means off, just like bits. (I/O, the letters i and o, stands for Input/Output, which is something completely different; see *I/O*.)

More fun from Steve: You can represent any number as a series of bits—you just have to use the "base 2" or *binary* counting system. In our accustomed base 10, or decimal, system, 7 equals 111. Reading from the right, each place signifies a power of 2 rather than a power of 10; in this case, the digit on the right means 1, the middle digit means 2, the one on the left, 4, for a total of 7. The expression 101 in binary equals 5 in decimal (1 plus 0 plus 4). In binary, three bits (digits) are enough to make any number between 0 and 7. Four bits cover everything from 0 to 16, eight bits take you up to 256, and 32 bits allow for numbers as large as 4,294,967,296. Whew!

bitmap, bitmapped

A **bitmap** is an image or shape of any kind—a picture, a text character, a photo—that's composed of a collection of tiny individual dots. A wild landscape on your screen is a **bitmapped** graphic, or simply a bitmap. Remember that whatever you see on the screen is composed of tiny dots called *pixels*. When you make a big swipe across the screen in a *paint program* with your computerized "brush," all that really happens is that you turn some of those pixels on and some off. You can then edit that bitmapped swipe dot by dot; that is, you can change any of the pixels in the image. Bitmaps can be created by a *scanner*, which converts drawings and photographs into electronic form, or by a human artist (like you) working with a *paint program*.

This painted image on the left, created with two brushstrokes, is bitmapped. On the right is an enlargement of a piece of the image, where you can see the individual pixels that have been mapped to the screen by the bits of information.

The simplest bitmaps are *monochrome*, which have only one color against a background (like black on white, as that shown above). For these, the computer needs just a single *bit* of information for each pixel (remember, a bit is the smallest unit of data the computer recognizes). One bit is all it takes to turn the dot off (black) or on (white). To produce the image you see, the bits get "mapped" to the pixels on the screen in a pattern that displays the image.

In images containing more than black and white, you need more than one bit to specify the colors or shades of gray of each dot in the image. Multicolor images are bitmaps also. An image that can have many different colors or shades of gray is called a "deep bitmap," while a monochrome bitmap is known as a "bilevel bitmap." The "depth" of a bitmap—how many colors or shades it can contain—has a huge impact on how much memory and/or disk space the image consumes. A 256-color bitmap needs 8 times as much information, and thus disk space and memory, as a monochrome bitmap.

The *resolution* of a bitmapped image depends on the application or scanner you use to create the image, and the resolution setting you choose at the time. It's common to find bitmapped images with resolutions of 72 *dots per inch (dpi),* 144 dpi, 300 dpi, or even 600 dpi. A bitmap's resolution is permanently fixed—a bitmapped graphic created at 72 dpi will print at 72 dpi even on a 300 dpi printer such as the LaserWriter. On the other hand, you can never exceed the resolution of your output device (the screen, printer, or what have you); even though you scanned an image at 600 dpi, it still only prints at 300 dpi on a LaserWriter, since that's the LaserWriter's top resolution.

You can contrast bitmapped images with *vector* or *object-oriented* images, in which the image is represented by a mathematical description of the shapes involved. You can edit the **shapes** of an object graphic, but not the individual **dots**. On the other hand, object-oriented graphics are always displayed or printed at the maximum resolution of the output device. But keep in mind that an object-oriented graphic is still displayed as a bitmap on the screen.

bitmapped font

All fonts (typefaces) that you see on the screen are **bitmapped.** That's the only way the computer can display the typeface on the screen, since the screen is composed of dots (pixels). Some fonts have no other information to them than the bitmapped display you see on the screen, while other fonts have additional data that is used by the printer to print the typeface smoothly on a page (*outline,* or *scalable* fonts).

Please see the definition under *font* for a discussion of bitmapped and screen fonts, outline and printer fonts, resident and downloadable fonts, PostScript and TrueType fonts, etc.

bleed

Bleed refers to any element on a page that is printed beyond the edge of the paper. Whenever you see anything (text, graphics, photographs) that is printed right up to the edge of the paper, it was actually printed onto larger paper over the margin guidelines, and the paper was trimmed.

Some press operators might tell you, "No bleeds allowed." Some might say, "We charge extra for bleeds." They might ask, "Does it bleed?" You might say, "Can I bleed this line?" An image that bleeds off all four edges of the paper is called a "full bleed." This page you are reading has no bleeds; for this page size and the press we used, bleeds were not allowed.

The reason the printer might not allow a bleed is because she plans to print the job on the right-size paper, without any trimming. If the image prints right up to the very edge of the paper, then the ink "bleeds" off the paper and gets all over the press and makes a mess.

bleeding edge

People who are on the leading edge in the computer field often feel like they are really on the **bleeding edge,** as they struggle with making incredibly complex new technology function properly, as they push the borders and find limitations, as they are overwhelmed with the acceleration of information, as the pace of their lifestyle increases with the pace of technological advances and what is expected of their machines.

s i g h . . .

blend

In many graphics programs you can choose to create a **blend,** where a color starts at one part of an object and blends into another color at another end of the object, like the sky at sunset. A good blend avoids *banding,* where you can visibly see the steps in the range of color.

blessed System Folder

With later versions of the Macintosh System, including System 7, the System Folder has a little icon of a Mac on its cover. This tells you that the folder is blessed (pronounced "bless edd," with two syllables). What that really means is that this folder contains the System and the Finder that are running the computer.

The blessed System Folder

Some people make the dreadful mistake of putting more than one System Folder on their hard disk. (Don't do that—the computer can get confused and check out. Like seriously crash.) Even though there may be more than one System Folder on a hard disk, only one of those folders should be blessed. Remove any System Folder you may find that is not blessed. (Look inside first to see if you want to keep any of the *extensions, control panels,* or other *files* that may be in there.)

block

In any kind of text editing, a **block** of text refers to any chunk you select, whether it is two pages, three paragraphs, a line, half a word, or even a single character. You can perform "block operations" on all the text in the block at once. When you take the block and move it somewhere else in the text, that's a block move. "Block move" accomplishes the same thing as a *cut-and-paste.*

In telecommunications, a **block** is a group of characters that is transmitted as one unit. Blocks can be different lengths, and the longer the block, the faster the transmission goes, generally speaking.

BMUG

BMUG used to stand for the Berkeley Macintosh User Group, but now their official name is just BMUG and it doesn't really stand for anything. But they are still the largest Macintosh *user group* in the world, with over 10,000 members in more than 50 countries. BMUG is an educational non-profit corporation dedicated to collecting, evaluating, and disseminating information about graphical interface computers (which is what the Mac is). They give users the information they need to use their computers most efficiently and painlessly.

BMUG's semi-annual newsletter is a valuable tome of 400 pages, offering reviews, tutorials, reference material, and commentary—and no advertising. They run a Helpline with high-quality technical assistance and expert data recovery, available in person and over the phone every business day. The BMUG *BBS* (bulletin board service) is a multi-line information service linking EchoMac, UseNET, the Internet, and local conferences, as well as the latest files from their massive software library.

BMUG holds meetings in Berkeley, San Francisco, and San Jose, California. They also hold a large number of meetings for special interest groups *(SIGs),* such as graphic design, beginners, networking, programmers, HyperCard, multimedia, etc. etc. etc.

board

A **board** (short for *printed circuit board),* or *card,* is a thin, rectangular sheet of fiberglass with *chips* and other electronic parts attached to it. Shallow metallic "traces" etched into the surface of the board serve as connecting wires between the parts. The main circuits of the computer are mounted on the *motherboard;* smaller boards that attach to other boards are called *daughterboards;* and boards designed to be plugged into slots inside your computer are called *add-in boards*. Please see Appendix A for several illustrations of boards and where they go.

boilerplate

A **boilerplate** is the part of a memo, newsletter, or any other regular communication or design that is repeated in every similar project. For instance, in your newsletter you use the same masthead across the top every month—that is the boilerplate. In an advertisement, the logo and the address are a boilerplate. In a mail *merge,* the entire letter is referred to as the boilerplate, except where the merge data goes.

A boilerplate is different from a template in that a template usually holds information that will change for each publication, while a boilerplate is exactly the same each time. For instance, you might have a template for a newsletter that has all the type set up and all the specifications prepared; each month you replace the existing type with the new text, and you change the volume and date. In this newsletter template, though, you probably have some boilerplate information—like the return address, the masthead, and perhaps a list of the Board of Directors—that stays the same every month.

bomb

Bomb is another term for *crash;* I suggest you read the definition for *crash* for more details on why this may happen to you and some tips on preventing it.

The Macintosh turns this figure of speech into a visual image, putting a picture of an old-fashioned bomb on the screen when the system crashes. I don't think this is very nice. It lays a guilt trip on us, too: we always think it's our fault. But when you see this little bomb icon on your screen with little fizzies popping off the wick, it does not mean that you need to jump under your desk to avoid the shrapnel as your machine explodes. It just means that the software suffered a malfunction of considerable dimension and decided to check out. Occasionally you may have a clue as to why this happened (such as you ran out of *memory* or you opened very old *software* in a new *System*), but sometimes you just have to accept the fact that you will never know why.

This is the bomb icon you will see.

~~If~~ When you get a bomb—the only thing you can do is turn your computer off for about fifteen seconds, and then back on again. If you have a *reset switch* (see *programmer's switch*) and you know where it is, use it. Everything you have not saved onto your disk will be destroyed, which is another good reason to save every few minutes. Really.

Boolean expression, Boolean logic

Boolean logic is a form of algebraic logic where every answer is either true or false. It's used to solve the kind of problems you buy a database program to handle: like, "show me the list of men who dance, *and* who cook *or* like washing dishes, *and* who are *not* married." (Boolean logic is big on **expressions** such as *and, or,* and *not.*) You use Boolean logic every time you search your database. Boolean logic comes in very handy on a *digital* computer (which is what you have) because each answer can be a *bit* that is either on or off, true or false. The computer uses Boolean logic in a great deal of its work.

boot, boot up

To **boot** or **boot up** means to start your computer system, usually by turning on the power and/or pushing the "on" button. It's called "booting" because the computer is going inside itself and turning itself on (doing a lot of preliminary checking and adjusting before it's ready to run your programs). Hence the machine is considered to be "pulling itself up by its own bootstraps."

A *cold boot* is when the power to the machine has been completely turned off and you turn it on with the power switch.

A *warm boot* is when the computer is already on but you want to *restart* it without actually turning off the power. You might do this because the system crashed or you want to clear out the *memory*.

On a Mac, you might warm boot to activate a new *extension* or *control panel*. To warm boot a Macintosh, choose Restart from the Special menu while at the Finder.

If you have a PC, maybe you want to activate a new *device driver* or deactivate a *memory resident program* that doesn't have an Unload command. Press *Ctrl Alt Del.*

Reboot simply means to start up the computer again, after it was already on. You can reboot by doing a warm boot, but sometimes you crash so bad that you have to turn off the power and then turn it back on with a cold boot.

bootleg, bootlegger

As an adjective or a verb, **bootleg** refers to illegally copied software. If you make a copy of a *non-shareware* program from a friend, you are **bootlegging** the program; you have a **bootleg** copy; and you're now a **bootlegger.** This term is synonomous with *pirate*.

Boston Computer Society

See *BCS*.

bps

The acronym **bps** stands for **b**its **p**er **s**econd, which is a measurement of how fast information (data) travels between two devices. When two modems converse via the telephone line, the speed of modem communications is measured in bps. (When people talk about modems, they often use the word "baud" to mean the same thing as bps, though technically the two terms are not synonymous. See *baud*.)

The modems most commonly used with personal computers run at 1200 and 2400 bits per second. At 1200 bps a modem can send about 1600 words through the phones lines per minute. At 2400 bps, it can send roughly 3200 words per minute.

Many vendors sell modems that run at 9600 bps, which is close to the maximum that standard metal phone wires can handle reliably. Much faster modems are available, however, and the newer fiber optic phone cables allow transmission rates of more than 500,000 bps. But no matter how good the phone line, be careful about buying a modem any faster than 2400 bps. It can only use the higher speeds if it's communicating with another modem that's equally fast, and if both modems use the same *protocol*, or system for communicating. (Watch out—many faster modems don't adhere to the "standard" protocols for high-speed communications.) Don't even think about installing a 300 bps modem, even if someone gives it to you for free. These things are obsolete now that you can buy 2400 bps models for well under $100.

You may also use a serial connection to link your printer to your computer, especially if you have a Macintosh (most PC printers are connected via the *parallel* port, but speed ratings for parallel connections aren't usually quoted). Typically, a computer communicates with a serial printer at 9600 or 19,200 bps. Under the right conditions, the *serial ports* on your computer can usually handle much higher speeds—57,600 bps for the Mac and 115,200 bps for the PC.

breadboard

A **breadboard** is a thin board, sometimes fiberglass and sometimes plastic, with lots of little holes arranged in a grid. Electronics engineers use breadboards to create prototypes of circuit boards by wiring chips, resistors, and other electronic parts onto the board by hand, with the connecting wires running underneath. The term sometimes refers to the finished prototype itself, but more often it's applied to the hole-filled board to which the electronic components are attached.

b-tree

The term **b-tree** refers to a way of organizing database information so that you can quickly search through it to find exactly what you're looking for. B-tree is a way of organizing database *keys* so you can quickly search them on disk.

A database contains collections of related information, similar to file cards. Each record contains several fields, such as name, address, date of birth, etc. We want to find something based on one of those pieces of information. We could search sequentially through the whole file, but imagine if you had to search through the entire database of California social security numbers one by one. A b-tree allows you to search efficiently for specific names or numbers even as you are adding or deleting from the database. It dynamically grows and shrinks and maintains an up-to-date index stored as a tree, hence the name. (The letter "b" is to differentiate from other types of previously existing trees.)

btw

This cute little acronym, **btw,** stands for "by the way." It's often used in online communications. Please see *baudy language* for more online terms of endearment.

buffer

A **buffer** is a section of *memory* set aside for temporary storage (if you don't understand what memory is, you should probably read that definition first). For instance, you've probably noticed that you can quickly type a bunch of commands on the keyboard, and then just sit back and watch your computer carry out the commands one after another. This is because the computer stores the keys you pressed in a buffer—after doing one command, it goes back to the buffer to get the next one in line.

Some programs use the term **buffer** to refer to the space reserved in memory for the information of a file you're working with. Actually, you can start a new buffer and put information in it before you save it on disk as a file, but the idea is that the buffer represents a collection of information in memory that is handled as a unit.

Printers, modems, and other devices can have **buffers** too. If your printer doesn't have a buffer, your computer can only send it a little information at a time, as much as the printer can output immediately. With a buffer, the computer can send a document at full speed, at least until the buffer fills up. The information waits in the buffer until the printer is ready to print it, but meanwhile you can go back to work with your computer.

In DOS, the word **buffer** refers specifically to the small buffers reserved for DOS's use when transferring information to or from a disk. DOS buffers are a crude kind of *disk cache;* when information is available in a buffer, it doesn't need to be re-read from the disk, and your system operates faster. DOS automatically sets up a standard number of buffers (usually 15), but you can change this by entering a command such as **BUFFERS=20** in your *CONFIG.SYS* file. Why do this? Because, up to a point, the more buffers there are, the faster the system will run (although there's no practical way to figure out the ideal number of buffers). On the other hand, if you are using a true disk cache such as SMARTDrive or Super PC-Kwik, DOS buffers are redundant, wasting memory and actually slowing things down. In this situation you should always set the number of buffers to 3 or 4.

bug

The B word. If something is wrong with a piece of software or hardware so that it stops working or destroys your data or just acts weird, the product is said to have a **bug,** or to be **buggy.** The term actually comes down to us from the real live crawling and flying bugs that used to get into those old giant-sized computers and wreak havoc, particularly a moth found squished in the Mark II machine at Harvard in 1945.

Bugs are not your fault; they are malfunctions accidentally built into the hardware or software. Of course, it can be hard to know whether the problem you're facing is a bug or whether you're just doing something stupid. And occasionally a *virus* can make a program appear buggy.

Bugs can be minor or major. Here's an example of a minor bug: In PageMaker version 4.0 you could search for the characters "f i" and replace them with the *ligature* "ﬁ." But in the dialog box in which you do this, when you typed the key sequence for the ligature "ﬁ," PageMaker displayed the ligature "ﬂ" (although it replaced the characters with the proper ligature). This bug was fixed in PageMaker 4.2.

Bugs give *beta testers* their mission in life, their raison d'être. Beta testers find and report the bugs in products that are under development so the bugs get exterminated (the item gets *debugged*) before we pay lots of money for the products. But almost always, some bugs are missed during the beta testing process, and they wind up in the version of the item that you buy anyway.

bullet •

In printed documents, a **bullet** is often used to indicate an item in a list, like this:

- Here's item one.
- Here's item two.
- Here's item old.
- Here's item new.

The little round mark (•) in front of each line is the bullet—at least the official one. Actually, you can use any character you like as a bullet. The Zapf Dingbats and Wingdings fonts contain lots of interesting symbols that can serve as bullets, and some fonts consist entirely of bullet-type symbols.

Zapf Dingbats ➞ ☆ ●❑❖❄ ✩❄▼❄❒ ✥✻❄❒❍❁■□✌ ❒❁▲▲✻❑■❁▼❄●❙ ❁■❄ ❄❄❄❒●❙✎

To type the official bullet on the Macintosh, press Option 8 (hold the Option key down while you tap the 8). It's the same principle as typing an asterisk, in which you hold the Shift key down and hit the 8—if you hold the Option key down instead of the Shift key, you get the bullet. Also see *Key Caps*.

On the PC, if you can type a bullet at all, the keys you need depend on which program you're using—consult the manual. In PageMaker, press Ctrl Shift 8. In other Windows programs, hold down the Alt key and type 0149 on the numeric keypad. When you let up the Alt key, the bullet should appear. Also see *character map*.

bulletin board, bulletin board service

See *BBS*.

bundle

When you buy a particular product and it comes packaged with another product for free, often from another vendor, we say that the extra product is **bundled** with the main product. For instance, when I bought my scanner from Microtech, it came bundled with ImageStudio software from Letraset so I could work on the images I scanned.

And similarly, when you buy two or more products for a single price, you're buying a bundle—the products have been bundled together, or one product has been bundled with another. Usually a bundle is a better deal than buying the products separately, but not always—sometimes they're trying to get rid of an older version that you might not really want.

burn-in

A **burn-in test** is often performed at the factory before a product is shipped, or at the store before you buy it. The product or system is kept running continuously so if there are any weak components they will be discovered before the system becomes a vital part of a person's life. The longer the burn-in time, the more effective the test.

Though the phosphors used in today's color monitors are pretty resilient, this burn-in test still isn't a good idea for monitors, because if you leave a bright image on a computer screen for an extended period of time, the image can get burned into the phosphor coating inside the screen, causing **screen burn-in.**

To prevent screen burn-in, you can turn the brightness level down on the monitor, or get a *screen saver,* which keeps a constantly changing image floating or flying across the screen.

Burn-in also refers to a period of time in which a person is so intensely involved in the project on her computer, such as writing a stupid computer dictionary, that she forgets to eat, sleep, or feed her children. They say excessive burn-in can lead to **burn-out,** but personally I think burn-in can lead to burn-out only if the time necessary for the luxury of burn-out is available.

bus

A **bus** is a combination of software, hardware, and electrical wiring that creates a way for all the different parts of your computer to communicate with each other. If there was no bus, you would have an unwieldy number of wires connecting every part to every other part. It would be like having separate wiring for every light bulb and socket in your house. For instance, you see the cables plugged into the back of your computer that connect it with the monitor or the printer or the hard disk or some other device—they all connect into the bus.

There are different kinds of buses. There's a **data bus,** over which data flow (data is plural) back and forth between *memory,* your disk drives, and the *microprocessor* that runs the show inside your system. If your computer has *expansion slots,* there's an **expansion bus**. Messages and information pass between your computer and the *add-in boards* you plug in over the expansion bus. And there are other types of buses as well, such as the "address bus" and the *local bus.*

The design of the bus determines how fast information can flow and how efficiently the various parts of the system can use the system. The details are too complicated for this book, but one important factor is the bus's *bandwidth,* the definition of which you might want to look at.

bus error

A **bus error** is one of the most commonly occurring error codes to appear on your Mac and is therefore respectfully designated as "Error code ID 1" or "Type 1 error." Although most of these error codes can only be made sense of by programmers and technicians, you can often cure a repeating bus error by allocating more *memory* to the application that was active when the error occurred.

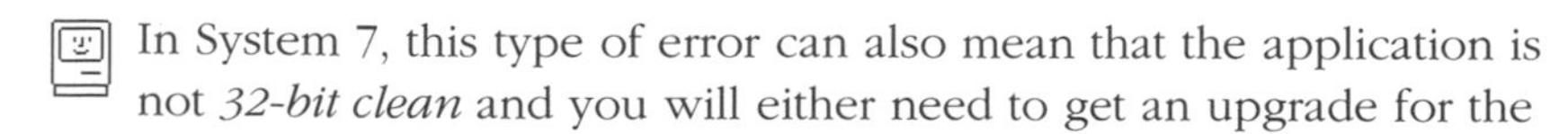

In System 7, this type of error can also mean that the application is not *32-bit clean* and you will either need to get an upgrade for the application or simply turn off *32-bit addressing* for the software to run properly. See *32-bit addressing*.

There are, of course, other reasons for a Type 1 error message on your screen, but the above-mentioned suggestions should be the first things you try. See *error codes*.

business graphics

Business graphics are those kinds of images like pie charts, bar charts, and scatter diagrams, etc., that are typically used by business people to show how many eggs they sold in Petaluma or how many days it takes to make a vacuum cleaner. Presentations that use slides during a speech are also lumped in with business graphics, even if they are straight text and not graphic at all.

There are several kinds of software that will draw charts and graphs for you—all you have to do is type in the information. You can then usually edit the graphic, either with the tools provided with the software or by exporting the chart to a graphic program. For instance, in most spreadsheet applications you can just tell the program what information to work with, and then hit a key or click a button. The application takes that information and creates a chart out of it. If you want really fancy graphs, or if you want an easy way to create text charts, you can buy separate business graphics, *charting*, or presentation programs.

bus width

If you think of the computer's bus as being a freeway that carries data, then it is easy to visualize the **bus width** as the number of lanes on this freeway. Naturally, the more lanes available, the more data that can be carried at any given moment. Bus widths are generally designated in multiples of eight; a *32-bit* bus width has the capability to carry four times the information of an *8-bit* bus in the same amount of time. See *bus* and *8-bit, 24-bit, 32-bit computer*.

button

A **button** on your computer screen is the visual equivalent of a button on a machine: it's a thing that you push and then you expect something to happen. Of course, you can't push a button on the screen with your finger, except on the rare computer that has a *touch screen*. Instead, you click on the button with your mouse pointer.

You can often use the keyboard to press a button, too. If a button has a dark border around it, that means you can hit the Return or the Enter

These are action buttons. If you press Return or Enter, the button with the dark border will activate.

key to make it work. If there's an underlined letter in the name of the button, you can press that letter (or a modifier key plus that letter) to activate the button instead of picking up the mouse and clicking.

In Windows, the dark border is kind of hard to see. And in Windows, you can activate a button by pressing Tab until the dotted rectangular highlight surrounds the button's label, and then press the Spacebar.

Buttons comes in several varieties. The different kinds of buttons give you a visual clue as to what to expect. Plain ol' buttons (sometimes they're called "action buttons") activate something as soon as you click on them (click once!). You might find an ordinary button in a tool bar (or button bar or palette) on your application's main screen.

Groups of buttons are used in dialog boxes to offer you choices among related items. In a group of *radio buttons*, you can only choose one, just like on your radio. In a group of *checkbox* buttons (or just checkboxes), you can check any number of the boxes from zero to all, just as on a list of checkboxes on a form.

Save as:
Publication ◉
Template ○

Radio buttons

Outline ☐
Shadow ☐

Checkbox buttons

byte

A **byte** is eight *bits*, eight individual electronic on/off signals, strung together to make a message that the computer can interpret. Although a bit is the smallest unit of information a computer can work with, most of the information it handles is organized into bytes. For example, most computers use one byte to represent each character (letter, number, or other symbol) that you see on your screen. A really small file, like a *batch file* on a PC or a one-page word-processed letter on the Mac, might consist of 20 bytes to 150 bytes.

The information-storing capacity of a computer is usually measured in multiples of bytes. A series of 1024 bytes strung together is a *kilobyte*. Lower-capacity *floppy disks*, for example, hold 360 kilobytes (KB or just K) or 800K of data; old PCs came with just 256K or 640K of *memory (RAM)*.

A *megabyte* (MB) is 1024 kilobytes. Your computer probably has 2 to 8 megabytes of RAM and a hard disk that can hold 20 to 260 MB of data.

C, C++

See *programming language*.

cables

A **cable** is that rubber- or plastic-coated big fat bundle of wires with huge plugs on either end that connects the computer box with the other parts of your system, such as a hard disk, monitor, printer, or a scanner. (You can't see the wires inside the cable, but they're in there.) Unfortunately, you need a different type of cable for each type of device you're hooking up to the computer—perhaps a *serial* cable or a *SCSI* cable or a *parallel* cable. Fortunately, each sort of cable has a different *connector* on the end that will only plug into a matching connector, so you can't really go too wrong. For instance, it is impossible to plug a SCSI cable into a serial port.

The connectors are either male or female (I'm sure you can could tell by looking at them which is which). And logically, you can only connect a male connector into a female connector, and vice versa. Of course, the cable connectors also have to be the correct size and shape to match the *ports* (the sockets) they plug into. Don't assume they will be—have the exact description of the ports and/or connectors handy when you order a cable (its shape, how many pins, etc.).

Also, be forewarned: when you buy computer equipment, they almost always charge you extra for the cable.

cache

A **cache** (pronounced "cash") is a temporary holding area in the computer's *memory* where the computer stores information that is used repeatedly. The computer can access the information much more quickly from the cache than if the information was stored in the usual place (which might be on a disk or in a part of the computer's memory that takes longer to *access*). The term cache can also be used as a verb—"to cache" means to place information in the cache.

disk cache

The basic idea behind a disk cache is that working with information stored in memory *(RAM)* is **much** faster than working with information stored on disk. A disk cache is a software *utility* that works by reserving a section of memory where it keeps a copy of information that was previously read from your disks. The next time your computer needs that same information, the data can be accessed directly from the cache, bypassing the slower disk. But if the *CPU* (central processing unit, the *chip* that runs the computer) needs data that is not in the disk cache, then it has to go to the disk and get it, in which case there is no advantage to having part of your memory set aside for the disk cache.

Since related data is often physically adjacent on a disk, the cache may also make a copy of information **near** the data that was previously used, in expectation that the extra data will eventually be needed. However, since there's limited room in the cache, only a fraction of the information on your disk is in the cache at any one time. But the caching utility manages things so that the information your software most often uses stays in the cache.

Some disk caching utilities also cache files you want to save, as well as other information your computer is trying to store on the disk. Because the cache sends this data to the disk in small drips and drabs instead of all at once, you don't have to wait until the saving process is finished before going back to work. Although you'll see ads saying that a disk cache "speeds up your hard disk," the disk doesn't actually run any faster—it just doesn't get used as often.

On a Macintosh, you can choose how much memory to set aside as a disk cache through the Memory Control Panel. In System 6 you can turn off the cache altogether; in System 7 you can reduce it to 16K. Keep in mind that any amount you set as disk cache takes away from your main memory. If you have 2MB or less in your computer, turn the cache off or down to 16K, since with less than a 256K cache you won't even notice a difference, and you really can't spare any if all you have is 2MB. With 4MB or more of memory, try setting your cache between 256K and 1,024K and see if you notice a performance improvement. If you don't notice an improvement, turn it down as low as possible because there is no sense setting aside that memory space if it's not doing you any good.

With IBM-compatible computers, you have to run your disk caching utility as a *device driver* (via your *CONFIG.SYS* file) or as a *memory resident* program (usually by including it in your *AUTOEXEC.BAT* file). Microsoft includes a decent caching utility called SMARTDrive with Windows and with MS-DOS versions 5.0 and greater. If you're willing to pay extra, you can get still faster caching utilities from other manufacturers; the most popular cache is Super PC-Kwik.

processor cache

This is a special area of very high-speed memory linked directly to the computer's *central processing unit* (*CPU,* the main *processor,* or chip, that runs the machine). The processor can access information in this cache much more quickly than it can get to data stored in the main memory area. The cache circuits monitor the data used by the processor, keeping a copy of the most recently and most frequently used information in the cache. Since programs are likely to access several items of related data that are near each other, the cache will even try to anticipate the processor's needs by copying data stored near the requested data in main memory. When the processor requests data from main memory, the *cache* first checks to see if it already has a copy, in which case the cache provides the data. Otherwise the main memory retrieves the data.

cache card

A **cache card** is a *board* you install inside your computer, specifically designed to temporarily store frequently used information that the computer would otherwise have to get from the disk (see *cache,* above). A cache card can dramatically increase the speed of many tasks because getting information from a cache card is much, much faster than getting the information from a disk.

Most computers built today either have a cache card built into them at the factory, or come with a slot inside the computer so you can add one at a later date. The beauty of a cache card is its price—often costing less than $200, this little jewel gives you "more bang for the buck."

CAD, CAM, CAE

The acronyms **CAD, CAM**, and **CAE** stand for **c**omputer-**a**ided **d**rafting (or **d**esign), **c**omputer-**a**ided **m**anufacturing, and **c**omputer-**a**ided **e**ngineering. These terms describe the use of the computer to make architectural or engineering drawings (blueprints), in planning and guiding manufacturing operations, and in developing and testing engineering concepts. Exactly what's meant depends on the person or company using the term. An engineering firm might hire someone to use CAE tools to develop their products.

Actually, you'll most often hear these terms used to describe software, as in "CAD package" or "CAE tool" (in computer talk, *packages* and *tools* are usually software products). There are numerous CAD, CAM, and CAE applications on the market. General-purpose CAD packages like AutoCAD get the most attention in computer magazines, but many CAD, CAM, and CAE products are tend to be fairly highly specialized for specific industries.

The acronym CAD, even though it stands for computer aided design, does not encompass graphics arts type desktop publishing design.

Calculator

Calculator, with a capital C, refers to the little electronic calculator you can use directly on the screen.

Macintosh Calculator

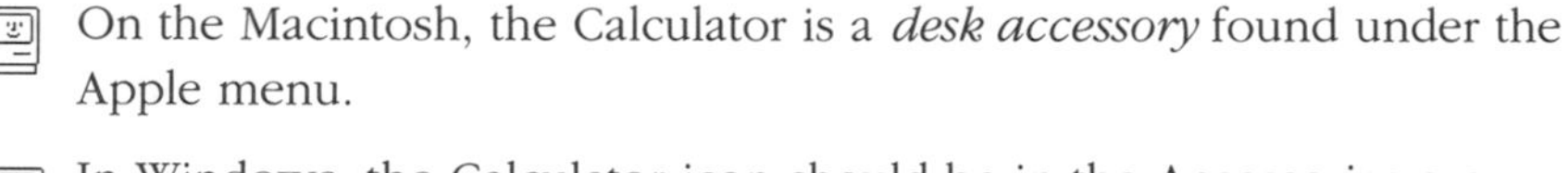

On the Macintosh, the Calculator is a *desk accessory* found under the Apple menu.

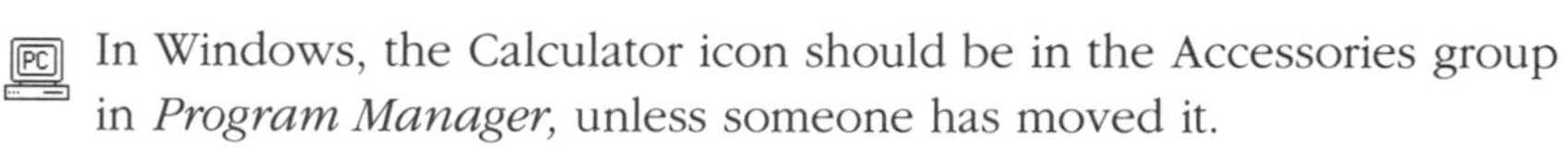

In Windows, the Calculator icon should be in the Accessories group in *Program Manager,* unless someone has moved it.

Once you open the Calculator, you can use it by clicking on its *numeric keypad* right on the screen, or by using the numeric keypad on your keyboard. You can even use the numbers across the top of the keyboard. The asterisk (*) is for multiplication and the forward slash (/) is for division (these two symbols work this way in any program that lets you do math).

Windows Calculator

Once you have an answer, if you press Command C (Mac) or Ctrl C (Windows) or choose "Copy" from the Edit menu, the answer will be copied to the *Clipboard,* and you can then paste it into your document.

If you copy a number from your document before you open the calculator, that number automatically appears in the little window of the calculator. (In Windows, you have to paste the copied number into the calculator yourself.)

call up

To **call up** a program is another way of saying to open the program, to get the program on your screen so you can use it. Or, when you're already using a program, you can call up (open) a particular file or document.

camera-ready art

Up until a few years, ago when you wanted to put an ad in a newspaper you went down to the newspaper office and gave the advertising department your scratch paper describing how you wanted the ad to look, and they took your information and prepared **camera-ready art.** That is, they turned your scratches into a presentable-looking ad that was ready for the camera person. The camera person took a picture of the ad with their special camera and then someone else used that picture to make a "plate" with which to print the page that included your ad. Now, with this desktop publishing revolution, you are creating your own camera-ready art. That is, now you probably march down to the newspaper office and tell them, "Print this," giving the ad department the page that you so carefully and lovingly prepared on your computer and printed on your personal printer (or perhaps you sent it to a *service bureau* for *output* on their *imagesetter*). Your ad, since it is all ready for the camera to shoot, is camera-ready art. This concept, of course, applies to any artwork to be reproduced on a press, not just newspaper advertising.

Cancel

Almost all *dialog boxes* provide a **Cancel** button. If you click it, the dialog box closes, but nothing you changed in that dialog box will take effect—the program restores the box's settings to whatever they were before you got there. This is quite valuable for those times when you realize you've made a mistake and you want to back out of it.

The Cancel button is also great for those times when you are just nosing about in a new application—choose any menu item that has an ellipsis (...) after the command (the ellipsis indicates that you will get a dialog box). You can check out the dialog box, enter any new numbers you like, pick different *radio buttons* or check other *checkboxes*, alter things, experiment, and you won't mess anything up as long as you remember to Cancel when you leave.

When you choose certain commands, such as Quit or Close, you also get a little box with the option to Cancel. In this case, click the Cancel button to cancel that command and return to where you left off.

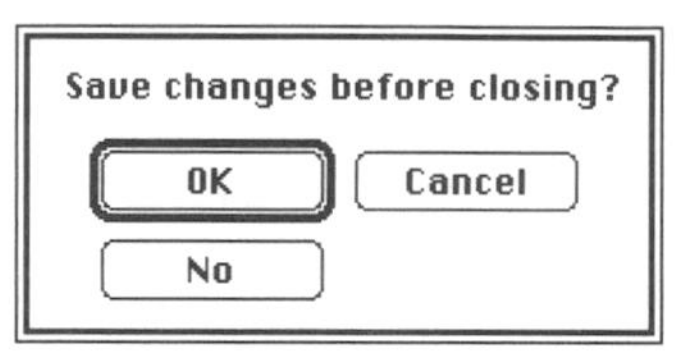

Typically, the safest option is the button with the dark border around it, sometimes called the "default button." You can press the Return or Enter key to select the default button.

In most applications for PCs, including Windows, pressing the Esc key does exactly the same thing as clicking the Cancel button.

In some Macintosh applications you can use the Escape key (esc) to activate the Cancel button, but a more common shortcut is to press Command Period.

Canon engine

The part of a laser printer that actually does the printing is called the "laser engine." Most laser printers sold around the world use a printer engine made by Canon of Japan, thus the term **Canon engine.** Several different Canon engines have been used in laser printers designed for personal computers, including the Apple LaserWriter series and the Hewlett-Packard LaserJet line.

Caps Lock

Caps Lock on the keyboard does the same thing as Shift Lock on a typewriter—it locks the Shift key down so you can type capital letters. There is one major difference, though, between Caps Lock and the typewriter Shift Lock: with Caps Lock down it **literally** types only caps; it does not type the characters above the numbers or any of those other characters you get when you hold the Shift key down, such as {, **?**, or *. With Caps Lock down you can still type numbers, and you still have to hold the Shift key down to get the characters above the numbers or the punctuation.

capture

Capture has at least two meanings in computer jargon. A **screen capture** refers to taking a quick picture of the computer screen. You don't do it with a camera; you do it with a screen capture *utility* program, or with a keyboard command (Command Shift 3 on a Macintosh; Print Screen in Microsoft Windows). Many of the images you see in the outer columns in this book were captured from the screen (then I took each image and pasted it into a *paint program,* erased the parts I didn't want to keep, and pasted it onto the page).

In most *telecommunications* programs, you can **capture** the **text** that appears on the program's screen or window, meaning that you save the text in a disk file. This is different from *uploading* or *downloading* a specific, complete file that already exists. When you turn the capture feature on, you'll be saving *everything:* whatever you type, and any messages sent by whoever or whatever is at the other end of the telephone line. For instance, you might want to capture messages from a *bulletin board* or an electronic information service like *CompuServe.*

card

A **card,** or *printed circuit board,* also known as a *board,* is a piece of plastic with *chips* attached to it. Chips are the tiny little circuits that run the computer. You buy a card and stick it inside the computer box. You can get accelerator cards (boards) that make your computer run faster, video cards that give your computer more graphic capability, and clock cards and printer cards or whole computers on a card. They range in price depending on what they do, who makes them, and so on. For instance, an accelerator card can cost $300–$1500.

Many cards you just order through the mail and then depending on what kind of computer you have you can open the back of the machine and stick 'em in, which I find to be a very frightening experience. Actually, once you take a few simple precautions, convince yourself (which sometimes you can only do if you have no choice) that you won't get electrocuted and you won't permanently destroy your computer, it can be a very satisfying, rewarding, and empowering experience to install a card yourself. I know.

You might also want to see the definition for *board,* which means exactly the same thing.

caret ^

A **caret** is a wedge-shaped mark, usually symbolized like this, ^, which you can type by pressing Shift 6. Some people call the ^ symbol a "hat." The caret is often used in search-and-replace commands to search for invisible characters or special items such as *tabs,* spaces, carriage returns, *hard spaces,* or index entries. For instance, in PageMaker you can search for tabs by entering ^t, or search for index entries by entering ^;.

The caret is also used in combination with other characters to indicate *control codes.* These are special codes within the coding system used to specify letters and numerals, set aside to control the computer system. Many computer systems don't have any visible characters corresponding to these control codes, so the caret is used instead. For instance, if you press Ctrl-B at the MS-DOS prompt, you'll see ^B. Some programming languages, including C, also use the ^ symbol.

Rarely, the greater-than symbol, >, used to indicate the *MS-DOS prompt* is referred to as a caret. On a PC screen a caret, symbolized by either ^ or > can indicate the text cursor.

carpal tunnel syndrome

One of the insidious side effects of working with computers is *RSI,* or repetitive stress injury, which is an injury that results from doing the same movement over and over again. **Carpal tunnel syndrome,** one of the most common and most debilitating forms of *RSI,* afflicts the wrists and hands with symptoms of numbness, tingling, pain, and weakness.

Although anyone who spends hours at the keyboard is vulnerable to carpal tunnel syndrome, you can make yourself much less vulnerable by keeping your wrists straight as you type, rather than tilting your hands up. In other words, don't rest your wrists and the base of your hands on the desk while typing. It also helps to position the keyboard lower than a standard desktop. Your typing teacher was right: your feet should be flat on the floor and your forearms should be horizontal to the floor. In fact, your whole body should be at right angles, either perpendicular to the floor, or horizontal to the floor.

carriage return

A **carriage return** on a computer is created, not by swinging that metal arm over to the right with a swat of your left hand, but by pressing the Return key, called the Enter key on some keyboards.

On a keyboard that has both a Return key and an Enter key, the two keys often do the same thing. If you have this choice, use the Return key for a carriage return at the end of a paragraph.

carrier, carrier signal, carrier tone

The **carrier** is what we have always called the "dial tone" on a telephone—that noise you wait for before you dial the phone (remember when we used to *dial*?). In its most basic role, the carrier is the modem's way of saying "I'm still here." It's something like the way you mutter "unh-hunh" every few seconds when someone's talking to you so they don't think you've fallen asleep. If the modem at the other end doesn't hear the carrier tone, it hangs up.

But the carrier is also critical to the modem's main job of transmitting information through the phone lines. A modem must encode information in a form that the phone lines can carry. It does this by rapidly altering or "modulating" the pitch (frequency), loudness (amplitude), or timing (phase) of the carrier signal. Faster modems use more than one of these techniques at the same time.

cart

Cart is an affectionate term for any kind of *cartridge*. For example, if you and your friend both have hard disk drives with removable cartridges, she might say, "Oh, just send me that file on a cart." See *cartridge hard disk*.

cartridge

Computers and related devices use many different kinds of **cartridges.** Some computers, typically portables and laptops, can run programs stored on *ROM* (read-only memory) cartridges, or let you add memory by plugging in *RAM* (random access memory) cartridges. If you have a tape *backup* unit, you may store your backup copies of files on tape cartridges. Many laser printers use replaceable toner cartridges so you don't have to measure out the messy toner and pour it into the "tank" yourself. Inkjet printers like the Hewlett-Packard DeskJet supply their ink in cartridges too. And fonts for many PC printers come in plug-in cartridges (these are actually a form of ROM cartridge designed for the printer).

cartridge hard disk

A typical hard disk is built right into your computer or is housed in a box nearby—and you never see the actual hard disk or take it out of its container. A **cartridge hard disk,** though, is removable. It works kind of like a giant floppy disk in that it slips into a slot in a special kind of removable hard drive case (actually, it's more like sliding a video tape into a VCR). A typical cartridge hard disk holds 44 megabytes (there are also 88s), costs as little as $40, and is about as big as a cheese sandwich with no lettuce. The *drive* (the case) that you put the cartridge into costs from $450 to $1000. But once you have the hard drive, buying a new cartridge is the cheapest way to increase the amount of hard disk space you have.

Some people feel cartridges are not as dependable for storing data, but others swear by them. I am one of those who swear. I have about 8 cartridges, each 44 megs. I can store backups of all my books on one cartridge and put it in a safety vault. I can keep System 6 on a cartridge so when necessary I can just switch back to 6. I keep all my teaching materials on a cartridge so when I trot off to the college I just bring the cartridge with me, pop it into the removable drive in the classroom, and use my own System and fonts and files. I do the same for the various presentations and workshops I speak at. Oh, I do think cartridges hard disks are a marvelous invention.

cascade

The **cascade** command in some applications automatically organizes all the windows on the screen in a tidy stack, making them all the same size, and placing them "on top" of one another but slightly offset so you can see the *title bar* of every window. You can then just click on the title bar of the one you want to work with and that window will come to the front of the stack. This is in contrast to the *Tile* command, which divides up the screen equally between all the windows, arranging them so they're touching each other on all four sides.

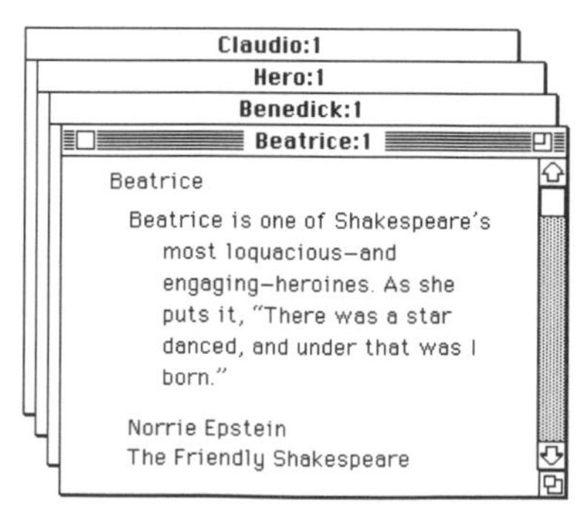

This is an example of what will happen to four separate windows when you choose the Cascade command. The window in front is the active window. You can click on the title bar of any other window to bring that one to the front. Each of these windows is probably also listed in a menu called something like "Windows," and you can choose to bring each one forward from that menu.

cascading menu

A **cascading menu** is a secondary menu that appears alongside an item in the original menu when you choose that item. Cascading menus are commonly found in both Windows and MacIntosh programs (although we call them *hierarchical* menus on the Mac).

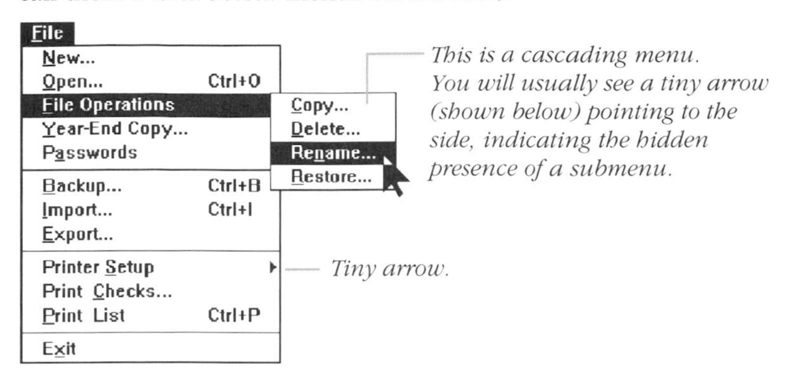

This is a cascading menu. You will usually see a tiny arrow (shown below) pointing to the side, indicating the hidden presence of a submenu.

Tiny arrow.

Most menu choices either do something immediately, like save your document, or bring up a *dialog box*. But some menus have so many related commands that they can't all fit in a menu of reasonable length. So rather than putting some of the commands on a separate menu, the program groups them under a choice on the same menu, For example, on the main File menu there are menu choices for opening, saving, and printing a file, but there might also be a choice that gives you a cascading menu for miscellaneous file-related functions you don't use quite as often, such as deleting or renaming files. Some programs even place cascading menus within other cascading menus, and once you open three or four levels of menus you can see why they call them "cascading."

Cascading menus can be a good way to organize a large number of commands so that no single menu is too cluttered. But they can be cumbersome to work with—the more cascading menus you have to go through to get to the command you want, the longer it takes you to get it done.

case

Case refers to whether letters are capitalized (uppercase letters) or not (lowercase letters). This term comes from the days when printers used metal type and stored the letters in wooden cases—the capitals were usually kept in the top, or upper, case; the small letters were kept in the bottom, or lower, case (see the illustration under *font*). Some applications have a command that lets you change the case of a word or phrase to ALL CAPS, lowercase, Initial Caps, or SMALL CAPS.

case cracker

A **case cracker** is the long, thin, metal tool you need to open the case of one of the small Macs, such as the Plus or the SE (on the bigger, modular Macs, you just take out the screw and pop the top open). If you open the case of the original Macintosh, you will see the signatures of the original development team embedded in the plastic.

case sensitive, insensitive

In this context, *case* refers to whether letters are uppercase (all capitals) or lowercase (small letters). When you hear that something is **case sensitive,** that means it matters whether letters are typed in upper- or lowercase. If it is **case insensitive,** it *doesn't* matter whether the letters are upper- or lowercase.

For instance, when you search for words in your word processor, you can choose to do a case-insensitive search: no matter what letters you capitalize when you type in the word you're looking for, the computer would find every instance of that word. For example, in a case-insensitive search, if you search for "John" with a capital J, the computer will find "John," "john," or even "jOHN."

If you click the button labeled "Case sensitive" or "Match case" or something similar, the computer will to do a case sensitive search—it will only find the "John"s that have capital Js.

cash

See *cache.*

CD-ROM

CD-ROM (pronounced "see-dee rom") stands for **c**ompact **d**isk, **r**ead-**o**nly **m**emory. A CD-ROM actually looks just like the CDs we play music with. To use one with your computer, you need a CD-ROM player, also called a CD-ROM reader. A CD-ROM can hold up to about 600 megabytes of information, which is the equivalent of about 700 regular floppy disks. There are CD-ROMs that hold the entire works of Shakespeare, complete dictionaries, histories, images of the works in the Louvre, etc., etc., etc. You can search the CD for the particular information you want to work with, copy it, then paste it into your own documents on your hard disk to do with what you will. You can only *read* from a CD-ROM, though—you can't store information onto it. The biggest complaint about CD-ROMs is that they are relatively slow (see *access time*).

cdev

A **cdev** (pronounced "see-dev" and properly spelled with lowercase letters) is a **control panel device,** now (since System 7) simply called a "control panel." A cdev is a *utility* (small program) that usually makes life easier for you. For instance, you can get note pads and fancy calculators and little calendars to show up on your screen by installing cdevs. Cdevs are similar to *extensions (INITS)* in that you just stick their icons in the System Folder, restart the computer, and they work. The biggest difference betwen extensions and cdevs is that you can control some of the functions of a cdev. Also see *Control Panel.*

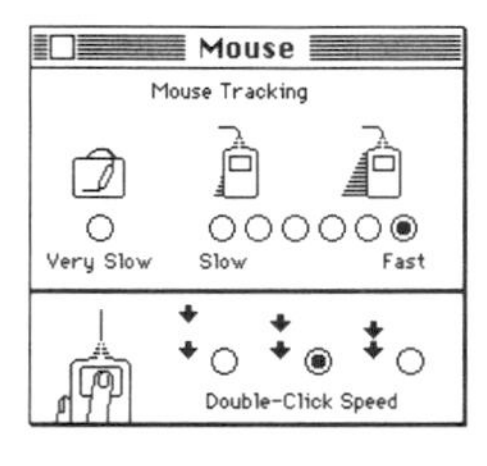

These are examples of cdevs.

cell

The visual layout of a spreadsheet consists of millions (sometimes billions) of little **cells.** Each cell can hold data or a formula. Each cell is identified by its vertical column and horizontal row. For instance, the cell B5 is in column B and row 5; in the example below, cell B5 contains the words "NET SALES." Many spreadsheet applications give you the option to *format* the data in the cell; that is, you can make it bold or italic or a different typeface.

C14 | × | ✓ | =SUM(C7..C11) — *Entry bar*

	A	B	C	D	E
1					
2					
3			*January*	*February*	*March*
4					
5		**NET SALES**	**$226,560**	**$260,478**	**$304,785**
6					
7		Cost of Goods Sold	111,328	116,024	130,239
8		Selling & Administrative	60,983	80,874	96,847
9		Shipping	4,265	4,604	5,047
10		Depreciation	2,148	2,350	2,245
11		Interest Expense	(2,458)	(6,360)	(5,600)
12					
13					
14		**TOTAL EXPENSE**	**$176,266**	**$197,492**	**$228,778**
15					
16					
17		Earnings Before Taxes	50,294	62,986	76,007
18		Income Taxes	20,118	25,194	30,403
19					
20					
21		**NET EARNINGS**	**$30,176**	**$37,792**	**$45,604**
22					

Each one of these spaces is a cell.

The cell with the border, C14, is the "active" cell. Its data (a formula) is shown in the entry bar, above.

or *centered,* is paragraph formatting because it is not possible to have just a few characters flush left or centered—either the entire paragraph is centered or it isn't. Tabs and indents are also paragraph formatting.

Character Map

Windows 3.1 comes with a program called **Character Map** which shows you all the characters (the letters, numbers, and punctuation marks) in your Windows fonts, one font at a time. If you need a special character such as ™ (registered trademark symbol) or ¥ (the Yen symbol) or an accented letter, start Character Map. The program will tell you if that character is available, and if so, which key or keys you need to press to insert it in your document. You can also use Character Map to see what the different fonts look like. The display is too small to be really helpful, but you can magnify one character at a time. In Windows, you can type characters that aren't represented on the keyboard by holding down the Alt key while you type a four-digit code; see *ANSI.*

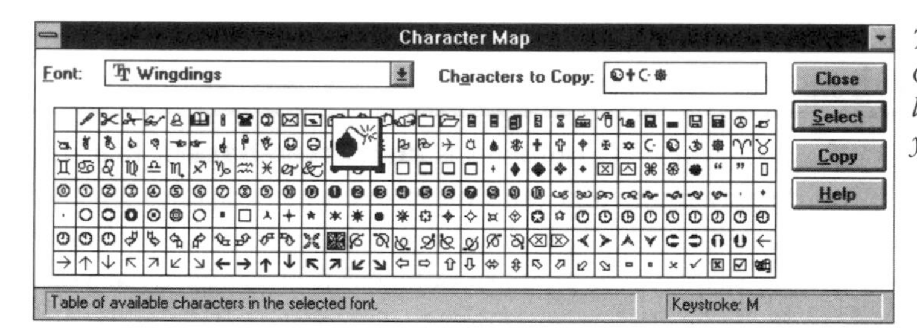

This is what Character Map looks like when you start it up.

character mode

Character mode refers to the operating state in which the IBM PC and compatible computers display only a fixed set of characters, in fixed rows and columns. This is different from *graphics mode,* where the computer is in an operating state ables to display *bitmapped* graphics. See *character-based.*

character set

When referring to fonts or typefaces, a **character set** is the group of letters, numbers, punctuation marks, and alternate symbols that are designed into the font itself.

When referring to computers, a **character set** is the particular group of pre-defined letters, numbers, and punctuation marks that the computer or printer can work with at any one time. Since a computer really doesn't know an A from a B, each character is represented by a code number. Macs and PCs were designed to allow for 256 code numbers, and thus 256 separate characters. That might sound like a lot, but it really isn't. Some of the 256 slots can't actually be used for characters you can see in a document (the system uses them for other purposes). And after you count all the regular letters, numbers, and punctuation marks, there aren't enough codes left over for all the characters people might want to use: legal symbols, foreign language characters, typographic marks, and so on. So the people who design each computer or each font have to limit themselves to a particular group of characters, and that's a character set.

The *ASCII* character set is the one most widely used—both PCs and Macs recognize it—but it only covers the code numbers from 0 to 127. Most sets with 256 characters are based on the standard ASCII set, filling the remaining code numbers with whatever other characters the creator of the set decided on. Examples of these include the PC-8 or "extended ASCII" set that's standard on IBM PCs, GEM International, ANSI, Roman-8 and the Macintosh's character set. Large corporate-type computers often use a completely different character set called *EBCDIC* that has nothing in common with ASCII. And some character sets consist entirely of special characters such as math symbols, bar codes, or musical symbols.

The term "character set" applies whether you're talking about the set built into a computer that operates in *character mode,* or the character set defined in a font that you install in your computer or printer. On PCs it's possible—and exceedingly frustrating—to have a software program that uses a different character set than your screen, and a printer that can only print yet another set. When this happens, what you see on the screen may not be what your software intends to print, and what you actually produce from your printer may look different yet. In Windows, everything uses the same set (unless you're working with special symbols for music, math, etc.). The standard Windows character set includes most of the professional typesetting marks such as *curly quotes,* *em* and *en dashes,* and the ©, ®, and ™ symbols.

When you want to use a character that isn't on your keyboard, how do you know whether it's in the character set, and how do you know what keys to press? You pop up a "map" of the character set. In Windows, run the *Character Map* program. In plain DOS, you can run a pop-up *memory resident program* such as SideKick to display a table of the standard PC extended ASCII character set while you're using any other program.

On the Mac, you can see the character set and find their keystrokes with *Key Caps.*

character string

A **character string** is a sequence of consecutive letters, numbers, punctuation marks or any other characters (a string can even include spaces) next to each other. While a word, a sentence, or a long passage are all strings, strings don't necessarily correspond to a unit of language—a string can be a part of a word, for instance, or the end of one word plus the beginning of another. If you search through your word processing document for the string "cat," you would find every instance of those three characters in a row, which would include "catalogue," "catatonic," "scathing," and "polecat."

checkbox, checkbox button

Checkbox buttons allow you to choose which features you want to apply to whatever you're doing. When the checkbox is checked so that an **x** appears in the square, the feature described by the box's label is turned on. When there's no **x** in the box, the feature is off.

A group of checkboxes are a visual clue that you can select any number of the boxes and each feature will apply. That makes them different from *radio buttons,* where you can select only one radio button, and thus only one feature, in a group.

Type style:	☐ Normal	☒ Italic	☒ Outline	☐ Reverse
	☒ Bold	☐ Underline	☐ Shadow	☐ Strikethru

You can usually choose from none to all of the checkbox buttons. In this particular example, though, if you click "Normal," all the other checkboxes will be emptied.

You don't really need to click right in the middle of the little square box—usually you can click anywhere on the label describing the button. Also try the *modifier keys*—in some applications you can press Command or Control plus the first letter of a button to turn it on or off.

In many PC applications, one of the letters in a checkbox label is bold or has an underline beneath it. This is a visual clue that you can type that letter, or perhaps press Alt or Control and the letter, to activate the button. In Microsoft Windows, you can also switch a checkbox on or off by pressing the Tab key until the dotted rectangle appears around the box and its label, and then press the Spacebar to put an x in it.

checksum

Checksum is a technique used by some *communications protocols,* such as *XMODEM,* to check for errors in the information that has been transmitted over the wires. The numbers of *bits,* or electronic units, of information, is added up (summed) before it is sent. Then the protocol sends that sum along with the data. When the receiving computer gets the data, it counts the bits and checks it with the sum that was sent along. If the two sums don't match, there was probably an error in the transmission.

Chimes of Doom

When you first turn on your Macintosh *(cold boot),* the computer performs a comprehensive memory test. If the test succeeds, the Mac validates, or okays, the *RAM (random access memory)* for use. When you do a *warm boot* (also known as a *restart*), the Mac does a similar but more limited test. Either way, if the computer discovers a problem before it can validate enough memory to run the monitor, the screen will go black (not a warm gray) and you will hear the **Chimes of Doom,** indicating that you have a serious problem. You might also hear the Chimes of Doom when there is a *SCSI* problem, a defective *board,* or perhaps loose or improperly installed *SIMMs* (memory chips). These chimes thoughtfully accompany your initial feeling of helplessness and desperation with an audio reminder of your dire predicament.

Does this guy look dead or what?

Supposedly the chords in the Chimes of Doom hold a clue to what the problem is, *if* you could tell which notes were in the chords and *if* those chords were documented in your manual. On the black screen, not placed safely and securely in the center of your screen but off to one corner, you will usually see the *Sad Mac.* On the black background and off-center, though, he looks more like a dead Mac.

If the computer finds a problem after the memory is validated, then you see the *Sad Mac* without the Chimes of Doom. Either way, you got trouble. The Chimes of Doom are sometimes referred to as the Heavenly Chords of Doom or the Chimes of Death.

I, personally, have heard the Chimes of Doom five times. I guarantee you, it is a memorable experience.

chip

This silicon wafer has been sliced into chips. A chip is then bonded to its package.

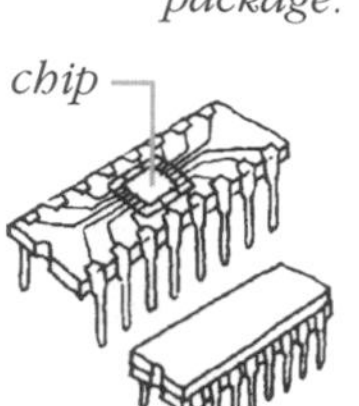

A **chip** is that truly amazing and remarkably tiny piece of *silicon* that has an entire integrated electronic circuit embedded within it. Chips are what make the computer. Chips *are* the computer. A tiny chip is one of the biggest pieces of human-made magic on earth. There are different kinds of chips, the most common being the *microprocessors* which run the whole computer, and *memory* chips, in which the computer holds and works with your information until you send it to a disk.

The chip that runs many PCs, the *80486*, has the equivalent electronic power—in its baby-fingernail sized wafer—of one of those room-sized mainframe computers from twenty years ago. The 80486 can be manufactures for just a couple hundred dollars; the mainframes of twenty years ago cost several million dollars. The impact on civilization and humankind of this mass-produced, inexpensive technology is going to be comparable to the impact of the technology of mass-produced, inexpensive books in the fifteenth century.

CHKDSK.COM

CHKDSK.COM (check disk) is a *utility* that comes with DOS and OS/2 that checks the *directory* and the *file allocation table* of a disk, fixing the problems it finds if you tell it to. When it's done, CHKDSK gives you a little report that tells you how many files are on the disk, how much free space there is, and how much memory (*RAM*) DOS has available to work with (CHKDSK only recognizes *conventional memory*, not "extended" or "expanded" memory; see *RAM*).

It's a good idea to run CHKDSK on each of your hard disks periodically, say once every day or two. You should also run CHKDSK any time the system starts acting weird, or if the computer stops working while you're using it, because of a power outage or a software crash. Be careful with CHKDSK — you're not supposed to run it when you have other software going, including *memory resident programs* or Microsoft Windows.

The reason you need CHKDSK takes some explaining, so bear with me. Because computers often split up all the information belonging to one file over different portions of a disk, parts of files wind up being scattered hither, thither, and yon. The *file allocation table* keeps track of all the parts of each file so that the whole file can still be used as a unit. Unfortunately, sometimes the information in the file allocation table gets altered so that DOS no longer knows where to find some of the file. The missing parts are called "lost chains," and CHKDSK can find them for you. CHKDSK can also detect files that are "cross-linked." This is another glitch in the file allocation table—this time, the table shows two files located in the same place on the disk.

CHKDSK can retrieve the information in lost chains and it can eliminate cross-linking, but don't expect it to restore the affected files to their original condition. To have CHKDSK do its partial repair job, you must run it with the /F *parameter,* by typing

CHKDSK/F

on the DOS command line. If it finds cross-links, it automatically tries to correct them. However, one or both of the files may no longer be usable, and sometimes CHKDSK just can't fix the cross-links. If CHKDSK finds lost chains, it asks you whether you want to retrieve them. If you type Y, CHKDSK turns the lost chains into files named FILE0001.CHK,
, and so on, placing these files in your *root directory.* You can then read them (use the TYPE command or a *text editor*) to see if they contain intelligible information you need. If not, erase the files to give yourself more disk space.

choke

A **choke,** or "shrink," is a traditional printer's technique for making an image or text of one color slightly smaller. It is often used in combination with the technique called a "spread," where an image or text of another color is made slightly larger. The purpose of chokes and spreads is to create a *trap* line, the thin line where two colors overlap a tiny bit to allow for paper stretching and imprecise registration on the press. Also see *trap* and *keep-away* for a more complete discussion on the concept and technique of trapping.

choose

When you read or hear the direction to **choose** a menu item, it means to select the command you want to put into effect. The most typical way to choose a menu item is with the mouse: position the tip of the mouse *pointer* on the menu name, press the mouse button, hold the button down and drag the mouse/pointer downward until the pointer gets to the menu item you want, and then let go of the button (don't try to click on the menu item!).

When the item is highlighted, as shown here, just let go of the mouse button.

See the following page for other methods of choosing items.

city-named fonts

Some of the *fonts* (typefaces) in your font list may be named after a city, such as Monaco, Geneva, or New York. This is almost always a clue that the font is *bitmapped* only, that it has no corresponding *printer font.* This means it is not *PostScript.* **City-named fonts** are designed to be printed on very *low-resolution* printers such as Apple's ImageWriter. **Unless you are using a *TrueType*** font with a city name, these fonts will not look nice at all on a *PostScript printer* such as the LaserWriter NT or GCC Business Laser Printer. A PostScript printer does not know how to make those city-named fonts (see *PostScript printer* to understand why not). A city-named font can stop an *imagesetter* (high-resolution printer) dead in its tracks.

The sizes you see in the outline style are the sizes that are installed.

City-named fonts, because they are designed for 72 *dots per inch* resolution, look pretty good on the screen, especially if you stick to the font sizes that you see outlined in your menu. These font sizes are the sizes that are installed in your System. If you print to a *QuickDraw printer* (a printer that is not PostScript, such as the Apple StyleWriter, HP DeskWriter, or the ImageWriter), the fonts will look pretty good (see *font*).

Some applications will automatically substitute a city-named font with another font when you print to a PostScript printer, most noticeably Microsoft Word printing New York or Geneva. Times gets substituted for New York, and Helvetica gets substituted for Geneva. But Times and Helvetica have very different letter and word spacing than New York and Geneva, so the printed page has big gaps between the words and the letters, numbers don't line up, tabbed indents don't even line up, lines that are supposed to be solid have spaces between them. That's because you tried to print a city-named font on a PostScript printer.

clamshell

The term **clamshell** describes the most common design for today's *portable* computers, in which the case is flat and shallow, and the screen folds up on a hinge from the back or middle of the computer. Like a clam. The keyboard is almost always an integral part of the case so that it can't be removed or detached. All true laptops and notebooks have clamshell designs. But somewhat less portable PCs may have a *lunchbox* design instead.

Clarus

Clarus is the name of the Dogcow that appears in the Page Setup dialog box. Please see *Dogcow.*

clean install

When you buy new application software or system software to replace an older version, sometimes you can install the new product right over the top of the existing one. Generally when you do this the software knows to remove the existing version as it installs the new one. But occasionally, especially if you run into trouble, your technical guide may tell you to do a **clean install.** This means to first *remove* the software you are replacing. Remove it completely and totally, including any files that may have become lodged in the System Folder, and start over again from scratch.

If you need to do a clean install of your System itself, you will have to start your computer with *another* disk that also holds a System Folder on it, because you cannot remove the System that is running the machine. (Well, you're not *supposed* to be able to, but I managed to do it one day while following the directions of a tech support person over the phone. No, it did not make me happy.) Before you re-install a new System, be sure you back up everything on your entire hard disk that is important.

click, click on

To **click** means to tap the button on the mouse—press the button down and let it up real quick. If a direction tells you to *point* to something and click, then just click *once.* Clicking twice when a direction has told you to click once can give you unexpected results. If you are supposed to click twice, the directions will tell you to "double-click."

Sometimes the directions just say to "click on" something on the screen; this is just a shorthand way to say "move the mouse pointer so that it's positioned over the item, then click the mouse button."

Many, many (far too many) directions tell you to "click" when they really mean "press." If you click, as the directions say, and what is supposed to happen doesn't, or perhaps you see something flash in front of your face and then disappear, then maybe they really mean to press: hold the mouse button down. The item (a menu, for instance) will be visible as long as the button is held down.

Some Macintosh mice and all PCs mice have two or three buttons. If you're using one of these mice and the instructions tell you to click but don't tell you which button to use, assume you're supposed to click the "primary" button. By convention, the primary button is the one on the left, but you can change the functions of the buttons in Windows (PC) or with your *mouse driver* (PC or Mac).

Also see *double-click, click-and-drag,* and *press-and-drag.*

click-and-drag

The term **click-and-drag** is misleading. When a direction or a person tells you to click-and-drag, what they really mean and what they expect you to automatically understand is that you are supposed to *press* the mouse button, *hold* it down, and *while* you hold it down, *drag* the mouse. A more appropriate direction is *press-and-drag.*

Very often you won't see the term spelled out as "click-and-drag," but merely as "drag." Drag is a shortcut term for "point, press, hold the button down, and drag the mouse."

clip art

Clip art has been around since long before computers were invented. Clip art came in books of art pictures and illustrations that you could clip out with scissors and paste onto your page. You can do it electronically now, but the concept is the same. There are many companies whose sole purpose in life is to provide you with computerized art created by professional artists and designers so you don't have to create it yourself. You can just order the disks from them and copy the graphics you want onto your page. You do need to order the clip art in a *file format* that your particular computer and software package can understand and use. To find clip art vendors, pick up any computer magazine and check the ads.

If you use a Mac, check out the book *Canned Art: Clip Art for the Macintosh* by Erfert Fenton and Christine Morrissett, from Peachpit Press. It's an 825-page indexed sample book of over 15,000 pieces of clip art available from 35 different companies. It comes with and includes tear-out coupons for over $1,000 in discounts on commercial clip art, plus two optional disks that contain 61 pieces of art. Great resource.

Clipboard

The **Clipboard** is an internal storage area that can **temporarily** store information you have copied or cut from a document. Both the Macintosh and Windows have Clipboards with a capital C. You can store text, graphics, spreadsheet cells, database lists, even sounds on the Clipboard (the Mac System 6 cannot store sounds).

If your application has a menu item called something like, "Show Clipboard," then you can see what is in the Clipboard. Otherwise, you have to just trust that you copied or pasted is really in the Clipboard.

Once an item is on the Clipboard, you can *paste* it somewhere else. You can paste it a million times, if you like, because the item will stay on the Clipboard until you turn off the computer, or crash, or until you cut or copy something else. The Clipboard can hold only one item at a time, so as soon as you cut or copy another item, the new item replaces the first one. Because the Clipboard is a system-wide feature, you can copy an item from a document in one program, close that program, open another, and paste the item into another document.

The Clipboard is often invisible while you're cutting, copying, and pasting, but some Macintosh applications have a menu item that lets you choose to see the contents (see above). If you don't have the option to view the Clipboard, you must simply trust that it is there and that the computer has placed your copied or cut item into it.

Windows comes with a *utility* called Clipboard Viewer that you can run to see what's currently in the Clipboard. Clipboard Viewer also lets you save the current Clipboard contents in a disk file for later use.

clock

A **clock,** as we all know, is a timing device. Most computers contain several different kinds of clocks.

real-time clock

All PCs and Macintoshes have a real-time clock that keeps track of the hours, minutes, seconds, and often the day and year. Except on early PCs, the real-time clock is battery-operated so the clock keeps track of the time even when you turn off the computer.

System clock

Inside the computer is an electronic circuit that generates a continuous, precisely spaced series of pulses. Like the beating of a conductor's baton, this "clock pulse" paces the electronic system in the computer, synchronizing all the circuits to its rhythm. Exactly how fast or slow this pulse beats is measured as the *clock speed* and is dependent on the *microprocessor* (the *chip* that runs the computer).

clock speed, clock rate

Clock speed refers to how fast the system *clock* drives the computer's *CPU (central processing unit,* the *chip* that runs the computer) which determines how fast the system as a whole can process information internally. Clock speed is measured in *megahertz;* a speed of one megahertz (1MHz) means the system clock is sending out its electric current one million times per second. The higher the clock speed of a computer, the faster the computer can operate, assuming all other factors are equal. However, clock speed isn't the only factor that determines your computer's overall performance, or even how fast the *microprocessor* (another term for the CPU) gets things done. Two different microprocessors may run at the same clock speed, and still take different amounts of time to finish a given job.

Although each microprocessor is rated for a maximum clock speed, the actual clock speed of your system is determined by the system clock, not the microprocessor. The original IBM PC and its *Intel 8088* CPU poked along at a torpid 4.77MHz; my Mac IIcx, with its *Motorola 68030* (pronounced "sixty eight oh thirty" or just "oh thirty" for short) runs at 16MHz; and Steve's *80486*-based PC *clone* barrels along at 33MHz.

The same microprocessor "model" is typically available in several versions, each capable of running at a different maximum clock speed. For instance, there are versions of the 68030 that run at 16MHz, 25MHz, 40MHz, and 50MHz, and of course they get faster all the time. The faster the processor can run, the more expensive it is, and that's one reason computers with a faster clock speed cost more.

clone

A **clone** of a computer isn't quite the same thing as a clone of a person, but the idea is similar—it's a replica so close that it's like having the same thing. Sometimes even better.

There are lots of IBM clones (which are not exactly the same as *compatibles*), but there are virtually no clones for Macintoshes (some people call Apple's *Performa* a Macintosh clone, but they're made by Apple). Because one clone is interchangeable with the next, the term also implies a kind of anonymity. The computer with the brand name you've never heard of that you buy from a little shop in the low-rent district is definitely a clone; well-known manufacturers hate to hear their computers referred to as clones. A clone is always cheaper than the original it was modeled after.

close box

Every Macintosh window has a little close box in the upper left. If you click in that close box with the very tip of the pointer, the window will (take a guess!) close, or go away.

Click in this little close box to make the window go away.

Toolbox

cluster, lost cluster

A **cluster** is the smallest single unit of the space on a disk (a hard disk or a floppy or even an *optical disc*) that your computer's *operating system* keeps track of separately. The operating system keeps systematic records of which clusters are occupied by each file stored on the disk (in DOS, this is called the *file allocation table,* or FAT). Clusters usually consist of more than one *sector,* a sector being the smallest unit of disk space that the computer can read data from or write data on. There are too many sectors on a hard disk to keep track of them all individually, so the operating system deals with them in groups called clusters instead.

Sometimes the operating system gets confused so it thinks a cluster is in use even though no particular file is using it. This is a **lost cluster.** You can free up the space occupied by a lost cluster, and maybe get back some missing data, by running a special disk repair utility (such as CHKDSK on the PC).

CLUT

CLUT stands for **c**olor **l**ook-**u**p **t**able. A CLUT is a software palette or set of 256 colors (it's actually a *resource*) that resides within the system software and most color-capable applications. On a computer with *8-bit color* (those that are only capable of displaying a total of 256 colors), a CLUT is a necessary reference to let the computer know which 256 colors out of the available 16.7 million colors *(24-bit color)* it can use at one time. If you think of all those 16.7 million colors as being a big (ok, very big) box of crayons, you can visualize a CLUT as being a small box of handpicked colors that someone has handed you to work with. Many applications give you the option of choosing which 256 colors you want to work with. You often can set up your own palette for each particular file. For instance, if you were painting a picture of a man's face, a palette of 256 different flesh tones would be more useful than a palette containing 256 colors found in the range between black and burgundy. Take the time to explore your particular application and its documentation for a variable palette feature.

CMYK

The acronym **CMYK** (pronounced as the individual letters: C M Y K) stands for the process colors **c**yan, **m**agenta, **y**ellow, and blac**k**. These four process colors are the transparent ink colors that a commercial press uses to recreate the illusion of a full-color photograph or illustration on the printed page. If you look at any printed color image in a magazine, especially if you look at it through a magnifying glass (a "loupe"), you will see separate dots of ink in each of the four colors. These four colors, in varying intensities determined by the dot size and space around the dot, combine together to create the wide range of colors you appear to see.

To get these four colors from the full-color image, the image must be separated into the varying percentages of each of the colors. There are several very sophisticated methods of doing this, and the result is a *four-color separation*.

Desktop color systems and the powerful page layout and art programs are now capable of making four-color separations for us. I can also create a color in my publication to match a color in the photograph. For instance, if I want to print my headlines in the same slate blue as in the model's tie, the computer could separate the headline color into the four different layers (sometimes called "plates") as follows: 91% Cyan, 69% Magenta, 9% Yellow, and 2% Black. The photograph itself would be separated into its variations of CMYK. When these four percentages of transparent ink are printed on top of each other, the colors combine to make the full-color photograph and the slate blue of the headlines, all at the same time.

This is different from *spot color,* where each spot of color is a separate, opaque ink color out of a can, such as red or blue or peach.

COBOL

COBOL (short for **co**mmon **b**usiness-**o**riented **l**anguage) is yet another computer programming language. COBOL was developed in the early sixties as a joint effort by several major computer manufacturers and the U.S. Department of Defense. This language is known for its English-like readability, which makes it (relatively) easy for another programmer to come along later and perform modifications or *updates*. COBOL has long been the most popular language for general business use, and most large- and middle-sized companies still use it to develop custom programs for their mainframes and minicomputers. PC versions of COBOL are available.

code

Code refers to the contents of a program for the computer. The term is used in two different ways. Code can describe the form of the program that a human being who knows the programming language can read or write. For example, a programmer might say, "I churned out a lot of code today" or you might say, "That guy writes elegant code," meaning that the programmer uses a minimum of programming commands to accomplish what he is trying to do. This kind of code, known as "source code," must be processed by a *compiler* or an *interpreter* until it is eventually turned into *machine code,* which is the form of the program that can actually be understood by the computer itself. Very few human beings can read machine code.

CODEC

CODEC is a shorthand way of saying "**co**mpressor/**dec**ompressor." It refers to a variety of software products that determine how a movie file, such as QuickTime, should be condensed, or compressed, to save space on the hard disk and to make the movie run faster. You might choose a different CODEC for video images than you would for still photography images. The different choices strike a different balance between picture quality and the size of the file (how many megabytes it requires to store it on the hard disk).

cold boot

The term **cold boot** refers to turning the computer on after it has been completely turned off. This may sound kind of dumb, since you would think that of course the machine is turned off before you turn it on. But you can also *warm boot* the computer, which means you can restart the machine without turning it off. See *boot.*

colon :

On both the Mac and PC, the colon (**:**)is an *illegal character* in a file name; that is, you cannot use a colon in the name of any folder, document, program, or any other kind of file. Have you ever tried to type a colon in a file name, like when you are saving a document? It either didn't show up, or another character appeared instead; maybe you got beeped at for your trouble. That's because the colon is reserved for the computer's internal use only. The colon tells the computer something—and it is not really what **you** want to tell it.

The colon usually indicates a *path,* as in **HardDisk:Newsletter:May:DogGraphic.** Translated (reading backwards), this means the DogGraphic document is in the folder (subdirectory) called May, which is in the folder called Newsletter, which is on the hard disk called HardDisk.

On a PC, a colon after a letter indicates a *disk drive*. For instance, **B:** means *disk drive B*.

color correction

Color correction is the art of matching the colors you see on your monitor to those that come off the press when you take it to be reproduced. Sounds simple? It's not. In the world of color, what you see on your screen is not what will print on your proof printer, which is not what will print on the commercial press. The video image you have spent hours perfecting still has to go through the lengthy process of color correction before the printed hard copy of that image matches the video image. The technician is going to have to adjust things such as brightness, contrast, mid-level greys, saturation, and hue. She has to make sure the flesh tones look like flesh and the bananas look like bananas. And she is going to have to attempt to get a basic color match between the video colors and the colors the printer is capable of producing (know as the printer's "gamut"). Sometimes the color on the screen has to be adjusted to it **appears** to be off-color on the screen, knowing that by the time the final product goes through the mutations of platemaking, inking, and dot gain, it will turn into the proper color.

In order to fully understand the difficulty involved with color correction, it helps to remember that video is created with light—something you can't really pick up, stir, or mix. Ink, on the other hand, is, well, ink. You can touch it, mix it, and smear it on your face. There have been books (big books) written about the science and art of color correction. There are literally thousands of technicians out there in service bureaus and print shops all across the county whose sole task is color correction—and you can bet that 50 percent of them are cursing at this very moment. The art of color correction requires talent, patience, and a whole lot of experience. Of course, modern technology has simplified the process of color correction a great deal in just the past few years. There has even been talk of a future system where all color correction will be unnecessary due to a new technology called "device-independent color" which promises to create the specified color, no matter the particular brand of hardware being used. But until technology catches up with promises, color correction is just one of those tasks that a person has to spend a lot of time learning, or leave in the hands of those who are already experts.

color look-up table

See *CLUT.*

color separation, color seps

Color separation is the process of separating artwork into the four basic *process colors* (or sometimes, for simpler work, into *spot colors*) for printing. From a desktop computer standpoint, this can take place at either end of the production cycle (and sometimes both). If you start with a color photograph that you want to place into a layout, for example, the photo must be scanned by a machine that can separate the colors (see *CMYK*). This machine will save a copy of the image in five files, with a separate file for each of the four process colors (cyan, magenta, yellow, and black), plus a fifth file that has a *low resolution* screen version for you to place in your document. This fifth file makes the image display faster on your screen and also takes up much less storage room on your hard disk. Then you add all the text and other elements to the page, and when it comes time to print the document to the imagesetter, the computer will print out one "plate," or piece of film, for each of the four colors. Most page layout programs are now capable of doing color separations right within the application, but some still rely on separate utility programs to do this chore.

color transparency

A common example of a **color transparency** is a slide, such as you shoot with your camera. A color transparency refers to any photographic image on transparent film of any size that shows a color positive (not a negative).

COM1, COM2, COM3, COM4

COM1, COM2, and so on are the names of the *serial ports* in an IBM PC or compatible computer. You can use these names in DOS commands to *configure* the ports (tell them how fast to transfer data, and what method to use in doing so). In most programs that utilize a serial port, you can pick the port you're working with by name from a menu. You probably have only COM1, or just COM1 and COM2.

COM file, .COM

For technical reasons you don't need to understand, DOS distinguishes between two kinds of programs you can run. Most programs are EXE (pronounced "E, X, E") files, but some are **COM** (rhymes with "tom") **files.** You can tell which is which in a list of files by the *extension,* the three letters that follow the period in the file's name. COM files are always relatively small programs, while EXE files can be almost any size.

You run either type of program, COM or EXE, by typing its name (just the first part, not the period or the extension) on the DOS command line. For instance, if the program file is named GO.COM, start it by typing GO and pressing Enter.

Comdex

Comdex in an acronym for **com**munications and **d**ata processing **ex**position. Comdex is just about the biggest trade show in the world on any subject, held in Las Vegas. It's spread out between the convention center and several hotels around town; in fact, they had to build two extra convention center buildings that are hardly ever used all year except for Comdex. Don't even try to get one of those fabulous deals on a Vegas hotel during the show because they figured out that Comdex folks don't gamble much so they have to make their money like any normal convention town: room rates. Rates Monday and Tuesday of the show can be twice the price of Wednesday or (better) Thursday. Take very comfortable shoes (this is one of the the biggest shows in the world, remember?) and plan on very long lines for buses to get from site to site, and even long lines for taxis. The crowds are much less beginning Wednesday. And unlike many shows, you can't buy stuff off the floor.

comma-delimited

See *delimiter.*

Command-click

When a direction tells you to **Command-click,** it means to hold the Command key down while you click the mouse button. This makes different things happen than if you click without the Command key down. Remember, sometimes when a direction tells you to *click,* they really mean *press* and hold the button down.

COMMAND.COM

COMMAND.COM is the program that serves as the DOS command processor, or the DOS *shell* if you prefer. Like any *operating system,* DOS itself is simply software, albeit software that has a very special role in running your computer. DOS consists of a conglomeration of *programs, utilities,* and *device drivers,* but at its core are three key pieces of software. They must be present on the disk you use to start your computer, or the computer won't work. Of these three pieces of software, the only one you're likely to run across is COMMAND.COM—you'll see it in the list of files on your screen when you display the *directory* of that start-up disk, by typing *DIR* and pressing Enter. (The other two essential DOS files are *hidden files,* so you won't see them in the directory list.)

Despite its importance, COMMAND.COM is an ordinary file, just like any other file on your disk. You can copy it to another place if you like, but whatever you do, don't erase (delete) the COMMAND.COM file, and don't *move* it to another directory or another disk.

COMMAND.COM is responsible for displaying the DOS *prompt* (like **C:>**). But its main job is to interpret the commands you type at the prompt (on the DOS *command line).* That is, when you type a DOS command and press Enter, COMMAND.COM "reads" what you typed. It figures out what you want to do—copy a file, display a directory, run a program, etc.—and then goes and does it. That's what they mean by "command processor." (By the way, it's possible to substitute another command processor for COMMAND.COM; see the entry for *4DOS.*)

In the technical sense, COMMAND.COM is a shell, in that it provides a *user interface,* a way you can get at DOS's functionality. But ordinary people use the term *shell* to refer to a program that lets you manage files and start programs without having to type commands—in other words, without interacting with COMMAND.COM.

command interpreter, command processor

A **command interpreter,** or **command processor,** is that crucial part of the operating system software that interprets, or processes, the commands you give, and then carries them out for you. In DOS, the command processor is usually *COMMAND.COM,* although DOS lets you substitute another command processor if you want. This sounds pretty technical and scary, but it really isn't difficult—see the entry for *4DOS* for the description of a better DOS command processor.

Command key

The **Command key** is on the bottom row of the Macintosh keyboard, the key with the California freeway cloverleaf symbol on it: ⌘. On some keyboards it also has an apple on it and you may hear it referred to as the Apple key or Open Apple. The key doesn't do anything all by itself; it is always used in conjunction with other keys to activate keyboard shortcuts. Don't call this key the Control key, because most Mac keyboards also have a Control key which does completely different things than the Command key.

command language

In a program you can control by typing out commands (as opposed to making choices from menus or clicking buttons), the commands you can type are collectively referred to as the program's **command language.**

command line, command line interface

In operating systems like DOS and Unix, and in many *text-based* or *character mode* programs, you control what's happening by typing commands on a **command line.** The command line is simply the line on the screen where you type your commands. The **only** way to control an operating system or a program that uses a **command line interface** like this is by typing commands—you don't get menus, dialog boxes, or buttons.

Commodore

Commodore Business Machines, Inc. is the manufacturer of several models of personal computers, including the Amiga, **Commodore** 64, and Commodore 128.

communications protocol, communications software

See *protocol* and/or *telecommunications software.*

comp

Comp, in the graphic design field, is short for "comprehensive rendering," which is a representation of the project as if it was printed. You create and show a comp to a client for their approval before you move forward to the completed project. There are loose comps, which just give a client a fairly good idea of the piece, and there are tight comps that don't leave much room for the lack of imagination of the client. For instance, if a designer is designing a package for perfume, she would actually create the three-dimensional package in the size and using the colors and the design and the typography of the finished piece so the client thinks they are looking at the product right off the shelf—that's a tight comp.

Compaq

Compaq Computer Corporation is a large manufacturer of IBM-compatible PCs. Compaq was IBM's first serious competition in the PC market, becoming a Fortune 500 company a few years after its inception. Its success was due partly to leading-edge technology—Compaq introduced the first portable PC (a *luggable* model with a small, built-in picture tube monitor) and the first *80386*-based PC—but a great marketing strategy deserves even more of the credit. Compaq PCs used to cost much more than most other PCs—as much as IBMs—even though they weren't really that much better. Lately, though, people haven't been willing to pay a big premium for the Compaq name, and the company has slashed its prices.

compatible

Compatible refers to the ability of two pieces of software and/or hardware to perform their advertised functions together without *crashing* the computer. For instance, you must make sure your new software is compatible with your *operating system;* your printer must be compatible with your computer; the two computers in your office must be compatible if you want to *network* them together and share files.

Compatibility has always been a major issue among computer users. Unfortunately, some computer and software makers still view incompatibilities as inconsequential. Be sure to read the enclosed documentation before installing something new (either hardware of software!). Often a manufacturer will issue a warning, in print, stating that the product you just purchased is incompatible with something you already have.

When speaking of an **IBM PC-compatible** computer, it means the computer is not manufactured by IBM but can *run* the same programs as the IBM computers.

compile, compiler

When a computer programmer writes a program, the programming language use has some resemblance to human speech—even we non-programmers can read some of the words, like **if** and **then** and **do,** and others, like **printf** or **struct,** almost look like real words. But a computer can't understand anything about a program written in a programming language, not even the plus signs. In order to run that program, the programmer has to first convert it into computer-ese, known as *machine code,* using a special program called a **compiler.** Usually, the compiler produces an intermediate form of the program which is then converted to the final, working form by a "linker," another special program.

As a verb, to **compile** a program is to convert it into machine code using a compiler. Contrast compiler with *interpreter.*

compile time

This phrase has two meanings. **Compile time** can refer to how long it takes to *compile* a given program (see the entry above). The compile time for a program depends on how long and complicated the program is, the speed of the computer, and the quality of the compiler.

The other meaning of **compile time** has to do with events that happen at the time a program is compiled, or "at compile time." A programmer may say, "Specific values are assigned to those variables at compile time."

compress, compression

When you compress computerized information, you make it smaller (taking up less space on the disk), meaning that less data is needed to represent exactly the same information. Using a *compression utility,* you can compress files stored on disk so that they take up less disk space and leave more space for other files. Some programs have the ability to compress data that's being held in memory, allowing the computer to keep more data in memory and thus spend less time retrieving data from the disk. And some *modems* and *communications* software can compress the data they send back and forth to one another. Since there's less data, it takes less time to transfer them.

Keep in mind that even when you are *archiving* files you never want to compress your original and only file. All it takes is to lose one bit, one electronic signal, from a compressed file and that file is destroyed. Bits get lost all the time. In an uncompressed file, it's a minor problem, but in a compressed file, it can be a catastrophe of considerable dimension.

compression utility

A *file* **compression utility** is a small program that *compresses* files for you, reducing the amount of space they take up on a disk. You must uncompress the files before you can use them. Examples of compression utilities include DiskDoubler, StuffIt, or CompactPro for the Mac, and PKZIP and LHA for the PC.

Some folks swear by compression utilities and others swear at them. Basically, this is how compression utilities work their magic: If you were to count the instances of the letter "e" on this page you would find, well, quite a few. At each instance of "e," this word processor is going to write a code on the disk that represents the letter "e"—each and every instance, and that's going to take up a relatively large amount of disk space. A compression utility will look at all those "e's" and say, "Why don't I just make a note here of how many there are and draw a tiny map of where they belong." This little notation is going to take up a lot less space than the normal method. Of course, it's more complicated than that but, for all practical purposes, compression can be thought of as a form of computer shorthand.

An "on-the-fly" *disk compression utility* compresses all the files you store on the disk automatically, while you work. You just "Save" in your application as you normally would, and the file gets compressed without any extra effort on your part. Likewise, the utility automatically uncompresses the file whenever it's used. Popular utilities include Stacker, SuperStor, and AutoDoubler. But **remember,** always keep one uncompressed backup of every important file.

CompuServe

CompuServe (whose formal name is CompuServe Information Services, or CIS for short and sometimes CI$ for a statement) is a commercial *online service* that you can access through your *modem*. It has an incredible array of forums (sections devoted to particular interests), conferences, facilities for *uploading* and *downloading* information, and all the other benefits a powerful online service offers.

CompuServe has been around for quite some time now and it's big. Vast. Cavernous. And getting bigger every day. *Navigating* around the many forums can be quite intimidating. It helps tremendously to invest in the available software specifically for CompuServe that provides automation of certain tasks, such as picking up *e-mail,* retrieving files and messages, etc., and for simply making navigation easier.

computer

A **computer** is a machine that processes data; that is, it takes information (data) that you put into it (input) and does something with it (processes it). Then it takes that processed information and outputs it, or puts it back out to you in a way you can understand. The output may be to the screen for you to view, or to paper (hard copy) for you to hold in your hands, or onto a disk or tape that you can store in a safe place. Basically, then, a computer does these three things:

Accepts input. Processes data. Outputs information.

A computer and all the parts that belong to it, the hard parts that you can bump into, are called the *hardware.* The instructions that tell the computer what to do are called the *software,* or software *programming.* Sets of instructions that allow you to perform certain tasks, such as balance your checkbook or design a brochure or draw an illustration, are called software *applications.* The people who run the computers are called the wetware, or liveware.

Computers come in all sizes and amounts of power and costs— from the supercomputers that can cost $20,000,000, to *mainframes,* minicomputers, and down to microcomputers (personal computers, like the kind you and I use) that can cost less than $1,000. See the illustrations in Appendix A.

computer on a chip

A *microprocessor* is a single *chip* that is the *central processing unit,* or the brains of a computer. To function as a complete computer, it also needs *memory,* a *clock,* and a *power supply.* Well, a **computer on a chip** has its own built-in clock and its own memory, so all it needs is a power supply to function. These tiny things are used in all kinds of things, from car parts to children's toys.

computer paper

Computer paper is the wide paper, 11 x 17 inches, that feeds through certain kinds of printers. Don't confuse this wide computer paper with the regular-sized *pin-fed* paper. Many home printers use pin-fed paper—you know, the kind that is hooked together and has those holey edges that you have to rip off. After you rip off the edges and tear off your page, the stuff you probably use at home or at school is 8.5 x 11, ***not 11x17.*** I'm stressing this because I see so many people choose "computer paper" in the Print dialog box, and then their image prints off to the side of the paper because the printer thinks you stuck 11 x 17–inch ("computer paper") in it.

COMSPEC

Most people don't need to know anything about the **COMSPEC,** so it's probably best if you just ignore this definition. I'm warning you, it's pretty technical. You're still reading? OK, here goes: In DOS, the "word" COMSPEC is an official term used with the SET command to tell DOS which program to use as the *command interpreter,* and where that program is located. (See the definition for *command interpreter.*) Actually, DOS already knows where the command interpreter is, because much of DOS *is* the command interpreter, and DOS can't run without it. But some other programs need to know where to find the command interpreter, and they ask DOS. The answer they get is the COMSPEC. You don't have to specify a COMSPEC because DOS automatically uses whatever command interpreter you started the system with as the COMSPEC. But if you want other programs to use a different command interpreter, you type a command like:

SETCOMSPEC=D:\DOS2\COMMAND.COM

concatenate

(Isn't this a great word?) **Concatenate** means to join separate things together. Often it is referring to *character strings,* or groups of letters, but you can also think of concatenating two files, such as two word processing documents.

An example of concatenating character strings: if you're working in *HyperCard* and you have set aside a variable called "my birthday" in which you previously put the day of your birth, you could concatenate the variable "my birthday!" with the sentence, "My birthday is." Using the ampersand (&) as the concatenating symbol, you might write it as

"My birthday is " & "my birthday" & "!"

The resulting sentence would read, after concatenation:

My birthday is October 9! *(HyperCard substitutes what was in the variable).*

Or in DOS you can concatenate, or combine, several files into one by using the COPY command and listing the files you want to combine with a plus sign between their names, followed by the name of the file for the combined files. Like this:

COPY FILE1+FILE2+FILE3 BIGFILE

conditional

Something that's **conditional** is an action that only happens if a specific condition is met. This term is mainly used in programming and similarly technical domains. For instance, "conditional compilation" refers to a section of a program that the *compiler* will compile if certain criteria (which are spelled out in the program) are satisfied; the compiler just ignores this section if the criteria aren't met.

CONFIG.SYS

If you have a PC you probably have a file named **CONFIG.SYS** on the disk you use to start, or *boot,* the computer. This file contains a variety of instructions used to customize the way your system works. Each time you start or restart the machine, the CONFIG.SYS file is read by the *operating system* (DOS or OS/2), which *configures* itself according to the instructions in the file. In DOS the typical CONFIG.SYS file starts with commands that tell the operating system how many files can be in use at any one time, and how many *buffers* to create, like this:

```
FILES=30
BUFFERS=20
```

The file may also have commands that *load* into *memory* one or more device *drivers,* which are little pieces of software that allow the system to communicate with and control some device like memory or a mouse. For example:

```
DEVICE=C:\DOS\HIMEM.SYS
DEVICE=C:\MOUSE\MOUSE.SYS
```

You can change the contents of the CONFIG.SYS with almost any word processor or *text editor,* since the file is just a *text file.* You do have to make sure to save your work as a plain *ASCII* file without any word processor-type formatting.

You don't absolutely need a CONFIG.SYS file to use your computer, because DOS makes assumptions about how to set things up if you don't tell it explicitly what to do. However, certain programs won't run unless the correct settings are specified in CONFIG.SYS.

configure, configuration

You can put a computer system together in an unlimited number of ways, depending on the kind of machine you buy, the kind of *monitor,* how big of a hard disk, which *operating system* you use, the amount of *RAM* installed, the software applications you need, and various other elements.

Your precise arrangement is your **configuration.** Once you know what you are doing, your friends may ask you to **configure** a system for them.

You may also need to configure a software program, meaning that you customize the program so it works with your particular computer or printer model, and that you choose the way certain features work.

connectivity

Connectivity is a true piece of jargon: a new word made up to make fairly simple ideas sound sophisticated. Connectivity can be used to describe anything having to do with connecting two or more computers together so that they can interchange information. You can use it to refer to networks, to modems, to connections between personal computers and larger minicomputers and mainframes, or what have you. This word was in vogue a few years ago, but isn't heard as much anymore—instead, you hear more specific, but equally pompous, words like *distributed computing.*

connector

Connectors are the parts on the ends of cables and cords that actually make the connection to another piece of hardware. Both the part on the end of the cord and the piece on the hardware that it plugs into are called connectors, and they are either male or female (take a look at any one and you can guess why).

NETWORKING

contiguous

If items are **contiguous,** they are next to each other, such as *sectors* on a disk, as opposed to being *discontiguous,* where they are not next to each other. In some applications you can select more than one item **only if** the ones you want to select are contiguous. For instance, in a spreadsheet you cannot usually select three cells here and two over there; in a word processor you cannot simultaneously select the first paragraph and the third paragraph (it seems like you should be able to).

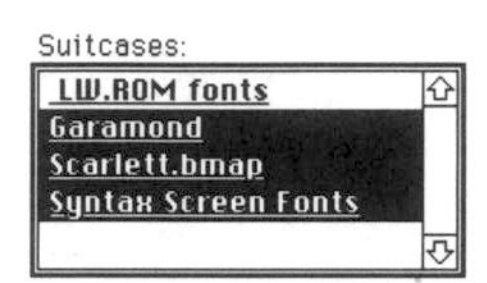

This is a contiguous selection, where all three selected items are next to each other.

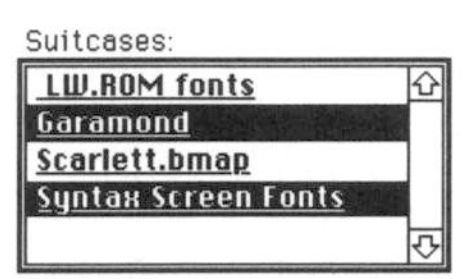

This is a discontiguous selection, where the selected items are not next to each other.

continuous tone

A photograph, a pencil drawing, a charcoal rendering—all of these kinds of images have **continuous tone,** where an uncountable number of shades of gray in the entire range between black and white blend into one another smoothly. You know what happens when you make a reproduction of a photograph on a black-and-white copy machine, right? A copy machine cannot print continuous tone; it is not capable of producing light grays and dark grays. Rather, everything comes out either black where the toner hit or white where the toner didn't. A laser printer cannot print shades of gray either, nor can a commercial press. All continuous tone images must go through a *halftone* process in order to be reproduced. Also see *halftone* and *grayscale.*

Control key, control key, Ctrl

A **Control key** is common on the keyboards of many types of computers, though the label on the key usually reads **Ctrl.** The Control key is a *modifier key,* meaning that it works in conjunction with other keys, and doesn't do anything (usually) by itself. You have to hold down Ctrl while you tap another key, like X or 2 or F3, to get the results you want. (More generically and confusingly, the term **control key** is sometimes used as a synonym for any modifier key.)

The special thing about the Control key used to be its location just to the left of the A, where you could easily reach it with your baby finger. This made it very easy to press Control key combinations like Ctrl S, Ctrl F and

Ctrl C (see *WordStar cursor diamond*). Recognizing the importance of the Ctrl key, IBM's engineers decided back in 1984 to give us two of them on the *enhanced keyboard,* which was good. But in an unfortunate triumph of symmetrical esthetic over function, both Control keys were placed down by the Spacebar. You can only reach them with your thumb, and you have to twist your hand uncomfortably to do it, which lifts the other fingers off the alphabet keys. Since IBM's the boss, just about every other keyboard manufacturer has followed suit—even the Mac's extended (enhanced) keyboard places the Command keys (the Mac equivalent) so they flank the Spacebar (which is where Robin likes them). I wouldn't be going on about all this if there wasn't a way to fix the problem—there is, at least if you have a PC. You can use any of several little *shareware* or *freeware* programs to alter the keyboard to swap the functions of the left Ctrl key and the Caps Lock key, which they put where the Ctrl key used to be, alongside the A. Presto, you have the old-style Ctrl key again. If you can't live with the mislabeled keys, you can also buy add-on keyboards with the Ctrl key in the right place, and then donate your current keyboard to science.

Control Panel

A **Control Panel** is a type of special *utility,* or small program that does something for you, like turns your screen into visions of flying toasters, or controls how many colors you see on your monitor, or determines who gets to share files. Each utility has its own panel for controlling the various features of the utility, which is why they are called Control Panels.

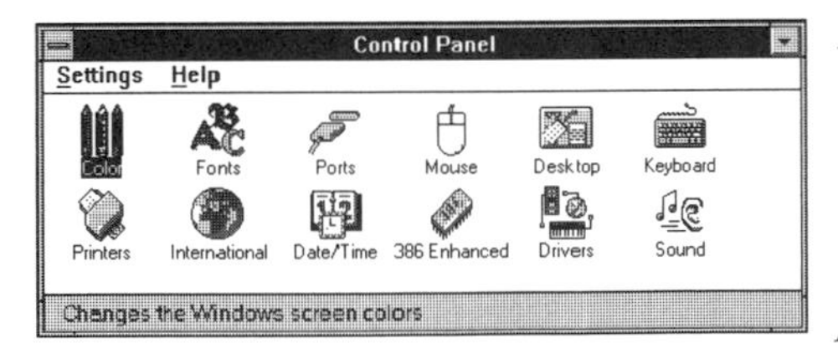

In Windows, Control Panel is the name of the program that lets you control the way Windows operates. You can pick the colors for each item on the screen, install or remove fonts, control the way the mouse works, pick which sound to play when Windows starts, and adjust all kinds of other settings.

Control panel devices used to be known as *cdevs* (pronounced "see dev") in System 6, and all the control panels for the cdevs were found in the one main Control Panel, accessed through the *Apple menu.*

In System 7, each item has its very own control panel, stored in the Control Panels folder within the System Folder. When you install System 7, the installation procedure automatically puts an *alias* of the Control Panels folder into the Apple Menu Items folder, which makes "Control Panels" appear in the Apple menu. You can, if you like, make an alias of any individual control panel and store that alias in the Apple Menu Items folder so you can get directly to the control panel through the Apple menu.

control unit

A *control unit* (or *controller,* same thing) is a piece of *hardware* that manages the activities of *peripherals* (separate devices attached to the computer, such as monitors, hard drives, printers, etc.) Control units found on personal computers are usually contained on a single printed circuit board. The control unit acts as a sort of "go-between," executing transfers of information between the computer's memory and the peripheral. Although the *CPU* (*central processing unit*—the "big boss" in the computer) gives instructions to the controller, it is the control unit itself that performs the actual physical transfer of data.

conventional memory

Without special help, DOS recognizes a maximum of 640 *kilobytes* of *RAM* (random access memory) for use in running your programs. That 640K is referred to as **conventional memory** to distinguish it from other types of memory *(upper, expanded,* and *extended)* that have been invented to overcome DOS's 640K limitation. See *RAM* for a comparison of the different types of memory used in PCs.

convert

To **convert** something means just what you think it would—to change it from one form to another. This term is most often used when you are talking about converting information stored in one kind of *file format* to another file format.

A typical application program creates files in its own special format; for instance, Microsoft Word creates "Word" files. Another word processing application, such as WordPerfect, can't *read* (understand) the Word format. So if you want to open a Word document in WordPerfect, you have to convert the Word document to a WordPerfect format.

Sometimes you need a special program to convert files from one form to another. Some applications can open more than one file format, so the application itself can do the conversion.

coordinate

Built into your computer is a mapping system, or grid, complete with the ability to pinpoint any location or **coordinate** in the application window. This grid is laid out in the common x,y format—x being the horizontal units of measure starting from the left side of the screen, and y being the units starting from the top of the screen. It's easy to see that 0,0 would be the upper left corner of the screen. Now, if you're only using your computer for word processing, then you have no real use for knowing exactly where your cursor is. But in the painting and drawing world, knowing these coordinates is very helpful—to say the least—and it's essential in a lot of instances. Nearly all graphic and page layout applications give you a separate window which shows the coordinates of where your cursor is located at any given moment. By watching your coordinates you can move, create, shape, or select objects or portions thereof with great precision.

coprocessor

A **coprocessor** is a *chip* that works side-by-side with the computer's main *processor* (the chip called the *central processing unit,* or *CPU*). The coprocessor handles some of the more specialized tasks, such as doing math calculations or displaying graphics on the screen, thereby taking some of the work load off the main processor so it can go on with the business of directing and keeping order over the whole show.

Math coprocessors, for example, are specialized for performing calculations on numbers, and they are much faster at it than the main processor in your computer. So if you have a program that does many math calculations, such as a *spreadsheet* or a *CAD* program, then adding a math coprocessor to your system can sometimes remarkably improve your computing speed.

There are video coprocessors that are used to speed up the display of graphics on your screen. Again, if you use any graphics-based application, including Windows, then adding a video coprocessor to your system on an *add-in board* can speed up your system even more than buying a faster computer.

One catch, however, with any kind of coprocessor, is that the software you use must be written so that it knows the coprocessor is there, otherwise your system will not recognize its existence and won't be able to use it.

copy

When you make a **copy** of something, whether it's a copy of a disk, a program, a document, or a paragraph or graphic image within a document, you are doing essentially the same thing as if you took a printed page and reproduced it on a copy machine—you're literally making a duplicate of the item while leaving the original item intact. Also see *Clipboard.*

If you make a copy of a **disk**, you aren't physically reproducing the disk. Instead, you're just copying all the files it contains onto another disk.

copy protection

Copy protection is any kind of technique used to prevent you from making duplicates of your software. The point, of course, is that without copy protection anyone can make as many copies of a software program as they want to. Obviously, software companies end up losing money if people copy software from others instead of buying it. The problem with copy protection is that it makes it impossible for you to make *backup* copies of software you have purchased. It is very important for you to have copies of your software in case something happens to your original disks or your hard disk fails. Because of this, market pressure has all but eliminated the use copy protection, at least with business type software. However, some games are still copy protected.

The technique used to copy protect software varies from product to product, but is usually based on special software tricks. Some games can't be played unless you answer questions, the answers to which are only in the manual—you only have a manual if you paid for the game.

Another form of copy protection, called "hardware copy protection," relies on a little device you stick onto the computer when you want to run a legitimate copy of the software. If your computer doesn't have this device, the software won't *run;* if the computer **does** have this device, any copy of the purchased software will run and so you can make backup copies. A problem with this technique is that you might lose the device. Also, it becomes rather ridiculous when you use several software programs that use hardware copy protection because you end up with a bunch of these little devices hanging off your computer.

core memory

See *RAM*.

Courier

Courier is the name of the *typeface* you are reading right now. It looks like the print of a typical typewriter, primarily because characters are *monospaced.* In monospaced type, each character takes up the same amount of space; that is, a period takes the same amount of space as a capital W. There are legitimate uses for Courier, but it is not a typeface that evokes a professional image, as shown in this paragraph.

You'll find Courier built into or installed in most printers and *graphical user interfaces.* If you use a particular font in your document and then take that document to another computer, (either to work on or perhaps to print) and the other computer does not have the typeface you originally used, that computer usually substitutes Courier for the missing typeface. That's guaranteed to make you irritable.

courseware

Courseware is educational software, applications that are designed to teach you some subject or to train you in some field. Courseware usually doesn't have much use outside of an educational institution.

cps, characters per second

The speed of a *dot matrix* or *daisy wheel printer* (or any printer that prints one character at a time) is measured in **characters per second,** or **cps.** A warning: the cps rating that a printer manufacturer puts in an ad is always much faster than the printer's actual speed.

Laser printers and similar machines that print an entire page at once are rated in *pages per minute (ppm)* instead.

CPU

The **CPU** is the **central processing unit,** the brains of the computer. In a personal computer it is the one tiny little *chip*, often called the *microprocessor,* that runs the show. Powerful magic. Sometimes people refer to the circuit board the microprocessor lives on, or the computer box (as distinct from the monitor and the keyboard), as the CPU.

The speed of a CPU is determined by several factors. One is in its "word length," or the number of *bits* it can process at one time. The more bits it can bite off in a chunk, the faster it can finish a calculation. For instance, the 80286 can process 16 bits at a time, while the 80386 and 80486 are 32-bit chips. Another important factor is the *clock speed,* which is measured in *megahertz* (millions of cycles per second). The Macintosh Quadra has a 25MHz microprocessor. The faster the clock speed, the faster the CPU can process the data. Finally, CPUs vary in efficiency—a newer processor can often accomplish a given task faster than an older one, even if both are running at the same clock speed.

Be aware that the performance of your computer depends on more than the CPU speed alone—factors such as the speed and "width" of the *bus,* the *memory* speed, and the presence of a *processor cache* make a big difference too.

cracker

A **cracker** is a *hacker* turned bad—a malicious meddler who likes to sneak into sensitive, secured information. Hackers are very interesting people, obsessively into their computers, particularly the programming of them (as opposed to *power users,* who just use the programs). Journalists got confused and starting calling the people who break into security systems "hackers."

crash

When your computer stops working or when part of the system suddenly breaks down, that's a **crash.** The term may sound overly dramatic, but the first time it happens to you, you'll understand. You'll say, "Oh. This must be a crash. I get it now." Perhaps the screen freezes, where typing on the keyboard or clicking the mouse on a menu doesn't do anything, or perhaps you see the bomb on your Mac's screen, or perhaps the system just decides to check out and tells you "There's been a system failure." Yes, everything you haven't saved to your disk is gone for good. Usually the only thing you can do now is turn the computer off and start over again (but first read about the *reset switch* on the Mac or *Crtl-Alt-Del* on a PC). If you turn off the power to the computer, wait at least a full minute before you turn it back on again.

Sometimes you may have a clue as to why your computer crashed; sometimes you just have to accept the fact that you will never know why.

A *hard disk* can also crash, meaning either that the files get *scrambled* or that something inside the disk actually breaks. Either way, this is a true disaster (do make your *backups*).

Cray

Seymour Cray founded a company called **Cray** Research, Inc., in 1972. Mr. Cray likes to build big, fast computers—"big" like double-refrigerator sized with 64-bit processing; fast like 165 *MHz*. The fastest Macintosh Quadra or IBM PC can process 15 *MIPS* (millions of instructions per second), whereas the **Cray** Y-MP/2 can process 8,000 MIPS.

creator, creator type

The **creator type** is a four-letter code found in all Macintosh files (they're written in by the original programmer) that identifies the application that created the file. When you double-click on a file or document, the computer will look for this code to tell it what application to open that file into. (Don't bother trying to find this code yourself; you need special utilities to be able to see it.) Often these codes are near-acronyms of the creator: RSED for ResEdit, MPNT for MacPaint, CPCT for Compact Pro. Canvas's creator code is DAD2, which could mean any number of things (a Father's Day gift from a loving programmer?) Who knows . . . it's totally up to the programmer to name the code.

crop marks

When you create a document in a *page layout program,* you can specify almost any page size you like. Everything will be printed onto standard letter-sized paper, unless your printer can accommodate a larger size. If you make an invitation, for instance, that will be a 5 x 7–inch finished page size, it will come out of your laser printer on 8.5 x 11 paper. It will be difficult to tell where the edges of the actual 5 x 7 page are unless you have let the printer apply **crop marks,** or unless you have drawn them in yourself. The crop marks appear as pairs of little lines at each of the four corners of the page. You can line up a ruler along the lines and slice the paper with a knife to cut the 5 x 7 page out of the 8.5 x 11 piece of paper.

Without the crop marks for this 5 x 7 page that is printed on an 8.5 x 11 piece of paper, you wouldn't be able to see the boundaries.

These are two of the eight crop marks on this page, two at each corner.

CRC

CRC (**c**yclical **r**edundancy **c**hecking) is a rather sophisticated *error checking* routine used to assure that two communicating computers are really getting the correct data. There are a whole bunch of these error correction and checking methods out there, but they all have pretty much the same objective: making sure that no data is lost or misinterpreted when two computers are exchanging information over the phone lines (through a modem, of course).

CRT

CRT is an acronym for **c**athode **r**ay **t**ube, the technical name for the picture tube inside a standard computer monitor or television set. See *monitor.*

Ctrl

Ctrl is an abbreviation for the *Control key.*

Ctrl-Alt-Del

On a PC, if you press the Control, Alt, and Delete keys (**Ctrl-Alt-Del**) at the same time, the computer stops whatever it's doing, the screen goes blank for a few moments, and then the entire system restarts or *reboots*—almost as if you'd turned the computer off and then back on. In the process, the system's *memory* gets a thorough flushing which removes all the programs and files it contains (only the data in *RAM* are erased—nothing on your disks is touched). In other words, if that letter you were writing before you press Ctrl-Alt-Del wasn't saved to disk, you'll never see it again.

So why would you ever want to press Ctrl-Alt-Del? The only good reason is if you get stuck, and stuck bad, while using a piece of software. For instance, you fire up your word processor, load a file, and start typing away, when suddenly, nothing seems to happen as you press the keys. The screen may freeze up, or you may see stray characters or splashes of dots. You might hear some fierce beeping as you type, or nothing at all. Anyway, your computer is out to lunch. Blotto.

In this situation, **don't** press Ctrl-Alt-Del until you've tried a couple of less drastic measures first:

Start by pressing **Esc** two or three times—this may be enough to bring your PC back to its senses

Next, wait for at least a minute. You may have pressed the *Print Screen*

(PrScr) key accidentally. If so, and if you don't have a printer hooked up, your system will keep trying to print for quite a while before it gives up.

If **Esc** doesn't work, try pressing Ctrl-C and then Ctrl-Break. In some programs, these keys interrupt whatever is going on—if you're lucky, you'll get back control of your software. If not, nothing will happen, or you'll be dumped unceremoniously back to the *DOS prompt.*

Pressing Ctrl-Alt-Del works differently when certain software is running, most notably Windows. In Windows 3.1 a screen will tell you if a program has blown its mind. If Windows detects the bad program you can terminate it by pressing Enter from this screen. Sometimes Windows doesn't recognize that a program has bombed. If this happens, you can't press Enter to terminate the offender. But you can press Ctrl-Alt-Del again to summarily shut down Windows and all your programs, restarting the whole system.

CTS

In the world of *asynchronous communication* (over Mac or PC *serial ports*), **CTS** stands for **c**lear **t**o **s**end. When a computer needs to communicate with an outside device physically connected to it (peripheral), the two of them have to go through this preamble of making sure each of them is ready to hear what the other has to say, and confirm that they are going to be speaking the same language. Somewhere near the beginning of this preamble, one device will send an RTS (Request To Send) message. The receiving device will then reply with a CTS message "Yes, I'm now ready for anything you want to send me." I know, the whole thing sounds like you may be sitting there drumming your fingers on your desk, waiting for these flipping machines to get done with their senseless checklist. Relax. Like most processes done by a computer, this one takes less than a millisecond.

curly quotes

Curly quotes refers to the true typographic quotation marks and apostrophe because the marks curl around the letters, as opposed to typewriter quotation marks, which are straight up and down. See *smart quotes* for illustrations and the keystrokes for producing curly quotes.

current directory

See *directory.*

cursor

The **cursor** is that little mark on the screen that moves when you move the mouse or perhaps press certain keys, such as the *arrow keys* or the *edit keys*. The term has different shades of meaning depending on the computer and the software you have.

On a *character-based* system, as in plain DOS or UNIX, the cursor is a flashing line or rectangle that shows you where the next letter you type will appear. By pressing the arrow keys, you can move this text cursor directly over characters you've previously typed.

In graphical software or *environments* such as the Macintosh or Windows, the cursor usually refers to the shape that moves when you move the mouse; this cursor can often be controlled with the keyboard as well.

Cursors come in different shapes, and each of these shapes has a corresponding available action or meaning attached to it. The cursor may be a pointer or a little light-colored rectangle on top of a letter or a crosshair or simply a tiny line.

The shape of a cursor can often give you a visual clue as to what it does. For instance, most paint or draw programs display a different cursor depending on what type of painting or drawing tool you have chosen.

There are also animated cursors, like little watches with hands that move or beachballs that roll around or hourglasses with dripping sand. While these cursors are doing their thing, it means the computer is busy performing a task (copying a file or disk, calculating an equation, opening an application, etc.) and you're just going to have to wait until it's done before you can take any new action.

cursor keys

The four *arrow keys* on the keyboard are sometimes called the **cursor keys** because in some programs they control the movement of the *cursor* on the screen. Sometimes a program (especially game programs) allows you to define which keys you want to use to control the cursor (or the laser beam, space creature, or rocket ship). These keys you define, then, are considered the cursor keys, even if they are not the arrrow keys.

cut

When you choose to **cut** an item of text or graphics from a document, you are accomplishing the electronic version of getting some scissors and cutting the item out of the page. Although there's no hole left behind in your screen, the cut item does disappear from its original place. When you *copy* an item, you reproduce it and the original item stays right where it was, intact.

You usually cut an item so that you can *paste* it in somewhere else. This "cut-and-paste" action is a very common form of editing, as in rewriting, and also of working with graphics and page layout. Also see *Clipboard* and *delete*.

cybernetics

Cybernetics is the comparative study of computers and the human nervous system. Whoa! Yes, this is the kind of stuff that used to scare a lot of us common folks away from computers. The fact is our own nervous system and the computer's electronic system have much in common, and cybernetics is the study of these comparable processes so we can better understand the similarities and differences. The hysteria and assumptions around cybernetics that arose in the sixties have long since been put to rest—thanks to the assiduous work of a group of scientists blessed with a working knowledge of reality. Today the study of cybernetics seems to be headed in a direction we can all support. As early as 1966, an ecologist named Romón Margalef began to see this, and wrote, "It is a basic property of nature, from the point of view of cybernetics, that any exchange between two systems of different information content does not result in a partition or equalizing of the information, but increases the difference. The system with more accumulated information becomes still richer from the exchange."

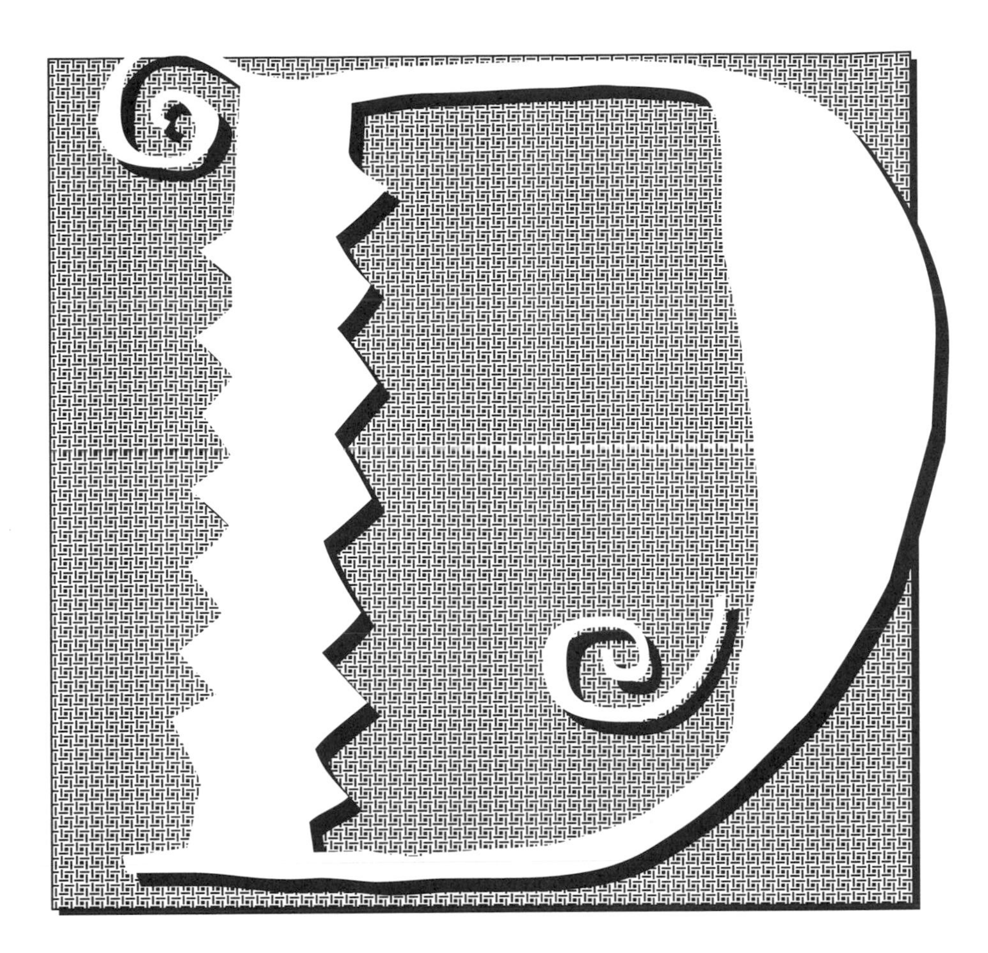

DA

See *desk accessory.*

daisychain

Sometimes, like in a school computer lab or in an office, you need to hook up several computers to one printer, or several computers need to connect to one *file server,* or maybe you have several devices (such as a scanner and a CD-ROM player) that you need to hook into one computer. Well, the only way to connect all these devices together, since there is only one *port,* or connecting place, on the back of the computer, is to connect each object (each device) to the next one in line, making a **daisychain.**

For instance, to connect the scanner and the CD-ROM player to your computer, you would have a cable going from the computer to the

CD player, and another cable from the CD player to the scanner. Through this "chain," the computer can use the scanner, even though it is not directly connected to it.

daisywheel printer

When personal computers first came out, **daisywheel printers** were the only type of affordable printer that could print sharp-enough text for important documents like business communications or college papers. Daisywheel printers work by pounding raised, fully-formed letters made of metal or plastic against the paper through a ribbon, just like a typewriter.

To be precise, a daisywheel printer has the characters mounted on the end of narrow projections arranged in a circle, like spokes on a wheel, or like petals on a daisy. If the printer has the raised characters mounted on a ball that spins around, as on an IBM Selectric typewriter, it's not technically a daisywheel, but some people would call it that anyway. "Formed element printer" is one generic term for any typewriter-like machine. We used to call such printers "letter quality" printers, but then many dot matrix printers came out with "near letter quality" modes, and it became very easy to confuse them.

dash

A **dash** is a typographic character, similar to but longer than a hyphen. There is an *en dash* and an *em dash*. Each has a very definite purpose and a very definite typographic character. Please see *en dash* and *em dash*.

Hyphen	-	(used to hyphenate words)
En dash	–	(used to indicate a range, show duration, or to hyphenate double compound words or adjectives)
Em dash	—	(used to indicate a strong break in thought, among other things [see your grammar book]; replaces the two hyphens we were taught to use on a typewriter)

DAT

Dat stands for **d**igital **a**udio **t**ape. See *DAT cartridge, DAT drive.*

data

The word **data** is actually plural for "datum," a single piece of information. You're supposed to say, "I analyzed these data," or "The data are conclusive." Most people have a hard time with this plural usage; it's natural to think of "data" as referring to a collection of computerized information as a whole. We won't hold it against you if your data is singular. (And please don't hold it against us.)

database

A **database** document is just a collection of information stored in computerized form. The simplest way to understand a database is to think of it like a set of 3 x 5 cards. Since the information is on your computer, though, a click of the mouse or the stroke of a key can alphabetize those "cards," or find just the names of the people on the cards who live in a certain town, or tell you who owes how much money, and so on.

Computer databases can be highly structured, storing the same kind of information about each item in the database in well-defined compartments. This works as if you printed a standard form on each of your 3 x 5 cards—perhaps with one space for a name, one space for an address, and one space for a telephone number. In a structured computer database, the "space" for a name, a part number, a price, is called a *field*. A *record* corresponds to one of the individual 3 x 5 cards. The record contains a complete set of fields, all filled with information corresponding to a particular item: if your database is a name-and-address list, each record represents a person; if your database is a parts catalog, each record represents one part.

A specific set of fields and records organized in a specific order, including the information they contain, is called a *table*. In fact, tables are often displayed on the screen with each item, or record, in a row, and each field as a column.

Structured databases can be either *flatfile* databases or *relational* databases. In a flatfile database, you can work with only one data table—one set of fields—at a time. In a relational database, you can use multiple tables (multiple database documents) at once. Flatfile databases are much easier to understand and use, but relational databases are much more efficient for many things you commonly do with data, especially in businesses.

A database can also be simply a free-form collection of information, without any particular structure. In this case, the analogy would be to a pile of notes you've written on whatever paper was handy at the time—the information on each piece of paper doesn't have to be organized in the same way.

The term **database** can also refer to the software package itself that you use to create the database. More often, the software is called a "database program" ("database application" is more specific) or a "database management system" (*DBMS*).

A database application is one of the most useful tools on the computer, and is actually an incredible amount of fun. If you've never used one before, read Guy Kawasaki's little book, *Database 101*. Everybody needs a database.

database engine

In a computer *database*, the **database engine** is the software that does the real work of *sorting* the information, finding specific data that you request, and so on. The term used to refer to a separate piece of software that ran on a central computer (in this case, it is more or less synonymous with the term "back-end"). Widely used database engines include Oracle, DB2, and Sybase. Separate *front-end* software running on your own computer lets you tell the database engine what to do (how to sort the data, what data to find), and displays the results of your commands.

If you're talking about database software that runs entirely on one computer, the term database engine is sometimes used to describe that part of the software responsible for manipulating the data, as distinct from the parts of the software that you interact with on the screen.

data cartridge

A **data cartridge** is a tape cartridge that looks like an oversized audio cassette. Data cartridges are used for backing up the contents of your hard disk. See *DAT cartridge.*

data entry

Data entry is the act of introducing information (data) into the computer. Although entering information by typing it in with the keyboard is the most recognized form of data entry, the term also applies to bringing in data via *scanners,* voice recorder, digital cameras, etc.

data fork

See *resource fork,* which explains *resource, resource fork,* and *data fork.*

data recovery

Data recovery is the art of restoring lost or damaged files. This damage can occur when your computer *crashes,* a *virus* infects, you accidentally reformat a disk that contains precious data, or you experience some other catastrophe of considerable dimension. And, at some point in your life, you're going to delete a file you really didn't mean to (believe me). The next time tragedy strikes, try running one of the many data recovery applications (powerful software written specifically for data recovery purposes) to see if it can correct the situation. Often these little jewels can work magic and save your day—and your files.

There are companies that specialize in restoring lost data. These people generally work with more serious situations, like when you lose the last three years of your business's financial records. Find them in the yellow pages under "Computers: Service and Repair." Check the ads in the back pages of computer magazines if you can't find someone local. It's very common to have to ship your hard disk or your computer off to a company that can fix it.

data transfer rate

Data transfer rate (sometimes just called "data rate") simply refers to just how fast data can get from one place to another—"one place to another" meaning between a computer and a *peripheral* (an external piece of hardware connected to the computer) or one *modem* to another modem, or even from one internal part of a computer to another internal part. This measured speed can vary a great deal, depending on just what kind of data is being transferred, what method is being used to transfer it, and what hardware and software is doing the transferring. Don't be too surprised by large figures when you see specific data transfer rates mentioned; the standard unit of measure for most transfers is *MB (megabytes)* per second!

DAT cartridge, DAT drive

DAT stands for **d**igital **a**udio **t**ape. It refers to a type of tape recording technology, originally developed for music, which represents music on tape with numbers (digitally) rather than as analog sound waves (see *digitize* and *A-to-D conversion*). High-fidelity digital music requires a system that can record a great deal of data at high speed, and DAT technology measures up. Since the same requirements pertain to backing up a hard disk, DAT technology has been adapted for use with computers.

A DAT drive is the equivalent of a tape recorder, except that it records computer data instead of music onto dat cartridges. The cartridges are only about as big around as a playing card, and not much thicker than a standard audio cassette tape. DAT drives are fairly expensive compared to more common kinds of backup tape drives, and they're mainly used by businesses which can justify their higher cost in return for faster recording speed and greater storage capacity per tape.

daughterboard, daughtercard

A **daughterboard** is a *circuit board* that attaches physically to another board in your computer. As an additional board, it provides additional functionality.

Depending on the daughterboard, it may connect directly to the *motherboard* (the main circuit board for a computer), or to *add-in boards* that fit into the *expansion slots* in your computer. For example, you can add a daughterboard containing additional *memory* to some add-in *accelerator boards*. See the illustrations of motherboards and add-in boards in Appendix A on "How to read a computer ad."

dBase

dBase is a specific software product used for creating and manipulating *relational databases*. The term can also refer to the dBase database *programming language* that first appeared in the dBase product, but which is now available in many other database forms.

As a software product, dBase was the first commercially successful database system for personal computers, and it remains one of the most widely used. Past versions include dBase II, III, III Plus, and several versions of dBase IV. In the early versions of dBase, to create a database or search for data or do anything else, you had to type out commands on a command line, much as you do in DOS. Nowadays, you can do much of the work using a system of menus, but you still need to be pretty savvy about databases to get very far. It's easier just to run programs that someone else has written using the dBase programming language. People who know dBase can write complete programs to accomplish customized tasks—like managing your company's inventory, invoicing system, and customer list. Once the program is working properly, anyone can run it without having to know anything about dBase commands.

Although the dBase programming language originated with the dBase software product, the language has taken on a life of its own. A variety of other software manufacturers have created their own versions. Since the name dBase is associated with a particular product, the abstract version of the dBase programming language is often referred to as **xBase** instead.

DBMS

DBMS stands for **d**ata**b**ase **m**anagement **s**ystem. A DBMS is a large, complex software product used to create and modify *relational databases* and to retrieve information from them. A DBMS is intended primarily for heavy-duty business use. Some people might use one for personal use, but a DBMS is overkill for most of us, and many of its capabilities would be wasted—a *flatfile database* is usually a better choice for things like filing names and addresses or recipes.

debug, debugger

A *bug* is a problem in software or hardware; to **debug** is to diagnose and correct said problem. Software programs inevitably develop bugs due to mistakes in planning or simply from accidentally typing the wrong command. Before a program can run properly, all bugs have to be found and corrected. A **debugger** is an application developed for the specific purpose of finding these problems; it lets the programmer run the program one step at a time so that she can see exactly where the mistake occurs. See *bug* for the origin of this entomological term.

debugger switch

See *programmer's switch*.

DEC

DEC is the acronym for **D**igital **E**quipment **C**orporation, a large manufacturer of computers and related equipment. DEC is best known for its VAX line of *minicomputers,* but it now has a successful line of mail-order PCs, too.

decimal tab, dec tab

See *tab marker*.

decrypt

To **decrypt** means to restore information that has been *encrypted* (turned into secret code for security purposes), making the information usable again. To decrypt the encrypted information you need to know the *password*. See *encrypt* for more details.

dedicated

If a computer or other device is reserved for one task, we say it is **dedicated.** For instance, in a college computer lab or in a office with several computers, there may be one or more computers that are dedicated *file servers,* which means no one can use them because their only purpose in life is to be the file servers (to give information to others).

de facto standard

If something is a **de facto standard,** it implies that the product or process has become an industry standard—not because a governing board ruled it as a standard, but simply by virtue of the fact that it is what most people use. Since most people use it, other developers create products to work with it, further embedding the product or process as a standard in the collective conscious. De facto standards become de facto standards because they are the best, at least at the time they appear. Better technology that comes along after a de facto standard becomes fully entrenched may not be able to establish a new standard.

Examples of technologies that have become de facto standards are the *PostScript page description language* for laser printers, the Hewlett-Packard *printer control language* (PCL) for PC-based laser printers, the *Hayes command set (AT command set)* for using modems, and the *XMODEM communications protocol.*

default

A **default** in any product is the automatic decision already made for you which will automatically be carried out unless you explicitly change it. For instance, when you open your word processor and begin a new document, certain parameters are set up for you, such as margins, tabs stops, the size of the type, etc. Your *telecommunications software* has defaults set for your modem; your database has defaults set for the fields; etc. You can always change these default parameters. Sometimes you can change the actual default itself, so that in the future you can start a new document with the parameters you prefer already in place, rather than having to change the existing ones every time. Check your manual to see if it's possible to change the defaults and how to do it.

When you change the actual defaults, you will affect only new documents created from that point on; any previously created documents will hang on to the parameters you specified for that existing document.

delete

To **delete** means to remove, erase, get rid of, trash, throw away. For instance, you can delete files from a disk if you know you won't need them anymore. (If you act quickly, you can usually get deleted files back—see *undelete.*) You can also delete items like text or graphics within a document.

If your application allows you to *cut, copy, and paste* (most Macintosh and Windows applications), it's important to understand the difference between *cut* and *delete.*

> **Cutting** an item removes the item from the page and places it on the Clipboard so you can paste it in somewhere else.
>
> **Deleting** will also remove the item, but the item does **not** go to the Clipboard—it's just gone. You won't be able to paste it anywhere. See *cut, copy,* and *paste.*

Remember, the Clipboard can only hold one item at a time. So if you have something holding in the Clipboard that you want to hang on to, don't cut or copy something else or that item in the Clipboard will be **replaced** with the item you cut or copy.

delimited, delimiter

A **delimiter** is a character that separates items within a database record or other set of data. The Tab character is a very common delimiter, as are the Return and the comma. Some programs accept a colon or a slash, and some programs let you decide on your own character to use.

Delimiters serve a very important function when you are *importing* or *exporting* information to or from a *database* or *spreadsheet.* For instance, say you're putting together an address book with your *database* file of names and addresses, and you want to add a list of names and addresses that someone else has already entered into a database on a different computer. If the other person didn't use the same database program as the one you have, they will have to *export* the list into a file that your program can *import.* In most cases, this file will use delimiters to tell your program which information goes where.

Database programs organize information into *fields* (such as Name, Address, and Phone Number) within *records* (in the address book, all the fields of information for one person constitute that person's record). Fields are sometimes referred to as columns, records as rows. In an import/export file, fields (columns) are typically delimited with commas or tabs, while and records (rows) are delimited with Return characters. If the fields in each record of your database are last name, first name, city, state, the address book file might look like this (you can't see the Return characters at the end of each person's record when you look at the file on your screen or in print, but the Returns are what makes each record appear on a new line):

Using a comma as the delimiter.

Kent, Clark, Metropolis, MI
Wayne, Bruce, Gotham City, NJ
Duck, Daffy, Universal City, CA

In the following lines I typed a Tab as a delimiter between each of the four items:

Using a Tab as the delimiter

Boop	Betty	Hollywood	CA
Boris	Badenov	New York	NY

Each of the examples above have four fields. Each field has a delimiter between it. So we could call them Fields 1, 2, 3, and 4. When this data is imported, the first name in each set would drop into Field 1, the second names into Field 2, etc.

Keep in mind when setting up data with delimiters that the computer doesn't know Kent is a last name and Betty is a first name—the computer just puts the first item into the first field and the second item into the second field, etc. I'm sure you can imagine what your new data would look like if you imported it into a database that did not have the same fields in the same order.

When you import or export you either have a choice of which delimiters to use, or the application will tell you which delimiters to use.

Also, **delimiters** are often used in DOS and other *operating systems* to separate one command from another; if you don't type the delimiter, the operating system won't understand that you mean two separate commands.

demo

A **demo** is a product demonstration, usually software, presented so you will buy the product after you've had this wonderful opportunity to see how dazzling the product actually is. A person who does good demo can sell anything and a person who doesn't do good demo can destroy a product.

The word "demo" can also be used as a verb, as in "they're demoing the latest version this afternoon."

demo disk

Disk vendors often give away **demo disks** at trade shows or by mail to demonstrate the program they sell. The demo software on the disk may just show you an "animated tour" of the program's highlights. Sometimes, though, the disk contains the complete application for you to try out yourself, with an important feature disabled—you can't save a document, or you can't print, or the document prints with a big "DEMO VERSION" diagonally across the entire page, right over your text.

density

When you're talking about *floppy disks,* **density** is one term used to describe how tightly you can pack data on the disk, and in turn, how much data the disk can hold. A double-density, 5.25-inch PC floppy disk holds 360K (kilobytes). A 3.5-inch double-sided, double-density disk holds about 800K. These days, 3.5-inch floppy disks that hold 1.44MB of data, and 5.25-inch disks for PCs that hold 1.2MB are very common. They're referred to as "high-capacity" or "high-density" disks.

descender

Letters of the alphabet sit on an invisible *baseline.* Any part of a letter than hangs below this baseline is a **descender,** such as the bottom parts of the letters g, p, or y. Also see *x-height, ascender,* and *baseline.*

descending sort

A **descending sort** is the opposite of an *ascending sort,* wood'nja know. To *sort* means to organize the information. If you use your application to do a descending sort on a list of numbers, the list would descend from the biggest number to the smallest, like counting backwards. A descending sort on a list of names would put them in alphabetical order backwards. See *ascending sort* and *sort.*

desk accessory

Desk accessories (often just called *DAs*) are little mini-applications, similar to the sort of accessories you might have on your desk. Your System installs several of these handy little tools automatically, such as the Alarm Clock, the Calculator, and the Note Pad.

In System 6, you need to install desk accessories directly into the system with the Font/DA Mover. In System 7, DAs no longer have to be installed—

they are treated (and behave) just like any other application. This means that you can put 'em where you want 'em, use 'em whenever you want to, and even make *aliases* of them to store in easily accessible places.

Desktop

The **Desktop** is the background on your screen when you're using a Macintosh, Microsoft Windows, and similar *graphical user interfaces*. The idea is that this screen background is sort of like the top of your real desk, and your program windows are all lying on the desktop in a pile.

Some programs may refer to their own "desktops." In this case, the desktop is what you see on the screen when the program is running but no document is open.

The Desktop is also commonly called the *Finder* (especially under System 7). Actually, the Finder is the software that runs the Desktop (among other things). But since you can only access the Finder while at the Desktop, and since whenever you are at the Desktop you are in the software program the Finder, the two terms are often used interchangeably.

The Desktop is where all your files are kept, all your applications and documents, the System Folder—all of your life on the computer is available here on the Desktop. You come back to the Desktop whenever you want to open another application, install more fonts, see what's on a disk, delete files, organize, etc. It's kind of like "home base."

Desktop level

The **Desktop level** is called a "level" because there is a *hierarchy* of different levels, kind of like different floors in a tall building. When you can actually see the Desktop, you can double-click to open folders and then you see that folder's window, right? That window is one level of the hierarchy. If you open another folder that is within that window, you go down one more level into the hierarchy. Well, the Desktop level is the very top of the hierarchy (or the root, some might say) because the first window (including the hard disk window) sits on the Desktop.

You will also notice this hierarchy in any Open, Save, or Import dialog box (among others). Click the Desktop button or press Command D to go directly to the Desktop level (System 7 only), where you can open other hard or floppy disks or get to any file that is currently stored on the Desktop level (directly on the pattern of the Desktop, rather than in any folder or window).

When you are at the Desktop and a window is *active* (meaning you can see stripes in its *title bar*), that window is the *level* that is active. If you choose to create a new folder, it will appear in that window that is active. If you want to create a new folder on the Desktop level, you must first *select* the Desktop level (System 7 only). You can do this by clicking once on the hard disk icon—notice not a single open window on the Desktop has stripes in its title bar now. You can also select the Desktop level by pressing Command Shift UpArrow.

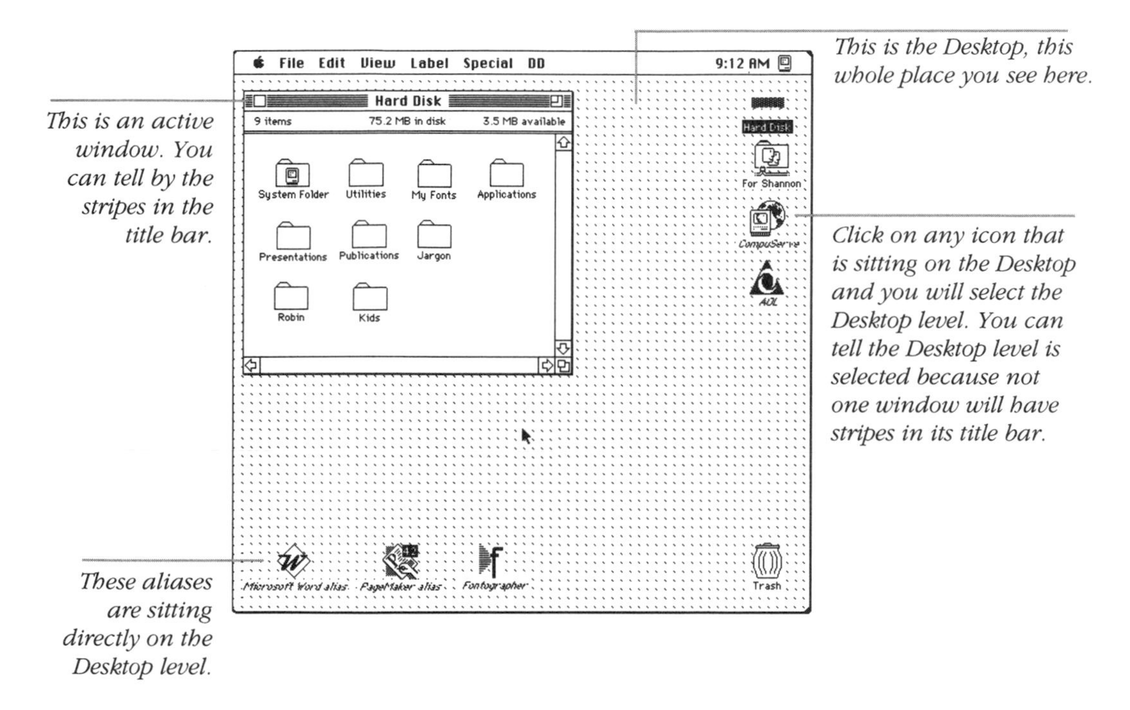

This is an active window. You can tell by the stripes in the title bar.

This is the Desktop, this whole place you see here.

Click on any icon that is sitting on the Desktop and you will select the Desktop level. You can tell the Desktop level is selected because not one window will have stripes in its title bar.

These aliases are sitting directly on the Desktop level.

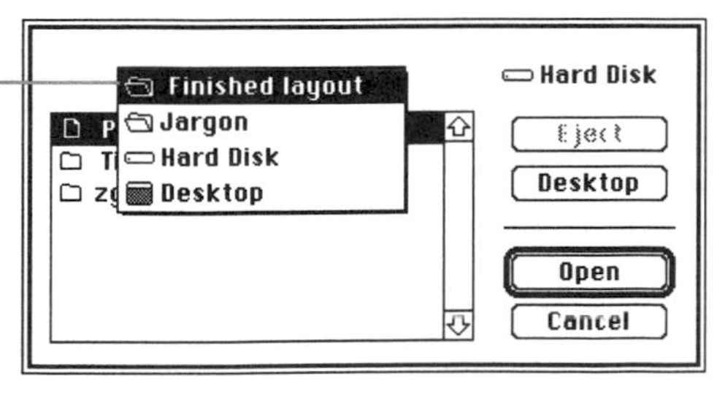

This hierarchy shows that at the Desktop is an icon to the Hard Disk. In the Hard Disk is a folder called Jargon. In Jargon is another folder called Finished Layout.

In an Open or Save As dialog box, you can see the hierarchy of levels. The Desktop level is always the top (or the bottom, if you prefer). If you choose to save onto the Desktop, the saved file will appear right on the Desktop, not in any folder. If you want to see the contents of another disk, choose the Desktop level to find the names of the other disks.

desktop publishing

Desktop publishing (DTP) is the process of creating printed documents that look professionally produced, using *page layout* software running on a personal computer, along with a high-quality, yet affordable, printer.

To publish something with the traditional method, you would send typed or handwritten text to a typesetter, who would turn it into typeset text called "galleys," which took a couple of days. If there were corrections, it took another couple of days to get those back. If you didn't know how to lay out the pages yourself, you'd take the galleys to a print shop, along with your art (illustrations and photographs). The people there would cut up the galleys with scissors and paste the pieces onto the pages along with the artwork. If something needed to be changed on the finished "paste-up" or "mechanical," it would be possible, but a lengthy and expensive process. Finally, the print shop would reproduce the document in quantity. Instead, you might pay a graphic designer to take the project from conception to completion, and the designer would go through this process, creating the mechanical herself and taking it to the print shop to be reproduced.

With desktop publishing, by contrast, you can create the entire document sitting at your own desk. You can think of the page layout software and the computer as the typesetting and layout area, and the laser printer as the printing press. You proof the project on your own printer; if it isn't right, you just turn back to your computer, make the changes, and print it again. Depending on how many final copies you need and the quality you want, you can print the job yourself, duplicate the masters you print on a photocopy machine, or have a professional press reproduce them. For the ultimate in quality, you can take your disk with your finished document to a *service bureau*, have it output on a *high-resolution imagesetter,* and then have the resulting pages professionally reproduced.

This book you are reading was desktop published. I wrote it (and Steve sent his contributions through the *modem* and on disk), created the illustrations and screen shots, designed the pages, helped Scarlett turn her type design into a PostScript font, laid out all the pages, proof-printed them on my laser printer, indexed it, had it edited and proofread, made all the corrections, took it to the imagesetter, and sent the final, camera-ready pages to Peachpit Press, who took them to the commercial press to be reproduced and bound into the final book. I did all this work at my Macintosh in my bedroom while my children were at school or asleep, or sometimes working at the desk next to me.

DESQview

Quarterdeck's **DESQview** is a popular software product that allows PCs to run multiple programs at the same time (this is called *multi-tasking*).

destination disk

The **destination disk** is the disk you send a file or other information to. If you copy the contents of one disk to another, the original is the "source" disk, and the disk you copy *to* is the destination disk (also called the "target disk").

developer

A **developer** conceives of, designs, and produces a product. The term is mostly used for those people or companies who develop software for commercial sale.

device

A **device** just means any kind of component that's part of or attached vto your computer. It can be located inside or outside the computer. A mouse, for example, is a device that sits outside the computer, while an internal disk drive (another device) is inside the computer.

Devices need instructions on how to communicate with the printer or the rest of the computer, and sometimes those instructions aren't part of the computer's standard *operating system*. In this situation, you have to add a little piece of software (a file) called a *device driver*.

The term *device* always refers to *hardware*, as contrasted with *device driver*, which always refers to *software*.

In DOS, the word "device" is used as shorthand for "device driver" in the *CONFIG.SYS* file (CONFIG.SYS sets up your system when you start the computer). To activate a DOS device driver, you add a line such as any of these to your CONFIG.SYS file:

DEVICE=CDROM.SYS	(CDROM.SYS is the device driver for a CD-ROM player)
DEVICE=VIDDRVR.SYS	(VIDDRVR.SYS is the device driver for a non-standard, large-screen monitor)
DEVICE=HIMEM.SYS	(HIMEM.SYS is an *extended memory* manager)

device independent, device dependent

Device independent components work right no matter what model of *device* you use them with. For example, if the *graphic file format* in your publication is device independent, the results you see on paper will look about the same whether you print to an HP DeskJet, an Apple LaserWriter, or a high-resolution Linotronic imagesetter (the graphic will be printed at whatever resolution the printer uses).

Device dependent components by contrast, work right only with a particular model of the device. A device dependent *bitmap* graphic looks the way it's supposed to only on a particular type of monitor—on other screens it looks funny or may not display at all.

All the programs that run under Windows and the Macintosh are device independent. A given program doesn't have to worry about how to work with every mouse, keyboard, printer, screen, or scanner on the market. Instead, as long as the environment has a software control module (or *driver)* for the devices in your system, every program will automatically produce the expected results. Of course, there may be some variations due to the differing capabilities of different models. For instance, text you print on a 300-dot-per-inch laser printer will not be as smooth as text from a 2540-dot-per-inch *imagesetter*. But the basic look of the text will still be the same if you have a device independent system.

device driver

A **device driver** is a piece of software designed for a particular device (printer, mouse, monitor, or what have you) and the particular application program or environment you're working with. The driver serves as a go-between for the program (or environment) and the device, translating the software's desires into commands the device understands.

For example, if you have a LaserJet printer, you'll need to have the LaserJet driver to print. If you buy a new or different type of printer, you will then have to change to the driver that matches the new model of printer.

DIALOG™

DIALOG is an *online* information service that you can dial into with your *modem* to search online *databases*. An online database provides access to information like journal abstracts (the summaries of the articles), financial data, or patent records. You can search for the particular information you need by subject, author, date, or whatever suits you.

dialog box

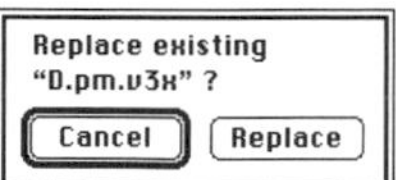

A **dialog box** ("dialog" for short) is any box, or window, on the screen that you have a dialog with. You can decide to pop up some dialog boxes to enter information in, or make choices about how a program works. Other dialogs appear automatically to give you simple messages (such as "Your disk is full") and ask you what you want to do about it.

Some people make a distinction between dialog boxes and *alert boxes* (where the computer warns you about something) or message boxes (where the computer just tells you about something), but that really only matters if you're a programmer. Everyone will understand what you mean if you call them all dialog boxes.

dictionary

Just as you use a dictionary to look up words to spell, some computer software uses a **dictionary** to look things up, too. Most often this term refers to spelling dictionaries in word processing or page layout programs. During a spelling check, the computer checks every word in your document against the list of words in its dictionary. If you have used a word it doesn't have in its list, the dictionary tells you the word is misspelled and/or gives you a list of words it thinks may be correct. Sometimes these suggestions can be pretty hilarious. You may, depending on the program, have the option to add words of your own to the dictionary. Many applications offer foreign dictionaries (useful, of course, when you are writing in another language).

The dictionary also controls the hyphenation of words (if your application automatically hyphenates). In page layout applications, where type and typography are very important, you can control many aspects of hyphenation. See *H&J* (hyphenation and justification).

DIF

DIF stands for **d**ata **i**nterchange **f**ormat, a standard *file format* for applications like spreadsheets and databases, where the information is structured in columns and rows. It was originally developed by Software Arts, the company that created the very first electronic spreadsheet program on earth, *VisiCalc,* created for the Apple computer.

DIF also stands for **d**ocument **i**nterchange **f**ormat, which is a standard file format developed by the U.S. Navy for interchanging documents between different computer programs.

digital

Information that is **digital** is information represented by numbers (digits) or more broadly, information that can be measured in discreet, exact values. The opposite term is *analog,* which describes information represented along a continuous range, where there are an infinite number of possible values. Trite as it may be, the best way to understand the difference between digital and analog is to compare a digital clock to a traditional round clock with hands. The display on a digital clock always shows one particular time, in numbers. A clock with hands, in contrast, is an analog device because the hands move along the entire circle of the clock face; at any one instant the hands can be anywhere on the clock, displaying an infinite number of moments in time. Water is analog; ice cubes are digital. All common computers work only with numbers and are digital devices. If you want to use your computer to work with information from the natural world, which is almost entirely analog, you have to convert that analog information to digital form. See *analog, A-to-D converter,* and *digitize.*

digital camera

A **digital camera** outputs images in *digital* form (real images converted into the digital numbers that the computer can understand) instead of onto photographic film.

Digital cameras come in different forms. There are digital video cameras (much like the standard video camera—but much, much more expensive) and there are still-image digital cameras that look and behave like your standard 35mm camera. In both cases, the images are either saved onto special floppy disks or sent directly to the computer's hard drive.

Today's digital cameras, because of their limited resolution, are used in situations which don't demand the highest quality, especially where the speed and convenience of filmless photography justify the trade-off. This may be for catalog work, timely news reporting, or when the resulting images need to be analyzed or manipulated—everything from medical science to specialized graphics to certain aspects of the motion picture industry.

digital signature

A **digital signature** is an *encryption* technique (secret code) used to verify that an electronic mail message was sent by the person the message claims sent it.

digitize

The qualities of the natural world—attributes like temperature, time, length, color, and the pitch of sounds—vary continuously over an infinite range of possible values. To **digitize** means to represent something that occurs in this infinitely variable *(analog)* form as a series of concrete, countable numbers. Since computers are digital devices—they can work directly only with precise numbers—information about the natural world must be digitized before it goes into your computer.

Say you want to add a photograph of your dog, Jane, to the family newsletter you're putting together. The only way the computer can deal with that photograph is if you digitize it; you need a *scanner* to digitize the picture. The scanner divides the image into an imaginary grid of small dots, and records the intensity of light reflected from each dot as a number—it turns the gray or color values of Jane's face into digital *bits* of information. This information can be reassembled into a recognizable picture by your computer. In digitized music (like the kind recorded on a CD), the sound wave is "sampled" every fraction of a second; the "height" of the wave (amplitude) at that moment is recorded as a number.

Any information that has been digitized contains only a sampling of the original information—much of the original is always lost in the digitizing process. Still, if there are enough samples, and if they are detailed enough, the reconstituted information can seem like the real thing. For example, the quality of a scanned, digitized photograph depends on how many gradations of light intensity the scanner can distinguish, and how small the dots on the imaginary grid.

DIN connector

DIN is an acronym for **D**eutsche **I**ndustri**N**orm (a West German standard). It is a type of round connector end on a cable, usually used for *serial ports*. It has little pins in it; some DINs are 5-pin, some 6-pin, some 8-pin. The Mac uses 8-pin DIN connectors for its keyboard, mouse, printer, and modem connections on the Mac Plus, SE, and Mac II (later Macintosh models use *ADB* cables and connectors for the keyboard and mouse). Most PCs use 5- or 6-pin DIN connectors to connect the keyboard and the mouse, if there is one.

dingbat

Cute as this word is, **dingbat** is actually a traditional printer's term for those little decorative type elements, like stars and hearts and tiny flowers and snowflakes. These are also sometimes called "ornaments," and the one that look like little pictures are often called "picture fonts." The dingbats shown below are from a font called Woodtype Ornaments.

Zapf Dingbats (pronounce both the p and the f in Zapf) is the name of one very popular collection, but there are quite a few dingbat fonts.

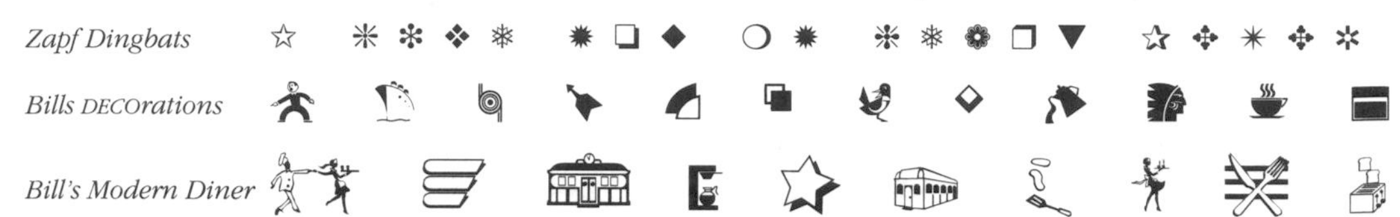

DIP

This is a DIP.

DIP is is an acronym that stands for **d**ual **i**nline **p**ackage and refers to the physical layout of most computer *chips*. The standard chip looks like a bug, or a little rectangular box with legs. Those two rows of metal legs are the "pins" that connect the chip to the computer's circuits. The whole chip is referred to as a "package," and it's a *dual inline* package because the pins are lined up in two rows. Also see *chip*.

DIP switches

DIP switches are those nasty, tiny plastic toggle switches that come mounted together in a row on a little box-like part attached to your computer's *motherboard* or on some of your *add-in boards*. Each individual switch in a DIP switch unit can be set either on or off, allowing you to control some aspect of your computer's function. For example, on the older IBM PCs and compatibles, you had to set the DIP switches to match the amount of memory installed in your computer, to tell it what kind of monitor you had, and so on. On the add-in modem board I just installed, I had to flip some of the DIP switches so the computer would know where to find the board electronically.

DIP switches are a monumental pain. For one thing, they're so small it's hard to switch them. Use the tip of a ballpoint pen, an unfolded paper clip, a baby screwdriver blade, or some other fairly pointy, narrow tool. Worse yet, they're always mounted in some out-of-the-way place, like on the back of your computer or inside it. Even if you can get to the switches without opening up the computer, you have to pull the machine out from the wall as far as it will go, cram your head back there, and just hope you'll switch the right switch. And the labels that tell you which direction is on or off are horrible. The writing is too small to see, and usually it says 1 and 0 instead of ON and OFF (yes, 1 [one] means on, 0 [zero] means off).

The point is, if two products are otherwise equal in features and price, choose the one that doesn't have DIP switches. It will set itself up automatically, or maybe it comes with a software utility that lets you set it up by choosing options on your computer screen. Also see *jumper.*

My kids think "dip switches" are the bad drivers on the freeway.

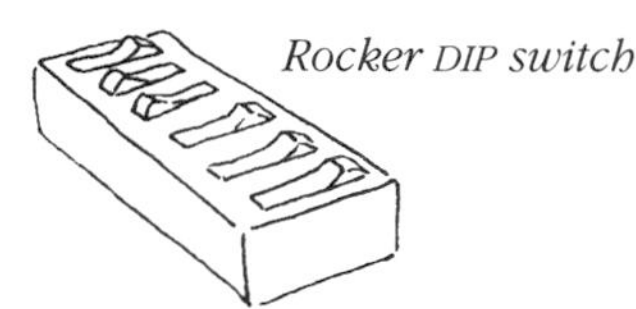

Rocker DIP switch

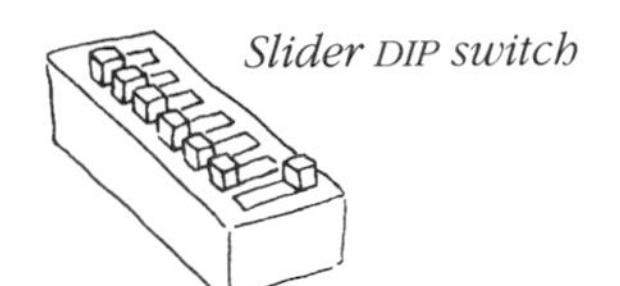

Slider DIP switch

DIR, dir

In MS-DOS and some other *operating systems,* **DIR** is the command you type to see a listing or directory of files on the screen. See *directory.* In MS-DOS, you can narrow the listing to show specific groups of files, like this:

DIR BEDTIME.TXT	Shows only the BEDTIME.TXT file
DIR .DRV	Shows all files whose names end in the .DRV *extension* (typing DIR *.DRV does the same thing)
DIR PROJECT1	Shows all files named PROJECT1, regardless of their extensions. For example, PROJECT1, PROJECT1.WKS, and PROJECT1.DOC would all be listed. Typing DIR PROJECT1.* does the same thing.
DIR PRO*	Shows all files whose names begin with PROJ, regardless of any other characters in the name or the extension: for example, PROJECT1. WKS, PROJECT2.WKS, PROFESSR, etc. The * character is a DOS wildcard—see the definition for *wildcard.*
DIR FUZZY.	With the final period in the request, this shows only the file named just FUZZY without an extension.

DOS also lets you add *command line parameters* to the DIR command to customize the way the directory list looks. Here are some good ones, but most only work in DOS 5.0 and later versions:

DIR /W	Lists the directory in five columns (instead of only one) so you can see more files. This works in all versions of DOS.
DIR /ON	That's the letter O, not zero, followed by N. Sorts the directory alphabetically by name. Highly recommended.
DIR /L	Displays the files in lowercase letters, which are easier to read.

You can combine a file specification and any number of command line parameters in a single DIR command. How about this one:

```
DIR .TXT /W /ON /L
```

directory

The term **directory** has several special meanings for people who use MS-DOS, Windows, OS/2, and Unix.

The first definition

For starters, you can think of a directory as a folder where you keep some of the files on your disk (in fact, the equivalent term in the Macintosh universe is *folder*). Directories organize and keep track of your disk files in a manner that makes sense. You might create a directory for all your financial records, one for your business correspondence, and one for your poetry. Each directory has a name, such as FINANCE, BUSLETRS, or POETRY. You can create directories inside other directories if you want. For example, your directory of financial records might contain a *subdirectory* for records from 1992, another subdirectory for 1993, and so on. Subdirectories can have subdirectories of their own, which can have their own subdirectories, and so on.

At any level of this hierarchy, when you're working with a directory and one of its subdirectories, you can refer to the directory as the "parent directory" and the subdirectory as the "child directory." The *root directory* is the top level of the hierarchy on a given disk, the level that encompasses all the files in all the directories and their subdirectories. In other words, the root directory is the way to refer to the entire file storage area on a disk.

In DOS, you identify the directory you want to work with, or a specific file in that directory, by its *path*—the path DOS has to tread through the disk and directory structure to find your directory. For example, the complete path for the TAXES.RPT file stored in your 1993 subdirectory in the FINANCE directory on disk D: would be

```
D:\FINANCE\1993\TAXES.RPT
```

In DOS, each *backslash* symbol (\) represents the next level in the directory structure, or *tree*. There's one exception: if the last backslash in a path comes after a directory name and before a file, as in our example, it simply acts as a separator between the two names. A backslash by itself refers to the root directory. So the path

\DISKINDX.TXT

refers to the DISKINDX.TXT file stored in the root directory.

In some DOS commands, you can use two other special symbols in path names: a single period (.) refers to the current directory, while two periods together (..) refers to the parent directory.

To see a diagram of the directory hierarchy on your screen, type TREE.

The second definition

A **directory** can also be a list of files located in a given directory—all of the files or some specific group of them that you define. Displaying, listing, or "doing" a directory means that you give the command that displays this directory list. In dos, you type DIR. If someone says, "I did a DIR," that's what they're talking about.

The third definition

When the talk turns really technical, **directory** can refer to the special file on your disk where the information about the files in a directory are stored. If someone says, "The directory got trashed" or something to that effect, they're saying that this special file got *scrambled* or *corrupted*. You can protect against serious loss of information from a scrambled directory by running the MIRROR program that comes with DOS 5 and later versions. *Disk repair utilities* like the Norton Utilities can sometimes restore a scrambled directory.

dirty ROMS

The term **dirty ROMS** is backwards-derived from the *32-bit clean ROMs* that were built into later Macintoshes. That is, they didn't **plan** to make dirty ROMs. A ROM is a *read-only memory chip* built inside the computer. It contains permanent and unalterable software to help run the computer. Well, in certain models of Mac computers—the Mac II, IIx, IIci, and the SE/30—the engineers and software programmers utilized a certain part of the ROMs for this or for that; it was a part of the ROM that nobody thought anybody would ever need . . . Lo and behold, technology developed that needed that certain part of the ROM. For all the gory details, please see the entry in the Symbols section, under *32-bit addressing*.

disc

The term **disc,** spelled with a "c," generally refers to *optical discs* such as laserdiscs or videodiscs. The term *disk,* spelled with a "k," refers to hard and floppy disks. We could make up reasons for this, but in reality that's just the way it is. (And who knows if "laserdisc" is really one word or two?)

dischy

Dischy is an affectionate term for a *discretionary hyphen* (see below).

discontiguous

The term **discontiguous** describes two parts of something (or two separate things) that are physically separated, as in the two halves of the state of Michigan.

With computers, dicontiguous sometimes means a *fragmented* file, one that is broken up into sections located on different parts of a disk. See *fragmentation* for more information.

There is another reference to **discontiguous**—have you ever tried to select two words or two paragraphs at the same time that were not next to each other? Or in your database or spreadsheet have you tried to select two rows or several cells that were not next to each other, avoiding the rows or cells in-between? To select two or more pieces of information which are not physically next to each other is called a "discontiguous selection," and very few applications allow you to do this. See *contiguous.*

discretionary hyphen

When you type a hyphen in a word to break it at the end of a line, the computer sees that hyphen just like any other letter in the word. That is, when you edit the text so that hyphenated word is no longer at the end of a line, that hyphen is still stuck in the middle of the word. Have you ever seen something like this: If you want the rain-bow, you gotta put up with the rain.

A **discretionary hyphen,** affectionately known as a "dischy," is a special hyphen you can enter within a word. If the word happens to fall at the end of a line of text, the software hyphenates the word at that point. But when you edit the text and that the word no longer breaks at the end of a line, the discretionary hyphen just disappears. It will appear again if necessary.

In most applications that can create a discretionary hyphen (such as page layout applications), you can usually create the dischy by holding down the Macintosh Command key or the PC Control key while you type the hyphen. If the word doesn't need the hyphen, you won't see it.

Disinfectant

Disinfectant is an anti-viral utility (an application that detects *viruses* and eradicates them). The first time you run it, it installs an *extension* in your System Folder that will quietly check any disk you put in your machine. If you insert an infected disk, Disinfectant sets off bells and whistles (literally). This program is *freeware,* written by a wonderful man named John Norstad at Northwestern University. He takes the time to write this program and is constantly updating it to catch new viruses. He does this out of the kindness of his heart, an intrepid soul battling the forces of evil. Write him a nice thank-you letter and tell him how wonderful he is.

Disinfectant is available from most *online services, BBS's,* and *user groups.* If you have no access to any of these places, you can send a floppy disk in a self-addressed, stamped disk envelope to the most wonderful John Norstad and he will send you a copy. But he says, "It's very important for people to keep their anti-virus tools up to date. Disinfectant was intended for people with access to electronic sources of information about new viruses and new Disinfectant releases. I normally recommend that people who don't have this kind of access use commercial anti-viral software instead." See *virus.*

John Norstad
Academic Computing and Network Services
Northwestern University
2129 Sheridan Road
Evanston, IL 60208 USA

disk

A **disk** is a thin, circular object used to store computer data (well, the actual disk itself is round, but it is stored in either a square, paper envelope or a square, hard plastic case). The disk goes into a *disk drive* which spins the disk very rapidly, allowing information to be located and transferred to or from the disk quickly. The surfaces of *floppy disks* and *hard disks* are coated with a magnetic material similar to recording tape.

Because hard and floppy disks store information magnetically, any magnet can destroy the data (information) on the disk. This means you must not get your floppy disks near a magnetic paper clip holder, the telephone, the stereo, a portable radio, or any other electronic device—and don't pin them to the filing cabinet with a magnet!

disk controller

The **disk controller** is circuitry on the computer's *motherboard* or on a plug-in *circuit board* that controls the operation of your hard disk drive, floppy disk drives, or both. When the computer wants to transfer data to or from the disk, it tells the disk controller. The controller in turn sends electronic commands to the disk drive making the disk spin and move its magnetic heads to the proper location on the disk. The controller then transfers the data between the computer and the disk drive.

There are several different types of hard disk controllers, based on the electronic method used to encode data on the hard disk. The type of controller must match the type of hard disk you have—an MFM drive must be hooked up to an MFM controller, an RLL drive needs an RLL controller, and a SCSI drive only works with a SCSI controller. If you add another disk drive to your computer because you need more disk storage, you **must** know what kind of disk drive and controller you currently have, and the new disk **must** match your existing controller.

disk drive, diskette drive

A **disk drive** (or **diskette drive,** depending on what you want to call the disk) is the part of the computer that takes the disks you insert and spins them. There are many different types of disks, so there are many types of disk drives. The one you will probably see most commonly is the *floppy disk drive;* actually, all you see is the slot where you insert the disk—the drive mechanism itself is inside the computer.

See the definitions for *drive, floppy disk drive, hard disk drive, CD-ROM drive,* and *WORM drive.*

diskette

A **diskette** is a *floppy disk*—the two terms are synonymous. Macintosh people always call theirs a "disk"; PC people are more likely to call theirs a "diskette." See *floppy disk.*

dismount

Dismount refers to the action of removing a disk's *icon* (the picture, or visual representation of that disk) from the Desktop, thereby making the information on that disk unavailable to the computer. In the case of an external device such as an external hard drive, CD-ROM drive, or a cartridge drive, you can dismount the disk even though the device is still physically connected to the computer. With floppy disks, the act of dismounting actually ejects the disk from the drive. Although there are many utilities out there that can dismount (and mount) a disk, the common method is to drag the icon of the disk to the Trash. See *mount.*

Display PostScript

If you don't know what *PostScript* is yet, I suggest you take a minute to read that definition, then come back to this one.

Normally in the Mac and PC environments, the PostScript code for a document or image you are working on is not generated until you send the document or image to a printer (or *print it to disk*). Instead of using the PostScript code to display the document or image on the screen, the computer uses a different *driver* (*QuickDraw* for the Mac and whatever driver your screen needs on your PC). The computer uses its own driver because its driver is able to draw the information much quicker than having to draw from the PostScript code. But this means that what you see on your screen is not exactly what will print out on paper. In Display PostScript, however, the image on the screen is displayed using exactly the same information (the PostScript code) that will be going to the printer—Display PostScript is the ultimate in *WYSIWYG* (what you see is what you get).

The NeXT computer is the only computer currently available that uses true Display PostScript to render on your screen exactly what will print. These machines ship with a minimum of 32 *megabytes* of *RAM,* and Display PostScript is slow even on them. Nevertheless, Display PostScript is what we all long for, because it is beautiful, because we can then design much more clearly on the screen, and because it cuts way down on the proof-printing process.

If you want to see the complexity of PostScript code sometime, print a page to disk (see *print to disk*) using a PostScript driver and then open the *printer file* in your word processor. You will be amazed at the pages and pages of code (like sometimes hundreds) it takes to draw a complex graphic page.

display type, display font, display typeface

Display type is text that's big enough to stand out as a headline, title, or other special element on a page, or at least large enough to draw attention. Traditionally display type refers to 14 point type or larger.

In some *expert sets,* fonts are redesigned as **display fonts.** These fonts are redesigned specifically for printing at larger sizes, usually over

24 point, with thinner strokes, a slightly different proportion of *x-height* to cap height, adjustments to where the curves meet the stems, more delicate serifs, etc.

distributed computing

Distributed computing is a quick but kind of stuffy way to describe the situation when a business relies on lots of small computers located throughout the organization, rather than a few big machines at some central location. They may still have a few big machines, but many important duties are assigned to the personal computers and workstations, too. All the computers are tied together in a *network,* communicating with each other so that different portions of an application run on different computers. The *front end,* the part of the application that the user interacts with to determine what information she wants to examine and how to organize it, runs on the user's own computer. The *back end,* the part of the application that actually finds and sorts the requested information, runs on a central computer somewhere else (see *front end, back end*). This type of distributed computing, also referred to as "client-server architecture," splits up the functioning of applications across a number of separate computers.

The advantages of distributed computing are increased speed and lower cost. It is currently quite popular and many businesses are converting to it as we speak.

dithering

Dithering is a trick many graphic applications use to fool your eye into seeing a whole lot more colors (or gray tones) on the screen than are really there. The computer achieves this optical illusion by mixing together different colored pixels (tiny dots on the screen that make up an image) to trick the eye into thinking that a totally new color exists. For instance, since pixels are so tiny, if the computer intermingles a series of black with white dots then you're going to think you're seeing gray.

Color dithering smooths out images by creating intermediate shades between two more extreme colors (called a *blend*). Dithering also makes the best use of the limited number of available colors, like when you open a *24-bit color* image (millions of colors) on a computer that's only capable of displaying 8-bit (256 colors).

There is a *halftone* effect for black-and-white images called dithering. Rather than dots of varying sizes, a dithered image has dots or squiggles all the same size, arranged in such a way as to create the illusion of gray values.

docking station

A **docking station** is a unit that attaches to a laptop or notebook computer, either directly or by cable, providing the laptop with more hardware. This arrangement increases the laptop's capabilities almost to that of a standard-sized desktop computer. A docking station typically includes standard *expansion slots,* as well as additional *ports* (sockets) for connecting a printer, modem, keyboard, etc.

So if you have a laptop computer and a docking station, you can take your laptop with you wherever you go, accepting its limitations as a laptop. Then when you come home or go back to the office, you can connect your laptop to the docking station (with some units this is like inserting a giant video tape into a VCR). Then your laptop virtually turns into a regular desktop computer, complete with large monitor, large hard disk, more *RAM,* etc.

docs

Docs is short for *documentation,* as in "I need the docs for this software."

document

The **document** is what you see on the screen, what you work with, and what you print, as opposed to the *file,* which is where the document is stored on a disk. I have a file called Play.March (prounounced "play dot march") that stores my document, a flyer for an upcoming play.

Database or spreadsheet documents are rarely printed out in full; in fact, a document doesn't ever have to be printed out on paper—"document" refers to the abstract version of whatever you're working on. Not everybody uses the word this way. Mac people talk about documents as files you have created yourself, as opposed to *applications,* which are what you use to create the document with. But PC people are just as likely to use "file" to mean either the abstract document or the disk file.

documentation

The **documentation** is the collection of written instructions and peripheral written material for a product, whether it's for your computer, a piece of *hardware,* or a software program.

Documentation usually comes on paper—in books, 3-ring binders, or just stapled sheets. This is often supplemented by information in a disk file (see *README files*). And in some cases (particularly with *shareware* or *freeware*), **all** the documentation for the product comes as a file on disk that you can either read on the screen or print out yourself.

(Michael Lilly says documentation is paper that is good for starting fires in your fireplace.)

Dogcow

The little creature shown below is the **Dogcow.** His name is Clarus. He is a trademarked symbol of Apple Computer, Inc. There have been technical notes written about the Dogcow that only real *hackers* can access. Clarus says Moof! You can find the Dogcow in the Page Setup dialog box in most applications, or even at the *Desktop*—click on the Options button.

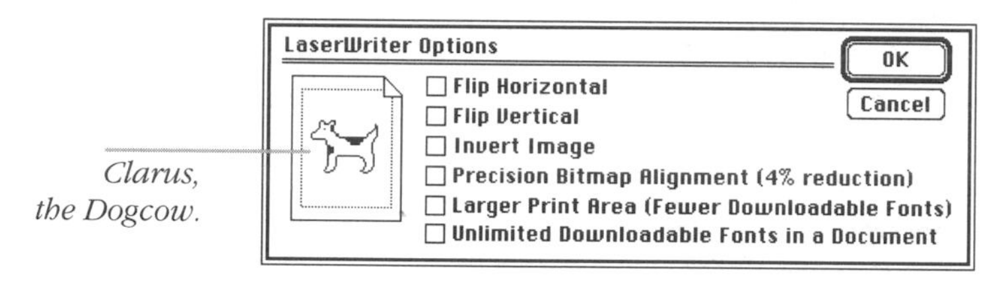

Clarus, the Dogcow.

Dan Druliner corrected me, however. He says, "You stated that 'Moof' is the call of the Dogcow. Actually 'Moof' would be the call of the Cowdog (Moo + Woof). This is a common syntactical error which can be corrected by selecting 'Flip Horizontal' in the Options section of the Page Setup dialog box." Others say, though, that when the Dogcow is in this position, he says !fooM.

So I suppose if Clarus did this: 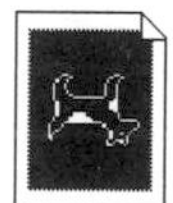 he would say: 

If you want to see the Dogcow, you'd better look quick. The new *PostScript printer driver* (PSWriter) from Adobe transmogrified Clarus into a boring letter a.

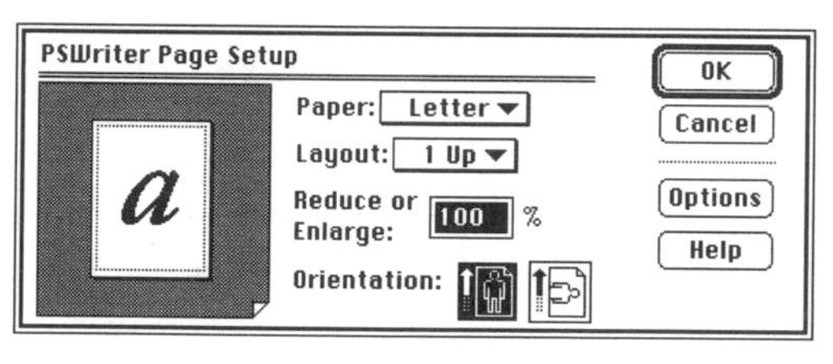

The PostScript printer driver dialog box. Boring ol' a.

DOS

Dos is an acronym for **d**isk **o**perating **s**ystem. Several different types of computers have had *operating systems* called DOS. These days, however, DOS almost always refers to the operating system (the master software) used by IBM PCs and compatibles.

There are several different versions of DOS, but all of them work essentially the same way and run the same software. For simplicity, we refer to all these products as DOS, unless there's a reason to be more specific. *MS-DOS* is the version you probably have unless you own an IBM-brand computer (MS stands for Microsoft, the company that created DOS). The version of DOS sold by IBM, originally called PC-DOS, is now called IBM DOS; it's just MS-DOS with a few minor added features. Then there's one called *DR DOS* made by another company, Novell. DR DOS is usually a step ahead of MS-DOS in terms of having better features and being easier to use, but it has captured only a small part of the DOS market.

Anyway, like all operating systems, DOS is the software required by your system to start up and run other programs. The name "disk operating system" is a little misleading, since an operating system does more than manage disks. Sure, DOS manages the hard and floppy disks, allowing the user to run programs, and allowing those programs to access files on the disk. But in addition, DOS lets your programs use printers, communicate with you via the keyboard and the screen, manage memory, and perform many other fundamental tasks.

Dos deserves its reputation for being hard to use. You have to type complicated commands yourself (how does DIR*.DOC/W/O-S look?), and there's nothing on the screen to give you a clue about how to proceed. Worse yet, a single typing error invalidates the entire command. And you certainly can't use a mouse to pick out options from a list or start a program by clicking on a picture. Dos has other problems too, such as the fact that it doesn't let your applications use more than 640 *kilobytes* of memory *(RAM)*, which these days is a meager supply (but see *DOS extender* and *expanded memory* under the *RAM* entry). Another weakness is it only lets you run one program at a time (it doesn't allow *multi-tasking.)*

But DOS has some advantages over fancier, prettier operating systems and *environments,* like the Macintosh and Microsoft Windows with their *graphical user interfaces*. Once you learn the commands, it's often actually faster to type a command than to fool around with all those windows, menus, and *icons* on the screen. Dos can execute a series of commands automatically (using a *batch file*), something neither the Mac nor Windows can do as well. And DOS itself is fast, just because it's so rudimentary. It needs much less memory, disk space, and processing power than a graphical user interface, which means it runs fine on older, cheaper, slower computers.

Besides, you can eliminate the pain of typing those yucky DOS commands by running a DOS *shell.* A shell is a program that displays the files and directories on your disk in a list on the screen. The shell lets you work on files (copy them, delete them, and so on), and run programs by picking them from the list with a cursor or the mouse. The MS-DOS Shell program comes with DOS 4.0 and later versions, and you can buy better shells such as XTree, the Norton Desktop for DOS, and the shareware programs QFiler and STS.

The core of DOS consists of three files that must be on your *boot disk* to start and run your computer. One is *COMMAND.COM,* which you can read about in the C section. The other two, "MSDOS.SYS" and "IO.SYS" are *hidden files* that you won't see in your regular directory lists, but which will show up if you have software that displays hidden files.

DOS extender

If your IBM-compatible PC has one *megabyte* or more of *RAM,* it almost certainly has *extended memory.* Despite the fact that extended memory is so common, DOS can't use it—not at all. But software engineers have found a way to use extended memory in their programs. A **DOS extender** is a software technology that can be built into other programs, giving them the ability to bypass DOS and recognize extended memory. See *RAM* for a complete discussion of all the types of RAM *(random access memory)* found in PCs.

DOS prompt

When you work with plain old *DOS,* the indicator that tells you where to type commands is called the **DOS prompt.** It might look like this: **C:>** or like this: **A:>** or like this: **C:\FILES>** or even something like this:

THE TIME IS 8:05:34>

(You can control the way the DOS prompt looks.) Immediately to the right of the DOS prompt, you'll see a blinking *cursor* which shows you where the next letter you type will appear. If you see more than one DOS prompt on the screen (which is common), the bottom one—the one that has the cursor—is the "real" DOS *command line,* the line where you can actually type commands.

DOS text files

Dos text files are files consisting of, yes, text. The key point here is that you can read them on the screen with DOS itself, using the TYPE command (as in TYPE READ.ME). On the screen, every word is complete, and the lines flow smoothly down the screen without any sudden jumps or bothersome beeping.

DOS text files are a type of *ASCII file*. They're different from Macintosh ASCII files in a couple of specific ways. For one thing, they contain DOS-type line breaks at the end of each line; Mac files don't have this type of line break, so DOS can't display them properly. Also, DOS text files contain only characters that display properly with TYPE. This can include the *high ASCII* accented vowels, boxes, and other symbols in the IBM character set, but not most of the "low ASCII" control codes.

dot (as in period)

The mark that looks like a period (cuz it is) is sometimes known as a **dot** and sometimes as a *point*. It's usually called a dot when you see it in file names, such as "frog.eps," which is pronounced "frog dot e p s."

The period is usually pronounced "point" when you see it in software versions—"version 3.1" is pronounced "version three point one."

dot bis

See *v.32bis*.

dot eps

See *EPS* (indicating a *graphic file format* of the Encapsulated PostScript type).

dot EXE

See *EXE*.

dot gain

Whenever a photograph, painting or drawing containing many colors or gray tones is printed, the colors and tones must be simulated with tiny dots (see *halftone* for details). **Dot gain** refers to an increase in the size of these dots when they are actually printed on the paper by the printing press. The dots can increase in size rather dramatically once the ink hits the paper, depending on the characteristics of the press, the absorbency of the paper, and the nature of the ink that is used.

The effect of dot gain on the final printed piece can be an increase in color intensity, because more ink is put on the paper (by the press) than was called for when the image was output, making the colors look darker than intended. Because the dots have gained in size, the final printed image can look not only darker, but muddy, low in contrast, and blurry.

The best of the *color separation utilities* offer a way to compensate for dot gain by adjusting the color curves when the film is imaged, but sometimes the artist has to compensate for dot gain manually within the software application used to create the image. It helps if the artist, the commercial press, and the service bureau work together closely so they can adjust for the quirks of each others' equipment and compensate for the dot gain effectively.

dot matrix printer

A **dot matrix printer** uses tiny metal pins striking the paper through an inked ribbon to print the text and graphics on the page. Each time a pin hits the ribbon, a little dot of ink gets deposited on the paper. As the *printhead* moves back and forth across the width of the page, electronics inside the printer tell the pins when to fire to create the correct pattern of dots. Because they have a built-in *tractor feed* mechanism, dot matrix printers work with "continuous paper"—the kind that comes in big fan-folded stacks with pinholes along either side. Apple ImageWriters and most Epson and Toshiba printers are dot matrix machines.

By rights, *laser* and *inkjet printers* also deserve to be called dot matrix printers, since they too compose text and graphics as a matrix of tiny dots. Be that as it may, the term just isn't used for lasers and inkjets. With a few exceptions, dot matrix printers print bigger dots than laser printers, so their *resolution* is lower. They're also slower and noisier—when a dot matrix machine is printing it sounds like a monstrous metallic insect. On the positive side, the typical dot matrix printer is much cheaper than a laser printer. And because the pins actually strike against the paper, you can print multi-copy (carbon or carbonless) forms with a dot matrix printer, which you can't with a laser.

dot PICS

See *PICS* (it's a *graphic file format*).

dot pitch

The dot pitch of a color monitor measures the size of the tiny individual dots of phosphorescent material that coat the back side of the picture tube's face. The dot pitch helps determine how sharp the image looks, independent of the *resolution* (which is measured in *pixels*). A smaller dot pitch is better.

Here's the technical scoop: Each point of light on a color monitor is formed from a triad of three separate dots of phosphor: one that glows red, one green, and one blue (the color you finally see depends on how intensely each dot in the triad is excited by the picture tube's electronic beam). The dot pitch is the vertical distance between the center of one dot and the next like-colored dot directly above or below it (the way the dots are arranged, pairs of like-colored dots are always two rows apart). The farther apart the centers of the dots are, the bigger the dots and the fuzzier the image. All other things being equal, a monitor with a smaller dot pitch is preferable to one with a larger dot pitch, though other factors are more important in determining image sharpness below a certain dot pitch threshold.

A dot pitch of .28 mm or smaller is ideal for 14- or 15-inch monitors; a dot pitch of .31 mm or less for 17- to 20-inch monitors. Resolution, in pixels, is determined by the video circuitry in your computer. Depending on the resolution and on the dot pitch, a single pixel may occupy 4 to 16 separate phosphor triads.

dot sea

See *sea* (it stands for *self-extracting archive*).

dot sit

See *sit* (it indicates a *compressed* file).

dot zip

See *zip* (it indicates a *compressed* file).

dots per inch

The *resolution* of a printer or a scanner is measured in **dots per inch,** abbreviated "dpi." If a printer has a resolution of, say, 300 dots per inch, that means in one inch there are 300 dots in a row across, and 300 dots in a row down. The more dots per inch, the smaller the dot has to be, of course. You can logically understand, then, that the higher the dpi, the smoother the printed image will appear to be.

300 DPI 1270 DPI

You may hear the resolution of a Macintosh screen, or monitor, measured in *pixels per inch,* or "ppi." Please see *resolution (printer)* and *resolution (monitor)* for the important difference between dpi and ppi.

double-click

To **double-click** means to use your finger to tap-tap (twice) on the button of the mouse. See *click.*

Whether you single-click (tap once) or double-click on the mouse button, you are telling the computer "do this" or "execute that." A single-click is very different from a double-click, though, and you want to make sure you use the appropriate action. For instance, if you have an *icon* for a word processing program and you use the mouse to position the little arrow directly over that icon and then single-click, the icon will *highlight* (change color) to indicate that it is *selected,* or chosen. If you double-click on that same icon, though, the word processing program will start up and run.

If you're using a mouse with more than one button and the instructions say to double-click but don't tell you which button to use, assume you're supposed to double-click the "primary" button. By convention, this is the left button, but you can change the functions of the buttons in Windows or with your *mouse driver.*

double-sided disk

A **double-sided disk** is a *floppy disk* on which can store information on both sides.

Double-sided is not the same as "double density"; some single-sided disks are double density on their one side. Neither "double-sided" nor "double-density" are the same as "high density." See *density.*

download

To **download** means to receive information, typically a file, from another computer to yours via your *modem.* For instance, you might go *online* and find some great *shareware,* such as *fonts* or games or perhaps a handy little *utility*. Or maybe someone sent you a long love letter as a separate file. Well, when you decide you want a copy of that item, you must download it from the computer of the *online service* or *bulletin board* to your own computer. In many cases, this is as simple as selecting the file and clicking the button labeled "Download."

The opposite term is *upload,* which means to send a file *to* another computer. Maybe you have a file you want to share with the world—you *log on* and upload it to the computer that runs the online service or bulletin board, and others will download it later. This is different from *e-mail,* where you send letters directly to your chosen recipient (although you can attach a separate file to an e-mail that your friend would then have to download).

Downloadable fonts, though, are a different thing altogether. If you want to use fonts (typefaces) that are not *resident* in your printer (meaning not permanently loaded into your printer), then you must send the fonts from your computer to the printer yourself. This is called **downloading** your fonts. See *downloadable font* and *font.*

downloadable font

A **downloadable font** is a *font* (typeface) stored on a disk (hard or floppy) that is sent, or downloaded, to your printer before you can use it to print. A resident font, in in contrast, is stored permanently (lives in) your printer.

Just because a font is called downloadable, though, does not always mean it has to be downloaded to use it. Whether the font actually gets downloaded depends on the type of font it is (PostScript, TrueType, etc.), what kind of printer you use, and whether or not you use *ATM (Adobe Type Manager).*

In the Mac world, any font you have paid money for (or unscrupulously acquired) is a downloadable font (whether or not it actually gets downloaded), just to distinguish it from resident fonts.

On the PC, examples of downloadable fonts are "soft fonts" for a LaserJet and disk-based *PostScript fonts* used with a PostScript printer. On the other hand, *TrueType* fonts and PostScript (Type 1) fonts used with non-PostScript printers via *ATM* are not downloaded.

downtime

Downtime is like when your car is in the shop—the computer is "down," or not running. It could be down for any number of reasons. On *networked* systems (where there are many users) downtime is often used to perform maintenance just as with your car, and preventive tasks that keep things running smoothly.

downward compatibile

Downward compatibile is the same as *backward compatible.*

dpi

See *dots per inch.*

draft

Draft refers to a *printout* you make for review or editing purposes. You might print a draft to edit the text, in which case the printout can dispense with any graphics that may have been in the document because you don't need to look at them. Or you might make a draft to check the look and layoutout of a document whose final version you intend to print on a high-resolution imagesetter; in this case, you do print your graphics, but you can use low resolution to save time. Some printers have no shame at all in using the worst possible typeface and spacing standards when printing a draft copy.

draft quality

Depending on your printer and the *printer driver* you are using, when you get ready to print a document you are usually given a choice of different qualities of printing such as **Draft,** Standard, *Near Letter Quality* (NLQ), or maybe something as simplistic as Good, Better, and Best. With a dot matrix printer, draft quality is by far the fastest method, but it also produces horrendously low-quality text and totally eliminates any graphics. Draft quality doesn't save much time when you use a laser printer unless your doucument contains a large variety of fonts and some complex graphics.

It's always a trade-off: speed for quality. Choose draft quality when you're in a big hurry or when you don't really care about the print quality (personally, I can never bring myself to use it under any circumstances).

drag

When an instruction says to **drag,** it's a shortcut for really saying this: Use the mouse to position the pointer over the object. Press and hold the mouse button down. ***While the button is down,*** drag the mouse across the desk, which will drag the selected item across the computer screen at the same time. The key thing here is that while you drag, you are holding the mouse button down.

You can drag icons around on your screen to organize them. You can drag *scroll boxes* to view the contents of a window. You can drag down a menu to choose a command. You can drag to select text, or cells, or rows, or columns. You can drag to draw shapes and pictures. Get the hang of dragging: you're gonna use it a lot.

drag-and-drop

Drag-and-drop is a feature in *graphical user interfaces* such as the Macintosh and Windows. It means you can perform tasks by using the mouse to drag an icon (representing a document, an application, a folder, or a disk, etc.) onto the top of some other icon. For example, to remove a document from your disk on the Mac, you drag the document's icon onto the Trash icon. When the Trash icon *highlights* (changes color), you release the mouse button and the document goes into the trash can.

You can just drag a document icon (indicated by the outline under the pointer) onto an application icon or its alias to open the document.

In System 7 you can *drag* the *icon* of a document onto the top of an icon of an *application.* If the application icon *highlights* (changes color), you can release the mouse button and that application will open the document, even if it was not the application in which the document was created. (If the application does not highlight, it will not open the document this way.) If you're lucky enough to have a big monitor, you can put *aliases* of your most frequently used applications around the edges of the Desktop; this will make those applications easily available for taking advantage of this drag-and-drop feature.

Some applications go one step further and perform a certain command on the drag-and-dropped file. For instance, some *compression utilities* allow you to simply drag-and-drop the icons of the files you want to compress onto the icon of the compressor, which then launches itself, compresses the files, and then quits. Cool, huh?

And in Microsoft Word, you can select some text, then point to the beginning of the selection, drag it to its new position in the text, and let go (drop). This is a shortcut to *cut-and-paste.*

DRAM

DRAM stands for **d**ynamic **r**andom **a**ccess **m**emory. DRAM is a technical term for a type of *random access memory (RAM)* that will not hold information for any real length of time, and therefore requires the computer to rewrite or "refresh" the information every few thousandths of a second (hence the term "dynamic"). DRAM is the most common type of memory found on SIMMs. See *SRAM (static random access memory)* for comparison. See *RAM* for the real low-down on memory.

draw program

You may think that a *paint program* is the same as or at least very similar to a **draw program,** but there is a major technical difference that is important to understand. Read the definition for paint program if you are interested.

A draw program creates objects (lines, circles, squares, etc.) by using mathematical formulas (rather than a configuration of *bits* on the screen, as in a *paint program*). Each *object* you draw is a separate entity—you can go back and change the object's size, the line thickness, the color, the shape; you can move it around, delete it, position it in back of or in front of another object and then change that position, etc.

Because the objects are defined by formulas rather than by turning certain pixels on or off, draw objects will print at the maximum resolution of your printer; that is, if you output the document to a 300 *dot per inch* printer, the images will be 300 dots per inch. If you take the same document and print it to an *imagesetter* with 1270 dots per inch, the image will be 1270 dots per inch. A bitmapped, paint image, on the other hand, will always print at the resolution in which it was created, and if you enlarge or reduce the image, it can become distorted. See *object-oriented, bitmapped,* and *paint program*.

DR DOS

DR DOS (pronounced "dee are doss," not "doctor doss") is an *operating system* (the master software that runs yor computer) for PCs. DR DOS is fully compatible with *MS-DOS,* the most common version of *DOS,* which means you can use all the same commands and run all the same programs, including Windows. But each version of DR DOS offers major enhancements over the comparable version of MS-DOS. In fact, DR DOS was probably the major reason Microsoft finally started a serious effort to improve MS-DOS over the last few years.

The DR stands for Digitial Research, the company that developed the product. Digitial Research is now part of Novell, the leading maker of *networking* sofstware.

drive

A **drive** is the part of the computer that takes the disks or tapes you insert into the slot and spins them to make them work. You probably have at least two different kinds of drives in your system: floppy disk drives are the ones with the little slots where you insert floppy disks; the standard type of hard disk drive comes sealed inside a case, which in turn is stuck inside your computer or inside a separate box—you never even see the disk.

On the other hand, there are *removable cartridge hard disk* drives. Each drive has a large slot in which to insert a cartridge hard disk. Other types of drives you might come across include *tape drives* (for backing up your files onto magnetic tape), *CD-ROM* drives (for "playing" CD-ROMs), and *WORM* or *optical disc* drives (for storing and retrieving data from WORM and read/write optical discs).

driver

A **driver** is a piece of *software* that tells the computer how to communicate with or operate another piece of hardware, such as a printer, scanner, or mouse.

Your *System Folder* has a *printer driver* in it (in System 7 they are stored in the Extension folder that is stored within the System Folder). If there is no printer driver in the System Folder, you can't print.

drop cap

A **drop cap** is the capital letter at the beginning of a paragraph that "drops down" and sits on another *baseline,* as in this paragraph. There are many variations of a drop cap—it may be taller than any other letters, it may tuck right into the paragraph, it may hang off the side, etc. An *initial cap,* as a subtle distinction, is supposed to sit on the first baseline. Whether you use a drop cap or an initial cap, please be sure to align the baseline of the capital letter with the nearest baseline of the text.

DSP

DSP stands for **d**igital **s**ignal **p**rocessing. DSP refers to the conversion and analysis of non-digital information by digital means. The idea is that we can take something, well, "real" such as music or the human voice, and *digitize* it (convert it into numbers—something the computer can understand) and then use the power of the computer to analyze it.

Since DSP chips are such heavy-duty number crunchers, they're used in many Macintosh *accelerator boards,* especially *JPEG compression* and Photoshop accelerators. Pretty amazing stuff, to be sure. Today this technology is also used in the analysis of everything from sonar and radar, to signals sent from earthquake monitors, and even in weather satellites.

DTP

See *desktop publishing.*

dual floppy drive

A computer with a **dual floppy drive** has two slots in which to insert a floppy disk. Although several years ago this meant you simply had two of the same kind of floppy drives so you could work from two different disks, it now usually means that one of the drives will take a 3.5-inch high-density floppy disk, and the other will take a 5.25-inch floppy disk.

dumb apostrophes

A **dumb apostrophe,** a fake one, a straight one, a typewriter apostrophe, looks like this: '. A true (sometimes called "smart") apostrophe looks like this: ' .

You may want to use a typewriter apostrophe intentionally as the symbol for the foot measurement. But typographically speaking, the correct symbols for inch and foot marks are called the "prime marks." They are similar to typewriter and quotation marks, but they are slightly angled, as if italicized. The Symbol font has true prime marks. ″ See *apostrophe.*

dumb quotes

A **dumb quote,** a fake one, a straight one, a typewriter quotation mark, looks like this: ". True (sometimes called "smart") quotes look like this: " ".

Using typewriter quotation marks is the single most visible sign of unprofessional type. See the note above in *dumb apostrophes* for info on "prime" marks, and see the definition for *quotation marks* for tips on typing real quotation marks.

dumb terminal

A *terminal* consists of a screen and keyboard which allow you to interact with a multi-user computer; the computer itself is often located in another room or even a different building. A **dumb terminal,** which is the most common type of terminal, has no computing capabilities of its own. You enter commands on the terminal's keyboard to tell the computer what to do, and the computer sends messages back to you on the terminal's screen. The computer does all the work (of running programs, for example), with no help from the dumb terminal. Please see *terminal.*

dump

Dump can be used either as a verb or a noun. As a verb it means to send a large volume of information somewhere. For example, you dump a file to a printer or to the screen. When you dump a file you are not manipulating or changing it, you are just sending the raw information that it contains.

As a noun, a dump is the result of something that has been dumped. For example, a screen dump is what you see after you've dumped a file to the screen. For more info on screen dumps, see *screen capture.*

duotone

A typical black-and-white photograph uses only one color. In a **duotone,** though, the black-and-white photograph (or other artwork) is reproduced using two colors. Perhaps it's black and brown, or black and grey, or dark grey and a rusty color. *Halftone* images are generated for the photograph, one slightly underexposed and one slightly overexposed, and the two are printed one on top of the other. The result can be an incredibly rich, powerful image—much richer and more interesting than the image with one color. The artist/designer has control over the values and percentages of the two different colors. It is also possible to make "tritones" using three different colors, and "quadtones," using four different colors.

A duotone (or tritone or quadtone) does not refer to the use of *spot color* for the second color; that is, if you take a photograph and color the lady's dress pink, that's not a duotone, even though there are two colors in the image.

duplex

Duplex is a *telecommmunications* term for communication channels that go in both ways at once. "Half-duplex" indicates that when you type a character on your keyboard, the character is displayed on your screen by your computer or terminal as it is sent to the remote computer or terminal. "Full-duplex" means that the character is sent to the remote computer, which then sends it back to display it on your screen. If your modem software is set for half-duplex, but the remote computer thinks you are full-duplex, each character you type will appear as two characters on your screen.

Dvorak keyboard

Dvorak is the name of a *keyboard layout* that is much more efficient than our present standard *Qwerty* arrangement. Typing speed can be increased dramatically and with less trauma to the hands. There is software you can install in your computer that will let your keyboard use the Dvorak arrangement (but your keys will still have the Qwerty letters on them). Compare with *Qwerty*.

Oh, and by the way, this term has nothing to do with John Dvorak, the Rush Limbaugh of computer columnists.

DWIM key

The acronym **DWIM** stands for **d**o **w**hat **I** **m**ean. The **DWIM key** is a mythological key that would make the computer do what you want it to when you can't otherwise figure out how to tell it what to do.

dynamic RAM

See *DRAM* (pronounced "dee ram").

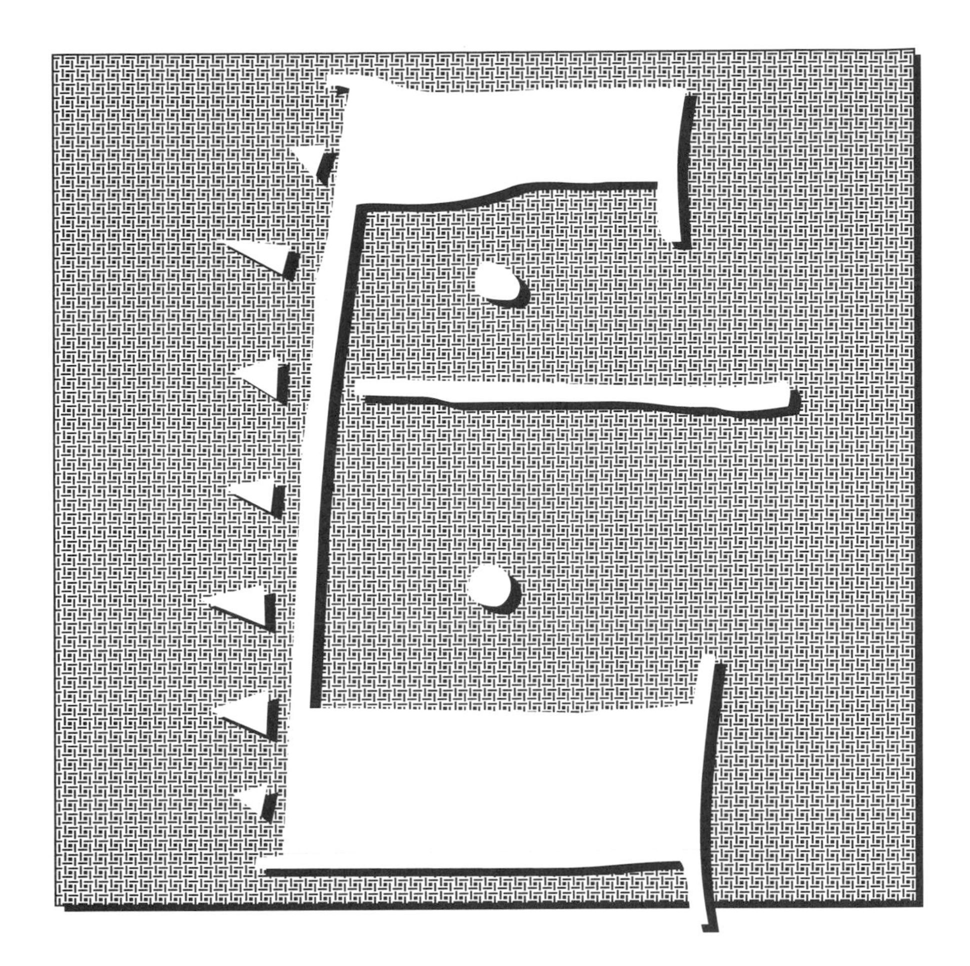

Easter egg

An **Easter egg** is a sweet little trick that a programmer has hidden somewhere in a program. Typically you do something normal, such as choose a menu item, but with a *modifier key* key held down—and something very unexpected happens. One of my very favorite Easter eggs is in WriteNow: Choose "About WriteNow" from the Apple menu; hold the Option key down, and click on the names in the box. Or open the virus detector *Disinfectant*, choose "About Disinfectant" from the Apple menu, and just wait a few seconds (also one of the best). In Word 5, hold down Command and Shift; from the Tools menu, choose Preferences; scroll to the bottom and click on the new "Credits" icon. You might accidentally find Easter eggs, but usually you hear about them through the grapevine. They are a great thing to share (oh, we are so easily amused).

EBCDIC

EBCDIC (pronounced "ebb see dick") is short for **e**xtended **b**inary **c**oded **d**ecimal **i**nterchange **c**ode. This is a coding system used to represent characters—letters, numerals, punctuation marks, and other symbols—in computerized text. The EBCDIC code allows for 256 different characters. It's widely used by *mainframes* and other large computers.

For personal computers, however, ASCII is the standard. If you want to move text between your computer and a mainframe, you can get a *file conversion* utility that will convert between EBCDIC and ASCII.

echo, ECHO

Just like a voice echoing off the walls of a canyon, a computer's transmissions may be reflected back to it. In *modem* lingo a computer **echo**es if it displays on-screen the characters being typed on its own keyboard. Some modem software offers the option of Echo on or off—if you can't see what you are typing, turn Echo on; if your screen displays every character in doubles, turn Echo off. In the case of two computers communicating, echo refers to the return of a transmitted signal to its computer of origin.

In DOS, the ECHO command is often used in *batch files*. Ordinarily when you run a batch file, DOS displays every command in the file, making a big clutter on your screen. But if the batch file contains the command ECHO OFF, then DOS stops echoing to the screen the batch file commands that follow. Other messages from DOS (like "1 FILE COPIED") and from programs you run from the batch file still appear. A line in the batch file such as ECHO PREPARE FOR TAKEOFF! displays whatever text follows the ECHO command, even when ECHO is off. This is a good way to send messages to yourself while a batch file is running, and still keep the screen uncluttered.

Eddy Award Winner

As you wander around a Macintosh trade show or conference or as you look through computer magazines, you may notice products claiming to be **Eddy Award Winners.** An Eddy Award Winner is a product that the editors of MacUser Magazine have deemed the best in its category that year. The awards are presented annually at a black-tie affair. As with any award, they mean a great deal to some people and are scoffed at by others, but in general it's a rather prestigious award and helps give you (as a consumer) some idea where that product stands in relation to others of the same ilk.

edit

To **edit** something just means to change it. You can edit text, for example, by typing letters on the keyboard, erasing existing letters, or moving them around. You can also edit pictures with a graphics program by changing colors or the sizes and shapes of the images or by applying special effects. Also see *editor.*

edit box

Edit box is the name of the little rectangle you see in dialog boxes where you can type something in. For instance, when you Save a file, you can always type in the name—you type it into the edit box. You can usually press the Tab key to move the selection (cursor) from edit box to edit box.

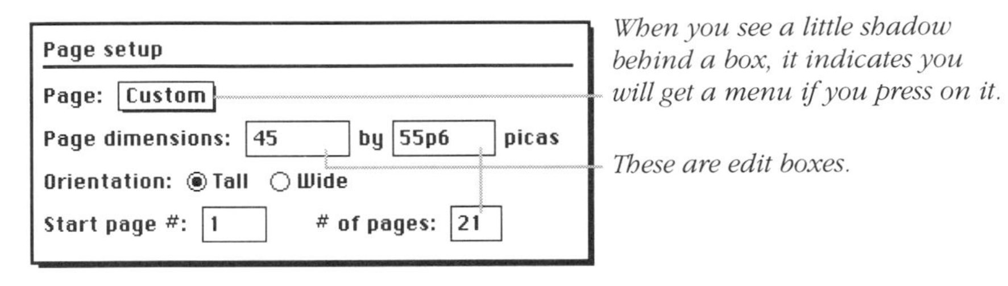

When you see a little shadow behind a box, it indicates you will get a menu if you press on it.

These are edit boxes.

In PC dialog boxes, you can press the letter in the label that is underlined to select that edit box.

edit mode

In some programs there is a separate **edit mode** for making changes to a document or whatever you are working with. In this type of program, you might use one mode to read a document, but you have to switch to edit mode to change it. Programs that have separate edit modes are usually more cumbersome to use than programs that let you do everything in at once.

editor

An **editor** is a software application that allows you to change the contents of a file or document, or to alter any other given collection of information. Most often the term is used in reference to a *text editor,* with which you edit text. (This is different than a word processor—see *text editor* for more information.) But you can also have a graphics editor to create and change graphical images; a music notation editor to place musical notes in a score and rearrange them; or a font editor to design and edit the appearance of the characters in a font. There are also a variety of editors that are used for very technical purposes, like the utility programs that let you edit the contents of a *directory* on your disk.

EEMS

EEMS stands for **e**nhanced **e**xpanded **m**emory **s**pecification, a type of *expanded memory* for PCs. EEMS was superseded by EMS version 4.0. See *EMS,* plus see *RAM* for the definition of expanded memory.

EGA

An **EGA** is a video *adapter* for IBM-type personal computers. This means it has the electronic circuits your computer needs to display images on the screen. EGA stands for **e**nhanced **g**raphics **a**dapter, but don't be fooled by the name: EGAs came out in about 1985, and what was "enhanced" then (compared to a *CGA*) is obsolete now. Yes, they still work, and if you get a computer that has one, don't throw the EGA *board* away. Most new graphics and Windows programs will function with an EGA. But don't buy one new—get a VGA instead, or better yet, a Super VGA. In graphics mode, EGAs can display a maximum of 640 by 350 pixels (dots of colored light) on the screen. That resolution is barely acceptable, and it produces squashed-looking images (squares look like rectangles). EGAs can display text too, but EGA text is a little too fuzzy to read for long periods of time. Though a breakthrough at the time, by today's standards EGA color is limited: 16 colors from a palette of 64 colors can be displayed on the screen at one time.

You'll hear people talking about their "EGA adapter," but that's probably reduntant, since the A in EGA stands for adapter. On the other hand, an "EGA screen" or monitor is a monitor designed to work with an EGA.

EISA

Extended **i**ndustry **s**tandard **a**rchitecture, or **EISA,** is the term used to describe the *expansion slots* and related circuitry (the *expansion bus*) found in some higher-priced PCs. If you see an ad for an EISA computer, it's talking about a PC with EISA slots.

EISA was developed by a group of PC manufacturers to compete with IBM's *Micro Channel* expansion slots, the kind found in most IBM *PS/2* computers. Like the Micro Channel slots, EISA slots can transfer 32 bits at a time at high speed, compared with 8 or 16 bits at relatively slow speeds for the standard slots in a regular PC. But EISA slots have the advantage of accepting standard PC *add-in boards,* which are by far the most common type you can buy, and which won't fit in the Micro Channel slots.

Practically speaking, though, hardly anybody needs EISA slots. That's because little of the equipment you can hook up to your PC through add-in boards can take advantage of the EISA slots' better performance. Unless you know you need EISA slots, save your money and buy a PC with ordinary *ISA* (industry standard architecture) slots.

eject

To **eject** means to pop a *floppy disk* out of the *floppy disk drive* (the slot you stuck it into). Ejecting only applies to floppy drives that are spring-loaded, so that the drive can forcibly pop out the disk. There are lots of ways to eject a disk.

On a Mac, you'll see an "Eject" button in every dialog box used for opening, saving, importing, or exporting files. You can press Command Shift 1 to eject a floppy disk from the internal drive, and Command Shift 2 to eject a floppy from an external drive. (If you have a Mac with two internal drives, Command Shift 1 ejects the bottom, Command Shift 2 ejects the top, Command Shift 0 ejects the external.) You can drag the icon out through the trash (won't hurt the disk; in fact, this is the most preferable way to eject a disk). You can select the disk and press Command Y in System 7, or Command Option E in System 6. If the floppy is stuck in your computer, hold the mouse button down when you turn the computer back on or restart. If all else fails, unbend a paper clip and poke it into the tiny hole next to the drive slot—push firmly and the disk will pop out.

Most 5.25-inch drives on PCs aren't spring-loaded, so you have to pull the disk out by hand. With a PC 3.5-inch drive, you have to press a little button on the drive to eject the floppy.

When you are using a *removable cartridge hard disk* and you want to eject one of the disk cartridges, it is called *dismounting* rather than ejecting. "Eject that floppy disk so you can insert another one." "I need to dismount this cartridge."

electronic

What's the difference between "electric" and **electronic**? Well, electronic devices—such as televisions, cellular phones, computers—all process signals of some sort, signals such as radio waves or bits of data through wires. If the device doesn't process any signals—if it just pops up the toast or gets hot or turns wheels—then it's electrical.

Electronic devices all used to have vacuum tubes that had a flow of electrons going through them, which is what made them electronic. Now *semiconductors* do what the tubes used to, and they can involve the flow of holes as well as electrons. "Holes"? Yes, a hole is a spot where there should be an electron but there isn't, and they really do flow the way electrons flow.

electronic mail

See *e-mail.*

electronic sex

Every night, the atmosphere around the world is buzzing with **electronic sex,** with people having some form of sex via their computers and modems. Sexy radio waves are flying through the air, whizzing past your ears, swirling through the trees. How does one have electronic sex? Well, it's simply a matter of connecting with someone over the *modem,* perhaps through an *online service,* or maybe a *bulletin board.* A friendship develops, one thing leads to another, just as in life, and instead of actually doing the physical thing, you type what you would be doing if you were together. You might type it in correspondences back and forth to each other, or if you're using a service like *America Online,* you can actually go into a "private room" together and be friendly. It's quite an interesting phenomenon, and a testament to the need for intimacy among humans.

elevator

Some people call the little box inside a scroll bar the **elevator.** Most people call it the *scroll box.* See *scroll box* for the details on what it does and how it works.

ELF

ELF stands for **e**xtra **l**ow **f**requency, one of the most insidious forms of magnetic radiation computer users are exposed to. (ELF is also generated by such common appliances as electric blankets and hair dryers.) When you hear about the dangers of radiation from computers, about the increase in miscarriages and cancer, it is ELF that is suspected of being the culprit. ELF fields are thought to cause fetal abnormalities and tissue changes, and even 20 hours a week in front of a monitor (or, worse, close to the back or side of someone else's monitor) is enough to increase the chance of a miscarriage. ELF has also been linked to cancer of the nervous system, lymphoma, and leukemia.

Don't sit within four or five feet of the back or side of a computer monitor. Don't sit closer than an arm's length away from the front of your own monitor. If you can't see at that distance, enlarge the view or the point size of the type until you are ready to print, then change to the printing size (I know—that's not always possible). Laptop and notebook computers that use an *LCD* or a gas-plasma display emit no ELF fields. See *radiation.*

Elite

On a typewriter, **Elite** is the name for the smaller size type which fits 12 characters in an inch (the larger size is called Pica and fits 10 characters in an inch).

On a PC, Elite is the name of a typeface that looks like the typewriter face. It's as lovely on a PC as it was on the typewriter.

ellipsis, ellipses

This is an **ellipsis: ...** The plural is **ellipses.** When you see an ellipsis at the end of a command on a pull-down menu or on a *button,* it's a visual clue that means you'll get a *dialog box* if you choose that menu command or click that button. Since dialog boxes always have a *Cancel* button, feel free to explore by choosing any command or clicking any button that has an ellipsis. You can look at the options in the dialog box, change them if you like, and then choose Cancel as you leave to restore the original settings. You can't hurt anything.

By contrast, if you choose a command that does **not** have an ellipsis, that command will immediately execute—so be sure that's what you want to do.

In typography, an **ellipsis** in text indicates a thought that's trailing off or a passage that has been edited out. In fine typography, the three dots should have a space between them, one before, and one after.

> Alas, when you type a space between the dots, you can end up with one or two of the dots at the end of a line, and the last dot or two at the beginning of the next line, since the computer does not recognize this series as a unit.
>
> Fortunately, you can type the ellipsis character (key strokes shown below), where the three dots are just one character, so they won't ever break at the end of a line.
>
> Alas, the ellipsis character has no space between the dots, so it is typographically unacceptable for professional text (it's fine for text where you're not concerned with fine typography).
>
> Fortunately, you can type a *hard space,* also called a *non-breaking space,* in all Macintosh programs and some Windows programs. So to create a tyographically acceptable ellipsis that will not break into pieces, type: hard space, period, hard space, period, hard space, period, hard space. (Actually, I usually type a thin space, or at least reduce the point size of the hard space to about half.)

On the Mac, type **Option ;** for the ellipsis. To type a non-breaking, hard space, press **Option Spacebar**.

In Windows, hold down the **Alt** key and type **0133** on the numeric keypad. Each PC page layout and word processing program has its own keystroke combination for a non-breaking space, so consult your software manual.

e-mail

E-mail is short for electronic mail, mail you can send or receive directly on your computer. Yes, with e-mail people can actually write you letters and send them to your computer, and you can turn on your computer and go pick up your mail whenever it's convenient. Many a love affair has begun through e-mail. I know. It's really fun, too. And useful, of course. E-mail can be a verb too, as in "I e-mailed my lover a letter from Lhasa."

The perfect love affair is one conducted entirely by e-mail.
Jon Winokur

E-mail may not be as personal as a handwritten note, but it's the quickest and most convenient way yet to communicate written information. If something absolutely positively has to be there in ten minutes, e-mail can get it there on time. I write for magazines and corporations all over the country, but I never leave my room. If an article has to be in by Friday at noon, I can finish it Friday at 11:45 and e-mail the computer file to New York before the deadline.

E-mail is also great for taking care of business without having to go through all the social pleasantries we endure in phone conversations—you can get straight to the point without having to ask how the kids are doing. "Attached is the file on the DeVere controversy. Tell me what you think." Contrast e-mail with *snailmail.*

To send or receive e-mail, you must have a *modem* or your computer has to be on a *network* (connected to other computers). Your company or organization may have its own e-mail system with its own software—in which case you'll run a version of that software on your machine. Lotus' CC:Mail is probably still the most popular e-mail software around for PCs. Alternatively, e-mail services are available on many *bulletin boards* and *online services,* and there are specialized services such as MCI Mail whose primary mission is to distribute e-mail.

You can also send and receive files by direct *telecommunications,* where your modem calls the other party's modem. But that entails the other person having to wait for your call and hanging around until they receive the entire file. They might not be able to use their computer for anything else in the meantime. With e-mail, you can send the mail anytime you want and they can pick it up anytime they want.

em dash

On your typewriter you were taught to simulate a long dash by typing two hyphens. Well, you need never fake it again with those two hyphens, because now you can have a real, typesetter's **em dash,** like this —. An em dash is used to set off parenthetical material that you don't want in parentheses, and to prepare the reader for an amplification or a shift in thought. See *en dash*.

On a Mac, to type an em dash, press **Option Shift Hyphen**.

In Windows, to type an em dash hold down the **Alt** key while you type **0151** on the numeric keypad. In PageMaker, press **Control Shift =** . In non-Windows PC programs you'll have to consult your software manual or *The PC is not a typewriter* because the keystrokes vary from program to program.

emoticons

When you read e-mail or have an online conversation (where you are typing at your computer and someone in another part of the country is typing back at you), there is obviously no way to interpret the tones of voice, facial expressions, and body language that we all use and depend on to get the meaning behind the words. So a substitute method has developed to communicate a joking tone, wry sarcasm, or genuine sympathy—or whatever you want the other person to think you're feeling. The trick is to use expressive little "faces" called **emoticons** in your writing. Emoticons are made up of ordinary text characters—mostly punctuation marks—that you can type no matter what kind of computer you have. Text sprinkled with emoticons is also called *baudy language,* which is really a much more fun term, so I have listed all the emoticons over there. Please see *baudy language*. (If you don't get the pun, look up *baud* while you're over there.)

EMS

EMS stands for **e**xpanded **m**emory **s**pecification. It refers to a combination of hardware and software that allows an IBM PC or compatible to use more *memory* than the 1 megabyte it would otherwise be stuck with. See *RAM* for a comparison of all the types of memory used in PCs.

emulate, emulation, emulator

When you buy something that claims to **emulate** another product, it means that the one you bought works just like the one it aspires to be like (emulates). For example, many laser printers emulate the Hewlett-Packard LaserJet, which means you can set up your software as if you had a real LaserJet. If your Brand X printer really does emulate the LaserJet, your printouts will look just like they would if a LaserJet printed them. If a product has an "emulation mode," this means you can turn a switch or change a setting to turn on the emulation—otherwise, the product has its own way of doing things. Some PostScript printers have a LaserJet emulation mode you can turn on when you're working with software that doesn't work with PostScript machines.

Emulation can be a feature of software, too—you can buy software that emulates the way other software works. And some software can make one device (like a printer) emulate another. For instance, you can buy software that lets a laser printer emulate a pen *plotter.*

Be forewarned that many products only partially emulate the original model. Some features may work perfectly, others might function incorrectly or at least differently, and some might not work at all.

emulsion

On a piece of photographic film, such as the kind you use to shoot photographs, one side of the film is coated with a layer of chemicals called the **emulsion.** This is the side that absorbs the light, and the emulsion is scratchable and dull. The non-emulsion side of film looks shinier and is more difficult to scratch. You can see the emulsion side on any negative you have hanging around.

The kind of film that comes out of *imagesetters* or that a pressperson uses to print your brochure also has an emulsion side. If you are creating something on your computer that will be output onto film (rather than onto plain paper from your personal printer or onto resin-coated paper from the imagesetter), you need to know whether the film should be output emulsion side up or down. The only person who can tell you the correct answer is the pressperson who will be printing the final job. She will say, "I need your film right reading emulsion side up" (RREU) or maybe "right reading emulsion side down" (RRED). This means that if you were to lay the film on a light table as if you were reading it properly (reading it right) the emulsion would be up, or facing you, or it would be down, on the side of the film away from you.

encrypt, encryption

To **encrypt** a file or other information stored in a computer means to convert it into a secret code so that it can't be used or understood until it is decoded or decrypted. You might want to encrypt a file if it contained a secret formula for a new invention, or some financial plans that your competitors would love to know about in advance. When you encrypt something, the computer will ask you to set up a pasword. After that, no one will be able to make sense of the information unless they have the same pasword.

en dash

An en dash is noticeably longer than a *hyphen*, but not as long as an *em dash*. If you grew up on typewriters, you never had to know what an en dash was because there wasn't one on the keyboard. But once you move up to a decent computer, you've grown beyond the typewriter.

Use an en dash instead of a hyphen to indicate a range, as a substitute for the word "to." For instance, you would probably read the phrase, "from 3–5 years" as "from three to five years." Right? So the proper symbol between these numbers is an en dash, not a hyphen. The same is true for other phrases showing duration, such as "October–December" or "7–9 P.M."

Another use for the en dash is when you are compounding two words and one of those words is already two words, such as "the San Francisco–Atlanta flight," or "the pre–World War II era."

To create an en dash on a Macintosh, type **Option Hyphen**.

In Windows, hold down the **Alt** key and type **0150** on the *numeric keypad*. In PageMaker, type **Control =** . In non-Windows PC programs, consult your software manual or *The PC is not a typewriter* because the keystrokes are not consistent.

End, End key

If you have a key labeled **End** on your keyboard, it usually functions to move you to the end of something in a file or document. In most PC word processors, pressing End by itself moves the cursor from wherever it is to the end of the same line. In some desktop publishing or page layout programs, End moves you to the last page in the document. If you're working with a graphics program, End might scroll the image so you can see the bottom. Pressing End in combination with another key like Ctrl or Alt often moves you to the very last item in the document or file.

On the Macintosh, you will have an **End key** only if you have an *enhanced keyboard;* it's in the little pad of "edit keys" between the alphabet keys and the numeric keypad. The End key isn't used much. In some word processors it will move the *insertion point* to the bottom of the screen, and if you hold the Command key down and press End, the insertion point will move to the end of the document. Other applications may or may not use it—experiment with it.

In a Finder window, you can press End to view the last item in the list (press Home to view the first item in the list).

end user

The **end user** is anyone who actually uses a product as it was intended. No one who thought of the product, or created the product, or who works for the company that produced the product, is an end user. And even the computer experts at a company who evaluated the product and decided to buy it for the sales clerks or secretaries or middle managers aren't end users. You and I are end users, the people who pay money for the product and use it as it was intended.

enhanced keyboard

The keyboards that came with IBM's first few generations of personal computers had several different layouts. Eventually—in the mid-'80s—the keyboard engineers at IBM came up with a layout that stuck. IBM called it the **enhanced keyboard**. Ever since, most other makes of full-size IBM-compatibles have imitated that layout. If you see an ad that mentions an "enhanced keyboard," it means one that has the same layout as IBM's. Features of the enhanced keyboard include 12 *function keys* in a row at the top; two *Alt* and two *Ctrl (Control)* keys, with the Ctrl keys down by the spacebar and the *Caps Lock* key next to the A key, where Ctrl used to be; a separate *numeric keypad;* and separate *arrow keys* arranged in an inverted T shape.

The enhanced keyboard may be popular, but I don't think the layout is so great. You can still buy keyboards that have the Ctrl key next to the letter A (see the entries for *Ctrl* and *Wordstar cursor diamond* for the reasons why) and with the function keys in two columns on the left where they're easier to reach.

The Macintosh version of the enhanced keyboard (usually called the "extended keyboard") is the same as above, except it has the standard Macintosh Option key (not an Alt key) and Command key (not a Ctrl key), and up to 15 function keys, plus the numeric keypad and the arrow keys.

There is also a little pad of "edit keys," above the arrow keys, that contain the Home, End, Help, Page Up and Down, and a backwards delete key. This edit pad is primarily used for PC applications used on the Mac.

ENIAC

ENIAC is an acronym for **e**lectronic **n**umerical **i**ntegrator **a**nd **c**alculator. The ENIAC, assembled in 1946, was the first operational digital computer. This monster occupied 1,800 square feet, used 18,000 vacuum tubes, and performed simple addition calculations at a rate of 5,000 per second (very, very slow by today's standards).

Enter key

On most PC keyboards, there are two **Enter keys:** one where the Return key would be on a typewriter (PCs don't have a key labeled Return), and one at the far right, along the *numeric keypad.* The Enter key is important, but it does different things depending on the software you're using and what you're working on at the moment. With most word processing software, you press Enter to start a new *paragraph,* which inserts a *Return character* into your text. Remember, you don't need to press Enter at the end of every line, just at the end of a paragraph (see *word wrap*).

When you're working with menus or dialog boxes, pressing Enter generally confirms whatever choices you've made. Once you've selected a menu choice with the keyboard (usually by moving a highlight over the choice), you have to press Enter to actually activate that choice. After you make changes in a dialog box, you press Enter to tell the program to accept your entries.

On a Macintosh keyboard, you'll find the Enter key along the right side of the numeric keypad. Sometimes the Return and Enter keys do the same thing, such as create a new paragraph in a word processing document. But sometimes they have different functions—again, it depends on the software package you're using. "To OK this dialog box, you can shortcut by hitting the Enter key."

environment

The term **environment** sounds both vague and pretentious, yet there isn't always a good substitute. Most often, environment refers to the system software that determines how your programs work. This system software may or may not be the actual *operating system,* which is one reason "environment" is the preferred term. For instance, Microsoft *Windows* isn't technically an operating system (unless you're talking about

Windows NT), but it creates an environment nevertheless. Another reason people use "environment" instead of "operating system" is because it emphasizes your experience—the way the system "feels"—rather than the technical specifications.

When some people talk about an environment, they mean a given combination of hardware and system software, but the term *platform* would be more accurate. An environment can also refer to the overall characteristics of a specific piece of software, usually software with a technical orientation. A magazine review might say, "This C language *compiler* offers a rich programming environment" (see *IDE*). In fact, sometimes environment is used in an abstract way to refer to the qualities of a programming language, independent of any particular compiler.

EPS, EPSF

EPS or **EPSF** stands for **e**ncapsulated **P**ost**S**cript **f**ormat, a graphic file format especially created for graphics that will be imported into other applications. The term "encapsulated" refers to the fact that an EPS file makes a tidy package for the graphic it contains, a capsule that can be easily incorporated into another document.

EPS files usually contain two parts. The essential part is the complex PostScript code that a *PostScript printer* can interpret and print in the highest resolution it is capable of. But since most applications can't display imported PostScript graphics, the EPS file usually also contains a *bitmapped* image of the same graphic that the applications can read and display on the screen. In a Mac EPS file, this bitmap is a 72 *dpi (dots per inch) PICT* graphic; in a PC EPS file, it is a *TIFF* file that can be of various resolutions. Anyway, if you print an EPS graphic to a *non-PostScript printer,* the printer will only be able to reproduce the *low-resolution* PICT image.

If you try to import an EPS graphic into another application and all you see is a grey box, that means you forgot to click the proper buttons when you saved or *exported* the file to make sure the file contained both parts of the graphic—the bitmapped image **and** the PostScript code. But even though you only see a grey box on the screen, a PostScript printer will print the graphic in all its splendor.

Like all PostScript files, EPS files are called "device independent" or "resolution independent." This means they will print at whatever resolution the PostScript printer happens to be. For instance, the same graphic will print at 300 dpi on a laser printer, or at 1270 or 2540 dpi on an *imagesetter*.

Epson printer

Epson has sold more *dot matrix printers* than anybody else, at least in the United States. Because of their popularity, the **Epson printers** established a *de facto standard* for the commands used to control the printer and other printer functions, and most other printer manufacturers still follow that standard.

erase

To **erase** something just means to remove it or throw it away. Erasing a file is the same thing as deleting it.

Paint programs provide an eraser *tool*. You drag the eraser over the part of your painting you don't want.

This is a typical sort of eraser; drag over the image to erase it.

In DOS, the ERASE command does exactly the same thing as the DEL (delete) command. (Everybody uses DEL because it's shorter.)

ergonomics

Ergonomics is the study of the human-and-machine relationship. It is the science of arranging your workspace so you can be comfortable working in it for long periods of time. Ergonomic adjustments include appropriate chair height, foot rests, wrist rests, height and angle of your monitor, etc., to put the least amount of strain on your body and eyes while you're working.

Ergonomics also deals with how you physically interact with your machine, such as the convenience or inconvenience of button and key placements, or whether or not the light from the lamp is harsh and tiring.

error message, error code

When something goes wrong with your computer or your software, you may see an **error message** or an **error code** number on your screen. Ideally, the message gives you a clue as to what caused the problem, and better yet, how to fix it. But you're just as likely to get a terse, inscrutable grunt such as DOS's "Error reading Drive A—Abort, Retry, Ignore?" If you're really unlucky, you'll just see a number; a typical message on the Mac might be "Error code ID 46." You might think that if you only knew what Error code ID 46 was you could fix this problem or at least prevent it from recurring. Wrong. The error code number is generally a message to the programmer (not the person who uses the computer) and even if you knew what it meant you probably couldn't fix it anyway. Try again.

For help with DOS's *Abort, Retry, Ignore* message, see that entry.

For a description of most of the alert messages (not really error codes) you may run across while using the Macintosh, try *The Macintosh Bible "What Do I Do Now" Book,* written by Charles Rubin and published by Peachpit Press.

Escape key (esc, ESC)

The **Escape key** (labeled **ESC** or **esc**) is usually located on the upper left of the keyboard. In most programs, pressing Esc lets you back out of commands or functions before you finally put them into effect. For instance, say you press the key that starts your word processor's *search and replace* function, but after typing in the word you want to search for you decide not to proceed; pressing Esc should cancel the command and return you to your writing. Sometimes you can even stop a command that's already been running by pressing Esc. In Windows and other PC programs that you control with menus and dialog boxes, pressing Esc closes the current menu or dialog box (the item disappears from the screen).

In some non-Windows PC programs you can even exit the program altogether by pressing Esc (the software usually gives you a chance to change your mind). Esc is almost always the key to press when you want to put away a pop-up, memory resident program (you're not exiting the pop-up program altogether, just deactivating it until you pop it up again).

The Escape key isn't really used much in Macintosh applications, and is only found on the Apple *ADB* and *extended keyboards.* In HyperCard you can use the Escape key to "Go Back" while in Browse mode, or "Undo" while in the Paint mode. The Escape key also comes in handy when you need to *force quit.* But the primary function of the Escape key on a Mac keyboard is to have it available when running DOS programs so your computer can be fully *compatible* with the DOS program. (Yes, you can do that.)

ESDI

ESDI stands for **e**nhanced **s**mall **d**evice **i**nterface and refers to one of the standards for PC hard disks and the circuits that control them (an ESDI hard disk needs an ESDI controller). ESDI devices were common at one time, but most new PC hard disks and controllers are the IDE type.

Ethernet

Ethernet (pronounced "eether net") is a *local area network,* connecting computers together with cables so the computers can share information. Within each main branch of the network, Ethernet can connect up to 1,024 personal computers and workstations. Also see *EtherTalk.*

EtherTalk

EtherTalk is Apple Computer software that allows the Macintosh to hook into *Ethernet networks.* To use EtherTalk and Ethernet, the Mac must have the Ethernet interface *card* installed inside the computer.

When you see a computer ad that says something like "8/80E," the **E** means the computer includes the Ethernet interface card. (The numbers tell you how many megabytes of RAM and hard disk space.)

evangelism

Evangelism is a way of selling (a product or a concept or a lifestyle) that involves communicating your personal enthusiasm to the point where another person or group becomes converted to your way of thinking. Apple adapted this concept to the personal computer marketplace as the cornerstone of their original development and marketing strategy. Basically, they sent Guy Kawasaki and his clones out to convince software and hardware developers to create products for a computer that had no installed base, no software, only 512K of memory, and no hard disk. They did it.

Evangelism also convinced people who were using IBM PCs and other personal computers that using the Macintosh was easier, more productive, more fun, and infinitely cooler. (Robin says, "That didn't take much evangelizing—it's just a fact." Steve says, "Now, Robin, be nice.")

Guy Kawasaki defines an **evangelist** as a person who sells dreams—as opposed to most people who dream of sales.

EXE, .EXE

Any file with a name ending in **.EXE** (pronounced "e x e" or "dot e x e") is an "executable" file: it's a program that can be run, or *executed.* The EXE file may or may not contain the entire program—sometimes other files are needed as well—but the EXE file is the one that starts, or executes the program.

If you're working from the DOS *command line* (rather than from Windows or a DOS *shell*), you start an EXE file by typing the first part of its name at the DOS prompt and then pressing the Enter key. So, to start a game program stored in a file named PILOT.EXE, you would type **PILOT**, then press Enter.

execute

You tell the computer what to do by typing in a command or choosing a command from a menu or list. The computer then **executes** that command; it follows the directions to complete whatever it was you wanted.

exit

To **exit** a program means to stop using it and to unload it from your computer's memory. Most software programs have an exit command that you use to leave the program and return back to your operating system or whatever it was you were *running* before.

In DOS, you use the EXIT command to terminate or quit the current version of the main DOS program, *COMMAND.COM.* But this only works if it is not the primary, original copy of COMMAND.COM—the one that ran when your computer started (when you *booted* it).

For example, when you're running DOS from within Windows, Windows has to run COMMAND.COM again in a secondary role. When you are done using DOS from within Windows you terminate that particular session of DOS by typing **EXIT** on the command line, and then pressing Enter. The DOS screen that you were using disappears and you go back to Windows. In a similar way, many non-Windows programs let you run a secondary version of DOS from within the program so you do not have to exit the program altogether in order to access or execute DOS commands. With such programs, after you are finished running this secondary version of DOS, you can just type **EXIT** to return to the program.

On the Mac, to **exit** means simply to leave the application you're using and go elsewhere. When you actually close up the application and put it away, it's called *quit,* rather than exit.

expanded memory

Expanded memory, also known as *EMS,* is a combination of hardware and software that allows an IBM PC or compatible to use more *memory* (normally, DOS limits PC-type computers to one megabyte). See *RAM* for a comparison of all the types of memory used in PCs.

expansion box

An **expansion box** is an *external* unit that provides a PC with more *expansion slots.* The box connects to the main computer through a special *port* on the computer, or via an *add-in board* that fits in one of the PC's existing expansion slots. Most commonly, expansion boxes are used with *laptop* or *notebook* PCs that lack slots of their own (see *docking station* also).

Once you've hooked a notebook up to an expansion box, you can use your little computer with any add-in board that works in a standard PC. You might, for instance, plug in a fax modem, an *adapter* for a large-screen monitor, or a sound board. Expansion boxes are sometimes marketed for people who've filled up all the slots inside a regular desktop PC and still need more add-in boards.

JUST AS GAIL HAD FEARED— HER PC WAS INFECTED WITH THE DREADED "SGV" (SALES GUY VIRUS).

expansion slots

Expansion slots are the long, narrow openings aligned in rows on your computer's main circuit board (the *motherboard*) where you plug in *add-in boards* to give your computer new capabilities (see the illustrations in the appendix on "How to read a computer ad"). The slots are openings within short plastic projections that stick up from the motherboard. The walls of an expansion slot are lined with metal contacts to match the contacts on the add-in board that you insert. Also see *full-length slot.*

expert system

An **expert system** is the type of software people usually mean when they talk about *artificial intelligence.* Software that uses an expert system comes up with recommendations for the best solution to a complex problem, based on information you (the user) feed it about the current situation. You ask the computer a question or pose a problem, and the expert system provides an answer, based on a "knowledge base" of human expertise.

The person programming an expert system analyzes the behavior of a human expert in the field, breaking down the expert's handling of the problem into a set of explicit rules ("if the molten steel contains ¼% copper, then increase the temperature to 675 degrees, unless the carbon content exceeds 1.5%, in which case . . ."). Once a good expert system knows all the relevant rules, its recommendations should match those of the expert. The catch is teaching the expert system *all* the rules it will need to deal with every possible scenario it might confront—no expert system can be as flexible as a person. When an expert system isn't programmed very well, some of its decisions can be pretty funny.

Expert systems are used in equipment repair, investment analysis, insurance planning, route scheduling, training, medical diagnosis, production control, and in other areas.

export

You typically **export** information from one *application* to use it in another. For instance, you might export information from a database so you could put into a spreadsheet. Or you might export information from one database so you could add it to another database. Usually, you export information by saving it onto a disk using a *file format* that the other program can understand. Then when you are using the other application, you *import* the previously exported file.

Sometimes you need to export information so another application can simply *open* it. For instance, if I write a story in PageMaker, which is a *page layout program,* MacWrite would not be able to open it because MacWrite is a *word processing program.* But I can, from within PageMaker, choose to export that story in a format that MacWrite or any other word processor could open.

extended keyboard

See *enhanced keyboard* and *keyboard.*

extended memory

In the vast majority of PCs sold these days, **extended memory** is the type of *memory* over and above the first *megabyte* (which is the *conventional memory*). A PC must have at least an *80286 microprocessor* (meaning that an *80386* or *80486* will work, too) to use extended memory. But even then, DOS doesn't recognize extended memory—to use it, your software has to incorporate a *DOS extender.*

Extended memory isn't the same thing as *expanded memory*—these two terms have specific meanings, and you have to know which one to ask for. See *RAM* for the complete low-down on the different types of PC memory.

extension (in file name)

In DOS and some other *operating systems,* the names of files are divided into two parts: the name proper and the **extension,** a few characters that are typically used to indicate the type of file you're working with. Like this:

LETTER.DOC

WP.EXE

SK.COM

TAXES.WK1

As you can see, a DOS extension consists of up to three characters, divided from the rest of the name by a period, called a "dot." When you're working with a file on the DOS command line, you must refer to the file by its complete name, including the extension. To delete (remove) LETTER.DOC from the disk, you would type DEL LETTER.DOC and press the Enter key. An exception is the DIR (directory) command—with that command you only need the first part of the name (see *DIR*).

You can use what looks like an extension to a file name on the Mac, but it's completely optional. It can be up to 31 *alphanumeric characters,* and the computer does not see the extension as anything special. An extension can come in handy to help identify files, which is the only reason you might see them or use them. For instance, the extension "*.sit*" after a file name gives you a clue that the file is *compressed*. In the files for this book, I labeled and relabed each one with version numbers (".v1," ".v2") to keep track of which file was at which stage of the process.

extension (in System Folder)

An **extension,** previously called an "INIT" (before System 7), is a little program that does things like make extra sounds, pictures, bizarre cursors, clocks, etc. Many extensions are extraordinarily useful commercial utilities and difficult to live without once you have them, such as *Suitcase,* Type-Reunion, and *QuickTime.* All the *printer drivers* are extensions. Some applications install extensions in your System Folder when you install the application, and the app uses these extensions when *running.*

Extensions

This is the folder where extensions are stored.

Extensions are stored in the Extensions Folder within the System Folder. In fact, that is the magical thing about them—they will **only** work if they are in the System Folder. Put an extension anywhere else and it doesn't do what it's supposed to do.

You rarely actually use the extension itself—it just allows things to happen, or makes them happen automatically. These are examples of extensions. Notice some are still called "INIT." An icon shaped like a puzzle piece is always an extension.

QuickTime 1.5

File Sharing Extension

multiple master INIT

Disinfectant INIT

(In System 7.1, printer font extensions like this one are stored in the Fonts folder.)

Symbo

PSWriter

PrintMonitor

StyleWriter

When you turn on your machine, all these little extensions get loaded into *RAM (memory)* and start working even before your System gets up and running. That's why they are sometimes called "Startup Documents," and why they were called INITs in System 6—because they work upon INITial-ization of your computer. As your computer starts up, you will see their icons appear in the lower left portion of your screen (some of those icons also represent *cdevs,* which are like extensions except you have some control over them through their *control panels*).

Most extensions from reputable sources will work invisibly and well. But many extensions, especially those written by individuals and sent out as *freeware* or *shareware,* are well-known sources of trouble to the System. If things start acting weird after you install an extension, take the extension out of the Extensions and System Folder, restart (from the Special menu), and see if things are normal again. Never install more than one extension at a time; rather, put one in, work on your machine a few days, then put another one in. That way you will be able to pinpoint the one that causes trouble.

You can't throw away an extension until it is disabled. You will have to remove it from the Extensions folder and take it to the Desktop or to another folder, restart, and then throw the extension in the trash (copy the extension to a floppy first if you want to save it). To disable all extensions, hold down the Shift key when you restart.

There are *utilities* available that help manage your extensions, enabling you to turn them on and off, disable them temporarily, make them load in a certain order, etc. Ask your local *user group* or on your favorite *online service.*

external

External refers to a device that's considered part of your computer setup, but that is in a separate case, outside the computer itself. Your computer probably has an *internal* hard disk that you can't see, but you may also have an *external* hard disk that is in a box outside your computer. You probably have a slot for your floppy disks, which is your internal floppy disk drive, but you may also have an external floppy disk drive in its own little box, next to your computer.

external command

In DOS, an **external command** is really just a *utility,* a separate little program, that comes with DOS. FORMAT, TREE, and BACKUP are external commands.

Ordinary DOS commands like DIR and COPY *(internal commands)* are built right into the main DOS software, so they always work—as long as you have a DOS prompt on your screen, an internal command will work when you type it on the command line. But an external command only works if the utility program for that command is available to DOS. That is, a copy of the program has to be stored either in the current *directory* or in a directory listed in your DOS *path*. See *DOS, DOS prompt, command line,* and *XCMD.*

eye-beam

See *I-beam*.

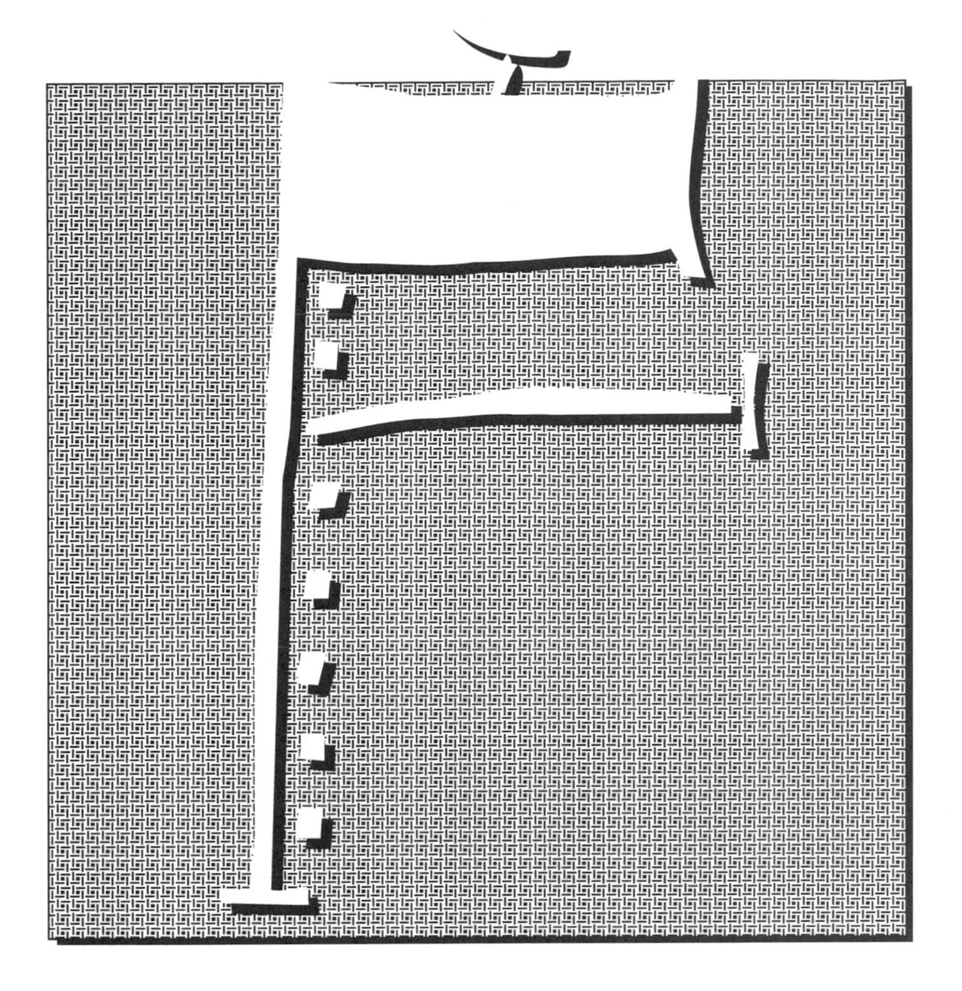

F1, F2 . . . F15

These numbers, **F1, F2,** etc., are the names of the *function keys.* If your keyboard has function keys, they will be in a row at the top, in two columns on the left, or in some minicomputer and workstation keyboards, in a cluster at the right. Please see the definition for *function key,* and if you are a Macintosh user, also see *FKey.*

facsimile transmission

See *fax.*

FAT

See *file allocation table* for the skinny on the **FAT**. *(oooh—Steve said that, not me.)*

fatal error

Fatal error!? Why do they do this to us? Why can't they just say, "Oh, excuse me, but your system has a serious problem. The trouble may just be temporary, and if you'll turn off the computer and try again, maybe everything will work fine."

But when you see this message, the word "fatal" tends to make your heart skip. You pause a moment, wondering how the fatality is going to occur—will it be a shot to the head from the person who owns the computer, or a long lingering disease caused by the radiation emanating from the screen? It conjures up dead bodies and grieving children. Then, when you realize it's not **your** condition that's fatal, but the **computer's,** you can just hear tiny little screams of agony from deep within the bowels of bits and bytes as the last threads of your hard work are devoured by the Fatal Error.

The computer may offer you a button that says Restart, but clicking it probably won't work. You can try a *force quit* (always after a force quit, save and quit all other applications and restart). If that doesn't work, the only thing you can do is turn off your computer, wait about fifteen seconds, and turn it back on again. Yes, it's true: anything you had not saved to disk before a fatal error occurred is lost for good.

Why did your System have a fatal error? Most of the time you may never know. It may never happen again. Or it may happen again if you completely recreate the scene just before it happened. It may be a particular graphic or event within your document that caused it. The best thing to do is always have a backup of your work and save every few minutes; then when a fatal error or some other milder form (ha) of a System crash occurs, you can just pick your chin up off the floor and continue on. Call your *power user* friend or ask your *user group* for suggestions. There are also several books available for troubleshooting; again, ask your user group for recommendations.

I do apologize for getting so dramatic here. I couldn't help it. The only thing worse than seeing the message "Fatal Error" is hearing the *Chimes of Doom*.

Some PC programs also display a "Fatal Error" or "Catastrophic Error" message if you do something that really ticks them off, like trying to run the program with not enough memory available. In this case, you don't have to turn your whole computer off. DOS itself isn't known for "Fatal Error" messages—it says mysteriously ominous things like "*Abort, Retry, Ignore, Fail?*"

FatBits

On the Macintosh, some programs let you edit *bitmapped* graphics as **FatBits.** In FatBits mode, the individual dots, or pixels, making up the image are blown up so you can work with them easily, one at a time. If you see stray dots in an image you've scanned, or if a line in a picture is just slightly too thick or too skinny, it's almost impossible to make precise changes working at normal size.

Traditionally, you can get into FatBits mode by selecting the pencil tool, holding down the Command key, and clicking on the image at the spot you want to see enlarged. You might also have a magnifying glass tool, or a menu command for FatBits or a command to Enlarge. Sometimes to get out of FatBits you can hold down the Option key and click.

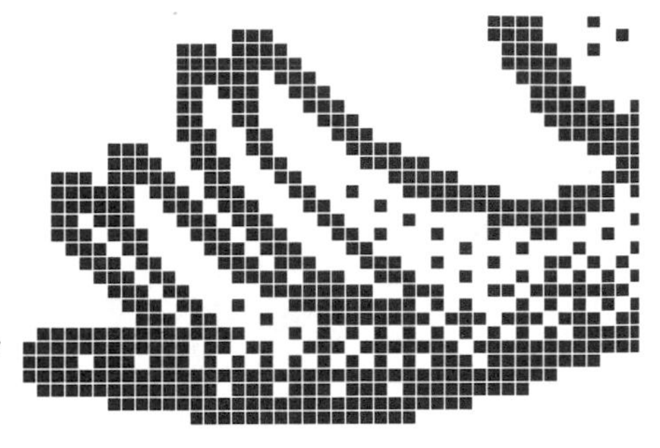

The image to the right is the dinosaur's tail in FatBits mode. You can clearly see each pixel and you can edit one pixel at a time.

fax

A **fax,** for that shrinking minority of people who don't depend on one daily, is short for "facsimile transmission." The word "facsimile" means a duplicate or copy of something. With a little machine not much bigger than an answering machine, you can send, or transmit, a fascimile of any document in your office. Just slip the piece of paper in the slot, dial the phone number of the other person's fax machine, and a copy of the document comes out, not of your machine, but out of their machine, even if it's across the world. Your fax machine scans the page, turns the information it sees into a series of electronic bits, and sends those bits through the phone line. The fax machine on the other end translates those electronic bits of information and reproduces them on a piece of paper. You can also use a fax modem (see next page), which even eliminates the step of having to print the document out of your computer. A fax machine is one of those things that you didn't know you needed until it got here, then once it got here you wonder how you ever got along without it. Like copy machines or post-it notes.

fax modem

If you don't quite know what a *fax* is, first read the definition on the previous page. And if you're shaky on what a *modem* is, it might be smart to read that definition too. Anyway, a **fax modem** is simply a modem designed for sending and (usually) receiving faxes via your computer. Using a fax modem, you can send a document to someone else's fax machine without having to print out the document and then stand there and make sure the pages don't jam in the fax machine. In fact, if the other person has a fax modem attached to *their* computer, they can receive the document directly into their computer and read it on their screen. Of course, they can print it to their own printer and then have a copy on real paper instead of that disgusting fax paper.

Apple's *PowerBook* (the *laptop* Macintosh) and many notebook PCs have fax modems built into them. You can take the PowerBook to your hotel, unplug the hotel room phone, plug that cord right into your PowerBook, and fax away. Incredible. In fact, if you need a *printout* and you don't have a portable printer, you can fax the document to yourself in the hotel lobby.

You can buy fax modems on *add-in boards* that plug into your computer, or in little boxes not much bigger than a deck of cards. Before you lay your money down, be sure you ask whether the fax modem can **receive** (some don't—they cost less, but you'll be stuck when someone wants to send you a fax). And ask how fast the fax modem is—the standard speed is 9600 bps, but some only run at 4800 bps. And find out for sure what software will work with it, because the modem won't do you any good if your software can't make it go.

FDD

FDD just means **f**loppy **d**isk **d**rive. Advertisements from cheapo computer stores often use this unintuitive abbreviation, as if the tiny print wasn't hard enough to read already.

FDHD

FDHD stands for **f**loppy **d**isk, **h**igh **d**ensity and refers to the floppy disk drive in current Macintosh models. This drive can read and write single-sided disks (400K), double-sided disks (800K), and *high-density disks* (1.2 megabytes). It can also understand 3.5″ DOS-formatted disks from IBM PCs and Apple II machines.

An FDHD is also known as a *SuperDrive*.

feature

A **feature** is any particular capability of the product you're talking about, be it hardware or software. You might compare the features of two software applications to decide which one is right for you—do you need the indexing feature found in one word processor, or is the local formatting *style sheet* in another word processor more important to your work?

Marketing people use the term "feature" for the Really Neat attributes they hope will convince you to buy the product. Some people call these fancy capabilities "bells and whistles." Sometimes salespeople try to convince you that a *bug* is a desirable feature.

female connector

A **female connector** is the *connector* (similar to a plug) that has, well, receptacles for the, er, prongs of the *male connector.*

fiber optics

Light beams can carry information. Enormous amounts of information can be transmitted along hair-thin strands of glass with the speed of light. **Fiber optics** is a communication system that utilizes dozens (or hundreds) of strands of glass (or other transparent material) that can each carry thousands of *digitized* voice conversations simultaneously. Light transmissions are not affected by random radiation like electrical impulses are, and being of a higher frequency (on an electromagnetic scale), can carry more information. If the telephone companies capitalize on it, fiber optics may bring *high definition television* into the homes of many, as well as video telephones, and will make other interactive services available to each home with a phone line.

FidoNET

Everybody likes to be part of something. **FidoNET,** founded in 1984, is a group of personal computer users all linked together sharing files, exchanging mail, leading discussion groups, etc. It originated as an IBM PC group but now includes users of all kinds of computers and represents thousands of linked *bulletin boards* worldwide.

field

A **field** is a specially prepared area in a document, usually in a *database* of some kind, into which you enter information. For example, in a mailing list database, you need to store names and addresses, right? So your database has a field set up where you input first names, another field for last names, another field for street addresses, and others for city, state, zip, and phone numbers.

You can *format* the fields according to the information that the field will contain. For instance, you probably want to format the zip code field as "numbers" so you can then *sort,* or organize, that information in numerical order. If you have a field that will contain dollars and cents, you can format the field so the dollar sign appears automatically. If the field will contain dates, you can format it for the way you want the date to display; that is, if you format the field for the "long date," then if you type 5/1/86, what will appear in the field is "Friday, May 1, 1986."

The similar concept in a spreadsheet is called a *cell.*

Each of these boxes where you can enter information is a field. The field titles will stay the same for each record, but the data in the fields can be different.

figure space

A **figure space** is a space that is the same width as a number. You see, a regular space created by pressing the Spacebar is proportional to the rest of the typeface, and each character (except numbers, also known as "figures") has a different width (an "m" is wider than an "i"). Numbers, however, must all be the same width or they will not align in columns. A few applications have a keystroke that will provide you with a way to create a figure space so you can enter a space within a column of numbers and still retain the alignment of the numbers (figures).

If your program does not have a figure space and you really need one, you can fake it: in the space where you need a figure space, type a number, then select the number and make it *reverse* (if your application allows you to make the text reverse).

file

A **file** is a particular collection of information you use as a unit. Files can hold just about any kind of information, including text, numbers, graphics, or software programs.

Files that hold the information you actually work with, such as a report you write or a graphic you create, are referred to as documents, document files, or data files.

Files that contain programs are program files; some program files are applications or utilities. Even the folders or directories on your disk are files.

Application files often have a lot of smaller files, such as *dictionaries* or *drivers* or a list of your *preferences;* the application needs these associated files to make certain features work, or work the way you want them to, or to communicate with other devices. Also see *file format.*

file allocation table

To save time, computers don't always store all the information belonging to one program or document in one place on a portion of a disk. Sometimes portions of files are broken up into pieces scattered hither, thither, and yon on the disk (see *fragmentation*). The **file allocation table** (FAT) keeps track of all the parts of the file so they can be linked together when the file is used again. Sometimes the information necessary to track these links gets lost, and then you have a "lost chain," or a *lost cluster.* See *CHKDSK.COM* for a way to find lost chains or clusters.

file association

A **file association** tells your system which application program you use to work on a particular type of document file. Let's say you use the Windows version of WordPerfect for word processing. Because a file association exists, you can double-click on the LETR2MOM.WP file in the Windows File Manager, and WordPerfect will *run* automatically, opening the letter document for you. In other words, you don't have to start WordPerfect first, then choose the Open command, and then find your document all over again through a dialog box.

File associations on PCs are based on filename *extensions,* not on the contents of the files. If you set up an association between WordPerfect and the extension .XYZ, WordPerfect will run and try to open any file that has the .XYZ extension, even if the file actually contains a Lotus 1-2-3 spreadsheet document.

In Windows, file associations are stored near the beginning of the WIN.INI file in your WINDOWS directory. Most Windows applications automatically set up file associations in WIN.INI for you when you install the program, but you can add to or modify these to taste. And Windows isn't the only place you can have file associations—they're available in some DOS *shells,* such as the Norton Desktop for DOS, and in *4DOS,* where you can just type the name of the file on the command line to run the associated application.

file conversion

Not all software applications can read documents created in other applications, so sometimes you have to do a **file conversion** to convert one file into a *file format* that another application can read. You can even convert PC files into Macintosh files. Some very smart programs can do a file conversion *on the fly;* they give you a message saying something like "This document is in a file format that MacWrite cannot read. Do you want to convert it into a MacWrite file?" On a Mac you can occasionally do a file conversion right within your program simply by saving under a different name and choosing a different format from the list available. And there are software utilities that can convert certain types of files for you, such as the *Apple File Exchange,* which converts PC/DOS files into Macintosh files.

file format

A **file format** refers to the particular structure that a *document* (also called a "data file") is stored in, whether it contains graphics, text, a spreadsheet, etc. For instance, in a word processing document, the file format would include the codes that represent each character; the codes for creating the text styles, such as italic or bold; and information such as the type of application the document was created in.

Each program has its own way of storing this information—its own file format. The MacWrite format is different from the Word format which is different from the WordPerfect format. To use a document created by another application, the program has to convert the foreign format into its own "native" format.

In addition to native file formats for every word processor, there are generic text file formats, such as ASCII (text-only) or RTF (rich text format).

There are many different file formats for graphics, as well, such as TIFF, PICT, PCX, MacPaint, WMF, DRW, EPS. Different programs can use different formats, and many programs can open and use more than one.

When you save a document to disk, the *default* (automatic choice) is to save it in the native file format. Most applications, though, offer a choice of alternate file formats. Choose another format when you know you are going to need to open the document in another application.

File Manager

In Windows 3.0 and 3.1, **File Manager** is the program that lets you organize your disk and files. It lets you copy, move, delete, and change the names of files. You can also create, remove, or rename *directories*.

When you open File Manager you are presented with a pair of windows. One displays the *directory* structure of the disk and lists each directory and *subdirectory*. For whatever directory is currently selected in that window, the other window shows the files that are in that directory.

In other words, File Manager lets you see the organization of your files on disk. You can have multiple pairs of windows open, so you can look at the organization of the same disk to see the contents of different directories. Or you can view the contents of different disks at the same time.

In Windows 3.1 you can copy or move files, or groups of files, from one of these windows to the other one just by dragging it, or them, with the mouse. You can also start programs by double-clicking on the name of the program in the window that contains the file list, or by double-clicking on a document associated with that program (see *file associations*).

The File Manager is important because it's the standard mechanism for dealing with Windows' files, although it can be augmented or replaced by a huge number of utilities available from other companies.

filename extensions

See *extension (file names)*.

file names

On the Mac, you can name your *file* (folders, documents, applications) anything you want, up to 31 characters. The only character you cannot use in a file name is the colon (see *illegal characters*). Although the Mac will allow you to start a file name with a period, don't do it; you run the risk of corrupting the file. And don't start the name with an exclamation point either. See *extension (file names)*.

Spaces, symbols, numbers, and punctuation marks are alphabetized before letters. You can take advantage of this fact to organize a list of files to suit yourself. For instance, you might want a particular folder or file to appear at the top of the list in a Desktop window or in any dialog box—put a space in front of its name. Two spaces will come in front of one space.

I wanted the template file to appear at the top of the list whenever I opened this dialog box, so I inserted a space before the name.

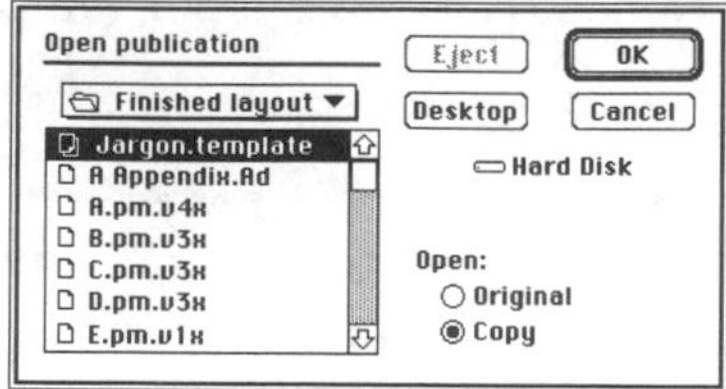

Avoid naming your files in all capital letters. Besides taking up too much space, they are much more difficult to read, especially when there is a long list.

On DOS machines, you are limited to eight characters, then a dot (period), then three more for the *extension.* But you can't use certain characters, including spaces or ^ * + = [ ] ; : " \ / , ? > or < and you can only use one period (between the main part of the name and the extension). And they are always capitals.

file server

In a computer *network,* the **file server** is the computer responsible for storing and retrieving the files used by all the computers connected to the network.

Let's say your computer at work is on a network (connected to other computers), and you want to look up the name of a customer in the company database. The information you need will be located in *database* files on the file server's hard disk. Database software running the server opens the necessary files and the information comes back to your screen over the network wires.

You can even run programs on your own computer that are stored on the file server's hard disk. Other people on the network can use the same files and programs. There is software running on the file server that controls who gets to use which files, and how many people get to do it at the same time.

Often, the file server is *dedicated,* meaning all it does is dish out files to the other computers on the network, and no one sits at the computer and actually uses it. In other cases, the file server also gets used as a working computer.

file sharing

File sharing is when a person on one computer can use the files from another computer **while the file is still on the other computer.** The computers that want to share files with each other must be connected via a *network.*

You can create a file sharing network on two or more Macintoshes as simply as sharing a printer through *AppleTalk.* For instance, my sister (Shannon) and I each have a computer in the same room, and our only connection to each other is that we print to the same printer—but that connection is enough to establish a network. She has a database that holds the names and addresses of the 500+ people on my *user group* mailing list. But when someone calls me and wants to be put on the mailing list, from **my** computer I can open the database that is on **her** computer and it appears on **my** computer and I can add the person's name and it instantly shows up on **her** computer also.

File sharing may be more complex, of course. There may be a *dedicated file server* in a school lab or an office that has applications and documents and fonts that everyone else on the network can use. See *file server.*

file system

Computers are asked to store massive amounts of information on *storage media* such as floppy disks and hard disks. Long ago, the computer just saved all the files on the disk, and if you weren't careful, you could end up with a list of files so long you could strain your wrist just trying to scroll through them all! It was kind of like putting all the papers into your filing cabinet without folders. As *operating system* software evolved, programmers acknowledged this problem and developed **file systems** so we can organize our electronic files. It's the same concept as the file system in the metal filing cabinet in your office: you store files in separate drawers, and each drawer can be separated into several categories, and each category can have any number of folders within it, and each folder can have other folders within it.

On the Macintosh, the file system is called the *hierarchical file system (HFS).* There are icons that look like manila folders in which you can store your files, and you can have folders within folders.

On DOS machines, the file system is composed of *directories* and *subdirectories* instead of folders, but they do the same thing.

fill

When you draw an object on the screen, you typically have a choice of the thickness and color of the line that borders the object; you can also choose the pattern and color of what is **inside** the object. This color and pattern **inside** the object is known as the **fill,** or perhaps the "shade." "Draw an oval and fill it with turquoise."

These two shapes each have a different fill.

film recorder

You have created a magnificent piece of computer art or you have edited a digitized photograph and no printing device can do justice to it. That's when you need a **film recorder**—an *output device* that captures your data and records it onto film. The film recorder creates a sharp, extremely high-resolution image, anywhere from 2,048 to 16,384 lines per **image** (not per inch) for film recorders that use internal picture tubes (CRT). Kodak LVT uses a rotating drum, not a CRT, to image the film, and can record up to 3,048 pixels per inch. High-end recorders like the LVT specify image resolution in pixels per millimeters. "Resolution 30," or "Res 30," for example, means 30 pixels per millimeter.

Depending on the film recorder, you can record anything from 35mm slides for presentations, to 4 x 5–inch or 8 x 10–inch film transparencies suitable for high-end color publishing or photographic prints.

filter

It would be wonderful if all programs could read each other's files. Unfortunately, though, most programs have their own ways of saving information, and unless your document can *export* data in a generic *file format* (such as *ASCII text*), you might have trouble reading a file in one program that was created by another. This became a real problem when desktop publishing became popular and people started using page layout software to combine text and graphics created by many different programs into one document. You could, of course, use a special *file conversion* program, but this extra step was cumbersome and time-consuming. To address this problem, the software engineers developed **filters** through which a given program can use a document created by another program.

For example, let's say you have a text file you typed in Microsoft Word, and you wish to import the file into PageMaker to use as part of a brochure.

The PageMaker filter for Microsoft Word will sift through the document and place it into PageMaker where it can be formatted and edited.

There are all kinds of filters for text, graphics, databases, and spreadsheets that will let you filter for *importing* and *exporting* to other programs.

find

In most word processing applications you can use the Find command to **find** a particular word, number, symbol, or even a *string* (group) of words. In some applications you can find text with certain attributes; for example, you can find all the words that are italicized, or perhaps any instance of the word "dogfood" that is bold. This process is also called *search* instead of find; both mean the same thing.

Often the find dialog box will include the ability to find a word and then change or replace it with another one; see *search-and-replace.*

All database software has a find, search, or "query" command that finds the information you need. If you type "Scarlett" in the name *field* of your Find or Search or Query screen, the database will find all the *records* that match your request (all the people in your database named Scarlett).

To *sort* is sometimes mixed up with to *find.* But sort means to organize. If you sort a list of names, they get alphabetized. If you sort a group of dates, they are put into chronological order. Please see *sort.*

Finder

The **Finder** is the part of the Macintosh *system software* that controls the *Desktop.* The Finder creates the Desktop, which is what you see when you first turn on your Mac—the trash can, the menu item "Special," the icons along the right side that display disks. The Finder is kind of like home base. It's always running, even when you don't have any other programs active; it's the top level of the *hierarchy;* it controls your main filing cabinet (your hard disk). Everything else in your computer can be found in this filing cabinet.

The Finder is actually a software program, and you can actually quit the Finder just as you quit any other program (but I don't recommend beginners do that). The Finder software is represented by an icon in the System Folder called Finder; if that icon is not in the Sytem Folder, the Mac cannot start.

Because the Finder and the Desktop are so closely related, the two terms are often used interchangeably.

firmware

Firmware is a category or class of *memory chips* which contain information that is permanent (meaning it isn't erased when you shut off your computer or when the power suddenly goes out on a dark and stormy night). The best examples of firmware are the *ROM chips* in your computer that contain programs installed at the factory. Firmware cannot be altered, per se, but in some cases the whole chip can be completely replaced by a technician when it becomes outdated or obsolete.

first-line indent

Remember when you were taught on a typewriter to set a tab stop five spaces into a paragraph, and then every time you started a new paragraph you hit the Tab key so the first line of the paragraph would be indented? Well, that's a **first-line indent**—the first line of a paragraph is indented.

On the computer we have an electronic version of the first-line indent that is incredibly convenient. In some Windows applications and in all Macintosh applications where you can *format* text (such as word processors and page layout programs), the first-line indent marker is in the *ruler* where you find the tab markers and indents. The left indent always splits into two parts. One part of the marker is where the first-line of each paragraph will begin; the other part of the marker is where the text in the paragraph will align.

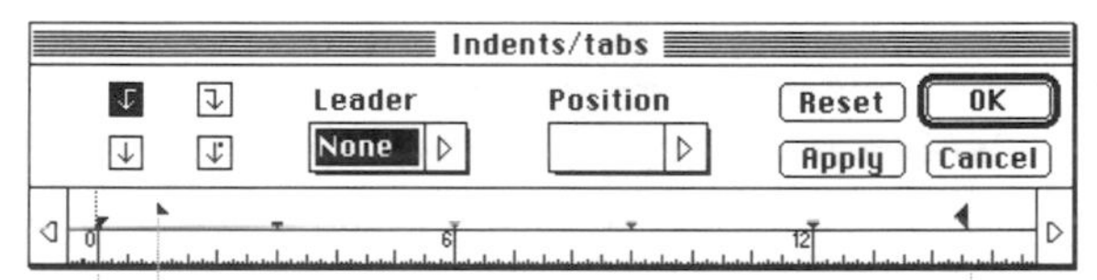

This ruler shows the indent setting to create the first line indent. The top half of the triangle is where the first line will begin. As you type, the text moves to the right, bumps into the triangle on the right, and bounces back to the bottom half of the triangle, the left indent.

You can experiment in your word processor. Separate the two parts of the left marker. Type several lines of a paragraph. Press Return. When you press Return, the computer knows that you are now starting the first line of the next paragraph, so the cursor will align at the first-line indent automatically—you don't have to press the Tab key.

fixed disk

A **fixed disk** is simply an ordinary *hard disk*, the kind without a removable cartridge. It's referred to as a fixed disk because you don't put it in or take it out as you would a floppy disk. Inside its case, though, the "fixed" disk actually spins (see *hard disk*).

fixed point

Fixed point refers to the way the computer stores and calculates numbers in which the decimal point is always in the same place. On paper, we are accustomed to working with fixed point numbers; we dutifully line up the ones in the ones column, the tens in the tens column, the decimal points all in a column, etc. But for working with very large and very small numbers, the computer uses *floating point* numbers.

Fkey

The terms **Fkey** and *function key* are often used interchangeably, although there are two separate meanings. "Function key" usually refers to the 10 to 15 extra keys along the top of an *extended keyboard,* numbered F1 through F10 or F15. Some applications use the function keys as keyboard shortcuts, and in almost all applications you can press F1 to *Undo,* F2 to *cut,* F3 to *copy,* and F4 to *paste.* You can also take advantage of the fact that few applications use all of these keys and you can program them yourself, usually with a *macro utility,* to perform certain tasks.

But Fkey (pronounced "eff kee") refers to a mini-program, or *utility,* that accomplishes a certain task using the keystroke combination Command Shift and a number. Several Fkeys are built into the Macintosh:

> Command Shift 0, 1, or 2 are Fkeys that will eject floppy disks from various drives (see *floppy disk*).
>
> Command Shift 3 is an Fkey that will take a picture of the screen (see *screen shot*).
>
> Command Shift 4 is an Fkey that will print your screen to an ImageWriter printer.

The other possible Fkeys (5, 6, 7, 8, or 9) and also the function keys can be programmed by you. Most Fkeys that have been created are not available commercially, but can be *downloaded* from *bulletin boards* or *onlines services* as *freeware* or *shareware.* They can do things like type in the current date or switch your monitor from color to grayscale—with just a keystroke combination.

Most Fkeys will come with their own installer, or you can use products like MasterJuggler or *Suitcase, ResEdit,* or FKEY Manager (*shareware* from Carlos Weber, found on most online services).

flag

A **flag** is a marker written into software or used in communications that tells the computer something in particular is happening, or that a particular condition prevails. For instance, the computer will find a flag at the "end" of a file telling it that this really is the end so there's no need to keep searching for more. Flags are also used to signal an event, such as an error or a result of two comparisons. Flags can be thought of as signs telling the computer to "Listen-up, something's happening here," or "When you get to this point, if such a condition is met, then do thus and so."

FLAM

FLAM is the **f**our-**l**etter **a**cronym **m**achine. It's very popular in the computer world. See *TLA* for more information on FLAs.

flame

To **flame** means to express your opinion *online* in a very strong, heated, emotional, and generally disagreeable manner. "Wow, did that guy flame when I mentioned armadillo rights." When a group of people get involved, posting flaming messages to each other and to the public, it can be considered a "flame war."

flat-file database

A simple, **flat-file database** has only a single *table,* which means it has one set of *records* with *fields*. Each record within the table contains the same fields with the same type of information in each field, although of course the specific information you enter into each separate field is different. For instance, you might have a "Last Name" field on every record, but the actual last name on each record is different.

This whole "card" is one record.

Each of these boxes is a field.

A database table consists of hundreds or thousands of individual records, each with different information in the fields.

First name
Last name
Address
City
State
Zip
Phone
Fax
Favorite book
Favorite play
Favorite quote

A database program that can only work with flat-file databases may itself be called a flat-file database, or sometimes a "file manager." This kind of application can record, organize and fetch information from only one file (table) at a time. This overall structure makes flat-file databases easy to work with. Also see *database.*

The other kind of database is the *relational database* in which information in different data tables is linked through established commonalities, such as a common field or column. When you input data into one table, it automatically (well, "automatically" after you program it to do so) gets recorded and acted upon in a related table. Relational databases are more expensive, much more complicated, and require much more time to learn to use. Of course, they can also do much more work.

floating palette

Paint and graphic applications have tools, patterns, and colors you can choose and use. Often these items are in a *pull-down* or a *pop-up menu.* To make them more convenient, many of the menus or tool boxes *tear off* and become **floating palettes.** They look like little *windows* (sometimes called *windoids*) and act like little windows—you can move them around with their *title bar* (which has no title in it and sometimes isn't even across the top, but down the side instead) and put them away with the tiny *close box.* You can move them around to get them out of the way, and yet they always remain available because they're always the front-most windows.

Floating palettes are very handy—for example, you can switch from using a paintbrush to using a pencil with just a click of the mouse. The palette also makes it easier to get those variable settings for the tool by double-clicking on the tool.

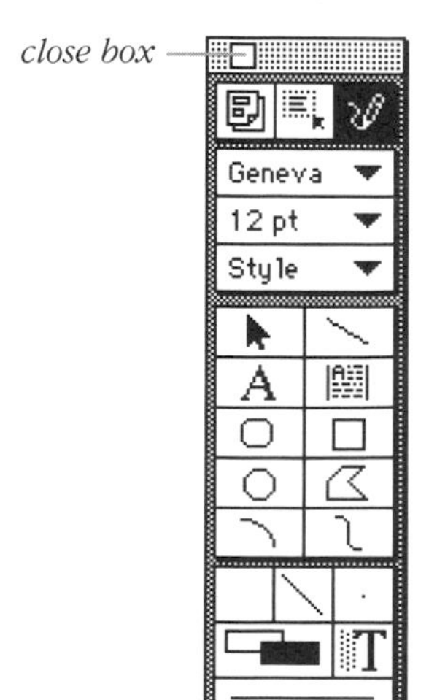

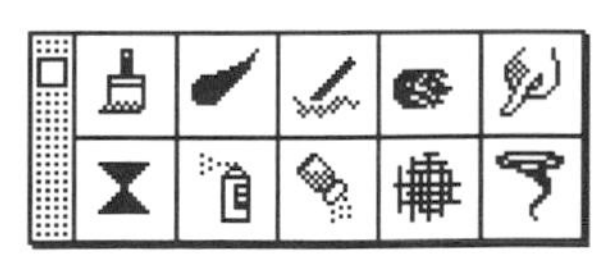

These are examples of floating palettes. Drag them around by the gray bar, and put them away with a click in the close box.

floating point, floating point coprocessor, processor

Floating point notation is another name for exponential notation (ha! remember that from high school?). Floating point notation is an efficient way to represent really large numbers (really large, like the kind with 10 or 15 digits) and very small numbers (like 10 or 15 digits to the right of the decimal point), without needing too many zeros. For instance, the real life number 10,953,000,000 is represented as 10953E6; the number .0000419 is represented as 419E-7. The point is, the decimal point can 'float' to wherever it's most useful, since the main part of the number is multiplied by the exponent.

Well, the reason you hear a lot about "floating point math" and "math chips" and "math coprocessors" is because many *microprocessors* (*chips* that run the computer) can't easily work with floating point numbers, and to do serious mathematical functions and high-end graphics (which use a great deal of math to represent the outlines and images), you need a computer with a special **floating point processor (FPU).** A floating point processor is also known as a "math coprocessor," a "numeric coprocessor," or a "floating point unit."

Adding a floating point processor (math chip) to your computer will speed up the math and graphics immensely, provided the application is written to take advantage of it. Some computers have a math chip already installed, or have it built into the main processor. Also see Appendix A, which has a little illustration.

floppy disk, diskette

A **floppy disk,** often called a **diskette** in the PC world, is a thin, round, flat piece of Mylar. It has an extremely thin coating of ferric oxide that is capable of storing magnetic fields. The *read-write head* in a *floppy disk drive* (the slot in the computer where you insert the floppy disk) stores data on the disk by altering the magnetic particles.

The large, 5.25-inch floppy disks that most PCs use are kept in a paper envelope. The smaller, 3.5-inch floppies that all Macintoshes and some PCs use is enclosed in a hard plastic case so you really don't even know it's floppy. It is. I understand that some people call the 3.5-inch floppies "microfloppy disks," although I have never heard anyone really say it.

Because floppy disks store information magnetically, any magnet can destroy the data (information) on the disk. This means you should never allow your floppy disks near a magnetic paper clip holder, the telephone, the stereo, a portable radio, or any other electronic device—and don't pin them to a filing cabinet with a magnet!

FLOPS

FLOPS stands for **fl**oating point **o**perations **p**er **s**econd. This is a measurement of *floating point* calculations, an indication of the number-crunching speed of the *CPU (central processing unit,* the *chip* that runs the computer) or the *FPU (floating point unit,* or *math coprocessor).* This speed also affects the graphics capabilities of the computer. "Flops" is usually preceded by mega- or terra-, as in 150 megaflops (which is 150 million floating point operations per second). See *megaflop.*

flush left, flush right

Flush left means the beginning of each line of type begins exactly on the left margin, while the right hand side is uneven (ragged). Flush left is also called "ragged right."

Flush right means each line begins exactly on the right margin, while the left hand side is uneven.

When both sides of the text align, it is *justified.*

folder

These are typical sorts of folders. You can name them anything you choose, of course. Use these folders just as you use folders in your office filing cabinet.

A **folder** on your Macintosh screen is similar to a folder in your filing cabinet—it's used for storing files. When you double-click on a folder, it opens to a window and you can see what is stored in the folder. You can create folders within folders within folders, ad infinitum (almost). See the illustration under *hierarchical filing system.*

To create a new folder, just select the window in which you want to store the folder, then press Command N. The untitled folder will appear in the *active* window, and you just type to name it.

To put an item inside the folder, just press on the item and drag it to the folder. When the tip of the pointer is on the folder, the folder will change color. Let go of the mouse button and the item will drop inside. **To take an item out,** open the folder by double-clicking on it and drag the item somewhere else.

To find a file if you can't remember which folder you stored it in, use the Find command from the File menu. **To put a file back into the folder it belongs,** click once on the file to select it, then press Command Y; the file will return to the folder you last stored it in.

fonts

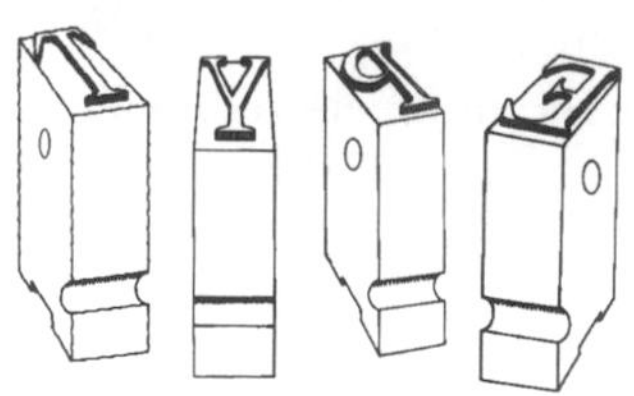

Example of metal type, rather large, as for a headline.

Computers have changed the definition of **font.** The terms "font" and "typeface" are often used interchangeably, but there is a subtle distinction that you'll need to understand when you're talking with a professional typographer or reading technical literature about type.

The term for the design of the characters, independent of their size, is **typeface.** Quoting Terry O'Donnell at Adobe Systems, "A typeface embodies the intellectual concept of the style, features, and parameters that make a particular design a unique entity. It is independent of whether the design is implemented in metal, wood, or digital form." Type designers design **typefaces,** not fonts.

A **font,** in traditional typesetting, was a specific collection of metal characters (letters, numerals, punctuation marks, and other symbols, and usually tiny, like those pictured here) that shared a common design theme, **all at a particular size.** That is, 9 point Times was one font; 10 point Times was another font; 10 point Times Italic was another font. With metal type, each font was kept in a separate case (see *uppercase* and *lowercase*).

Digitally, though, Terry says, "if the implementation is done as a *scalable outline* font in the digital medium, the font is scalable to any size." Type foundries digitize a typeface design and create a **font,** and a digitized font can be any size.

Font technology has become very complex over the last few years. What follows here is a necessarily brief overview of this technology. There is a lot of overlap between Mac and PC terminology, but there are some important differences also, so be sure you read the section for your computer. Macintosh users: It will help your comprehension if you understand the difference between *PostScript printers* and *non-PostScript (QuickDraw) printers,* so I suggest you read those definitions before you move on.

Until otherwise noted, the following information does not apply to TrueType or to systems with ATM installed—this first part is just the stripped-down layer. You should read it, though, and then the difference between TrueType and Type 1 will make more sense, and what ATM does for you will be clearer.

screen fonts, bitmapped fonts

A screen font is what you see on the screen. "Screen font" is synonymous with "bitmapped font"—the only reason you see it on the screen is because the shapes of the characters have been "mapped" to the grid of *pixels* (dots) on the screen. It has been mapped because a pixel has been turned on or off according to the *bit* of information that your keystroke sent to the screen.

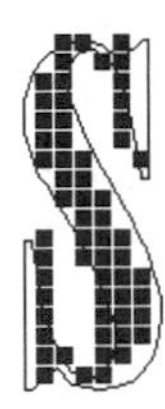

This shows the pixels, the bitmapping, for the screen version of the letter.

Other...
6
8
✓9
10
11
12
14
18
24
30

Bitmapped fonts come in "fixed" sizes. For instance, sizes 9, 10, 12, 14, 18, and 24 point screen fonts are commonly installed. You can tell what sizes are installed by looking at your font size menu—the numbers that are in the outline style are the installed sizes, as shown to the left. You can see that sizes 10 and 12 of the selected font are installed. (Word doesn't show this, though.)

Once you have installed those sizes, the computer knows how to display them on your screen. If you use a size that is **not** installed, the computer has to fake it and it usually looks pretty rough.

serendipity

The computer knows which pixels to turn on and off when you use an installed size. This is 12 point.

serendipity

But when you use a size that is not installed, the computer has to fake it. This is 18 point.

If you print bitmapped fonts to a **non-PostScript printer**, your type will look as good or as bad as you see it on the screen. That is, if you use a size that is not installed and it looks jagged and chunky on the screen, it will print jagged and chunky.

If you print to a **PostScript printer**, however, the type will print smooth and clean, no matter what it looked like on the screen. But—this is only true if the font does not have a city name.

serendipity *on the screen*

serendipity *printed to a PostScript printer*

city-named fonts

Almost every font with a city name (such as San Francisco, Geneva, Monaco, Chicago, Cairo) is a bitmapped font **and nothing more**. Fonts not named after a city (such as Bookman, Palatino) also have an *outline,* which means they can be *scaled* (resized) to print at any size with smooth edges. Fonts with a city name, though, have no corresponding outline, so it doesn't matter whether you have a PostScript printer or not—city-named fonts will never be more than jagged and clunky. (Unless they are *TrueType*—sigh . . . it's always somethin'.)

outline fonts, scalable fonts

In a *scalable font,* each character is stored as a mathematical *outline* of the character's shape. Because the outline is scaled to the correct size and then filled in with dots, a scalable font can print at any size with no loss of quality. Depending on which outline font technology you work with, it may also display smoothly on the screen. Scalable font technologies on the Mac are *PostScript* and *TrueType*.

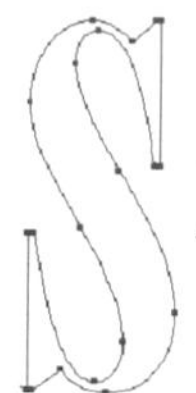

This is the outline for the letter S. The bitmap for this character is shown on the previous page.

This is the letter that will appear on the printed page.

PostScript fonts

Type 1 fonts: Each Type 1 font has two separate parts: a bitmapped, screen portion (that appears in your menu and on your screen, as mentioned previously) and an outline **printer font** (that the printer uses to create the font on the printed page). Both the screen font and the printer font must be installed.

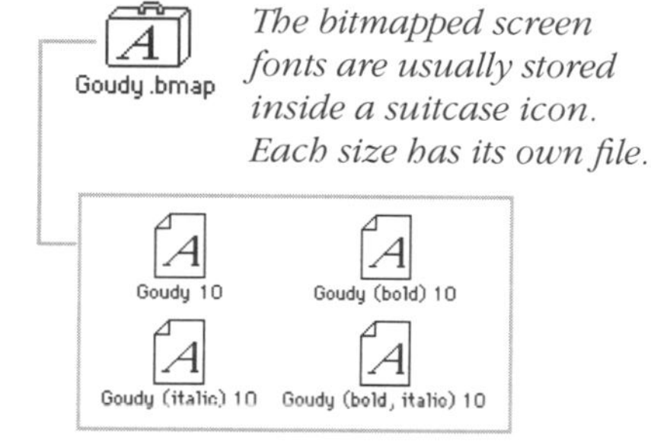

The bitmapped screen fonts are usually stored inside a suitcase icon. Each size has its own file.

For each member of the typeface family (but not for each size), there is an outline printer font.

The printer font contains the outlines of the characters. But printers cannot print lines or curves; they can only print dots (series of dots create the lines and curves) and, besides, the outline needs to be filled in with a solid color. A PostScript interpreter in the printer takes an outline and *rasterizes* it—that is, the interpreter tells the printer which dots it must print to create and fill in the outline. A Type 1 font can be printed in whatever *resolution (dots per inch)* the printer can produce.

Type 2 fonts: There is no such thing as a Type 2 font. It was a proposed font technology that was abandoned early in its life.

Type 3 fonts: Type 3 fonts can be more ornate than Type 1 fonts, with shaded insides and fancy shadows. Type 1 fonts use a *hinting* technology that originally was not public information, so no other vendor could create a real Type 1 font, although anyone could use the PostScript language to create Type 3 outlines. When Adobe made their proprietary hinting techniques public, all type vendors began converting their Type 3s to Type 1s. Very few vendors create Type 3 fonts anymore.

Type 3 fonts have several disadvantages: they take up more disk space than Type 1, do not print as well at smaller sizes, take longer to print, and *ATM* cannot *rasterize* their outlines.

resident fonts

Almost all PostScript laser printers have certain *printer fonts* built into their *ROMs (read only memory chips).* These printer fonts are usually Helvetica, New Century Schoolbook, Palatino, Times, Symbol, Zapf Chancery, Zapf Dingbats, Avant Garde, Bookman, and Courier. If you read the description of Type 1 fonts, you know that each Type 1 font has two parts: the screen, bitmapped information and the outline, printer information. When you buy a laser printer, you get a disk with the screen fonts to install in your computer; their corresponding printer fonts are already living (resident) in the printer. So these fonts, named above, are called **resident fonts.**

When you buy any other fonts, though, you will get both the screen and the printer fonts on the disk. See *downloadable fonts,* below.

downloadable fonts

First read the definition of *resident font,* above. You can apply the term *downloadable* to any font you buy (or acquire), because if you bought it, its printer font obviously does not live in your laser printer. And if the printer font doesn't live in the printer, then it must be downloaded to (loaded down into) the printer's *memory* so it can be used.

LaserWriter Font Utility

You can also use this utility to make your printer stop producing that page that prints every time you turn it on.

The downloading happens automatically as you print, although it is possible to use a font utility, such as Apple's LaserWriter Font Utility, to manually download printer fonts (it's on one of your system disks). The advantage of manually downloading is that your pages will print significantly faster. Turn off the printer for a minute before you download to clear the space in *RAM,* and only download the couple of fonts that are used the most in your document, unless you have more than two or three megs of RAM in your printer. Manual downloading is temporary—as soon as you turn off the printer, all downloaded fonts disappear from the printer's memory. If you have a hard disk for fonts attached to your printer, you never need to manually download.

If your printer is not PostScript, then even if you have a downloadable font, you cannot download it to your printer.

TrueType fonts:

TrueType is a *scalable* font technology developed by Apple. As with PostScript fonts, TrueType uses a mathematical outline to describe characters (but it's not the same math outline that PostScript uses). TrueType fonts appear smooth on the screen at any size, and they print smooth to any printer, PostScript or non-PostScript, even if they are city-named.

Chicago *on the screen (even though it is a city-named font!)*

Chicago *printed*

TrueType fonts do not have two separate parts, like PostScript Type 1 fonts. The screen information and the printer information is all rolled into one.

This is a TrueType font file, indicated by the 3 As.

Font size menu for a TrueType font.

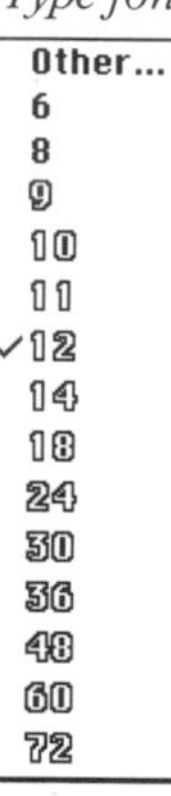

The TrueType technology is similar in results to the ATM technology (described below)—the outline is *rasterized* to the screen and is also rasterized straight to any printer, at whatever the resolution of the printer. You may find bitmapped versions of your TrueType fonts in the sizes 10 and 12; they are only to speed the drawing of the fonts on the screen at those common sizes. You can throw them away if you need the space and are willing to sacrifice a tiny bit of speed.

How do you know if your font is TrueType? Look at the icon, for one thing (see above). Or select some of the text on the page and check the font size menu—if all the sizes are in the outline style (which has nothing to do with outline technology), the selected font is TrueType (example to the left).

ATM (Adobe Type Manager)

Above, the letter s created from a bitmap; below, the same letter created from an outline.

Adobe Type Manager is a *utility* (little software program) that *rasterizes* the outlines of Type 1 PostScript fonts. What does that mean? Well, if you read the description of Type 1 fonts, you know that the PostScript printer interprets the *outlines* of the characters and turns the outlines into dots that the printer can print. Adobe Type Manager does the same thing—it gets the outline and rasterizes it into dots—but instead of doing it in the printer, ATM outputs the dots to the screen. So your type on the screen appears almost as smooth and clean as it does from the printer (except that the *resolution* of your screen is much lower than the resolution of your printer). ATM uses the same size pixels on the screen as the system does to create the letterforms, but because ATM uses information from the outline rather than the bitmap, it can do a better job (shown to the left). See the example on the next page.

This is a screen shot of the letter Q, size 72 point, in the font Shelley Volante, without using ATM.

This is a screen shot of the same letter, but while using ATM. Do you think that's worth the $7.95 cost of ATM? ooh la la!

If you have a non-PostScript printer, ATM will also rasterize the outline directly to the printer, and the type will print at the printer's full resolution of 360 or 400 *dpi (dots per inch),* or whatever the resolution is.

If you already have ATM, do you wonder why it doesn't work on fonts like Avant Garde, Bookman, Zapf Chancery, etc.? Remember that definition for *resident font?* Well, ATM rasterizes the *outline.* Resident fonts have no outline stored on your computer—the outlines are stored in the printer and ATM cannot reach them, so ATM can't do its magic. You can buy the resident printer fonts from Adobe so ATM will work on them, but personally, I would rather invest that money in new fonts altogether.

TrueType vs. Type 1

In the war between Type 1 and TrueType technology, it seems to have evolved into this:

> People who use type extensively and professionally, or who are involved in graphics, use PostScript printers, ATM, and Type 1 fonts. *Service bureaus* (for high-end output) prefer PostScript Type 1 technology.
>
> People who have PostScript printers, for any reason, can use either TrueType **or** Type 1 fonts, but it is generally preferable and more reliable to use PostScript font technology (Type 1) with the PostScript printer. And ATM, of course.
>
> People whose only involvement with type is to get words on the page, and who do **not** have PostScript printers, can be very happy with TrueType.
>
> People who do **not** have PostScript printers can use ATM with Type 1 fonts and get the same results as people using TrueType, and have a wider choice of fonts.

You can mix TrueType and PostScript fonts in one document. The big rule to remember is never install two fonts from the two different technologies with the same name; that is, don't keep both TrueType Times and Type 1 Times in your system.

fonts (PC)

PC fonts can be classified in various ways. For instance, they come in different physical forms. Some you buy as files on floppy disks or *CD-ROM*. **Cartridge fonts** are stored in circuits inside cartridges you plug into your printer. A printer's **internal fonts** or **resident fonts** are the ones that are built into the printer's circuits.

Another division is between **screen fonts,** which display text on the screen, and **printer fonts,** which work with a printer. With most **scalable fonts** and the right software, the same fonts will work for both the screen and the printer; otherwise, you need separate screen and printer fonts.

In a **bitmapped font,** the size of the font is fixed—the font looks bad, or doesn't print at all, at any other size. That means you need a separate font for each and every size you use. A bitmapped font represents each character as a collection of dots, the dots needed to create the character on the screen or the printed page (see the discussion of Macintosh *bitmapped* fonts for an illustration). Bitmapped screen fonts won't work on a printer, and bitmapped printer fonts won't work on the screen. Windows screen fonts are bitmapped. Other common bitmapped fonts are the ones that work with Hewlett-Packard LaserJet and DeskJet printers.

By contrast, a **scalable font** can print or display at any size with no loss of quality. Scalable fonts are much more convenient than bitmapped fonts, since you need only one font for all sizes you use. In a scalable font, each character is stored as the outline of the character's shape, described mathematically. When you decide to print a character at a given size, this master outline is scaled to the requested size. Only then is the outline filled in with the necessary dots so that it's ready to be printed or displayed (again, the Mac section illustrates the idea of outline fonts). Scalable fonts are also called **outline fonts.** The most widely used scalable fonts are Type 1, TrueType, and Intellifont.

The process of converting a scalable font into actual characters may take place in your printer, in your computer, or both. Let's talk about printing first. If you have a PostScript or PostScript-compatible printer, or a LaserJet III or LaserJet 4, the printer itself can do the job. The typeface outlines are stored permanently or temporarily inside the printer, which has circuitry that scales them to whatever size is needed.

But any laser or dot matrix printer will do if you have font scaling software that runs on your computer. In this case, the typeface outlines are stored on your computer's hard disk, and are scaled by the software to the requested size as needed. Once dot patterns representing the scaled characters have been prepared, the software sends them to the printer. Windows 3.1 has the ability to scale TrueType fonts so every program has access to scalable fonts. To scale Type 1 fonts with Windows, you need *ATM,* a font scaling utility program from Adobe Systems. Other font scaling utilities are available for such DOS programs as WordPerfect.

As for text on the screen, TrueType and ATM can generate screen characters at any size from the same fonts you use for printing. This alone is a great reason to use Windows 3.1 or ATM. Before they came along, even if you had a PostScript printer to scale your printer fonts, you needed separate bitmapped screen fonts. That is still true for PC programs that don't use Windows.

All scalable fonts work in roughly the same way, but there are different ways or formats for storing the information about the character outlines. The most popular scalable font formats are:

Type 1: This is the high quality format originally designed for PostScript printers but that now works with nearly any printer if you have ATM software. There are probably more Type 1 fonts available than any other kind.

Type 3: Type 3 fonts are another PostScript format. They look rougher than Type 1 fonts, especially when printed by a laser printer, and they only print on PostScript printers (not with ATM). The only reason you might be interested in Type 3 fonts is that the characters can be more ornate than with Type 1 fonts.

TrueType: TrueType fonts work automatically with Windows 3.1. Lots of inexpensive TrueType fonts are available.

Intellifont: Intellifont fonts only work with LaserJet III and LaserJet 4 printers. If you have Windows, you can get a special utility that works much like ATM, scaling these fonts for display on the screen (the printer still handles the scaling for printing).

font cartridge

For certain printer models, you can buy **font cartridges,** which are plug-in modules that contain *fonts* (typefaces) embedded in electronic circuits *(ROM)* within the cartridge. Once you've plugged the cartridge into your printer, you can use any of the fonts it contains just as if those fonts were built into the printer—you don't need to *download* the font first. Depending on the printer, you can buy both *bitmapped* and *scalable fonts* on cartridges.

Adobe makes a PostScript cartridge for non-PostScript Hewlett Packard printers that allows the HP printers to print PostScript type. A Postscript cartridge for a LaserJet is a cartridge, but it's not a font cartridge.

font conversion

For a variety of reasons, you may want or need to convert a *font* from one format into another format—for instance, from *Type 3* to *Type 1,* or from *PostScript* to *TrueType.* To do this you need special software to perform a **font conversion.** Software such as FontMonger, Metamorphosis, and Evolution perform font conversions of one sort or another.

Font/DA Mover

In System 6, the **Font/DA Mover** is a little Apple *utility* for moving *screen fonts* and *desk accessories (DAs)* into and out of *suitcases* and the System. You can open suitcases with the Font/DA Mover, then copy fonts into the System or remove them from the System, or create new suitcases for storing other fonts or desk accessories, or for transporting on floppy disks. In fact, in System 6 there is no other way to deal with fonts except through the Font/DA Mover. With the release of System 7, the Font/DA Mover has become a rather obsolete tool.

If you have the current version of the utility, you can still use the Font/DA Mover in System 7, although there is no need to. For some reason, everybody hated the Font/DA Mover. I'm not sure why. It made perfect sense to me and it did what it was supposed to do.

font ID conflict

A **font ID conflict** ("eye dee," identify conflict) occurs when your fonts get confused about who they are. This can happen easily when you create a document on your computer, then take it to someone else's computer. On your machine, the font was, say, Bembo. But when you take the document to your friend's machine, it gets confused and thinks Bembo is Franklin Gothic. It can even happen on your own computer—you might choose the font Janson and the font Syntax shows up.

Each font has a name and an identifying number. Some applications look for fonts by their names, some by their numbers. Font ID conflicts occur when two or more fonts have the same identification number or the same name. How can this happen, you say? Oh, it's easy. You might have bought several fonts from several vendors and their numbers are the same. And the identification numbers can also be changed by the font management software, whether it is an outside *utility* or the System itself.

If the identifying number or name on your machine is changed to prevent a conflict with another font, then when that document goes to another computer and the application calls for that particular font by name or number, the application finds the wrong one.

Solution? Don't change the names or numbers of your fonts without knowing clearly what you are doing. And use a font management utility, such as *Suitcase.* In Suitcase you can create a file that contains the ID numbers your computer is using. You can take this file to the other computer, temporarily *load* the file so the other computer uses your numbering scheme, then remove the file.

If the font you originally used isn't in the other computer, then the application can't find any font with a matching name or number and it "substitutes" a font it likes (usually Courier or perhaps Geneva). This is not really a font ID conflict, but just a matter of a missing font. If you install the font the document needs, the correct font will show up.

font metrics

The **font metrics** for a typeface are the detailed design specifications of the font, such as the width of characters, the thickness of the underline, how far below the *baseline* the underline will sit, how tall the capital letters are, the *kerning pairs,* etc.

On the Macintosh, these font metrics are contained within the *screen fonts,* the ones that are stored in *suitcase icons,* or in the TrueType font file itself.

On the PC, where the font metrics are stored depends on what type of font you're working with and which program you're using. Font metrics for Type 1 fonts used with Windows are stored in PFM files (PFM stands for PostScript font metrics); the metrics for a TrueType font are stored in the TrueType font file itself.

Type 1 fonts usually arrive on the disk with corresponding AFM (Adobe font metrics) files. Although the applications you use don't need the AFM file, you can actually read its information: open it in a word processor. The metric information looks like this:

```
StartFontMetrics 2.0
Comment Generated by Fontographer 3.5 3/12/93 1:04:49 AM
FontName Scarlett
FullName Scarlett
FamilyName Scarlett
Weight Medium
Type 1 by Fontographer 3.2.
ItalicAngle 0
IsFixedPitch false
UnderlinePosition -133
UnderlineThickness 20
Version 001.000
EncodingScheme AppleStandard
FontBBox -13 -217 1033 893
CapHeight 794
XHeight 394
Descender -212
Ascender 788
StartCharMetrics 77
C 19 ; WX 420 ; N DC3 ; B 0 205 341 550 ;
C 33 ; WX 170 ; N exclam ; B 7 12 169 645 ;
C 35 ; WX 574 ; N numbersign ; B 18 -34 548 472 ;
C 36 ; WX 547 ; N dollar ; B 23 9 494 579 ;
C 65 ; WX 917 ; N A ; B 28 13 898 861 ;
C 66 ; WX 824 ; N B ; B 54 6 809 853 ;
C 67 ; WX 824 ; N C ; B 1 -24 771 850 ;
C 68 ; WX 817 ; N D ; B 47 6 852 815 ;
C 69 ; WX 660 ; N E ; B 57 -56 643 780 ;
```

Do not install the AFM file on your hard disk! Keep it on its original disk on the off chance you ever in your life need it.

footer

Also known as the "running foot," a **footer** is the identifying text that appears at the bottom of document pages. Footers can be inserted in most word-processing programs. They may include the author's name or document title, and may contain automatically changing info, such as the page number. Most applications that can create footers can even include things like the date of printing, which is picked up automatically from the computer's internal clock. Usually you can choose the alignment, style, and other formatting for the text in the footer.

If the information appears at the **top** of every page, it is called a *header* or a "running head." Word processing programs can also insert headers automatically.

footprint

Footprint has two different meanings, depending on whether you are referring to software or hardware.

A hardware **footprint** refers to the amount of actual space a device will take up on a desk. A small footprint means the dimensions of width and depth are relatively small. Sometimes a small footprint indicates the lack of room for expansion, as in a restriction on the number of devices you can add on later. A "zero footprint" indicates the potential to stack the chosen hardware under the computer, on top of the computer, or perhaps hang it from the side.

A software **footprint** has an altogether different meaning. It is a trace that software leaves behind that lets other software know it's there, or that allows the software to spring into action when a particular key is pressed. A footprint is a kind of minimal presence of the software that enables something else to happen.

force quit

When your screen *freezes* or when you *crash,* you can sometimes **force quit** to avoid turning off the computer altogether. You will still lose everything you had not saved in that application, but you may be able to salvage documents from other applications that were open at the same time.

To force quit, hold down the Command and Option keys. While they are down, press the Escape key (esc) once. You should see an alert box that says something like this:

Click "Force Quit." Your current application should quit, and you should find yourself back at the Finder.

If the force quit **does** work and you end up back at the Finder (the Desktop), immediately go to your other open documents, save them all, and *restart* (at the Desktop, choose "Restart" from the Special menu). It's important to save and restart because you never know just what was lurking around that caused the freeze or the crash in the first place, and you can eliminate a lot of trouble by just starting over.

A force quit doesn't always work, though. If clicking the Force Quit button does nothing, or if you never even see the Force Quit button, then try the *reset switch* (see *programmer's switch* for details). If the reset switch doesn't work, then you have to turn off your computer, and yes, you lost everything on your screen that you had not previously *saved to disk.* That's why you must **Save Often, Sweetie (SOS).**

fork

All files on the Macintosh contain at least one of two **forks:** a **data fork** and/or a **resource fork.** The data fork usually contains things like the text, numbers, and graphics. The resource fork, on the other hand, contains "parts" of applications, such as the menus, icons, dialog boxes, and sounds.

Not all files contain both a resource fork and a data fork. For instance, a simple word processing document will usually not contain a resource fork, as the only item it really needs to display is text (the fonts to display the text are found elsewhere—not in the text file itself).

There are many utilities on the market today that let you edit either of these two forks. The most popular is Apple's own *ResEdit,* a powerful resource editor that allows you to change or alter things like windows, icons, and sounds. Data fork editing is much more complicated than resource editing, and is generally not used except for security or data recovery purposes. In either case, always do your editing on a **copy** of a file—never edit an original. Then, if your editing doesn't turn out as planned, you've still got the original to fall back on.

format (a disk)

Before a disk can be useful in a computer, the disk must be **formatted.** Formatting, also called *initializing,* organizes the storage area on the disk—it magnetically marks the disk with *tracks* and *sectors,* each with indicated boundaries, so that the information you store can later be located easily. The process involves erasing all that is on the disk, testing the disk to make sure all of its sectors are reliable, and creating a directory—an internal address system used for locating information later.

Do be aware that it is easy to **mistakenly** format a disk you didn't mean to, even a hard disk. I mean, you can **re**format a disk that already has valuable information on it. Reformatting makes any existing information invisible so your computer doesn't know it's there. Software such as Norton Utilities can often salvage the information on the disk if you accidentally reformat it—but not if you reformat the disk using a different type of formatting. For example, if you stick a *high-density* floppy disk in a low-density drive, the computer will insist the disk is unreadable, even if that high density disk contains all your inventory information for the past three years. If you go ahead and reformat the disk for low density, the information will be completely and permanently wiped out.

Most hard disks need special software for formatting. You can also buy both floppy disks and hard disks pre-formatted; they cost more, but it may be worth the savings in time.

On the Macintosh, floppy disks can be formatted by just sticking them in the floppy disk drive. If it's unformatted, the computer will ask if you really want to format the disk. If it's really empty, click yes.

On a PC, the computer knows when you put in an unformatted floppy disk, but it won't automatically format the disk. You have to run a DOS program called FORMAT, or use a disk management utility that can do the job.

format (text), formatting

When you **format** text, you are choosing the typeface, type size, alignment, tab settings, style (bold, italic, etc.), or any other options. Once you apply any option to text, you have applied some kind of **formatting.**

You may hear people (or yourself) complain that the formatting changed on the document. This can happen very easily when you open a document in an application other than the one in which it was created, or if you open a document on another computer that does not have the same *fonts* as you have on your computer. The length of the document may be longer or shorter, the lines end at different places, and you may even end up with the dreaded Courier font. There is software designed to take care of exactly this problem. On a simple level, *SuperATM* solves most lost formatting problems. On a larger, *cross-platform* scale, the new technology from Adobe Systems called *Acrobat* will solve formatting problems for documents that need to travel to and from different computer systems.

formfeed

The **formfeed** button (also known as FF) is a printer command that advances continuous paper to the beginning of a new page, provided you did everything else right. Formfeed will send one complete page size rolling through the printer—it does **not** roll the paper through to the next perforated section to ensure that when you print you start at the top of the page. No. It's **your** job to make sure the perforation in the continuous paper is properly aligned.

FORTH

FORTH, short for **fourth**-generation programming language, is a programming language developed by an astronomer named Charles Moore. Moore created FORTH in 1970 for the sole purpose of controlling equipment at the observatory where he was working. Now, applications written in FORTH are often used to control other hardware devices such as robots, arcade games, and even musical contrivances.

forum

An electronic **forum** is the equivalent of a traditional Roman forum—a place *online* to meet other people, to exchange information, to get on a soapbox and expound your philosophies, to discuss ideas and opinions. There are hundreds of separate forums for separate topics.

Many *online services* and *bulletin boards* have forums or something similar. Depending on your computer system and the software, it is usually just a matter of clicking buttons to get into the forum of your choice and participate.

four-color process

See *CMYK.* CMYK stands for the four "process" colors: **c**yan, **m**agenta, **y**ellow, and blac**k**. Also see *four-color separation,* below.

four-color separation

Four-color separation refers to both the process and the final result of separating the colors in a photograph or illustration into the four process colors that will be printed: cyan, magenta, yellow, and black. Special hardware and/or software (depending on how the procedure is accomplished) views the full color image and electronically decides how much of each of the four colors is in each part of the image. Then a separate page is created for each of the colors: the cyan (blue) tones, the magenta (kind of red) tones, the yellow tones, and the black tones. The four separate pages are not output in color—they are black dots on clear film. A printing press can only print solid ink, so the colors are simulated by dots—the combination of the size of the dot, the space around the dot, and how many dots in an area determine how intense each color is.

When the four separated pages (made into printing "plates") are printed in the four transparent process inks, the combination of inks creates the illusion of full color. You can easily see the four separate colors in any printed piece with a magnifying glass—check it out. Also see *CMYK* and *color separation.*

FPD

See *full-page display.*

FPU

FPU stands for **f**loating **p**oint **u**nit, a small *chip* built into some computers. Sometimes referred to as a *math coprocessor* or a *math chip,* an FPU can calculate complex mathematical problems and various graphic tasks much faster than the "general purpose" *CPU (central processing unit)* found in all computers. It's called a *floating point* unit because its speed really shows when calculating equations involving numbers having decimal portions, such as 67.9345 x 0.00345. See *floating point* for more details.

fractal

Fractal is a term coined by mathematician Benoit Mandelbrot to describe a category of shapes characterized by irregularity. Although fractals are, in fact, irregular and seemingly chaotic, they create a distinct pattern. If you magnify any portion of a fractal, you find that every portion has the same characteristics as the entire fractal. The concept behind and the complex mathematics involved with fractals can get very confusing and should probably be avoided by those of us concerned with our mental stability. But if your computer has an *FPU* (a *chip* that can handle all those complex mathematical calculations) there are several low-cost applications on the market that can create very beautiful fractals for printing or just viewing on your screen.

fragmentation

Fragmentation occurs when you save a file containing more data than can fit into one *contiguous* (connected) space on the disk. When this happens, the drive splits the data into many pieces small enough to fit wherever it can find the next empty space. If that first space isn't large enough, it puts down the next piece at the next empty space, and so on and so on. This is actually quite an efficient method of making use of the entire disk, as you can imagine how quickly a disk would fill if the computer needed a contiguous space for each file.

What causes fragmentation? Change. An example: In the very beginning, when you first saved or copied data to a newly formatted disk (an "empty" disk, for simplicity's sake) all the data you saved was laid down in a contiguous fashion—one file abutting the next file, that file abutting the next file, and so on. Let's say several days later you decided to add a few more chapters to that novel you've been working on. When you save these new pages, they simply are not going to fit in the same space that the novel file previously occupied. Therefore, the disk is going to have to fragment the newly expanded file by putting these new pages somewhere else on the disk. Fortunately, the computer keeps a record of just where everything resides on the disk, so there's no extra, inherent danger in losing your data due to fragmentation.

What does occur, though, is now the disk heads (the physical part of the drive that reads and writes the data—picture the arm on a record player) have to travel a greater distance to retrieve that data next time you need it. This extra travel causes two things to happen. First, it's going to take more time to *access* that information. Not a lot of time, mind you, but if your disk is severely fragmented, you may notice large files taking longer to open and/or save. The other thing that's going to happen is this extra movement is going to put extra wear and tear on the drive heads themselves. As with any mechanical device, the less unnecessary use the better. See *defragmentation* for tips on what you can do about fragmentation.

freeware

Freeware is software made available for public use by the author, and it's free. You're not under any obligation to pay for it. Freeware is usually distributed in the same places you find *shareware* and *public domain* software: on *bulletin boards,* at *user groups*, and by commercial shareware distributors.

The difference between freeware and public domain software (which is also free) is that the freeware author retains the copyright to the software. Because of this you cannot do anything with the software that the author doesn't allow you to. For example, you can't distribute it for profit, and you can't incorporate it into other programs, unless the author gives permission. Public domain software, by contrast, is given over in its entirety to the public domain with no strings attached. As for *shareware,* you should see that entry.

frequency

Frequency refers to how often something happens. For our purposes, frequency is measured in *hertz,* which is one occurrence or cycle per second. See *hertz.*

friction feed

Friction feed is a paper-feeding mechanism in the printers that print a line or a character at a time. This is the same sort of paper-feeding mechanism that is in typewriters, where you just stick the paper between the rollers and there is enough friction to pull the page through. In the case of a printer, the roller is turned by a motor, just as in an electric typewriter. Standard *dot matrix printers, inkjet printers,* and old-fashioned *daisy wheel* printers all have a friction-feed mechanism. A printer might also use the *tractor feed* mechanism.

front end, back end

You can divide the functioning of some computer programs into two main sections: the **front end,** the part that you, the user, actually interact with; and the **back end,** the part that does the "real" work of the program.

The distinction between front and back ends is most commonly made when talking about *distributed* or "client/server" applications running on a network. Many businesses are moving to systems in which each person has her own personal computer or workstation at her desk, with all of these little machines connected via a network to a central computer, which holds the data for the whole company.

In a distributed application, the front end part of the application runs on the user's own computer, giving you an easy way to request, say, a list of all the company's customers sorted by sales volume and state. The front-end software translates this request into a series of programming commands which it sends over the network to the back end part of the application running on the central computer. The back end performs the dirty work, processing the commands and finding the requested data. It returns that information to the front end, which then displays it on your screen.

Another similar sense of the term **front-end** is when one program uses another program as the way to view or use the information. For instance, an interactive training setup may use a program like *HyperCard* as a front end, as a way for trainees to use interactive training materials. It's the same as the above definition, except that the back end is right there in the training program, not in some other computer.

Frontier

Program your computer to do your work without you! **Frontier** is software from UserLand that allows you to write *scripts* (small programs in the UserTalk programming language) that tell your applications what to do. It's rather complex and not for the faint-of-heart, but Frontier scripts can reduce enormous amounts of repetitive work down to a few keystrokes.

frozen, frozen screen

Frozen means your computer is totally unresponsive to any action on your part. You can click or drag the mouse, tap on the keyboard, stomp your feet, scream and yell, and nothing happens on the screen. Your cursor might roam around the screen, but it doesn't activate anything.

A **frozen screen** is usually caused by some incompatibility problem (when two pieces of software or hardware can't live together comfortably), or there's a bug in the software you're using at the time of the freeze, or, commonly, from too much static electricity in the air.

If you have regular trouble with static electricity in your geographic area or at certain times of the year, like during thunder storms, you can invest in an *antistatic device*. And always avoid those activities that build up static, like shuffling your slippers across the carpet or rubbing balloons on your head.

full color

When you see a color photograph (or painting, illustration, etc.) that has been printed in a magazine or brochure and you think it has been printed in **full color,** you are actually being fooled. The photo has really been printed with only four colors. People who know what they are talking about call it a "four-color" print job, but people who don't understand think they heard this person say it was a "full color" print job. The four colors are a very special group, called "process" colors: cyan, magenta, yellow, and black. See *four color process*.

The four process colors cannot produce every color designers want, so there are five- six-color jobs, etc. Whenever you see a printed photograph or illustration with a rich, beautiful, deep red, you can be quite certain that was a five-color job, because the four process colors cannot create that rich red.

full duplex

Full duplex is a term used in networks and communications *(modems)*. It refers to transmitting and receiving data at the same time. Technically, this feat is achieved by either using two pairs of wires or by splitting the *bandwidth* of one pair into two *frequencies*. See the *duplex* definition under *modem*.

full-length board, full-length slot

A **full-length board** is an *add-in board* that's long enough to take up all the space available from the front to the back of the computer. The original IBM PC established a standard for the physical characteristics (as well as the electronic characteristics) of PC add-in boards. If a board fills up the whole space alloted, so that its front slides into the plastic guide which holds it there, it's a full-length board. If it's shorter than that, it's a half-length or three-quarter length board, or whatever.

A **full-length slot** is an *expansion slot* (an electronic connector) which is set into the computer so that full-length boards will fit, as opposed to a half-length slot. Actually, the slots themselves are the same length—it's the room in front of them that is different.

Obviously, if you have a half-length slot in your computer, you can't put a full-length board in it. The IBM PC/XT, for instance, had one short slot, which you could only fill with half-length boards. See also *8-bit slot* and *16-bit slot*.

full-page display

A **full-page display** is a monitor that is physically large enough to display a full 8.5 x 11–inch area so you can view your document at actual-size.

Keep in mind, though, that standard monitor sizes refer to the screen size measured **diagonally**. Although a standard 13-inch monitor (640 pixels by 480 pixels) sounds like it should display a full page, it will really only show an image slightly under 9 x 7 inches. If you need a true full-page display, make sure you're getting a monitor with an active screen size of at least 640 by 870 pixels (or roughly 9 x 12 inches).

function keys

The **function keys** on your keyboard are the ones labeled F1, F2, and so on. Depending on the keyboard and the computer, you may have 10 function keys, 12, 15, or even more. Many Macintosh keyboards have no function keys at all (if your Mac has function keys, you have an *extended keyboard*). If you have them, the function keys are arranged either in a long row at the top of the keyboard, in two rows at the left side, or in a cluster on the right side.

Many software programs have special uses for these keys. If a particular function key isn't otherwise in use, you can use a macro utility (such as SmartKey or QuicKeys) to program it to perform some common task. In a word processor, for instance, you could program one of your function keys to fix all those typos where you transpose two lettesr.

On the Mac, pressing F1 will usually *undo;* F2 will *cut;* F3 will *copy;* and F4 will *paste.*

On a PC, pressing F1 will usually display an online *HELP* message on the screen.

fuzzy, fuzzy search

Fuzzy is another way of saying inexact. One common use of this word is in the term **fuzzy search.** This is a feature in some software programs that allows you to search for text that is similar to, but not necessarily exactly the same as what you tell it to look for. For example, you might type in something like "phonics," and the fuzzy search might find "phonics" or "telephone" or "Phoenicia," or even "corn pone." You can even type in phrases and the fuzzy search will come up with similar sections of text. This is great for finding all the "Boadecea's" in your story, when you can't remember how many different ways you spelled her name.

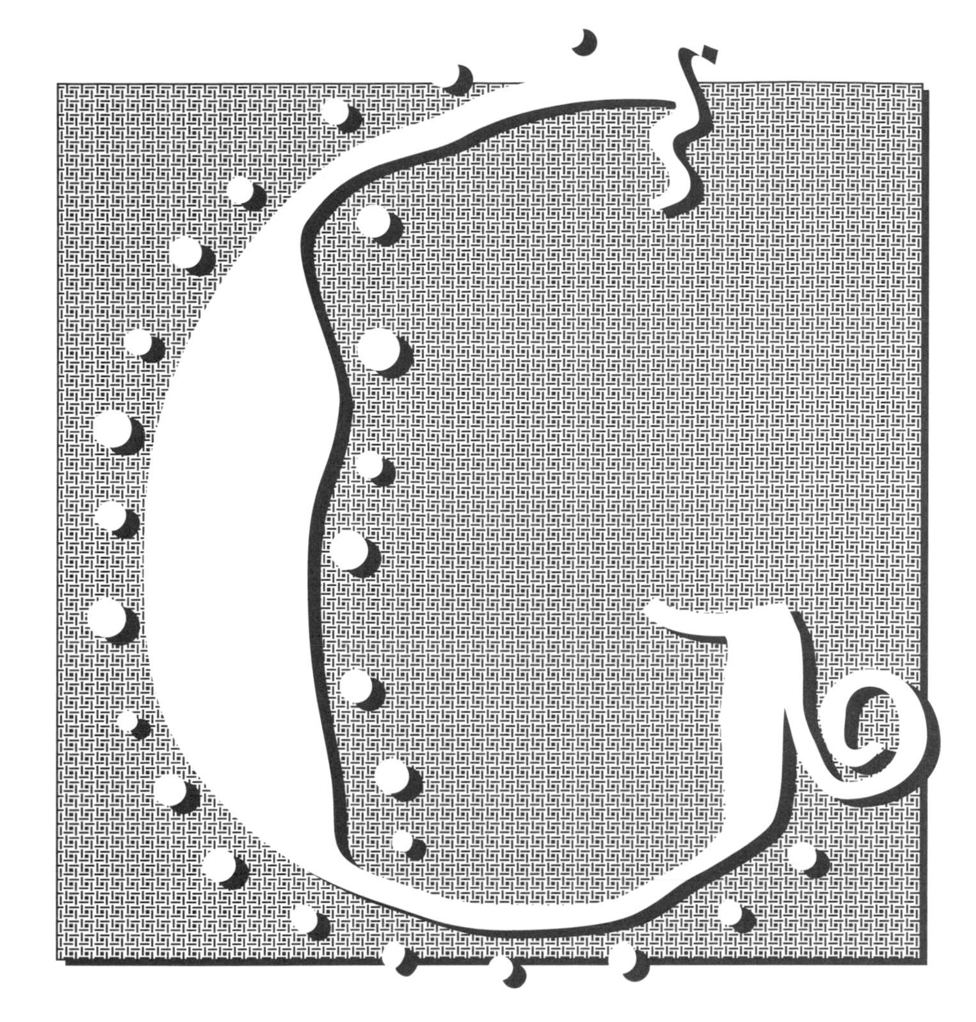

G, GB

G and **GB** are abbreviations for *gigabyte;* please see that definition.

garbage

Garbage is junk you don't need, like old software that's of no use any more, or an obsolete computer. More specifically, the term refers to information that is unintelligible (at least to you) in a file or memory. For example, if you open a file and you can't make any sense out of it, then you would refer to it as garbage. Garbage on the screen is a hodgepodge of characters that you can't read. Frequently you can also find garbage in *memory*. Be careful, because some things that are look like garbage to you may actually be very important to someone else or to certain software.

geek

The New Hacker's Dictionary says a **computer geek** is "one who eats (computer) bugs for a living. One who fulfills all the dreariest negative stereotypes about *hackers:* an asocial, malodorous, pasty-faced monomaniac with all the personality of a cheese grater." It's one of those personality labels, though, that can't be used by outsiders without implying an insult to all hackers. A geek may be the larval stage of a hacker, or simply a fundamentally clueless individual.

GEnie

GEnie stands for **G**eneral **E**lectric **n**etwork for **i**nformation **e**xchange, which is a large *on-line* information *service.* With a *modem* you can call the service GEnie and become informed. In fact, GEnie has a well-stocked library of downloadable public domain and shareware programs, and their non-prime time (6 P.M.–8 A.M.) rates are cheaper than *CompuServe's.* Unfortunately it is not as easy to use as CompuServe; GEnie has yet to put together a good software package (MacGenie is in the works), relying instead on *command-line interface.*

Get Info

On the Macintosh, there's a **Get Info** window for every file on the disk: when you are at the *Desktop,* click once on any file, then press Command I (or you can choose "Get Info" from the File menu). This window supplies information such as when the file was created or last modified, what kind of file it is, where it's stored, etc. There is a little box in which you can type your own personal notes or quips about the file. (Note: anything you type in the Get Info box will disappear when you *rebuild the Desktop.*)

There are several other important or playful uses for the Get Info window. From the Get Info window you can:

- Increase the *memory* allotted to a particular application (see *application heap*).
- *Lock* a file (see *lock*).
- Find the original of an *alias* (see *alias*).
- Customize the *icon* of any file (see *icon*).
- Disable the warning on the *trash can* (see *trash can*).
- Create a stationery pad (a *template*) out of any document (see *template*).

GIF, giff

The acronym **GIF** (pronounced "jiff") stands for **g**raphics **i**nterchange **f**ormat. This is the *file format,* developed by CompuServe and H&R Block, that is used to *compress* and store graphics that get *uploaded* (sent) onto *online services*. When you go online, you can *download* (get) the GIF file.

The GIF format is not tied to any particular computer or operating system or screen resolution, so it provides a useful way to exchange files between different systems. One of the nice things (actually, the critical thing) about the GIF format is that it is compressed: it takes up considerably less space in memory and on disk. This is convenient because when you upload or download GIF files to another computer via a modem you spend less time on the telephone, and in turn you save money.

The only problem with GIF is that many graphic software programs don't recognize it. This means that after you get your GIF graphic image you may have to convert it into a form that your own software can understand and use. Or you can also download "Giffer," a *utility* that allows you to view GIF files, and save them into other formats for placing into documents created in other applications.

gigabyte

A **gigabyte** (G or GB) is a unit of measure for such things as file size or hard disk space. It's very large.

Technically, one gigabyte is 1024 megabytes, which is the same as one billion bytes, which is really 1,073,741,824 bytes. Impress your friends with this useful knowledge.

GIGO

GIGO stands for **g**arbage **i**n, **g**arbage **o**ut, meaning that if you put meaningless or incorrect information into a computer, then your results will also be meaningless or incorrect. You can also apply this to life.

glare, glare screen

A **glare screen** is a clear, mesh panel that slides or clips on in front of the glass on your monitor to keep light from reflecting back in your eyes. These are said to reduce eyestrain that can sometimes result from long hours of keyboarding. Some of the newer ones also claim to reduce harmful emissions (see *radiation*) that may or may not be coming from your monitor.

glitch

Glitch has become a popular word to explain any kind of malfunction. A "power glitch" can crash a computer; a "software glitch" can drive you crazy, and a "hardware glitch" can be temporary and/or random. Hardware glitches can be the result of power glitches, software glitches can be the result of hardware glitches—and *Murphy's laws* prevail. Some glitches are the result of supposedly simultaneous signals arriving at their destinations at different times—a computer slip-up.

Generically, glitch can refer to any problem of any sort.

global, global search

Global describes anything that applies to an entire file or document or whatever it is you are working on. For instance, if you make a global change to the margins of a word processing document, then the margins change in the same way throughout the entire document.

A global *search-and-replace* means that, with one command, you search for and alter a particular sequence of characters everywhere it appears throughout a document. This is extremely convenient and takes a lot less time than going through the document and finding each instance of the sequence and changing it manually. Say you have the phrase "fuzzy red slippers" sprinkled throughout a document; with a global search-and-replace you can change the phrase to "shiny blue shoes" wherever it appears.

"Local" is the opposite of global. When you make a local change it applies only to the particular portion of the file or document that you have selected.

glossary

A **glossary** is a wonderful time-saving feature in some word-processing programs. A glossary is a place where you store entries for frequently used text, and then you can use a keyboard shortcut to enter that text into your document. For instance, you can have your name and address listed in the glossary. Whenever you need to type your name and address in a document, you type the keyboard shortcut that refers to that information, and your name and address appears on your page. In some applications, a glossary is called a "library" (although in some applications "library" may refer to a collection of graphic images).

gooey

It sounds like **gooey,** but it's spelled *GUI.* Please see the definition for *graphical user interface.*

GPPM

GPPM stands for **g**raphic **p**ages **p**er **m**inute. It refers to the non-text output speed of a laser printer. Most manufacturers brag about the number of "page per minute" that a printer can output, but they are talking about pages with simple text. If you do intensive graphic work, you'll be more interested in the number of "graphic pages per minute" the printer can output. It takes quite a bit more time to output graphic images than it does to output plain text. The only problem is, the number of graphic pages per minute that a printer prints will vary wildly with the type of graphics being printed, so a GGPM rating may not be applicable to your own printing projects.

gradient fill, graduated fill

A **graduated** or **gradient fill** is when the color that fills or shades an object makes a gradual shift from one color to another, or from a dark tone to a light one. The ideal result in a graduated fill is to avoid *banding,* where you can see bands of the varying shades as they shift.

grammar check, grammar checker

A **grammar checker** evaluates your writing to see how grammatically correct it is. A grammar checker tries to find mistakes in syntax, tense agreement, word use, and so on. It may suggest stylistic improvements, such as avoiding passive constructions, non-standard expressions, or overused phrases. It can make a qualifying statement about the document's readability for a particular audience.

A grammar checker may be a separate software program or it may be a feature within a larger application.

Grammar checkers have come a long way since they were first introduced, but the final word is still the same: Don't fire your human copy editor! Grammar checkers are not intelligent creatures. They can, however, make the work of a copy editor easier. **Grammar check** the document first, then go through the document yourself, or have someone else do it who knows more about all those rules of grammar than you do.

graphical user interface

A **graphical user interface** is fondly called "GUI," pronounced "gooey." The word "graphical" means pictures; "user" means the person who uses it; "interface" means what you see on the screen and how you work with it. So a graphical user interface, then, means that you (the user) get to work with little pictures on the screen to boss the computer around, rather than type in lines of codes and commands.

A graphical user interface uses *menus* and *icons* (pictorial representations) to choose commands, start applications, make changes to documents, store files, delete files, etc. You can use the mouse to control a cursor or pointer on the screen to do these things, or you can alternatively use the keyboard to do most actions. A graphical user interface is considered *user-friendly*.

The Macintosh is a purely GUI computer. PCs can use programs such as Microsoft *Windows* or the OS/2 *Presentation Manager* to create a GUI *shell* to work with.

graphics-based, graphics mode

If software is **graphics-based,** then your computer must be in **graphics mode** to display everything on the screen, even text. Technically, graphics mode means that each *pixel* (dot) on the screen is independently controlled. This is in contrast to a "character-based" or "text-based" mode, where entire characters have fixed shapes and are displayed in fixed rows and columns. The Macintosh has always been only graphics-based. PCs and some Unix computers can run either text-based or graphics-based software.

The advantage of graphics-based software is that theoretically it can show anything you want to see on the screen, within the limits of the screen's *resolution* and color capabilities. How the text looks on the screen is the way it will appear in print; you can see pretty pictures, draw diagonal lines that connect things, and so on.

The big drawback of graphics mode is that it requires a fast computer. Say you want to display a letter like A on the screen. In graphics mode, the computer would have to store and manipulate a separate piece of information for each of the individual dots that make up the character on the screen. When you multiply that by a whole screen's worth of text or information—especially if you have actual illustrations that have complicated lines or colors—it slows down the display of anything on the screen. Text mode is inherently much faster, because a single small unit of computer information will display an entire character.

But with the speed of computers being produced now, graphics mode is clearly becoming predominant over text mode. And once you have worked with graphics mode, it is difficult to return to using a text-based display.

Graphics Interchange Format

See *GIF*.

graphics tablet

A **graphics tablet** is a flat piece of plastic or similar material that is electronically sensitive to touch. By pressing on the tablet with your finger or with the *stylus* (a pen without ink) that is usually supplied with it, you can control the cursor on the screen. Graphics tablets are most often used in conjuction with graphics software, such as illustration and paint programs, though they can substitute for a mouse with most any program.

Some graphics tablets are sensitive to pressure, as well as touch. Combined with certain graphics software, the lines you draw are thinner or thicker depending on how hard you press.

grayed, grayed-out

A **gray** or **grayed-out** item in a menu or dialog box indicates that the particular option or command isn't currently available or appropriate to use. For instance, the commands *Cut* and *Copy* will only be available (they will be black, not gray) if you first select the text or the object that you want to cut or copy. These are described as "grayed-out" because they appear in gray rather than black text.

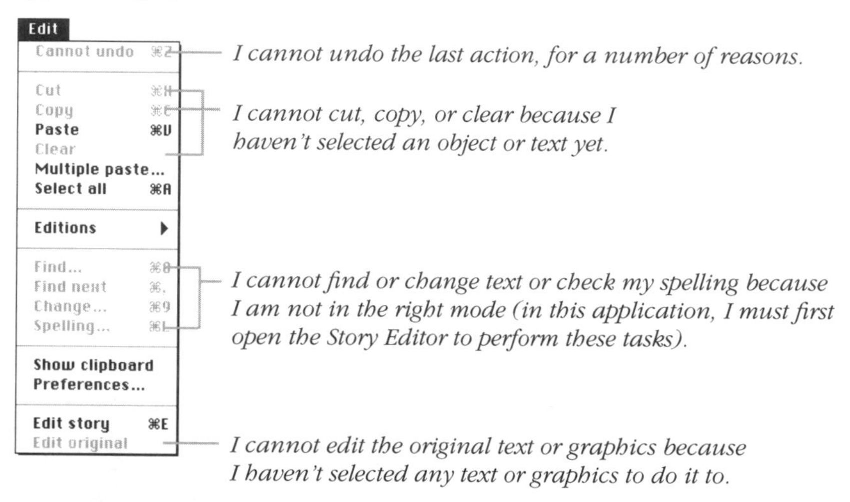

Gray font names in the menu imply that the open document contains fonts not presently available in the computer's system.

Gray icons on the *Desktop* indicate that the file is open. Even though it's gray, you can double-click the icon to bring its window or document or application forward.

A **gray floppy disk icon** means one of two things.

If the icon is gray and there is no detail, its window is open on the screen somewhere. If you can't see its window because of the other windows on the screen, double-click the gray icon and the disk window will come forward.

If the icon is gray and there is some detail showing, the disk has been ejected. If you double-click on the icon, the computer will ask you to insert the disk. To avoid leaving the gray icon on the screen, don't press Command E or use the menu to eject the disk—instead, press Command Y or drag the disk icon down to the trash can (don't worry—it won't erase anything on the disk).

grayscale

Some computer screens are **grayscale,** rather than plain ol' black-and-white *(monochrome).* On a black-and-white screen, there is only one *bit* of information being sent to each pixel (dot), so the pixels on the screen are either on (white) or off (black). On a grayscale monitor, anywhere from 2 to 16 bits of information are sent to each pixel, so it is possible to display gray tones in the pixels, rather than just black or white.

The gray tones are the result of some of the bits being on and some being off. If the monitor uses 4 bits, there are 16 possible combinations of on and off, so there are 16 possible shades of gray.

A grayscale is also one variety of *TIFF* (tagged image file format). When you scan an image as a grayscale, each dot on the screen can register a different gray value. A grayscale tries to approximate the continuous gray tone of photographs.

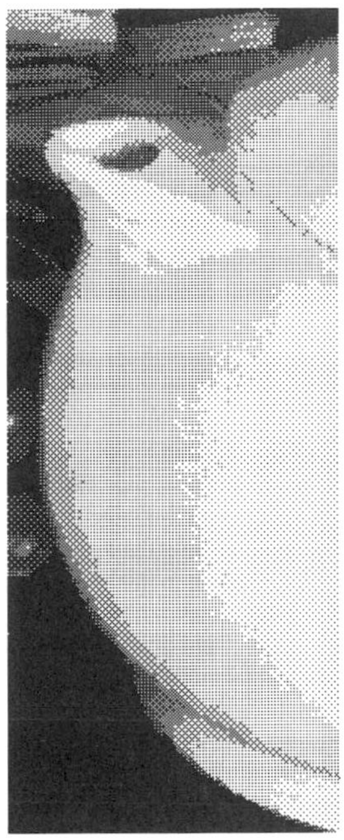

This is a black and white image; there is only one bit of information being sent to each pixel. That one bit tells the pixel to turn on or off (white or black).

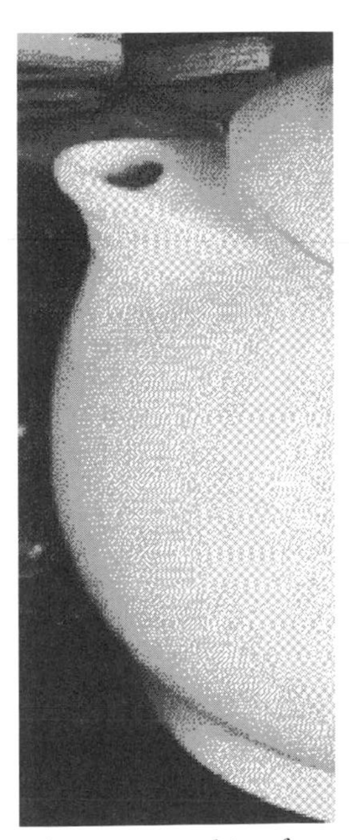

There are two bits of information going to each pixel in this image, creating four possible shades of gray.

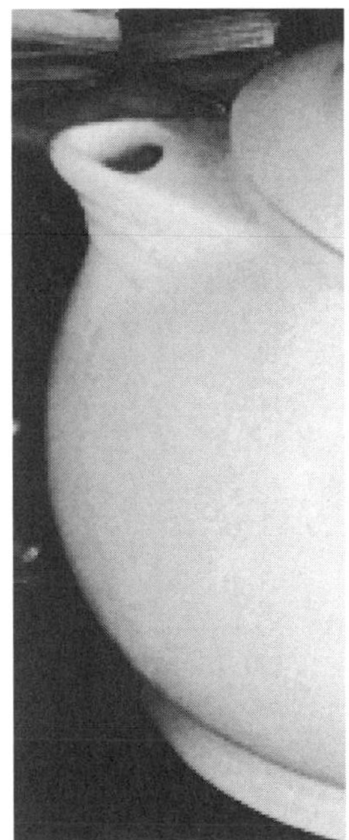

This full grayscale image is created with all 256 levels of gray, from 16 bits of information per pixel.

greek, greeked, greeking

Greek text is nonsense text used to indicate real text. This term has two variations of meaning.

The more traditional use of the term is when you are just filling a document with fake text to see how the layout will eventually look. You can use any old text file and then later replace it with the real words. The text isn't really Greek, because the Greek letterforms don't give us the same visual interpretation as our Roman letterforms. It often appeared to be Latin (but wasn't really). "Lorem ipsum dolor sit amet" is an example of greek text. An advantage to using greek in a *comp* is that your client doesn't get wrapped up in critiquing the copywriting but can focus on the design instead. We still use greek text in this way on the computer.

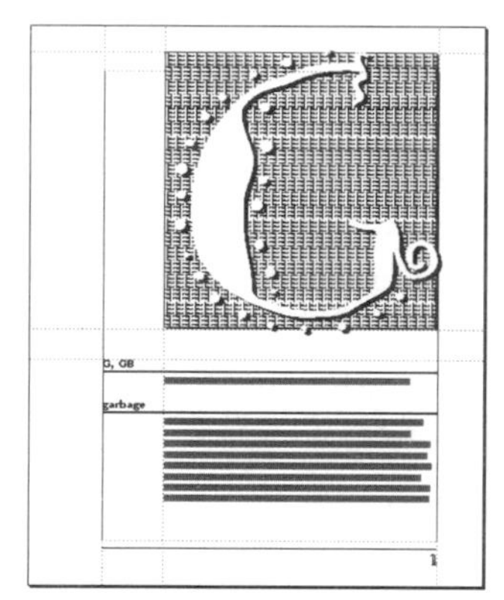

Another very similar meaning of the term pops up in software applications that can enlarge and reduce the view of the page. When the text in the reduced view is too small to display on the screen (as shown to the left), you may see squiggles or gray bars that represent the words. If you enlarge the view of the text, the real words will appear. This form of greeking is useful because it speeds up the display of text on your screen—it takes much less time for the computer to display little squiggles than it does to display the shape of each individual character.

Some applications greek automatically when the text gets so small that it would be hard to read anyway. Other applications let you pick the size of text at which greeking takes over.

GUI

Gui (pronounced "gooey") stands for *graphical user interface.*

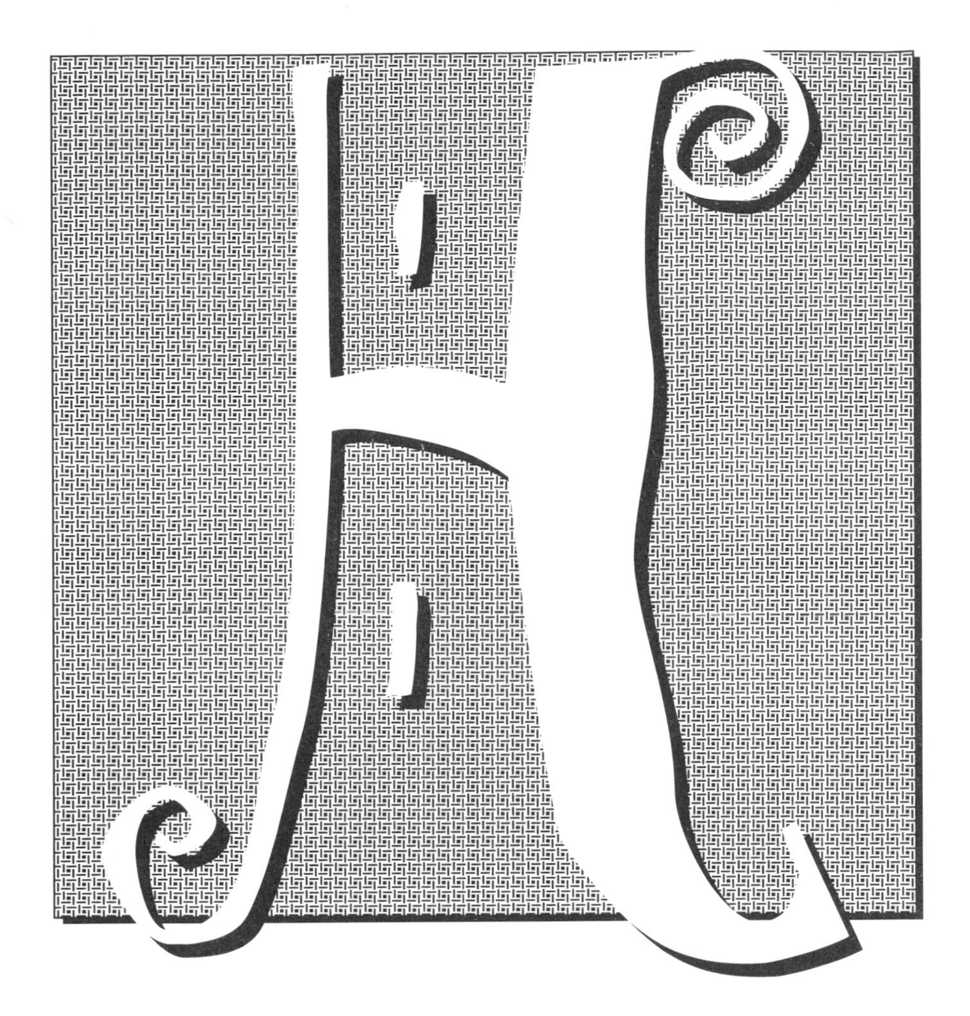

H&J

H&J stands for **h**yphenation and **j**ustification. Typographers (people concerned with the finer aspects of type) are always comparing the H&J controls in the various page layout applications, because how often and where words hyphenate is a sign of good or bad typography. The hyphenation used is directly related to the justification of the type, and the two controls interact with each other.

hacker

A **hacker** is a computer enthusiast, one who is willing to hack away tediously at understanding and programming computers. In the news media, hacker was mistakenly given derogatory overtones, implying that hackers were people who corrupt and/or steal data from unsuspecting victims by gaining unauthorized access to their systems—a more correct term for this kind of person is "cracker." A hacker is technically sophisticated, dedicated to, and perhaps obsessive about computers. Compare with *power user.*

half duplex

Half duplex means data can be relayed in only one direction at a time. Two-way transmission is possible, but the transmissions must be alternate. A walkie-talkie is half duplex—when one person is speaking, she cannot also listen (remember that little button you press to talk?). A telephone is *full duplex*—information can go both ways simultaneously; both ends can talk and hear all the time—so does a modem.

Your communication software may have a half duplex setting, but this doesn't actually cause your modem to alternate between sending and receiving. Instead, it just sends the characters you type directly to your screen and to the other modem, too. In this sense, half duplex is the same as "local echo" or "local echo on." If you can't see what you type when you're telecommunicating (because the other modem isn't sending the characters back as you type them), switch to half duplex, or turn local echo on.

If you see two copies of every character you type, switch to *full duplex,* or turn local echo off. In full duplex mode, your modem sends the characters you type only to the other modem, not to your screen. But you will still see the characters because the other computer is sending them back to you.

handle

Objects in a document, whether text or graphics, can usually be selected by positioning the pointer on the item and clicking the mouse. Graphic objects, such as a drawn circle or an imported image, display an outline around the object with two, four, or eight small square boxes attached, called **handles.** Handles only show on-screen and never print.

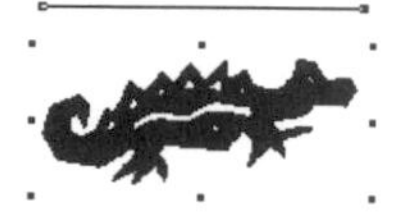

A selected line will have two handles, one at each end (as shown to the left). A two-dimensional object will have up to eight handles, one on each corner and one in the middle of each side (as shown to the left). In certain software applications, a block of text can be selected as an object.

Handles allow you to resize the object. Drag (press-and-drag) on a corner handle to resize in both dimensions; drag on a side handle to resize in only one dimension. Experiment with holding down a *modifier key;* for instance, what happens when you resize while the Shift key or the Command or Control key is down?

handshake

A **handshake** is an electronic exchange between two devices; the handshake establishes that they are communicating with each other. It's a formalized greeting-and-response routine that the two devices use when they first make contact to determine if they are talking to each other in the proper way; additional communication can occur after that point. For example, when you use your *modem* to dial another modem, the two modems begin their electronic communication with a handshake.

A **hardware handshake** is the actual signal over wires, while a **software handshake** is the information transmitted that establishes the communication agreement. Handshaking is also a method for controlling the flow of data—two devices use a handshaking signal to stay in sync with one another. For example, a computer sending information to a printer watches the printer's handshaking signal in order to know when to hold back data and let the printer catch up.

H and J

See *H&J,* the first entry in this section.

hanging indent

An *indent* is that space before the first word in the first line of a paragraph (as in this paragraph) that makes the first sentence begin to the **right** of the rest of the paragraph. If a paragraph has that space, that paragraph is indented.

A **hanging indent** is when the first line of a paragraph hangs out to the **left** of the rest of the paragraph, as you see here. Both a hanging indent and a regular indent can be created automatically in almost every word processing or page layout application.

Happy Mac icon

The little icon you see to the left of all these Macintosh definitions is the **Happy Mac icon.** When you turn on your computer, you should see the Happy Mac. If you don't see the Happy Mac, you got trouble—you might see the *Sad Mac,* the flashing *X,* the flashing question mark, or, worst of all, you might see the Dead Mac and hear the *Chimes of Doom.*

I have read that you must never turn off the computer while the Happy Mac icon is on the screen. I'm not sure why, but I'm not gonna experiment.

hard copy

Hard copy is the printed version of the document you create on the computer. It may be printed onto paper, onto film, or onto any other "permanent" thing. While the document is in your computer or on a *disk* of some sort, it is a *file,* sometimes called "soft copy." When you print the file, you get hard copy.

hard disk

You can think of your **hard disk** as a filing cabinet for the information you store in your computer. On the hard disk (in the "filing cabinet") you'll keep copies of all the software applications you use, such as your word processor and your spreadsheet program, plus copies of all the documents or files you create.

Some computers don't absolutely need a hard disk for storing things—they use floppy disks instead. But a hard disk drive is much, much faster than a floppy disk drive, and it stores much more information. And as software programs and *operating systems* get more sophisticated, they also take up more disk space—so much disk space that many programs and systems can't even fit on a floppy disk. These days, almost every computer has a hard disk.

The hard disk itself is a flat, round platter; actually, most hard disks have from two to eight platters. Each platter in the hard disk has a special coating of magnetic particles. Each side of each platter has a *read/write head* that floats over the surface of the rapidly spinning disk and picks up (reads) magnetically stored data, or records (writes) data onto the disk. It's just like a tape recorder, except that in a hard disk, the head doesn't touch the disk surface. Instead, the read/write heads float on an extraordinarily thin cushion of air, about the height of a fingerprint. Really. That's why it's important not to jostle your hard disk or kick it while it's doing its work.

Hard disk sizes are measured by how much information they can store; for instance, my Macintosh has an 80 *megabyte* hard disk internally, and I have a *removable cartridge drive* that holds cartridge hard disks of 44 megabytes each. By comparison, a 3.5-inch floppy disk holds around 1 megabyte of information.

Hard disks are also rated for speed, for how quickly they can find the information you need. See *access time.*

It is a fact of life that hard disks die. They all eventually die. Crash. Burn. Bite the dust. A typical hard disk has a life span of about three years. **So you must be responsible** and aware of this and create your life-saving (and job-saving) ***backup!*** Of course you won't get serious about backing up until you go through your own catastrophe of considerable dimension—but I warned you.

hard Return

In almost every application where you type words, the words will get to the right margin, bump into it, and automatically bounce down to the next line. You do not have to hit the Return key as you did on a typewriter! When you do press the Return key (Enter on a PC), you create a **hard Return.** When you create this hard Return, the line will always end at that point. This might be what you want, as in a return address—of course you want a hard Return at the end of each line. But if you accidentally create a hard Return in a paragraph of text and then *edit* the text, like add or remove a few words, the line will still end at that point where you hit the hard Return.

So the point is, let the text *word wrap* (bump into the margin and automatically bounce to the next line) unless you really do want the line to end at that point.

hard space

When you hit the Spacebar between words, the computer knows it can break the line there—it knows that one word ends and another is about to begin. A **hard space** looks just like a regular space, except that it **connects** two words rather than separates them. A hard space is seen by the computer as another character and so it thinks the two words are actually just one word. This prevents words from separating that should stay together, such as Mr. and DeVere. Instead of typing a regular space between the two words, type a hard space and those two words will never separate. A hard space is sometimes called a "required space" or a "non-breaking space."

On the Mac, you can almost always type a hard space by typing Option Spacebar instead of just Spacebar.

hardware

Hardware refers to those parts of the computer that you can bump into, such as the printers, drives, modems, etc. (Clay says hardware are the parts you can kick.) The *software* is the invisible stuff that is stored on the disks in *digital bits* and *bytes*. (Clay says software is the stuff you can swear at.)

Hayes compatible

The term **Hayes compatible** refers to a *modem's* ability to understand the Hayes AT Command Set. Hayes created this set back in the early seventies and today it is the *de facto* standard by which all other modems are designed. Although it is rare to find a non-Hayes compatible modem today, Hayes compatibility is the first thing you should check for when purchasing a modem.

HDTV

HDTV, which stands for **h**igh-**d**efinition **t**ele**v**ision, has twice the *resolution* of normal television, giving you a very clear, vibrant picture. Europe, Japan, and the United States are all competing to bring forth the best technology in this field. There are no standards set yet.

head crash

The slightest bit of dust, dirt, even smoke, can cause a crash between the surface of a *hard disk* and the *read/write head*—called a **head crash.** After all, the read/write head hovers just a breath above the spinning hard disk, and although the head touches the surface of floppy disks, its contact with a hard disk is a big no-no. It can mean the loss of data, and if the directory gets damaged, the computer won't be able to read the file and you will need to replace the hard drive and/or the read/write head. Contact between the head and disk damages the magnetic coating of the disk. If any particles are set loose and encountered by the head later, even more damage occurs. (The lesson to learn is to use a dust cover when the computer is not in use and to never, ever move or bump a hard disk while it is running.)

head, read-write

A **read/write head** retrieves (reads) and stores (writes) information on a disk. It's a magnetic device that moves back and forth just micrometers above the disk. The read/write head can also erase data.

Information is stored magnetically on the disk, and the head has the ability to sense (read) the series of magnetic dots on the disk, and to change the orientation of the dots (write or record), creating new information on the disk. The head is also responsible for the conversion of this magnetic information into electrical impulses the computer can understand. Also see *hard disk* and *head crash*.

header

Text that is repeated at the top of each page in a document is called a **header** or "running head." The words at you see at the top of these pages are running heads—headlines that run through the entire book. Good word processing programs allow you to create different headers for even and odd numbered pages, to insert variables such as page numbers and dates, and the ability to suppress or change a header within a document. You usually have options for formatting the heads, also.

In data processing, a **header** is the first record in a file that contains the essential information on the file to follow—file name and date, and other identifying data. In other types of communication a header plays this role of identifying the information to follow. Without a header a file may be misread.

heap

The **heap** is a section of the computer's *memory* that is parceled out in different sized blocks according to the needs of a program. The heap is "free" (available) memory that a computer relies upon when it needs to load and run programs; how much of the heap the program will need isn't known until the program is running. Some programs know how to borrow from the heap when they need it.

A heap can become *fragmented*, requiring compaction to allow the memory to be used more efficiently.

Help

Help, with a capital H, generally refers to *online help*. Please see the definition for *online help*.

Hercules, Hercules graphics

Hercules is a company that manufactures *add-in boards* (mainly video *adapters*) for PCs. By far its greatest success came in the early days of the PC when Hercules developed a board that allowed the IBM *monochrome* monitor to display relatively sharp *(high resolution)* graphics. IBM's own monochrome video board could only display text, so the Hercules alternative caught on in a big way and became a *de facto standard* for monochrome graphics. Soon after, other companies began to make boards that worked like the Hercules version, and **Hercules graphics** became a generic term describing any such board.

hertz

A **hertz** (abbreviated as Hz) is a measure of electrical vibrations per second; one hertz is one cycle (vibration) per second. You'll usually hear people talk about *megahertz,* which is the definition to read.

(My sister just looked over my shoulder and asked if I would like a Hertz donut, as she punched me in the arm. 'Member that stupid joke?)

heuristic

Heuristic refers to a trial-and-error method of finding an answer or a solution to a problem—the rules of thumb we learn as we go through life. It's the self-learning model, where we learn from what we just experienced and apply that to the next experience. Heuristic is kind of the opposite of algorithmic—an *algorithm* is a fixed set of rules that (supposedly) always leads to the correct solution. Obviously, life does not play by algorithms; computers do.

Artificial intelligence programs try to use heuristic methods to solve problems.

Hewlett-Packard Company

The **Hewlett-Packard Company** (known as **HP**), is a major manufacturer of electronics, mini- and personal computers, *laser printers, plotters,* and other technical instruments. Like Apple computer, HP was founded in a California garage by two guys, William Hewlett and David Packard.

hex, hexadecimal

Hexadecimal is a number system in base 16. It utilizes the numbers 0 through 9 and the letters A through F. The hex system is convenient to use in programming because it is compatible with the binary system and is easier to read and more compact. Two hexadecimal numbers can represent one byte. For example, 2B7D equals 0010 1011 0111 1101 in binary.

HFS

See *hierarchical file system.*

hidden files

A **hidden file** is a *file* (such as a document or software program) that is stored on the computer, but you can't see any record of it on the computer—it's invisible. Files are sometimes hidden to prevent accidental corruption or damage, as in a school lab where you don't want students to trash important things. Sometimes files are hidden from prying eyes. Since a hidden file can't be viewed in the normal directory list or window, it can't be read, changed, copied, or deleted.

To hide files, you need special software. On the Mac, you can use *ResEdit.*

hidden text

Hidden text is text within a document that is visible or not, depending on whether you choose it to be. Some word processing applications let you create hidden text so you can make notes to yourself that you can read as you work, then hide when you print. The software application itself might create hidden text to mark indexing or words that are to be added to the table of contents.

hierarchical file system

The **hierarchical file system,** also known as HFS, is the system the Macintosh uses to organize all your files on the computer. The HFS is displayed on the screen as folders within folders.

Originally, the Mac used what was called MFS, the Macintosh file system. That was long ago before Macs had hard disks and no one had many files to keep track of. At that time, you could not store folders within folders—there was only one level. You may occasionally run across old, single-sided floppy disks that were formatted with MFS.

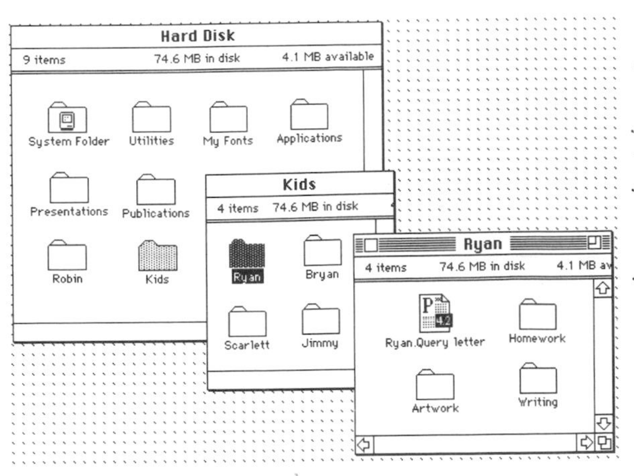

This is an example of the hierarchical file system. The hard disk has a folder called "Kids." Inside "Kids" is a folder for each child. Inside each child's folder they can create other folders for storing their files, as shown by Ryan's folder.

hierarchical menu

A **hierarchical menu,** affectionately known as an "h-menu," is a menu with subsets (sub-menus) to it. The subset is usually indicated by an arrow, and to get to the subset you need to drag the pointer off to the side of the menu item—not always in the direction of the arrow! It's possible to have hierarchical menus several layers deep.

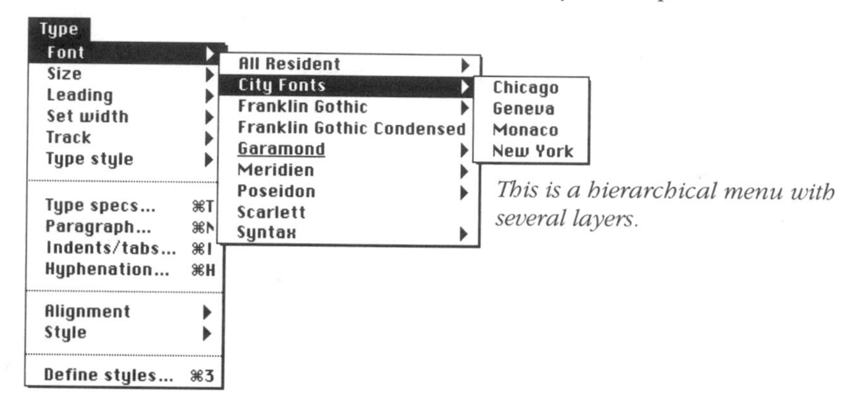

This is a hierarchical menu with several layers.

high ASCII

ASCII, the standard computer coding system for text characters, defines code numbers for only 128 characters (the upper- and lowercase lettters, the numerals, and the common punctuation marks). In most computers, however, 256 "slots" are available, and the extra codes are used for foreign language letters, math symbols, typographic marks, and miscellaneous characters like the trademark and yen symbols. The term **high ASCII** is sometimes used in the PC universe to refer to characters with code numbers from 128 to 255 (the count starts at 0). Actually, there is no ASCII standard for these characters—DOS, Windows, and the Macintosh all assign different characters to the high ASCII code numbers.

high-density disk

Density is a term used to describe the type and amount of storage area on a *floppy disk.* A disk is covered with particles of magnetic 'dust' that can store pieces of data. The larger the particles, the fewer the disk can hold and therefore the less information it can store. On a **high-density disk,** the particles are smaller; the disk can store more information because more particles can be packed in more densely. High density is not the same as "double density," which is an earlier technology still currently used that packs only about half as much data onto a floppy disk.

high-level disk format

When you *format* a hard or floppy disk, you are performing a **high-level disk format.** Formatting creates "housekeeping" *sections* on a disk so information can be stored and later retrieved from known locations.

Low-level formatting is a process the hard disk vendor puts your hard disk through before they send it to you. If they don't do a low-level format, you need special software to do the low-level formatting yourself, before you can do a high-level format. Usually the software performs both the low- and the high-level format in one procedure.

If the hard disk is sold to you completely unformatted, it is sometimes called "raw."

high-level language

A **high-level** programming **language** is one that resembles (albeit crudely) humans' *natural language,* the kind we talk with. BASIC, COBOL, and Pascal are examples of high-level languages. Programs can be written more quickly in a high-level language, but they are often slower to run. See *programming languages*.

high memory

This term refers, confusingly enough, to two different types of memory on PCs. Some people call "upper memory," which is the region between 640K and 1024K, **high memory.** But the term is also used for the High Memory Area (HMA), the first 64K of *extended memory* after 1024K. See *RAM* for a full discussion of all the types of memory on a PC.

High-Performance File System

See *HPFS*.

high resolution

The **higher** the **resolution,** the more information there is in a given amount of visual space (in a square inch, for instance). It may mean that each pixel in that square inch of the screen is providing more information for resolving the image, or it may mean there are more printed dots per inch on the page. Either way, the image appears to be in finer detail and with smoother edges. Please see *resolution*.

highlight, highlighting

When you *select* a menu item or an object or text, it becomes **highlighted.** The highlight may look like a colored bar over the words, or the words may reverse (become white on a dark background), or the object itself may reverse. Selecting another item or clicking the mouse button anywhere (even in the highlighted area) will deselect (and thus un-highlight) the selection.

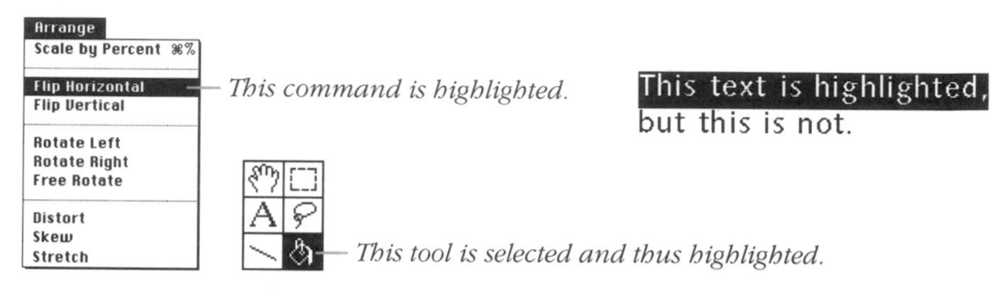

This command is highlighted.

This tool is selected and thus highlighted.

hints, hinting

Hinting is a technique developed by Adobe Systems that improves the appearance of small-sized characters when displayed on computer screens and printed on low-resolution personal printers.

Remember that a character in a *scalable font* is represented mathematically as an *outline* of the character's shape, while the characters you actually see on the screen or on a printed page are made up of dots. The job of font scaling software is to decide which dots to "turn on" in order to best match the ideal outline of the character. At smaller font sizes, it commonly happens that the ideal outline would cut a dot down the middle, so that part of the dot is inside the outline, and part of it outside. In this case, if the font scaler turns on the dot, the character will look too bold or chunky; if it doesn't turn on the dot, the character looks too thin, or a line may even disappear.

Hints, which consist of extra information encoded in the font along with the character outlines, help the font scaling software to make this decision. Ideally, the resulting characters look more recognizable and less jagged, even if their outlines have been altered significantly.

Hints can be applied using set mathematical formulas which work with all the characters in a font. However, getting the best possible results requires hints chosen individually for each character by an aesthetically-minded human. The problem is that hinting characters one by one takes lots of time, and many font vendors don't bother.

Whether or not a font has hints, and how good those hints are, makes a difference when you're working with small type (around 5 to 14 point type) on low-resolution devices such as the typical laser printer or the computer screen. The higher the resolution of the printer and the larger the size of type, the less important hints become: the dots are smaller and/or there are more of them to work with, so the characters automatically look better.

histogram

A **histogram** is a type of graph. In Adobe Photoshop, the histogram is a graph that gives you a very clear idea of the balance of the distribution of brightness and darkness levels of the file you're working on. It's a gauge, for all practical purposes, that tells you just how much variance there is between the darkest and brightest pixels of your image. Given this information, you can make various adjustments to "equalize" the image to suit your particular needs.

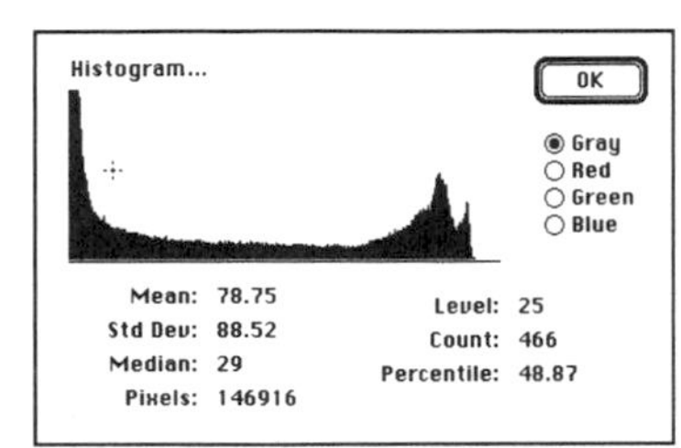

Example of a histogram from Photoshop.

HMA

HMA is an acronym for **h**igh **m**emory **a**rea, a type of *memory* on PCs. See the definition of *RAM* for details.

h-menu

The term **h-menu** is a shortcut way of saying *hierarchical menus.*

Home key

All PC keyboards have a **Home key,** but with a Macintosh, Home is only found only on *enhanced* (extended) *keyboards.* Most word processing applications will respond to a press on this key by taking you to the beginning of the current line or the start of the document you're working on. If pressing the Home key by itself doesn't do anything, try using a *modifier key* with it; that is, try holding down the Option, Alt, Command, or Control key in addition to the Home key.

horsepower

Horsepower is just a picturesque way of referring to a computer's speed. To measure computer performance quantitatively, use a *benchmark* test.

hot link

See *link*.

HP

See *Hewlett Packard*.

HPFS

OS/2, IBM's *operating system* for PCs, can use one of two *file systems,* or methods: either the DOS method (which limits the names of files to a total of 11 characters), or a new method called **HPFS** (**h**igh-**p**erformance **f**ile **s**ystem). HPFS is more efficient and lets you have long file names (up to 254 characters), including spaces and punctuation marks.

HPGL

HPGL stands for **H**ewlett-**P**ackard **g**raphics **l**anguage, a set of commands originally used for controlling HP *plotters* (a plotter is a printing device that makes images on paper by actually drawing the picture with pens). More recently, HPGL was incorporated into *PCL* Level 5, so that HP LaserJet III and 4 printers can understand the same commands for printing sophisticated graphics. Anyway, HPGL includes commands that tell the plotter or printer where to draw a line or a shape, how large to make it, and what color it should be. You don't have to know any of the commands, since they're generated for you by your software (if it's capable of controlling an HP plotter or a PCL Level 5 printer).

HP plotters are the most popular brand, so more software knows HPGL than any other plotter language. For this reason, it helps if your plotter is HPGL-compatible (understands the commands). HPGL comes in two major versions, plain HPGL and HPGL-II, so a plotter that understands only the standard version won't be able to process HPGL-II commands.

HyperCard

HyperCard is a software application, originally from Apple, that is difficult to explain because it is unlike anything else on earth. A HyperCard document, called a *stack,* is a series of "cards" (which is why a document is called a stack, as in a stack of cards). Each card can hold text, graphics, sounds, and animation.

This is the icon for the HyperCard application.

As a user, you don't really notice the cards—you just notice what is going on. The user controls the flow of information. Rather than having to read the information in a linear format, as in a book where you start at one end and go to the other, in a HyperCard stack you can read the information in any way you choose.

When you create a document in HyperCard, it's called a stack. This is a stack icon.

HyperCard stacks can be *interactive, multimedia* events. For instance, you could create a stack that takes a person through a portion of American history, say the Lewis and Clark expedition. On one of the first cards you could choose to go on this expedition as Lewis, or as an Indian guide, or as a woman, or as a mountain man carrying the packs, or perhaps even as a horse. Along your path you can click buttons to hear sounds, see animated clips or little movies of events, select text to get more details on points of interest, answer questions, or switch to another point of view.

There are thousands of ways to take advantage of HyperCard. The one point to remember is that HyperCard stacks are meant to be used while they are inside the computer; that is, rarely do you actually print *hard copy* of a HyperCard stack, except maybe an address collection or the result of your journey through the stack. Also see *HyperTalk.*

HyperTalk

HyperTalk is the programming language that mortals like you or I can use to control HyperCard stacks. Using HyperTalk, it is so easy to program a button to do what you want—activate an animation sequence, play sounds, ask questions. You can make dialog boxes appear, you can request information and act on it, you can create tutorials. This is an example of HyperTalk:

on openCard *(which means, when the user gets to this card)*

answer "On whose tombstone is engraved the epitaph, "It was a grand adventure. I am content"?" with "Shakespeare" or "Dwiggins" or "Who cares" *(this is what the dialog box will have in it)*

end openCard *(this is the end of this command)*

On the next page is an example of what you created. You could then take the answer and do things with it. Oh, it's too much fun, which is the biggest problem with working in HyperCard and HyperTalk—it's so much fun and so empowering that you'll stop cooking dinner and sleeping and instead spend all your time "working" on your new HyperCard stack.

This is the result of that short script on the previous page.

hypertext

Ted Nelson invented **hypertext** in the 1960s as a database that allowed the user to follow associative trails of thought through a document. Rather than a linear structure of information, hypertext is a non-sequential, complex web of information, all creatively linked—you can click on a word or phrase and instantly get more information on that word, or perhaps find the next or previous instance of that word, or watch an animated sequence explaining the word, or be transported to another part of the document that contains related information. *HyperCard* is a hypertext system. *Online help* screens are now often done in hypertext. See *link.*

Hz

See *hertz.*

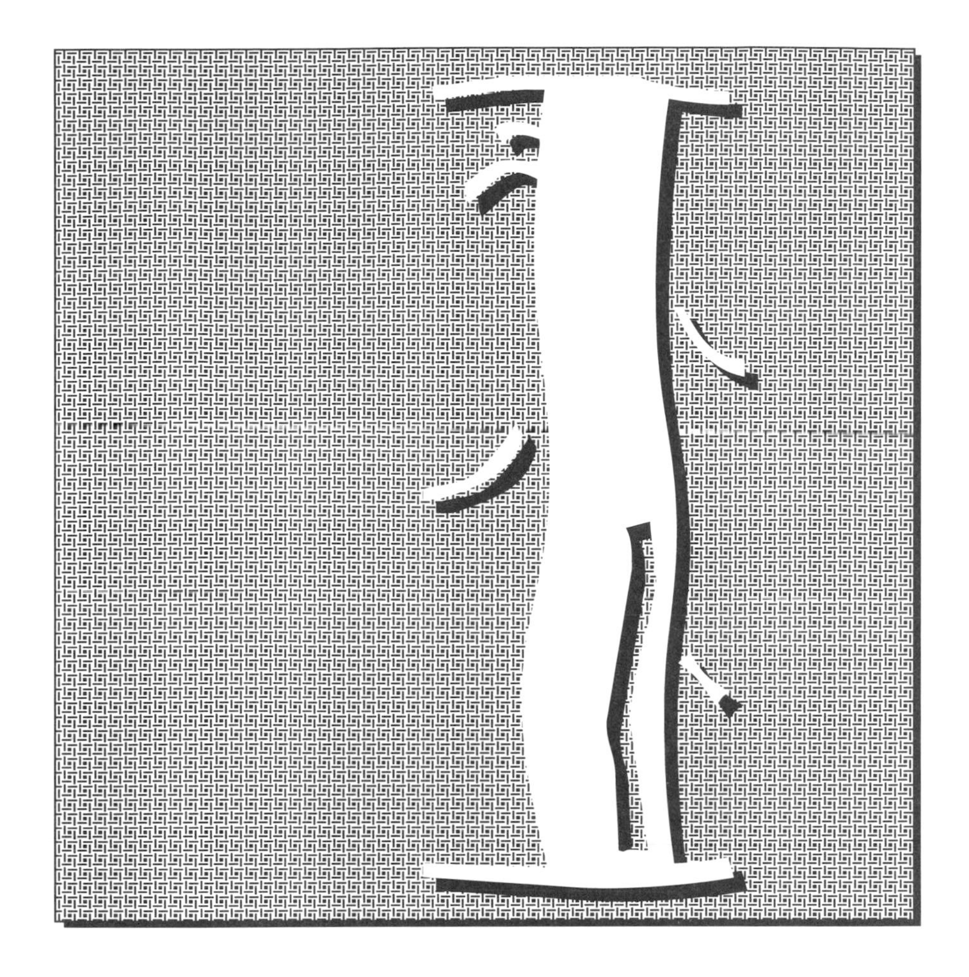

IAC

IAC stands for **i**nter**a**pplication **c**ommunication—a process in System 7 where one application can send a message to another application, either to exchange information or perhaps to make something happen. It does this by sending an "AppleEvent," a set of messages that tells the other application what to do.

IBM

IBM is short for **I**nternational **B**usiness **M**achines—until very recently the largest and most powerful computer company in the world. It began in 1911 as the Computing-Tabulating-Recording Company, a merger of three smaller companies. Thomas J. Watson, Sr., became general manager in 1914 and by 1924 had turned the New York City business into an international enterprise and renamed it IBM. The company set standards with advanced technology several times in its history.

IBM PC, IBM PC compatible

IBM PC refers to a personal computer from the IBM corporation. An **IBM PC compatible** or an **IBM compatible** computer is a computer made by another company that can run the same software as an IBM computer and use the same *peripheral devices.* These compatible computers are often called *clones.* **Every** software program will not necessarily run on every machine, though—there are hundreds of different compatible computers and there are so many variations of hardware and software that it's impossible for everything to run smoothly on everything else. But the term "IBM compatible" certainly distinguishes computers in that category from computers in the Macintosh category.

I-beam

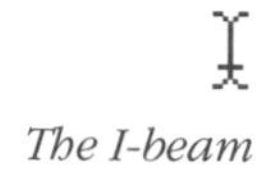

The I-beam

The **I-beam** is a mouse *pointer* or *cursor,* shaped kind of like a capital I. It moves when you move the mouse. The I-beam is the tool you use to tell the computer where you want to type when you're entering text. The I-beam is thin enough to position it accurately **between** letters; when you click the mouse button, the I-beam sets an *insertion point* (a vertical flashing line like this: | but it flashes) in the position you clicked. When you type, the text appears at the insertion point. **Text does not appear at the position of the I-beam—text appears at the insertion point.** The I-beam is just a pointer—use it to insert the insertion point or to select text, then move the I-beam out of the way!

To *select* a range of text, *press-and-drag* with the I-beam (that is, position the I-beam at the beginning of the text you want to select, hold the mouse button down, and drag the I-beam across the text).

In many applications you can position the I-beam at the **beginning** of the text, then click; next position the I-beam at the **end** of the text you want to select, even if it is pages away, **hold down the Shift key**, then click—everything between the two clicks (click and Shift-click) will be selected.

Or if you have some text selected and you want to add or delete a few words at the end of the selection, hold the Shift key down and position the I-beam where you want the selection to end; the words will be added or deleted from the selected range.

icon

Icons are those little pictures on the screen that represent the files stored within the computer. An icon can represent your hard disk, an application, a document, a font, a utility, etc. Icons are one of the features that makes a computer system *user friendly;* instead of typing a memorized, obscure command, you find the icon.

These pictures are mnemonic—they visually remind you of what they represent. This makes it so much easier for you to work with all your files—rather than having to guess what the cryptic name of a file alludes to, you can instantly tell what an icon will do by the picture.

A folder to store things in.

A paint application.

A font.

A printer driver.

Notice that a document icon (on the right) matches the application it was created within (on the left).

If you click once on any icon, it changes color, indicating you have *selected* it. If you double-click on a document or application icon, that document or application opens on the screen. If you double-click on a folder icon, the folder opens to show you what is inside. You can press on any icon and drag it to another place—into another folder or directory, into the trash can, or perhaps just to clean things up.

ID

ID stands for **id**entification. If you want to know about font IDs, check the definition *font ID*. If you want to know about a *SCSI* (scuzzy) ID, check the definition for *SCSI*.

IDE

Every computer needs special circuits to control the hard disk, either on an *add-in board* plugged into an *expansion slot,* or circuits built right into the computer's main *motherboard*. In the PC world, there are several different standards for these controller circuits, including ST506, ESDI, and *SCSI*. But these days the most common type is called **IDE** (for **i**ntegrated **d**rive **e**lectronics). In practice, all of them are equally fast. IDE wins because it's simple and inexpensive.

Still, the main thing to remember is that your hard disk has to match the controller—if you have an IDE controller, you need an IDE hard disk, and vice versa. On the other hand, if you're buying a new hard disk for an old PC that has an ST506 or ESDI controller, you need that type of hard disk instead (or you have to buy a new controller too). SCSI hard disks and controllers are relatively complicated and expensive, so don't bother with them unless you really need SCSI's special capabilities (such as disks bigger than 600 megabytes, and the ability to hook up other types of devices, like CD-ROM drives, in series with your hard disk).

IGES

The initial graphics exchange standard, **IGES,** is an interchangeable *file format,* a type of file that can be used by many different applications and/or types of computers. IGES is probably the most popular file format for transferring CAD (computer-aided design) files between different computer systems. CAD systems on micros, minis, and mainframes all support this well-established file type.

illegal character

An **illegal character** is a letter, number, or punctuation symbol that you are not allowed to use in a particular situation. This term crops up most often when you're talking about naming files, because there are always some characters you can't use in a file name. Which ones are illegal depend on the computer and the operating system. See *reserved character* for details and a list of illegal characters.

image editing, image processing

Image editing or **image processing** is what it's called when you edit, or change, a graphic on the computer. Most typically it refers to the changes you can make to a photograph that has been scanned in and opened in an application like Adobe Photoshop or TimeWorks' Color-It. If you scan a photograph of an old boyfriend and replace his face with that of an iguana, that's image editing. When you scan your own photograph and remove the wrinkles from your eyes and neck, that's image editing.

imagesetter

An **imagesetter** is a high-quality printer that *outputs* (prints out) your files. At home you probably output to your personal printer, either a laser printer or perhaps a StyleWriter or DeskWriter or DeskJet. You know the kind of quality you get with your own printer. When you want higher quality print-outs, you can put your file on a disk and take it down to a *service bureau* that has an imagesetter. The service bureau follows the same basic process you follow to print your file, except their printer costs about $100,000. It doesn't print onto "plain paper" (like the kind you buy at Costco) with toner; it prints onto resin-coated paper with a photographic process. The *resolution* on your home printer is probably around 300–400 (possibly 600) *dots per inch;* the resolution on a typical imagesetter is 1270 or 2540 dots per inch.

Many people refer to an imagesetter as a "lino" (pronounced "lie no"), which is short for Linotronic. Although the Linotronic is the most popular brand of imagesetters, it is not the only one.

Because the quality of output from an imagesetter is so much higher than that from a personal printer, and because they can do things like output onto film and do *color separations*, service bureaus tend to say they "image" your pages, rather than merely "print" them.

imho, IMHO

This little acronym, **imho,** stands for **i**n **m**y **h**umble **o**pinion. It's often used as a typing shortcut in *online* communication. When it is capitalized, you are *shouting*. You might also see the term "imnsho" on the screen, which stands for **i**n **m**y **n**ot-**s**o-**h**umble **o**pinion. See *baudy language* for more of these online acronyms.

impact printer

Impact printers create marks on paper with a mechanism that actually smacks up against the paper, usually with an inked ribbon in-between. The ImageWriter is a dot matrix impact printer—it uses tiny pins to print dots in the form of characters. Impact printers are great for printing through all the parts of a multi-part form, which you can't do with a *laser printer.*

import

When you **import** information, it means you bring information from one document or computer system into another document. For instance, you might create text in a word processing application and graphics in a graphic application, then import both the text and the graphics into a page layout application to put them all together into one document. Or you might import data from one database table into another, or import spreadsheet information into your word processing document.

To make importing work properly across different *platforms* (computer systems) and applications, *file formats* have been developed. Each application *supports,* or accepts, certain file formats. For instance, the file format *TIFF* is a common graphic file format specifically created for *scanned* images. If your application accepts TIFFs, you can import them. If your application is not set up to accept a certain file format, you cannot import that kind of file into that application.

indent

An **indent** is the amount of space between the margins of a page and the actual text. Don't confuse an **indent** with the ***margin.*** The margin sets up the actual "live" area within which all the text is confined—the indent is the extra space that the text moves over.

These are the margins.

This is the indent.

For instance, in this book the margins are out to the **left** edge of the jargon you see in bold, and over to the **right** side of this paragraph. But you can see that the left side of this paragraph is **indented**; I've added extra space between the actual margin and this paragraph.

A right indent indents a paragraph on the right side; a left indent indents a paragraph on the left side. A *first-line indent* indents just the first line of a paragraph. An "outdent" or *hanging indent* leaves the first line of a paragraph at or near the margin, and the left side of the rest of the paragraph is indented. See the examples in *hanging indent.*

information bar

The **information bar** is the line or two of information you see above a window that is open on the Desktop. What you see in the information bar depends on your view of the window.

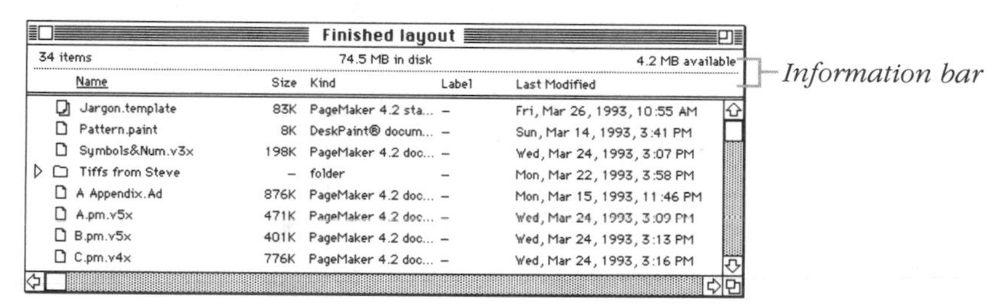

Information bar

Do you see only one line of information visible when you view your window as a list (By Name, By Kind, etc., rather than By Icon)? If you want to see both lines, as in the example above, go to the Views Control Panel. Click the checkbox button, "Show disk info in header."

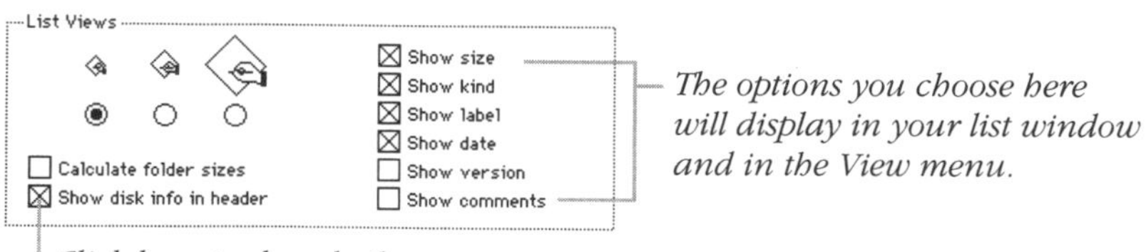

The options you choose here will display in your list window and in the View menu.

Click here to show both lines of information.

INIT, INIT conflict

An **INIT** (short for **init**ialization) is now called a *System extension* under System 7. Whether you call them INITS or extensions, they act the same and they conflict the same. Please see the definition for *extension*.

initialize

To **initialize** is to prepare something for use. You must initialize your hard disk and floppy disks before their first use. On disks, this process is also called *formatting* (although there is a slight technical difference between the two processes) and includes erasing any data already on the disk, creating *sectors,* testing the disk, and writing start-up information.

A computer initializes each time it starts up—it clears and sometimes tests memory, then gets the information it needs from *ROM* and the hard disk and puts it in *RAM*. When a computer sends info to a printer, it first initializes the printer to prepare it for the data that will be coming down the line.

inkjet printer

An **inkjet printer** prints by spraying tiny streams of quick-drying ink from little nozzles onto the paper. Circuits in the printer tell it which nozzles to turn on and when, timing the jets of ink precisely to produce a pattern of dots that form the desired text and graphics.

Inkjet printers are quiet, lightweight, and relatively cheap, and many print almost as sharp as a laser printer. The most popular models are HP's DeskJet (for PCs) and DeskWriter (for Macs).

inner cap

Many titles of companies, software, and other products are written with an **inner cap,** a capital letter in the middle of the name, as in HyperCard. Usually the name is a combination of two words, and since there is no space between the words, the inner cap serves to identify the second word, making the phrase easier to identify. I'll bet that if Disneyland had been built in the past five years it would have been named DisneyLand.

input

The word "input" can be a noun or a verb. As a noun, **input** is any information fed into a computer from the keyboard or from some outside device, such as a modem or maybe a scientific instrument of some type. As a verb, to **input** data means to feed information (data) into the computer.

input device

Any device that can be used to enter information into a computer is an **input device.** The best example is the keyboard. Sometimes the mouse is called an alternate input device. Other assorted *peripherals* (trackball, modem, graphics tablet, and scanner) are also known as input devices.

input/output

See *I/O,* which stands for input/output.

insertion point

If you use an application that has an *I-beam* (as in all Macintosh text applications and many Windows applications), you also have an **insertion point.** You see the insertion point when you click the I-beam between two characters: it looks like a flashing vertical line (like this: | but it flashes). The insertion point marks the spot where characters typed on the keyboard will now appear. The I-beam remains independent of the insertion point—you can move it out of the way by moving the mouse. The insertion point can be controlled with the mouse, the I-beam, or the arrow keys.

You'll see the flashing insertion point in many other places besides text applications: when you change the name of a file or folder, when you name a document in the Save As dialog box, when you open any dialog box where you can change the information in the *edit boxes,* in the message box in the *Get Info* window, and a few other places. The insertion point—wherever you happen to run into it—is a visual clue that you are in a typing mode—anything you type on the keyboard will appear at the insertion point.

install, Installer

When you **install** an application, a font, a system, any other kind of program, etc., that means you copy it onto your hard disk so you can use it. Sometimes the installation process is just that simple—you simply copy the program to the hard disk. But as programs have gotten more complex, the installation process has gotten more complex. Now many applications arrive on several disks and you have to run the **Installer** to properly load everything onto the hard disk. The Installer is a file that comes on the first disk in the series; typically you open the Installer and it does its business, which includes making folders or directories and installing items into the various places they need to go. For instance, on the Mac, a program may install items in several of the folders within the System Folder.

You may have a choice of an Easy Install or a Custom (or Expert) Install. The Easy Install is the one to use if you haven't a clue as to how or why to pick and choose between the multitude of supplemental files that the application wants to run. An Easy Install will install everything, whether you need it or not. This will take up more space on your hard disk, but someday you can ask your *power user* friend to come help you delete superfluous stuff. In a Custom, or Expert, Install, you can choose just the things you want installed. For instance, maybe you know you will never use the templates or the tutorial files, so you uncheck those from the list of files. Perhaps you have a choice of millions of *printer drivers* to install—a Custom install lets you choose just the ones you need.

integrated software

The term **integrated software** applies to one *package* that includes several major kinds of application programs. Typically, an integrated package contains a word processor, a spreadsheet, a database, and usually *telecommunication* capabilities. Some integrated software products include other modules too, say for graphics or for scheduling appointments.

The obvious advantage of integrated software is that you get all these features in a single product that costs a lot less money than you would pay for a collection of separate programs. Another plus is that all the modules use the same commands to do the same things. For example, the command for saving a file is always the same, whether you're working on a letter, a database, or a financial worksheet.

The disadvantage of integrated programs is that they're not as strong in any one of the categories as a specialized program would be. Still, the good ones provide all the features that the typical user will need, making an integrated product a great first choice for anyone just getting started with computers.

Integrated software packages usually have the word "Works" in their name, such as ClarisWorks, GreatWorks, or Microsoft Works.

Intel

The **Intel** Corporation is a *semiconductor* manufacturer. They make the *microprocessors (chips)* in the *8086* family (which includes the *80286*, *'386*, and *'486*), which are the central processing units (CPU) in IBM PC and compatible computers. You'll often hear people talk about these particular microprocessors as **Intel chips.**

intelligent terminal

A "terminal" is simply a screen and a keyboard. There are "dumb," "smart," and **intelligent** terminals. A *dumb terminal* has to rely on a central computer to do its processing. A "smart terminal" has **some** processing capabilities. An intelligent terminal has its own *memory* and *processor,* but not (necessarily) a disk or storage area. It can *download* information from the main computer and perform certain operations independent of the central processor.

interactive

If a program, game, presentation, or any other product is **interactive,** it means the user has some control over what is going on. Perhaps you can click buttons to make things happen; perhaps you can type in questions or answers; perhaps you can point to the screen to change the items. Broadly, everything on a personal computer can be considered interactive.

We now take this interactivity for granted. But not too long ago "batch computing" was the norm—data was fed into the computer and the answers came out later. In between there was no interaction on the part of the user and the computer. Nowadays working on the computer is almost conversational. This back-and-forth dialogue is the essence of interactivity.

Inter-Application Communications

See *Apple Events* and *IAC*.

interchangeable file

An **interchangeable file** is one that is recognized by many different applications, and in some cases, can even be used by different applications on different types or brands of computers. For instance, I can create a file in one application, save it as one of the many interchangeable file formats, and then take it over to a friend's house (who may not have the same application) and use it on her computer.

interface

Generally speaking, an **interface** is the connection between two things so they can work together. This may apply to the *user interface,* which describes how the user works with the computer (see below). This may apply to the "hardware interface," referring to the connectors, cards, plugs, etc., that allow the computer to work with the other parts. This may apply to the "software interface," referring to the programming that enables an application to work in a particular *operating system.* This may refer to the *protocols* of *networking* and *telecommunications* that allow computers to "talk" with each other.

There are three basic sorts of user interfaces on personal computers.

The *command-line* interface requires the user to type lines of text to command the computer to do things.

The *menu-driven* interface uses *menus* of one sort or another, where the user can choose from a list of commands. You can make the choice by typing a character, by using the arrow keys, or perhaps by using a mouse.

The *graphical user* interface (*GUI,* pronounced "gooey") displays icons and menus and uses metaphors such as the "desktop" so the user can easily understand what they see and can control the computer by pointing and clicking with the mouse, rather than by memorizing and typing in coded commands.

interference grating

See *moiré.*

interlaced/non-interlaced monitors

In a standard television-like computer *monitor,* an image is produced on the screen by a beam of electrons sweeping rapidly across the surface of the picture tube, lighting up the screen as it passes. Starting at the top, the beam traces one horizontal row across the screen, shifts down a bit and does another row, and so on, until the full height of the screen has been covered.

In an **interlaced monitor,** the electron beam takes two passes to form a complete image: it skips every other row on the first pass, and then goes back and fills in the missing rows. A **non-interlaced monitor** does the whole job in one pass, tracing each row consecutively. Interlaced monitors are easier to build and therefore cheaper, but as you can guess—they aren't as good as non-interlaced monitors. The problem is that all things being equal, it takes twice as long to create the complete screen image on an interlaced monitor. That's long enough to spoil the illusion that you're looking at a steady picture, and the image on the screen flickers annoyingly. Please see *monitor* and *refresh rate.*

interleave

A hard disk's **interleave** (interleave factor) should be set to the optimum value by the manufacturer or dealer when you buy it. You can take it on faith and ignore the rest of this definition—or, if you need to be sure you're getting maximum performance from your system, you can buy special disk *utilities* (little programs) that test the system and reset the interleave to the best possible value—ask your dealer to recommend one.

You really want to know what's happening? Okay, okay. The data on a disk is laid down in concentric rings called tracks, which in turn are divided into sectors—arc-like chunks of equal size around each track. The numeric order of those sectors is called the interleave and it dramatically affects how quickly your computer can read data off the disk. When the interleave is 1 (or 1:1), the sectors are numbered sequentially: if the first sector is sector 1, then sector 2 comes next, followed by sector 3, and so on. This arrangement sounds logical, but it's a good idea only if your computer can keep up with the disk.

Remember, the disk is always spinning very rapidly. If the computer can transfer the data from one sector into *memory* almost instantly, so that it's ready as soon as the next sector moves into position, an interleave of 1 is ideal. But if the next sector has already spun past, the computer has to wait for it to revolve all the way around again. Those fractions of a second add up, and your system seems to slow down. It would be better to increase the interleave. With an interleave of 2, the numbering sequence skips every other sector—if the disk has 4 sectors, they would be numbered 1, 3, 2, 4. The computer reads sector 1, stuffs that data into memory while sector 3 spins by, and is then ready for sector 2. With an interleave of 3, every third sector is next in the sequence (as in 1, 4, 7, 2, 5, 8, 3, 6, 9).

internal

Something that is **internal** is inside a computer or some other device. For instance, an internal hard disk is mounted inside your computer, an external one is in a separate box on your desk. An internal *modem* is built into the computer, whereas an external modem is a little box that sits next to the computer.

internal command

Some DOS commands that you type are carried out without running a separate programs. These are called **internal commands.** They include all the most commonly used commands, like DIR, COPY, and DEL. See also *external command*.

Internet

The **Internet** is a worldwide *network* of about half a million computers belonging to research organizations, the military, institutes of learning, and corporations of all sizes. The Internet is not a destination; it's a means by which you get to your destination. To access the Internet, you *log on* to a nearby computer (a "site" or "host") that's been linked into the *network*. Then you can run a set of programs to process mail and files.

CompuServe and *America Online* (two of the most popular *online services*) use one huge central computer to which everyone logs on to leave and pick up mail and files. But on the Internet, mail and files move from computer to computer until they reach their final destination. Andy Ihnatko says the Internet functions like one massive living, breathing organism.

You can send personal mail, as well as access many of the services you'd expect from any other commercial online service. There are more than 2,000 "newsgroups," places where anyone can post messages for all to see. You can also download piles of free or almost free software, but because the process is rather complex and unfriendly, downloading isn't the most popular aspect of the Internet.

Using the Internet is *free*. You might have to pay a slight fee (like a dollar) to your local Internet host computer, but there's no sign-up fee, no monthly charge, no per-message charge, no hourly rate. If you're in school or if you work for the government or a big corporation, the big computer you have to deal with every day is probably on the Internet. Ask your system manager for details on how to access Internet services. Many companies rent time on their computers to people who want Internet access. Check with your local *bulletin board* and ask for the number of a PAU (public access UNIX) system in your area.

interpreter

An **interpreter** is special software that converts commands written in a programming language that humans can understand into the machine code that a computer can understand. An interpreter does this conversion every time you run the program, whereas a *compiler* does it only once, creating a "stand-alone" version of the program that can run without any other software. Though compiled programs run faster, interpreted programs are easier to write and test. That's because the interpreter checks each line of the program for errors as you write it, and because you can make changes in the program and run the new version immediately, without having to recompile it first. Many computers come with a *BASIC* interpreter. DOS *batch files* are actually simple programs, which are run by DOS's built-in batch language interpreter.

interrupt switch

See *programmer's switch.*

intuitive

If a program or computer system is **intuitive,** that means you can manage to bumble your way through it without having to rely on a massive manual or weeks of intensive training. On a computer system like the Macintosh, most programs are intuitive once you have learned the basics, because all the different programs work so consistently—*cut, copy,* and *paste* are always in the Edit menu; Save, Close, and Quit are always in the File menu. *Objects* in every program are always selected the same way; text in every word processing program is formatted in a similar way.

inverse video

Inverse video, sometimes called "reverse video," is when the image you see on the screen, whether it is a graphic or text or a menu command, is shown in the opposite pattern of light and dark than you are used to. Inverse video is often the visual clue that lets you know an item is selected. That is, if the text you see on the screen is normally black letters on a white background, then when you select a portion of the text, that portion is displayed in inverse video — it is displayed as white text on a black background. A menu command may display in inverse video when you choose it.

When text is selected, like this, it appears in inverse video.

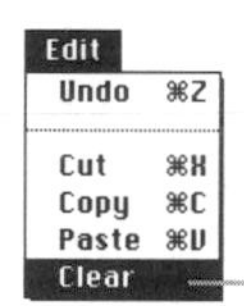

When a menu command is selected, it displays in inverse video.

The graphic on the right is in inverse video.

invisible file

An **invisible file** is one that, although it's physically present, doesn't show up on your computer. Most invisible files are made invisible for a reason: somebody (usually the original programmer) doesn't want you to accidentally delete or modify it. In school computer labs, certain files may be made invisible so students can't harm them. There are *utilities* available

that let you make files invisible and that let
made invisible (by changing their file attrib
file was made invisible for a good reason,
are also called *hidden files,* especially on

An example of an invisible file on the M
can't really "use," but its presence is ess

An example of an invisible file on a PC
contains the *drivers* for the keyboard, t

IO, I/O

IO, or **I/O** (pronounced "eye oh" because it is the letters "i" and "o"), stands for **i**nput/**o**utput. A computer basically does three things:

It takes **input**—the information you put into it, using devices like the keyboard, the mouse, a scanner, etc.

It **processes** that information—that's the job of the *CPU,* the *central processing unit,* which is the *chip* that runs the computer.

It **outputs** the information—in the form of displaying the information on the screen, printing it to a page, or even just storing it on a disk.

The term I/O can describe anything that has to do with moving data between your computer and something else. An I/O *port,* for example, is a connector on the computer where you can plug in a cable from another device, and through which information can flow between the two units. The I/O *bus* consists of circuits that let you plug other devices (such as add-in boards or *PCMCIA* devices) into the computer itself.

Yes, I/O looks very similar to the 1 and 0 (one and zero) that symbolize on and off in digital technology. They look similar, but they don't mean the same thing at all.

iron

This is a macho way of referring to big computers like *mainframes.* The guys and gals who run these big machines love to talk about their **iron.**

ISA

ISA stands for **i**ndustry **s**tandard **a**rchitecture and is just a fancy way of referring to the *expansion slots* (and the associated circuits, the *bus*) in the IBM PC/AT and the myriad of AT-compatible PCs. An ISA slot transfers 16 bits of data at a time between the *add-in board* plugged into the slot and the computer proper. They also accept the *8-bit* add-in boards that work in the original IBM PC. See *EISA* and *Micro Channel* to read about other types of PC expansion slots.

The **ISDN** acronym stands for integrated service digital network, a global system for *digital telecommunications* that is gradually replacing conventional *analog* telephone systems. By ensuring standardized, high-speed digital links, ISDN will permit the transmission of voice (people-talk), images (both still and video), and computer data—all at the same time—between any two points connected to the system. ISDN sounds like the "information highway" envisioned by our politicians, but it may turn out to be too slow for that role—we may wind up using a modified version (BISDN, broadband ISDN), or a faster alternative.

italic

Italic refers to the style of type where the letters slant to the right, somewhat indicative of handwriting. True-drawn italic letterforms, though, are not simply slanted; they have been completely redesigned (see below).

It is sometimes possible to slant a typeface on the computer, but please don't think you are making the typeface italic. You are merely slanting it, which is actually a distortion. On some *sans serif* typefaces you can sneak in a little bit of a slant without most people noticing it.

Many sans serif faces don't use the different letterforms of the *serif* italics, but instead have that simple, slanted look, called *oblique*. A high-quality typeface, though, still has true-drawn obliques and does not rely on a mere computerized angle.

a f g *a f g*

Notice that these sans serif letters are oblique—they have simply been slanted.

a f g *afg*

Notice how the italic of these serif letterforms have been completely redesigned.

Type that is not italic is "roman." Roman does not mean the typeface is serif or sans serif—it simply means that the type does not slant, just like the type on the Roman columns.

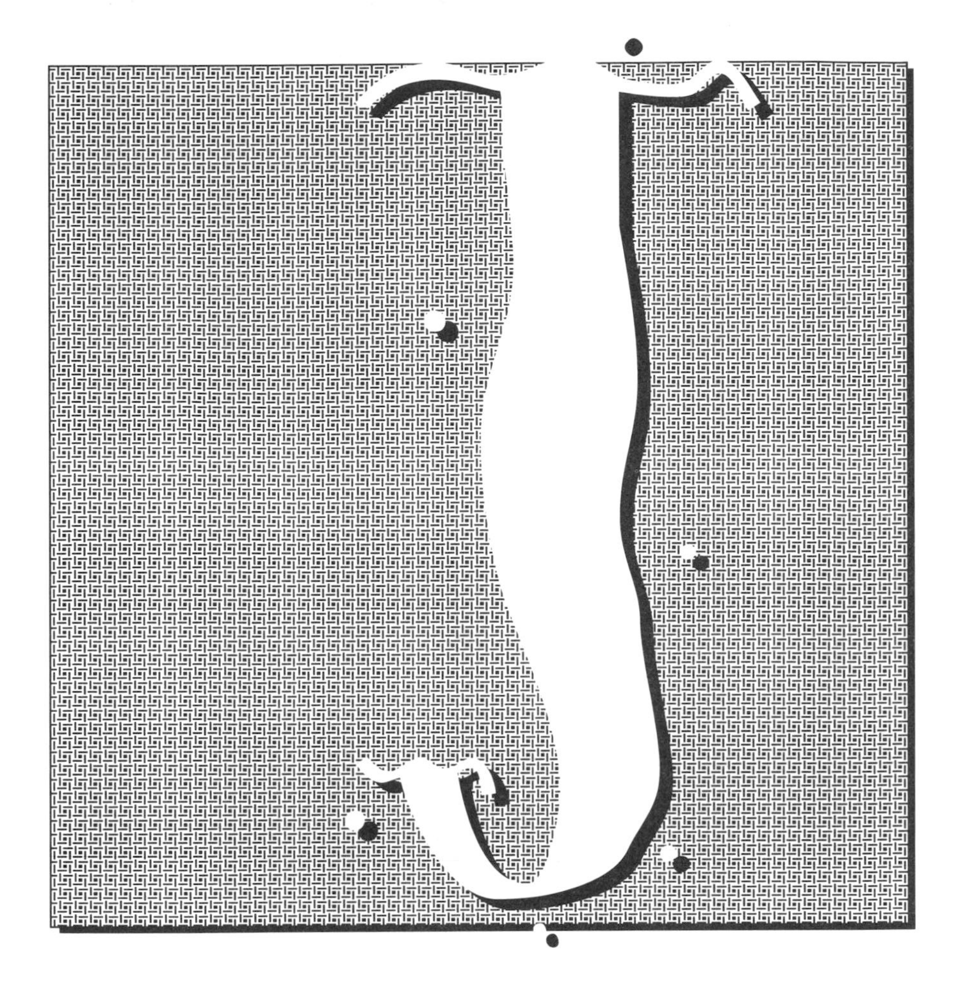

jaggies

When the diagonal or circular lines in a graphic image or text look as if they were built of bricks, displaying a "stair-step" appearance, we say it has the **jaggies.** This effect is also called *aliasing,* and there are several *anti-aliasing* methods that help cure the jaggies.

This text has the jaggies.

JIF

See *GIF (graphic interchange format).*

joystick

The **joystick** is the lever in video games that controls the movement of the cursor or other object on the screen. It's "omnidirectional," meaning you can move it in all directions, not just up or down, left or right. Occasionally people use a joystick with their computer, particularly in *CAD* systems.

JPEG

JPEG (pronounced "jay peg") stands for **j**oint **p**hotographic **e**xperts **g**roup, and is a *graphic file format* used for compressing large, color image files. JPEG works in connection with the *QuickTime extension* to reduce the size of graphic files down to about 1/20th of their original size. It does this by coding the information in the graphic. For instance, if there is a large area of blue, JPEG will apply a code to that area instead of storing all the color information. When you *uncompress* the graphic, all that blue goes back in and replaces the code. JPEG compresses large, solid areas of color more extensively than it compresses detailed areas because you are less likely to notice the change in the detailed areas in case it doesn't come back just the way it was. What?? Well, it's true: JPEG is considered a *lossy* compression scheme because you don't get back exactly the same image that you compressed—some graphic information gets lost along the way, depending on how much compression was applied, the image itself, and how many times you have compressed the file.

Julian date

The **Julian date** expresses the date of the year as a number; the number is the actual number of the day of the year, starting from January 1. For instance, the Julian date 59 is February 28, because February 28 is the 59th day of the year. Computers convert the regular dates into Julian dates when necessary for calculating.

jump

In the Microsoft Windows *on-line help* system, a **jump** refers to a term or other item in the help window for which more information is available. To display the full information about the jump (which can be a word, phrase, or graphic picture), first select it by either clicking on the item or *highlighting* it using the *cursor keys*. Then press Enter.

The jump appears in a different color than the main help text.

> If the jump in the help window is underlined by a solid line, when you select it you'll see a completely new help topic, which will replace the original help screen.
>
> If the jump is underlined by a dotted line, when you select it you see a small box containing a sentence or two, usually giving a brief definition of the term.

To select jumps with the keyboard, press Tab to move the highlight from one jump to the next down the help window; press Shift Tab to move one jump at a time in the reverse direction. This only works for the jumps that are displayed on the screen—if there is more help information than will fit in one window, *scroll* until you can see the jump, then press the Tab key to select it.

jumper

On some circuit *boards,* you control the way the board operates by fiddling with little devices on it called **jumpers**. Jumpers are even worse than *DIP switches* as a means of controlling circuit board options.

A jumper is a set of two or three tiny metal pins mounted side by side, plus a tiny electrical connector mounted in a thin plastic block that fits over the pins. Each jumper controls a particular option on the circuit board. If you want to change the setting for that option you have to slide the little plastic connector over a certain pair of jumper pins or remove it all together. To do this you have to read the manual for the circuit board. Unfortunately, the manual diagrams are often very hard to figure out, with a poor-quality sketch of the board, and it's hard to tell which jumper is which and how the connectors should go. Sometimes the drawings are plain wrong, because they have changed the way the whole board is oriented since the diagram was made. Another disadvantage of using jumpers is that it is really hard to remove and reseat the connector without dropping it. I strongly recommend you use a pair of tweezers to do the job. Better yet, look for a board or a computer that doesn't have jumpers.

justify, justified

To **justify** means to arrange text so the ends of each line, both right and left, end at the same place. You might hear text set like this called "fully justified," or "quad left and right," or just plain ol' "justified." The paragraph you are reading right now is justified.

In **left justified** text, the ends of the lines align on the left side, but not on the right (called "rag right," "ragged right," "flush left," or sometimes "quad left"). This paragraph is left justified.

In **right justified** text, the ends of the lines align on the right side, but not on the left (called "flush right" or "quad right"). This paragraph is right justified.

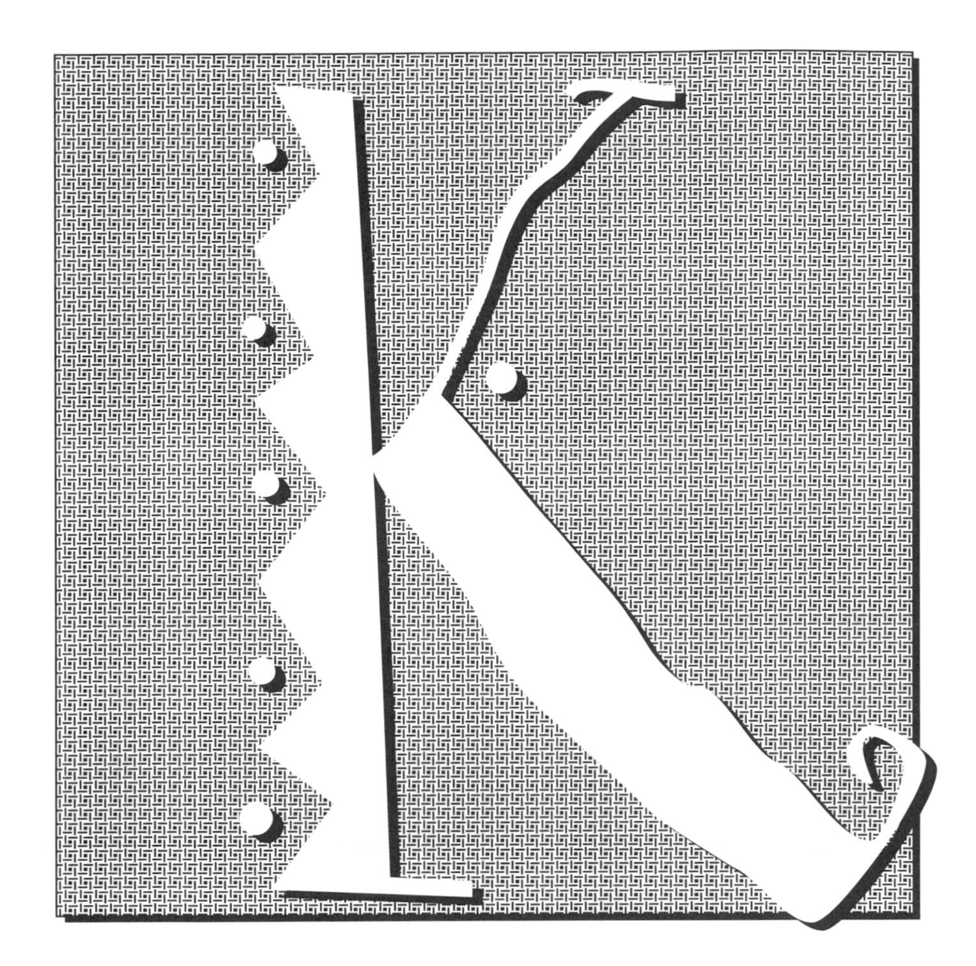

K, KB, Kb

The initial **K** stands for *kilobyte,* as in "an 800K floppy disk." A kilobyte is a unit for measuring the size of things like hard disks, documents, applications, or *memory*. Specifically, one kilobyte is 1,024 *bytes*. **KB** means the same thing, as in "640KB of system memory." To get a clear picture of kilobytes, please see *byte* for a complete explanation in relation to *bits* and bytes and *megabytes*.

Technically, **Kb** (with a lowercase b) is not the same as KB (with a capital B), although many people use it interchangeably. Kb refers specifically to the smaller unit of measure, kilo*bits*.

keep-away

In color printing, a **keep-away** is the opposite of a *trap*. In a trap, two colors that butt up against each other are set to overlap a tiny bit to allow some leeway for the flexibility of paper going through a press. If the colors did not overlap, there would be little slivers of white paper showing through where they abut (as in the example below).

This is an example of what a trap is meant to avoid—those slivers of the white paper showing through between the two colors.

A keep-away is often used when type or some other object is *reversed* out of a solid made of two colors (see below). For instance, to get a good clean black, high-quality printers usually print another color under the black so the black appears darker and richer. But the same problem can occur with these two colors that are supposed to butt up against each other to leave the white image—one color may slip into the white space a tiny bit more than the other, showing a sliver of color where it shouldn't be. So a keep-away is prepared for the under color: it **backs away** from the white image a tiny bit so although the dark solid is made of two colors, only one color actually meets the white area. This is also called a "cut away."

This is an example of what a keep-away is meant to avoid—those slivers of the under color along the edge of the reversed text.

Kermit

Kermit is a *telecommunications protocol* (a standard for sending files over a *modem* from one computer to another). It is mainly used to transfer files between a *mainframe computer* and a *microcomputer* (such as a *personal computer*). Don't use Kermit unless you have to—it's too slow and with some versions you can only send plain *text files*. If you're transferring files between two personal computers, *XMODEM* or *YMODEM* are better than Kermit, and *ZMODEM* is better still. If you're downloading from CompuServe, the protocol "COMPUSERVE-B" is best. And if all this confuses you, don't worry—most software has a *default* (automatic choice) already set with the appropriate protocol selected. The modem at the other end must use the same protocol or it won't work; so if it doesn't work, call the people at the other end and ask which protocol you should use.

kern, kerning

Kerning is the process of adjusting the spacing between two adjacent letters. Traditionally kerning meant to *decrease* the amount of space, but the term has come to mean either increasing *or* decreasing the spacing between the letters.

Each letter in a typeface is designed with a predetermined amount of space to its right. But certain combinations of letters, such as Ta or Wo, look awkward when you set them with this standard spacing. And when you use a badly-spaced pair, the awkward letterspacing can detract from the *legibility* of the entire word. So the bad combinations must be kerned. The ideal is to create **visually consistent** letter spacing—when you look at the text, you should not notice any letters that are set too tight or too loose **in relation to** any other letters.

Won't you?	**Won't you?**
Unkerned text	*Kerned text*

Most commercially available typefaces are designed with predefined kerning instructions for certain pairs of letters, those most commonly known to cause problems. Based on this information, some applications will automatically use these "kern pairs" as you type. In some applications, you can "manually" adjust the kerning between any two characters or over a range of characters. Typically you will find kerning capability only in page layout *(desktop publishing)* programs, and in some top-of-the-line word processors.

key

You know what a **key** on a *keyboard* is—any of those little buttons you press. There are *alphanumeric* keys (letters and numbers) and punctuation keys. There are special keys like the *function keys* and the *arrow keys*. There are *modifier keys* that may do nothing all by themselves when you press them, but they modify or alter the result of other keys (modifier keys include the Shift, Option, Command, Control, or Alt keys). For explanations of special and modifier keys, please look up the name of each key separately, and also check the Symbols section at the beginning of this book.

In a *database* application, **key** also refers to an identifying *field* (some applications call this an "index" field). A typical key might be the name of a client, account number, or product code, where the information in the key is unique to each *record*. These keys or composite keys are often saved in key tables (also called "index files") and specially indexed so you can get to them quickly.

Also in a database program, a **sort key** is the field you use to *sort* information (organize it in alphabetical or numerical order). For instance, if you want to organize your database by age, then your key for sorting is the age field.

When a file is *encrypted* (the information has been scrambled for security reasons), the **key** to unscrambling the data is the equivalent of a secret decoder ring. It takes the same key to encrypt the file in the first place. The key is the same as the *password*.

Key Caps

Key Caps is a *desk accessory* on the Macintosh that displays all the available characters in each *font* (typeface). Open Key Caps from the *Apple menu* (under the apple on the far left of the *menu bar*). What you see is a keyboard layout, right? Notice in the menu bar you have a new menu item called "Key Caps." Press on that menu to see a list of the fonts that are available in your system (good trick to know). The font you choose from this list is the font that will appear in the Key Caps layout.

There are four different keyboard layouts for each font (some fonts have more characters than others). The layout you initially see displays the standard keys you know and love. Hold down the Shift key and you see another set of keys that you also know and love. Hold down the Option key and you will see a new keyboard layout, one that holds many characters we did not have on our typewriters, such as ™, £, ¢, or ❏. Hold down the Option and the Shift keys together and you see a fourth keyboard layout.

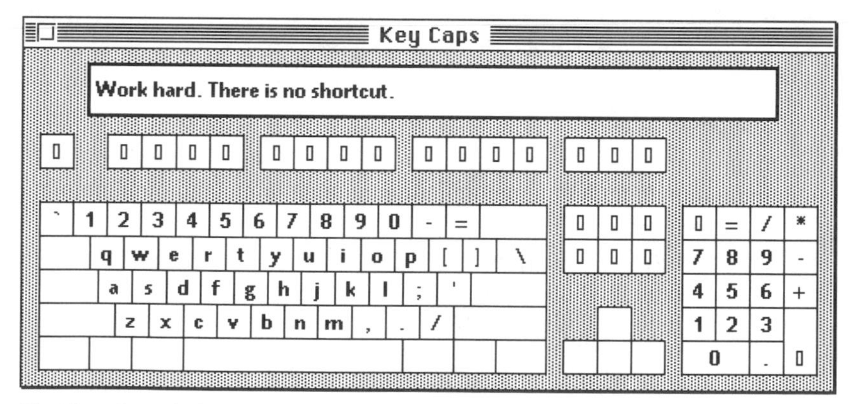

The Key Caps desk accessory.

Key Caps is only meant to act as a reference. Since it is a desk accessory, you can open it while you are working in any application. You open Key Caps only to find out which keys to press to get the character of your choice. For instance, say you want to type the © symbol. You open Key Caps and press the Option key or the Option-and-Shift keys until you see the character ©. Then you poke around on the keyboard until you discover, in this case, that the © hides under the letter **g** key with the Option key pressed. You close Key Caps (click in the close box in the upper left, or press Command W)—since it has served its purpose—and you go back to your document. Click the insertion point at the point in your document where you want the © to appear, hold down the Option key, and type **g**. Et voilá, a © appears.

keyboard

The **keyboard** of a computer is the piece of *hardware* that has all the keys such as you are accustomed to seeing on a typewriter. A computer keyboard, though, has more keys than a typewriter. It usually has a *numeric keypad* off to the right, that looks like a little calculator. An *enhanced* (or *extended*) *keyboard* has keys across the top with the letter F and a number, called the *FKeys* or *function keys.* An extended keyboard usually also has a little *edit pad* between the main alphabet keys and the numeric keypad, with keys like "home," "end," and "help."

For descriptions of each of the individual keys (besides the letters and numbers), please see their names. If you don't know their names, try looking in the Symbols section of this dictionary, before A.

keyboard layout

One definition of a **keyboard layout** is the arrangement of the keys on the *keyboard.* The standard layout is called *Qwerty* because of the first six letters across the top alphabetic row. Your keyboard is probably Qwerty.

The Qwerty layout was purposely created to be inefficient. The mechanical keyboards of the late 1800s jammed easily because the typists' fingers could move faster than the keys, so the typewriter designers arranged the characters according to letter frequency. They separated the commonly used keys to minimize jamming, essentially slowing down the typist.

The *Dvorak* layout has a very different keyboard arrangement, one that is supposed to be incredibly more efficient than the Qwerty. Some personal computers can be switched to the Dvorak keyboard. See *Dvorak.*

The *Maltron keyboard* refers to a different shape of keyboard, which can use either a Qwerty or Dvorak layout. See *Maltron*.

Keyboard layout also refers to the way the characters are "mapped" to the keyboard you use; that is, what characters will appear when you press a certain key or combination of keys. Most typefaces have a fairly standard layout up through the regular characters and numbers. Then, depending on what *font* (typeface) you are using, and particularly depending on what kind of computer you are using, there may be other keyboard layouts with alternate characters for each typeface. Mac users, see *Key Caps*; Windows users see *Character Map*. You may also be interested in *expert fonts*, as well.

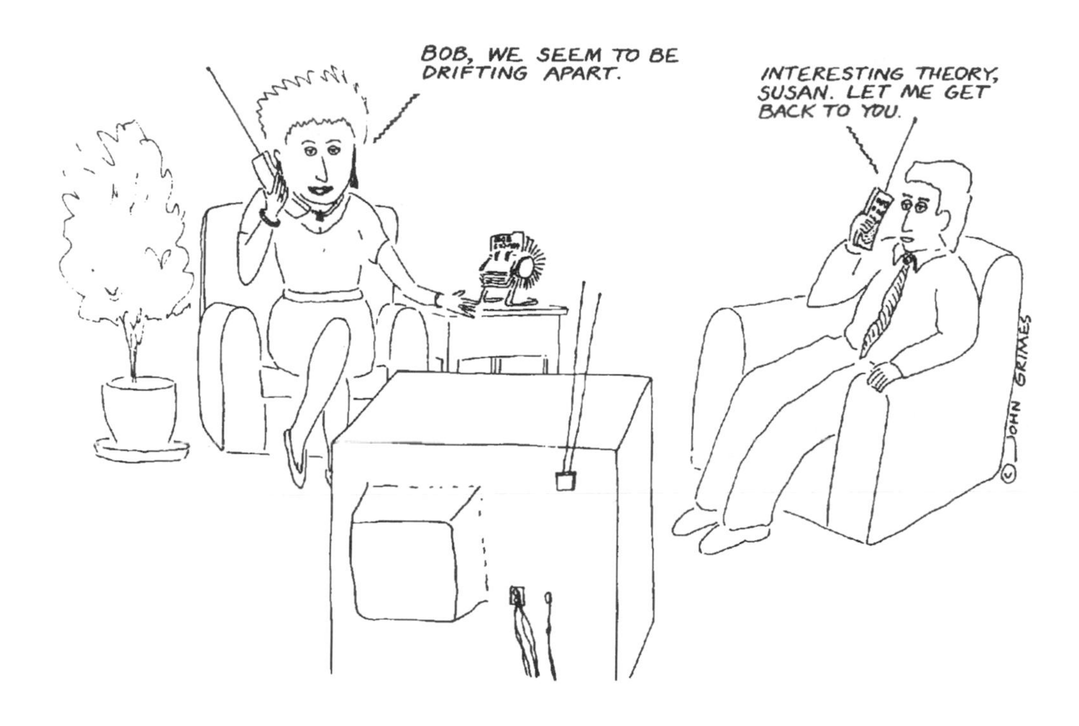

keyboard shortcut

On some computers, everything you do has to be done from the keyboard—to save a file, you press a key or type a command; to search for a word in your document, you press another key. But other computers let you use a *pointing device* (such as a *mouse*) and *menus,* so you can just choose the commands from a list.

Although using a mouse is great when you're first learning a program, it can slow you down once you learn the ropes—you have to lift your hand off the keyboard, move the mouse to the menu, open the menu, and choose the command. For people who like the efficiency of the keyboard, menu commands often have **keyboard shortcuts.** These are just keys you type to accomplish the same thing as the menu command, **instead** of using the mouse. Keyboard shortcuts usually consist of two or more keys: one or more *modifier keys* that you **hold down** while you type one or more ordinary letter or number keys. For instance, instead of using the menu to choose the command to print, you can just press the Command key and the letter P.

keyboard symbols

See the Symbols section in this dictionary, before the letter A.

keypad

When you see a reference to **keypad,** as opposed to *keyboard,* it means you should use the number keys that you see arranged in a square on the right side of most keyboards. It is important to note whether a direction tells you to type a number from the key*board* (from the top row) or from the numeric key*pad*. In many *keyboard shortcuts* it can be critical that you use one as opposed to the other. For instance, in Microsoft Word (Mac) you can use the keyboard shortcut Command Option + (the plus symbol from the key*board*) to add an item to your menu; you can use Command Option + (the plus symbol from the key*pad*) to add a keyboard shortcut to a menu item.

More generically, a **keypad** can refer to any small group of keys used for a special purpose, such as the *edit keys* between the main keyboard and the numeric keypad on an *extended keyboard.* Also see *numeric keypad.*

kilobyte

A **kilobyte** is a unit of measure on a computer. It measures such things as the size of your documents, how much space you have in your hard disk, or how much *memory* your computer has. Each kilobyte, or K, is 1,024 bytes (often rounded off in conversation to 1,000 bytes). To see where kilobytes stand in relation to smaller and larger units, and to get a more detailed view of the way things are measured on the computer, please see *byte*.

kiss-fit

Kiss-fit refers to printing different colors that are touching each other on a commercial press with no *traps*. A trap is a tiny overlap of the two colors to allow for shifting or stretching of the paper during the printing process. Kiss-fitting is not appropriate for all printing jobs, but it certainly can be done, and when it's done well the result is cleaner and clearer than when using traps. If you have any doubts, take a look at the brilliant monthly magazine by John McWade, "BEFORE & After."

kluge, kludge

Kluge (pronounced "klooj") describes a makeshift, temporary, inelegant, or poorly planned solution to a problem with hardware or software.

For instance, I have a *scanner* and a *cartridge hard disk drive* attached to my computer. Because I haven't been able to figure out how or which device to *terminate,* and also because I haven't bothered to get the right *cable,* I cannot have both devices working at the same time. So when I want to use the scanner, I have to turn off my entire system, switch the cables, and turn everything back on again. This is definitely a kluge, a dorky solution that happens to work, albeit inefficiently and temporarily. I wish I'd fix it.

Most software kluges are invisible to the user and may actually result in faster or more efficient execution. The bad thing about them is they're hard for a programmer to understand during any later modification of the code.

Kluge is usually a noun but can be mutated into any other form of speech. Some people *["Me among them," says Steve]* insist on misspelling *["Spelling," says Steve]* this word as "kludge." But if you saw the word "kludge," wouldn't you pronounce it to rhyme with "fudge"? *[Nah.]*

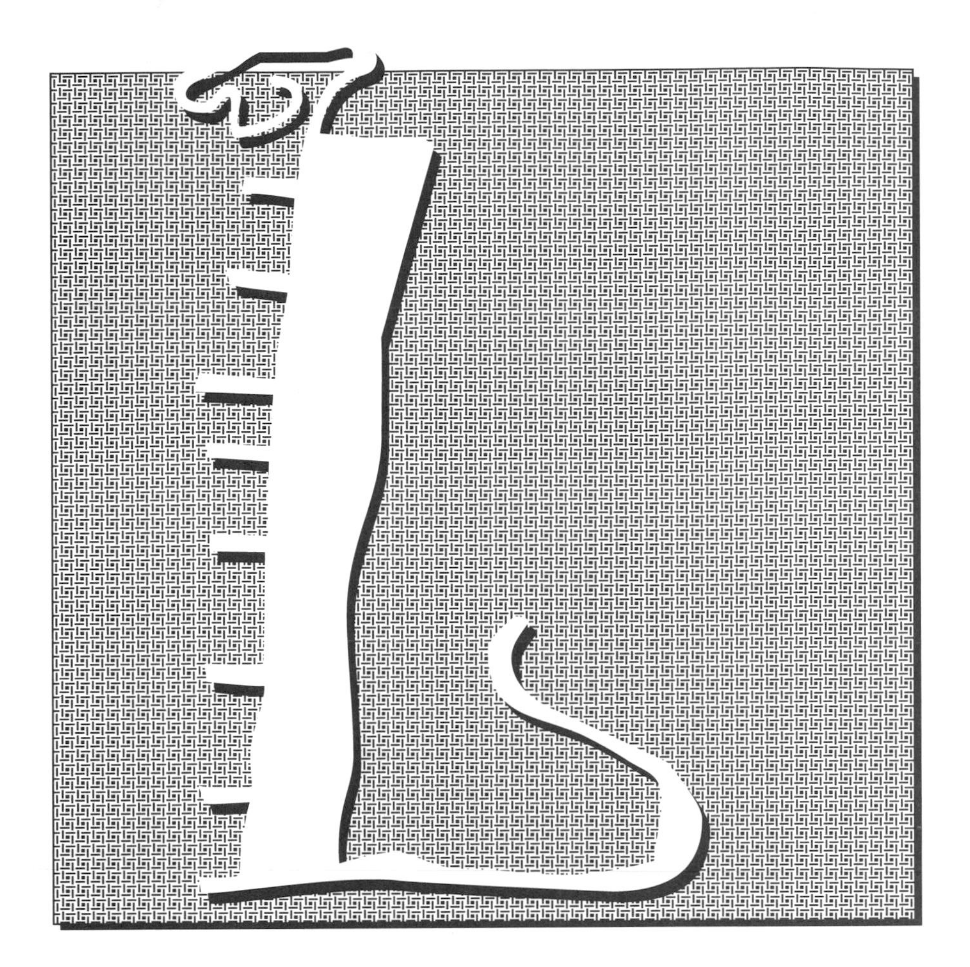

Lady Lovelace

Lady Lovelace is known as the first programmer in the world. Some of her programming notes written in the mid-1800s for Charles Babbages's "inference engine" have survived. Lady Lovelace, also known as the Countess of Lovelace, was Augusta Ada Byron, born in 1815, the daughter of Lord Byron, the English romantic poet (or was he the romantic English poet?). The Pascal-based programming language that was developed as a standard for the U.S. Department of Defense is named Ada in her honor.

LAN

LAN looks like an acronym that a board of directors spent a lot of money and time trying to create, but it actually stands for any generic **l**ocal **a**rea **n**etwork. A *network* is a group of computers and other devices connected together so they can pass information back and forth. LAN is a generic term for any network in which the devices are in close proximity (like in the same office or building).

For instance, in a computer lab in a school or office there may be several computers that share a printer or two, that *access* information from one main computer, or that are set up to be able to send files to each other. The entire conglomeration of the computers, the printers, and whatever software and hardware they use to communicate with each other is the LAN. Compare with *WAN,* wide area network.

The term LAN can also refer just to the hardware and software that allows you to connect all the devices together. In this sense, *LocalTalk* is one kind of LAN, *Ethernet* is another. (*AppleTalk* is the *protocol* for LocalTalk.)

landscape

Landscape refers to the orientation of a page when you create or print a document. In a landscape orientation, text runs across the long direction of the paper, so that you read the page with the short edges at the sides. (Most of us would call this "sideways.") By contrast, the usual way to look at a page (or to create one or print one), with the short edges at the top and bottom, is called *portrait* orientation. Most applications let you set up your documents in either portrait or landscape mode.

Landscape orientation is sometimes called "wide" or "horizontal," and portrait is sometimes called "tall" or "vertical."

This is the tall orientation, also known as vertical or portrait.

This is the wide orientation, also known as horizontal or landscape.

language

There are many different kinds of **languages** relating to computers, including many different programming languages, page description languages, machine languages, etc. Basically, the languages are used to tell the computer or the printer what to do and how to do it. The languages can be as different from each other as Chinese is from Swedish, and they each have a specific purpose in life. Also see *programming language.*

laptop computer

A **laptop computer** is a computer that is small enough and light enough (generally less than twelve pounds) to use on your lap. Laptops are primarily intended to be used when you travel. They're the most wonderful invention for those of us who get jittery and nervous if we are away from our computers for too long.

Most laptops run on batteries so you can actually use them on an airplane or at the beach. But to spare the batteries and to recharge them, laptops also plug into the wall socket when you're at home or in your hotel room. In the most common design, the laptop folds open, with the keyboard on the bottom and the screen flipping up to a vertical position (see *clamshell*).

A distinction is sometimes made between *portable* or *luggable* computers and laptops. Although laptops are definitely portable, a "true" portable is larger, weighs more (like maybe thirty pounds), and can't run on batteries. "Portable" simply means that it's fairly easy to move from one place to another without a dolly. But most people have come to think of a portable machine in the sense of a laptop or *notebook computer*. That's okay.

laser

Laser is actually an acronym that stands for **l**ight **a**mplification from the **s**timulated **e**mission of **r**adiation. Normal, or incoherent light, has multiple *frequencies* so it kind of spreads out all over the place, sort of like water pouring over a dam. Lasers, however, create coherent light that can be focused very precisely, because it has a single frequency and phase. It travels to its destination in a tight beam, kind of like water rushing through a narrow pipe. *Laser printers* use low-powered lasers, while surgery and welding use high-powered lasers.

laser disc

A **laser disc** (also known as a "video disc") is similar to a music CD, but it holds visual images as well as music. In fact, laser discs can store entire movies, concerts, operas, recordings of live theater, and a wide variety of educational material. Its signal gets fed right into your television or video monitor, just like the video tape movies you rent. Laser discs are typically 12 inches wide, just like a standard long-playing phonograph record (remember those?). Like a CD that you buy for listening to music, you can't record onto laser discs—they are *read-only,* meaning you can only listen to and watch them. The difference in quality and ease of use between a video tape and a laser disc is similar to the difference between a cassette music tape and a music CD. Prices for laser discs run from $20 to $500, with the most expensive ones being for the educational market. Some consumer-level laser disc players can play both laser discs and your collection of music CDs.

For techies: Laser discs store information in *analog* form, not *digital* form. There is no compression required, as the output of laser discs is analog video *(NTSC)*, not digital data.

LaserPrep

The **LaserPrep** icon you may see in your System Folder in System 6 (leave it there!) represents the information that tells the Mac how to print documents to a LaserWriter printer. It works in conjunction with the *LaserWriter printer driver*, represented by an icon that should also be in your System Folder.

LaserPrep icon in System 6

Printer driver in System 6

Printer driver in System 7

If you are using System 7, you no longer have a LaserPrep icon because all the information is now contained within the *driver*.

laser printer

By allowing ordinary people (that's me) to produce printed documents that look almost like they were professionally published, and to do it all in their own home or office, **laser printers** represent a true technological revolution. The first relatively affordable laser printers were the Apple LaserWriter and the Hewlett-Packard LaserJet. Prices have dropped like a rock and quality is way up since those machines came out, but the basic technology remains the same.

Laser printers work by laying down an array of tiny, evenly spaced dots of "ink" on the paper. The dots are so small and they blend together so seamlessly that text looks very nearly as clean as what you get from a traditional typesetting machine. Yes, there are little jagged edges along curved lines, but they're hard to see. Graphics sometimes don't turn out quite as sharp, but they still are clear enough for a professional look in many situations.

The quality you get from your laser printer depends mostly on the resolution, the fineness of the dots it uses to print the images. Resolution is measured in *dots per inch (dpi)*, which is how many dots it can print along a line, either vertically or horizontally. Most laser printers have had 300 dpi resolution, but 600 dpi is just now becoming the new standard. If you're shopping for a new printer and are willing to spend around $2,000, make sure you buy a 600 dpi machine—it's really tough to see any jagged edges.

Although the output of a laser printer is sometimes described as "near typeset quality," don't mistake the mechanical resolution with the aesthetic quality of good typography. No matter how high the resolution, if you still use typewriter conventions (such as two spaces after periods, straight quote marks [" instead of " and "] two hyphens instead of a dash, underlines, and other typographic faux pas), the text will never really look "typeset quality," even if you use 2540 dpi. If you are still using typewriter conventions and trying to produce high-quality output, you should check out the books *The Mac is not a typewriter* or *The PC is not a typewriter.*

Behind the scenes: a laser printer relies on a process something like that used to generate images on a television. In a TV set, a beam of electrons rapidly scans across the video tube, building an image out of tiny dots of light. In a laser printer, a tightly focused laser beam does the scanning. Although the laser source itself is stationary, a rapidly spinning mirror directs the laser beam so that it quickly traces a narrow, precisely horizontal line from one side of a special light-sensitive drum to the other. Initially, this drum carries a uniform static charge. As the laser beam moves across the drum, however, it flashes on and off, and wherever it touches the drum it reverses the static charge of a minute dot.

When the beam reaches the far edge of the drum, the drum rotates precisely. The laser beam then scans back across the drum, continuing to flash on and off. This cycle repeats itself until the laser has etched the drum with a pattern of charged dots corresponding to the entire page to be printed.

Even before the laser beam has finished its work, the drum's slow rotation brings the portions already etched into contact with the toner, a black plastic powder that serves as the printer's ink. The toner itself is charged so that it sticks to the dots etched by the laser. Eventually, all the etched areas of the entire drum receive a coating of toner. Meanwhile, a sheet of paper has been drawn into the printer by a system of rollers. In the process, these rollers have given the paper a static charge of its own, but one stronger than the charge on the drum. As the drum presses against the paper, the paper's charge attracts the toner away from the drum and onto the paper. More rollers move the paper into the fusing system, where heat and pressure affix the toner permanently to the paper. That's why the paper feels a little warm when it emerges a few moments later from the printer. As this is going on, a thin corona wire restores the drum's original homogeneous charge, readying it for the next page. Whew.

LaserWriter

LaserWriter, one word with a capital L and W, refers to any of the *laser printers* made by Apple (LaserWriter Plus, IINT, IINTX, IISC, etc.). The definition of laser printer is on the previous page.

Most of Apple's LaserWriters are *PostScript printers,* meaning they understand the PostScript language (which you should read about if you don't know what it is because everybody talks about PostScript and it's nice to know what they're talking about).

LaserWriter 35

Most *LaserWriter* printers have 35 *fonts* installed in the *ROMs* (the *read-only memory* built into the printer). These 35 fonts have become a standard of sorts for computer printers and are referred to as the LaserWriter 35. Actually, you get only 11 *font families,* but some of the families have four different styles. The first 32 of the 35 fonts are Courier, Times, Helvetica, Helvetica Narrow, Bookman, Palatino, New Century Schoolbook, and Avant Garde, all of which have four styles each (regular, italic, bold, and bold italic). Then there are Symbol, Zapf Dingbats, and Zapf Chancery, each with one style only (regular).

This is Avant Garde. **This is Bookman.** This is Courier.
This is Helvetica. **This is New Century Schoolbook.** This is Palatino.
This is Times. Τηισ ισ Χουριερ (Symbol). *This is Zapf Chancery.*
☆ ●❑❖❄ ✸✯✳✌ ✰✣▼❄❒✁ (Zapf Dingbats).
This is Helvetica Narrow, which cannot be printed to an *imagesetter!* Don't do it!

lasso

The **lasso** is a tool found in many *paint* and *image editing* programs. It's used for selecting any randomly-shaped part of an image so you can make changes in what you select. When you click on the lasso tool, the cursor becomes a lasso. The *hot spot* of the lasso (the one spot that has any power in it) is the tip of the dangling rope (see illustration). With the mouse button held down, you drag the hot spot so it lassoes (encircles) the area you want to select. When you draw around an object or group of objects, you don't really need to connect the two ends: just let go of the mouse button and the end will snap straight over to the beginning point.

The hot spot of the lasso tool.

Once the area is lassoed, the image appears to shimmer so you can tell it's selected. You can then make your changes by cutting or copying the selected area to the clipboard, clearing it out of your picture altogether, moving it, or changing its colors, etc.

Most applications also provide other selection tools, such as a rectangular tool. What you can do with a selected area sometimes depends on the selection tool you used. For example, you may not be able to flip or stretch a selection you made with a lasso, but you can if you made the selection with the rectangle tool.

launch

To **launch** a program simply means to start it or to open it (same thing).

layer, layering

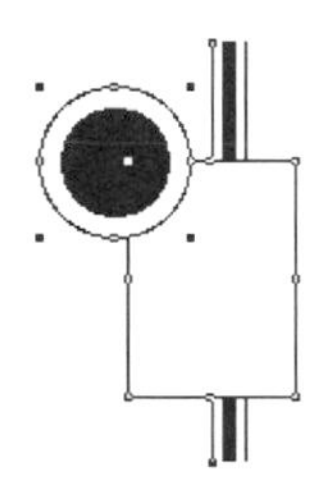

Each of these drawn objects is on a separate layer. Their positions on top of or behind another object can be changed. You can see the handles on each separate object.

The concept of **layers** has a couple of different meanings, depending on the type of application you are working with. Simple *bitmapped paint* programs have only one layer; that is, if you put one item on top of another item, the first item underneath no longer exists—it has been replaced by the second item because there is only one level of existence.

In *draw* programs (also known as *vector* or *object-oriented* programs) every drawn object is on its own **layer.** Thus any object can be placed "on top of" or "behind" any other object, and you can pull it off of or out from under any other object. In addition, many draw programs separate the drawing as a whole into completely distinct layers, each of which has its own collection of objects (each object still occupies its unique layer within larger layer).

From another viewpoint, many sophisticated graphics programs (both *vector/draw* and *bitmapped/paint*) let you work on different layers of your picture separately. In a draw program of this type, such as Aldus FreeHand, Altsys Fontographer, or Micrografx Designer, each distinct layer has its own collection of objects; the objects on each layer have their own front-to-back position within that layer. Fancy bitmapped graphics software—*image editing* programs such as Adobe PhotoShop—also have different layers. Each layer has its own bitmap, and you can apply different effects to the various layers using *alpha channels.*

Layering also applies to the different ways and levels you can work with and view a project in programs such as Aldus FreeHand or Altsys Fontographer. The various levels, or layers, can display background, foreground, and guides, for instance. Each layer has a specific purpose in the creation of the image.

Apple's HyperCard has *background* and card layers, each fulfilling a different but interrelated role in the construction of a *stack.*

LCD

LCD stands for **l**iquid **c**rystal **d**isplay. Your *digital* watch uses an LCD to show you the time, and most portable computers use an LCD to display the screen. There is actually a liquid compound, liquid crystals, sandwiched between two grids of electrodes. The electrodes can selectively turn on the different cells or *pixels* in the grid to create the image you see.

Steve's more technical explanation: An LCD consists of a layer of gooey material—the liquid crystals themselves—between two polarizing filters. These filters are sheets of plastic that let through only those light waves travelling parallel to a particular plane. Between the filters and the liquid crystal layer runs a thin grid of transparent electrodes.

The two polarizing filters are arranged so that their polarizing planes are at right angles. That setup would block light from passing through except for the fact that the liquid crystal molecules are "twisted." They pivot the light coming through the first filter, aligning the light with the polarizing plane of the second filter. Since the light makes it all the way through both filters, the screen looks light in color. However, the liquid crystal molecules that are controlled by a particular electrode become untwisted when a current is applied. Light no longer passes through the second filter, and you see a black or colored dot on the screen.

Most LCDs are passive matrix designs, in which each dot, or pixel, on the screen shares electrodes with other dots. *Active matrix* designs, which produce much brighter, more colorful images, have a separate transistor for each pixel, which allows greater control over the current for that pixel.

In "supertwist" LCDs, the liquid crystal molecules have a more pronounced twist than in the ordinary screens, improving contrast. The chemist's term "nematic" refers to the molecular structure of the crystals—all LCDs use nematic crystals, so this term is used in ads just to impress you.

Although you can read an LCD screen in room light, the contrast is mediocre at best. Today, the LCD screens on most computers are illuminated by backlighting or edgelighting (fluorescent-type lights mounted behind the screen or along either side).

leaders

Leaders (pronounced "leeders") are the periods, dots, or other little symbols that lead your eye from one column to the next, as in a table of contents. Most applications that use words have a tab that automatically applies the leaders.

Dot leaders .... **page 304**

leading

Leading (pronounced "ledding") is the amount of space between lines of text. The term comes from the little strips of lead that used to be inserted between the lines when all type was made of metal—little strips that were just a few points thick—so they added just a few points of space between the lines (each point is 1/72 of an inch).

So if leading is just a few points, why do you have to enter amounts like 12, 14, and 18 point leading? Well, the way it works is that you take the point size of your type, say 12 point. You add a couple points of lead, say 3 points. You add the 12 and the 3 together and we say you have a leading value of 15.

A standard amount of leading is an extra twenty percent of the point size of the type. For instance, if you are using 10 point type, your standard leading would be 12 point. This is the norm from which you can deviate, depending on whether you want tight lines or loose lines, whether your type has a large *x-height* or a small one, whether you're trying to conserve space or fill it, whether your type is bold and black or light and airy, and so on. And in some situations, you may need "negative leading," where the leading value is *less than* the point size of the type. This is very common with large type, especially when there aren't many *descenders*. The example below is 24-point type with 20-point leading.

EXPECT
MIRACLES

(Traditionally, a leading value of 15 always meant there were 15 points from the *baseline* of one line of text to the baseline of the next. Electronically, there is no standard as to how leading is applied. Some applications apply the extra 3 points above the line of text, some apply it below the line of text, some divide it up and apply a little above and a little below, etc. You need to check your manual if you are concerned.)

The term "leading" was just being phased out for the more appropriate term "line spacing" (since we no longer use lead) when desktop publishing hit the streets and hung onto the outdated term "leading." Now it looks as though we're stuck with "leading" for a while longer.

learning curve

When you hear people talk about an application's **learning curve,** they are referring to how long it takes and how hard it is to learn to use the application. A high, steep, or long learning curve indicates the program is difficult and that the person who uses it will probably require training. A short learning curve implies you may not even need to read the manual (but you should anyway!).

The learning curve is often an indicator of the power of the program. For instance, the *relational database* called Fourth Dimension has an extremely high learning curve and it can do extremely complex things. The personal contacts database called TouchBASE has a very low learning curve—like it will take you about maybe ten minutes to figure it out—but it doesn't do anywhere near what Fourth Dimension can do. Although the learning curve is a good indicator of the complexity of a program, it does not indicate its relative usefulness to a person—for instance, I personally have no use for Fourth Dimension; I would not be very happy without TouchBASE.

LED

LED stands for **l**ight **e**mitting **d**iode. You know those little lights on your computer, usually near the hard disk, that flash while the computer is working? Those are LEDs. They work on the principle of electroluminescence, which refers to substances that glow when you apply electricity. LEDs were used in digital watches, but now all digital watches use *LCDs* because LCDs take less power.

LEDs are ordinary diodes (the most basic electronic component) that, due to their composition, happen to glow red, green, or amber when energized by a couple of volts. They use less power than incandescent bulbs and last over 100,000 hours.

LED printer

An **LED** (**l**ight-**e**mitting **d**iode) **printer** uses an electrophotographic process similar to laser printers, where an electrostatically charged drum transfers toner to the paper. The most significant difference between a *laser printer* and an LED printer is that the laser printer uses a single light source, sweeping back and forth, whereas an LED printer uses a stationery array of lights that go on and off. See *laser printer*.

legible, legibility

Typographically, **legibility** is different from *readability.* Legibility refers to how easy it is to tell which letter is which: Is that an **n** or is it an **h**? Readability, however, refers to how easily a great deal of text can be read without tiring your eyes or driving you nuts.

There have been extensive studies on legibility and readability. The studies show that *sans serif* typefaces are more legible than *serif* typefaces. That is, it is easier to instantly recognize a short burst of text set in a sans serif font. (Serif type, on the other hand, is more readable; it is much easier to read large amounts of text when it is set in a serif face.) Now, this doesn't mean that just because a typeface is sans serif that it is legibile. There are many design features that create or destroy legibility. Take a look at these three words for a quick comparison of their legibility.

STOP STOP STOP

letter quality

It used to be a big deal if the printer for your computer could print **letter quality,** which meant that the quality of the type looked similar to the quality you could get from a very nice typewriter, instead of looking like it was being made out of coarse dots from a computer. Now with *near typeset quality* printers available and in widespread use in homes and businesses (like laser printers), if you print a document with mere letter quality people snicker at it.

LHA, LHARC

LHA is a fast, efficient PC program for *compressing* files so they take up less room on a disk and less time to send by *modem*. This one program does everything that *PKZIP, PKUNZIP,* and *PKSFX* do all together, and arguably as well. Plus it's free from *bulletin boards* and *online services*—you don't even have to pay a *shareware* registration fee. LHA used to be called **LHARC.**

license

Guess what: You actually do not own the software for which you paid all that money. What you bought was a **license** *to use a copy of the software.* The original—and actually all the backup copies you make—belong to and are technically owned by the software company, forever and ever. You can, if you like, "transfer" (what you might think is "selling") the program and its license to someone else, as long as you include all the documentation and the license. Some software companies, though, do not like to transfer the license and thus the privilege of *upgrading*— this can create problems for the person to whom you "transferred" the software.

Sometimes people sell their used computers loaded with all kinds of software and jack up the price of the computer to include the software, making you think you are getting a "good deal." Keep in mind that it is completely illegal to sell the software unless it comes complete with all *documentation* (like the manuals) and *serial numbers*. If they can't give you the docs and serial numbers, it may mean they *bootlegged* it from someone else and they shouldn't be charging *you* for it. The nerve.

license agreement

The **license agreement** is the small print on the outside of the envelope that contains the software for which you paid a lot of money. If you open the envelope, you agree to the conditions in the license agreement. See the definition above for what the *license* means to you.

LIFO

LIFO (pronounced "lie foe") stands for **l**ast **i**n, **f**irst **o**ut. When data gets processed, it is often lined up in a queue (orderly line) of some sort. In a LIFO arrangement, the last item in line is the first one out to be processed. (Another sort of arrangement is FIFO, which is **f**irst **i**n, **f**irst **o**ut.)

ligature

A **ligature** is a special character that combines two or more single characters into one. For instance, in many fonts when an **f** is next to an **i**, the hook of the **f** bumps into the dot of the **i** (shown below). So traditionally typesetters have used a ligature for those two characters: ﬁ. There is also a ligature for fl: ﬂ.

flying fish *without ligatures* *ﬂying ﬁsh* *with ligatures*

On a Macintosh, you can type the ﬁ ligature in any non–city-named font by pressing Option Shift 5. Type the ﬂ ligature by pressing Option Shift 6.

In Windows with *TrueType* or *ATM*, you have to use a special *expert font* to get ligatures. Or you can use BitStream's FaceLift with Speedo fonts. For other PC programs, consult either the little book, *The PC is not a typewriter*, or your software/font manual to find the keystrokes, which may be an ANSI or ASCII code.

LIM

The acronym **LIM** stands for **L**otus-**I**ntel-**M**icrosoft, three of the biggest companies in the PC industry. They got together to work out the standard for *expanded memory,* or EMS, and the term LIM is used only to describe the EMS standard. "This memory board provides 4 megabytes of LIM EMS 4.0." See *RAM* for a definition of *expanded memory.*

line feed

Some printers have a button for **line feed.** Each time you press the button, the printer pulls the paper through the printer one *line* at a time (usually 1/6"). This is different from *form feed,* which feeds one entire sheet of paper through (or the equivalent length of paper, typically 11 inches). The printer must be *off-line* to use this (either push the "off-line" or the "select" button).

On an ImageWriter printer, if you hold the line feed button down, after four lines it will start to form feed until you let up on the button.

lines per inch, line screen, line frequency

Lines per inch (abbreviated *lpi*) is sometimes confused with *dots per inch,* but the two terms mean very different things. Dots per inch refers to the *resolution* of a graphic image or a printer, or how many dots the printer uses to create the image on the page. The higher the number of dots, the smoother the edges of the image. The resolution of a printer is fixed (I strongly suggest you read the definition for *resolution).*

On the other hand, when you print a photographic-type *grayscale* or colored image, you can choose (if you use the right software) the lines per inch setting, also known as the **line screen** or the **line frequency.** This measure refers specifically to *halftone screens* or *tint screens.* (They are called "screens" because we used to make halftones or tints by shooting the original image through a "screen," a piece of mylar with halftone dots or tint dots embedded in it.)

So what is a **line screen**? A black-and-white photograph has gray tones that blend into one another, right? But you know what happens if you reproduce a photograph on a copy machine. The copy machine cannot reproduce the actual shades of gray, because it can't put different values, or shades, of gray toner on the page—it only knows how to put black toner *on* the page or leave the toner *off* the page. The same is true of printers that generate output from a computer, from a laser printer to a commercial *imagesetter*—they can only put black on the page or nothing

If you look closely, you can see the halftone dots.

on the page. So if you try to print a grayscale photograph or gray lines or gray text, the printer has to imitate those gray values. It does this by breaking the image into dots; some dots are larger and surrounded by a small amount of white space; some dots are small and surrounded by a lot of black space; some dots are small and surrounded by a lot of white space; and every variation in-between. When we look at an image that has had its tones broken into dots (a *half* tone), our eyes see the **illusion** of gray.

Anyway, these dots are lined up, usually diagonally, in rows with a certain number of lines per inch, such as 80 or 133. The number of lines per inch is also known as the "line screen" or "line frequency." The coarser the line screen (the lower the number, the fewer lines per inch), the bigger the dots (see the example).

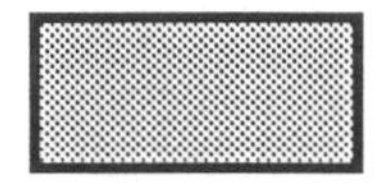

This box with a 20 percent shade has a line screen of 53 lines per inch.

All *PostScript* and some non-PostScript laser printers apply a *default* (automatic) line screen value to gray tones as they output the document. That means that if you select a box and apply a gray value to it, like 20 percent, the text will not be printed with gray toner, but with black and white dots (I'm sure you've seen those dots in a box before). Most personal laser printers apply a default screen ruling of 53 lines per inch (see the example).

This is the same box as above with a line screen of 105 lines per inch.

High-resolution i*magesetters* (large, expensive printers) have their own default line screen values. If you plan to send your file to an imagesetter at a *service bureau* for the final printout, you need to find out what the imagesetter's "default line screen value" is (call and ask). You can ask the service bureau to change the imagesetter default for your job, which they should be happy to do. You can use your software application, such as Aldus PageMaker, QuarkXPress, or Adobe Photoshop, to change the line frequency of an image, in which case **your** specifications will override the imagesetter's defaults for that particular image. Thus you can actually have several different images, each with a different line screen.

If you can specify the number of lines per inch you need, how do you know what you need? You ask your printer. No, not your laser printer—you ask the press person who will be reproducing the job. Newspapers like to use screens of about 65 to 85 lines per inch. Quick presses like about 85 to 100 lines per inch. Standard high-quality such as you see in a good magazine is 133 lines per inch. A high-quality press can reproduce up to 150 lines per inch, but you'd better ask first. Richard Benson can print 300 lines per inch.

Now this is the trick statement: Whether you print an image with a line sceen of 25 lines per inch or 150 lines per inch, *each of the dots of the screened image will be created with the resolution of the printer.* In the example on the next page you can see the large dots from a laser printer at 300 dpi as compared to the same large dots from a Linotronic with 1270 dpi. The halftone dot is the same size, 25 lines per inch (lpi), **but the resolution of the dot itself is different.**

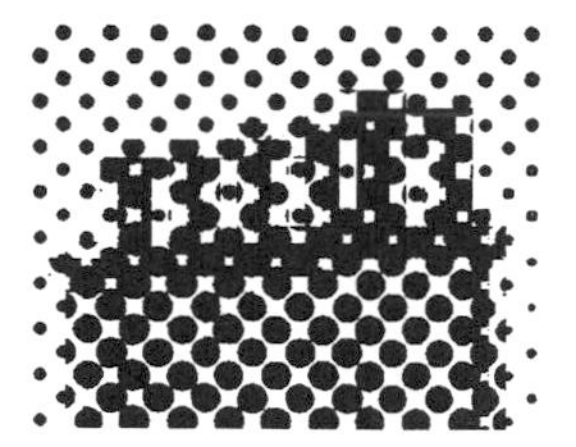

This is a 12 line screen, printed at 300 dots per inch.

This is a 12 line screen also, but it was printed at 1270 dots per inch.

This is the same image with a 105 line screen, printed at 1270 dots per inch.

line printer

A **line printer** is a high-speed printer that is used with large computers like *mainframes,* rather than with personal computers used by people like you or me. A line printer prints an entire line at a time, sometimes hundreds of lines per minute, whereas a *dot matrix printer,* such as many people have at home, prints one character at a time. Line printers usually use *computer paper,* which is 11 x 17 inches wide.

line spacing

Line spacing is the space between lines of type, which you may also hear referred to as *leading*. It was properly termed "leading" when all type was made out of molten lead, but "line spacing" is really a more appropriate term at this point in history. Unfortunately the computers don't always follow typographic trends (as evidenced by the fact that every Mac arrives with Helvetica installed). Please see *leading*.

link

When an application claims to retain **links** to objects outside of itself, it means that the program keeps track of elements that have been incorporated into the document, elements that were created by other applications. For instance, graphic images can get quite large, taking up a lot of disk space. When you put graphics in your document, the document gets larger and more complex, and it becomes slower and more cumbersome to work in. But if a program is capable of retaining links to the graphics, then you have the option of placing just a *low resolution* version of the graphic in your document for placement and manipulation, thus keeping the file size smaller. When you print, the application follows the link to the original graphic and gets all the data it needs to print the actual, *high resolution* version of the graphic.

Another advantage to linked files is that if you change the original, say the original of that graphic we just talked about, then you can update the graphic on the page of your document to reflect those changes without having to replace the graphic—you just click a button and the image updates. The same is true of *word processing* files: Usually when a large publication is being created, several people write and edit the stories and someone else arranges them on the page in a *page layout program*. If the page layout program can link to the original word processed file, then when the writer makes any changes to the original story, you can choose to update the text on the finished layout page with those changes.

Hot links, or **live links,** go a step further. In a program that provides hot or live links you can use a keystroke combination, such as Option double-click, to pop you directly into the program (or sometimes a comparable program) that created the object you double-clicked on. You can change the graphic or text of the original, then pop back into your original program and instantly see the changes. Oh, it's too amazing.

In PC usage, "hot links" are the same as "links," as described above. *OLE* does what is described as "hot links" here. Same thing; slightly different terminology.

So the only difference, really, between a **link** and a **hot link** is that a plain link knows where the original is and updates the document in which the link appears if **you** ask it to, whereas a hot link **automatically** updates the document.

Hypertext links are links in a text document that are connected to related information in the same or in other documents. When you click or press on a word in a hypertext document, it might lead to such activities as a search within a document, a pop-up menu with choices for finding related information, or definitions and annotations for that word or concept.

Another kind of live link, often called "file sharing," is the connection between two running programs that exchange information. In this way, another user can make a change using one of the programs; the change is immediately seen by other running programs.

Programmers use other kinds of links, but since this book is not written for programmers, we won't talk about those.

lino

Lino is short for *Linotronic,* the most popular brand of *imagesetter* (see the definition below). Although Lino is technically a brand name and there are other popular imagesetters, the term is often used to mean any output (printed pages; see below) from any imagesetter. The term **lino** can be a verb, as in "Are you going to lino this job?" or a noun, as in "Here's my lino," or an adjective, as in "Let's get lino output for this." In each instance, though, lino refers to the high-quality output.

Linotronic

You're probably accustomed to using your computer to print to your personal printer. Printers in the home or office usually print onto letter-sized plain ol' paper and have a *resolution* of anywhere from 74 to 300 or 600 *dots per inch*. (This resolution, or dots per inch, determine how smooth the edges of the text and graphics appear; see illustration.) If you want very high resolution, *typeset quality,* you can print to an *imagesetter,* which has a resolution like 1270 or even 2540 dots per inch, uses a smooth, resin-coated paper, can print onto film or negatives, and can print onto larger sizes of paper. A **Linotronic** (affectionately called a "Lino") is a high-quality imagesetter, one of the most common and popular. It is not the only kind of imagesetter, although some people mistakenly use the word to mean any imagesetter (like the way we use the words Kleenex or Xerox).

Lisa

The **Lisa** was a computer introduced by Apple in 1983, a predecessor to the Macintosh. It was the first personal computer with *integrated software,* and was the first personal computer to use an adaptation of Xerox's *user interface* with *icons* and a *mouse*. It didn't get wide acceptance (partly because of its $10,000 price), and when the Mac was introduced a year later with a lower price tag, greater dependability, and a more compact shape, the Lisa was doomed.

LISP

LISP is an acronym for **lis**t **p**rocessing, which is a high-level programming language that is one of the standards for *artificial intelligence* applications. Also see *programming language*.

list box

In *dialog boxes* such as the kind you get when you want to save a document or open a new one, you usually see a list of the files that are already in that directory or folder. This is a **list box.** An *edit box* is, for example, the little space where you type the name of the file you want to save.

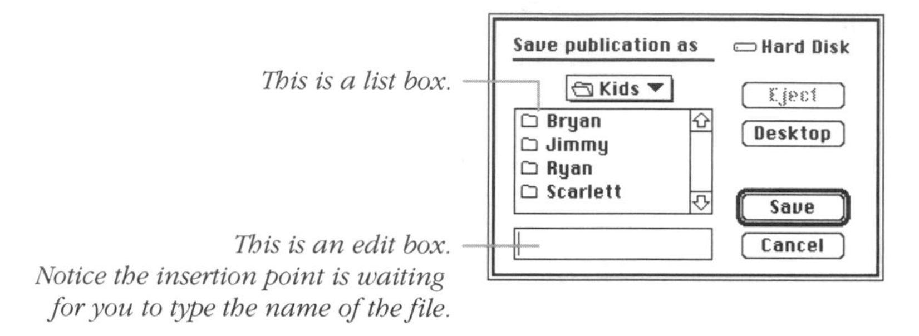

live link

Also known as a *hot link.* See *link.*

liveware

Well, *hardware* is the stuff you can knock on. *Software* is the invisible magic that makes the computer do what we want it to. And **liveware** is us—the living and breathing and often basically intelligent human beings who push the buttons. Other slang for humans is jellyware or wetware, but those sound a bit naughty.

load

Load means to get something ready to use, typically by transferring the disk, application, or file from wherever it is stored to where you can use it. You can load a hard disk by copying programs or documents onto the disk so you can work with them. "Before you open that document, let me load these new fonts," means to load the font files into the system so they can be used. You can load a floppy disk, which simply means to stick the floppy disk in the disk drive slot. Compare *download* and *upload.*

Most often you load software or information by moving it from disk into memory. For example, you load your *memory resident* programs *(TSRs)* into memory so you can pop them up later.

load high

PCs have a section of *memory* called "upper memory" that used to be off limits for your programs, but which you can now use under certain conditions (see the long definition of *RAM* for a description of upper memory). In particular, if you have a computer with an *80386* or newer *microprocessor,* you can load *memory resident programs (TSRs)* and some parts of DOS into upper memory, or "**load** them **high.**" That way, you can use as many TSRs as will fit into upper memory without reducing the amount of *main memory (conventional memory)* available to run your main applications. To do this, use a special utility called *LOADHIGH* that comes with DOS 5 or DOS 6, or a similar utility that comes with a memory management product such as QEMM-386.

LOADHIGH

LOADHIGH is a *utility* that comes with DOS 5 and DOS 6. It allows you to load your *memory resident programs (TSRs)* into upper memory (to load them high). See *load high*.

local bus, local bus video

If you're buying a new PC with an *80486 microprocessor,* get one with **local bus video** for high-speed display on the screen. Here's why: Traditionally, *VGAs, EGAs,* and other video circuits for PCs have been placed on standard *add-in boards,* the kind that insert into the main *expansion slots* inside the computer. The problem with this is that the standard expansion *bus* (the circuits of the expansion slots) can't transfer information from the computer to the video circuits quickly enough for graphical software like Windows. It takes lots of information to generate a graphics screen, and the amount of information required goes up geometrically as you increase *resolution* and the number of colors you can see. Even if you have a super-fast PC, your screen may respond sluggishly because the expansion bus acts like a kink in a hose, constricting the flow of information to the display.

The answer is local bus video. A **local bus** is one designed to match the speed of the microprocessor, the computer's main electronic brain. While a local bus could speed up other types of high-speed devices, video benefits the most. Some PC manufacturers have come up with their own local bus video circuits, but proposed standards such as VESA's VL-Bus and Intel's PCI are now emerging for local bus expansion slots. With a standard slot, you can pick a corresponding video board from any manufacturer you choose.

Local bus video can't be added to existing PCs. But you can still get a major boost in the screen speed by adding a *video accelerator board.*

localization

Localization means to adapt a program to suit the locale of where it's going to be used. For instance, the words in the menu bar may have to be changed for use in a different language, or the alphabetizing order in a database may need to be adjusted, or the decimal points and commas in numbers need to be switched, or the date format needs to be changed.

LocalTalk

The *AppleTalk network* (a network system is a way for computers, printers, and other *devices* to share information and services) is built into every Macintosh model. **LocalTalk** is one of the types of network hardware available in the AppleTalk system. For instance, in a school lab situation there may be several computers that need to print to one printer. So you buy these little LocalTalk connector boxes and a cable for each computer and string 'em all together in a *daisychain*.

LocalTalk is the physical connection. *AppleTalk* is the *communications protocol,* or the rules the data must follow as it flows through the wires so the information doesn't get mixed up or go to the wrong destination.

PCs can be connected to a LocalTalk network with the right *adapter board.*

lock, locked

On a Macintosh, you can **lock** a disk, an application, a document, or any other file on your computer. Typically when a disk or file is locked, you can look at it, you can open it, you can read it, but you cannot make changes to it. You might be deluded into *thinking* you can change it, perhaps because you can type into the document or move something off the disk, but it is fooling you. As soon as you try to close the document or take the disk out, all the changes you thought you made will disappear. If you try to save the changes to a locked document, you'll find that the Save button is deactivated (gray). You can save the document with another name, which makes a second version of it, or you can unlock the document or disk.

To lock a 3.5 inch disk, just make sure the little tab is snapped into the open position—if you can see through the hole, the disk is locked. (Yes, you have to take it out of the *drive* to lock or unlock it.) On *high-density disks* there are two holes, but only one hole has a movable tab. It's a good idea to always lock the original disks of your software to prevent any accidental changes and to prevent *viruses* from infiltrating.

To lock files (including applications) you need to be at the *Finder* (also known as the *Desktop*). Click once on the file *icon,* then from the File menu choose "Get Info" (or press Command I). Click in the little box that says "Locked." Put this window away by clicking in its close box in the upper left corner (or by pressing Command W). Unlock a file by unchecking the "Locked" box. [illustrate]

You can also use the "Sharing..." command in the File menu to limit a person's access to a file more specifically. It isn't really "locked," but the effect is the same.

In the PC universe, the term "lock" isn't used much. The term for a locked disk is a *write-protected* disk. A locked file is called a *read-only* file.

logic, computer logic, program logic

The sequence of instructions in a software program is called **logic,** or **program logic.** The chips and circuits that process data and control the operation of the computer is the **computer logic**. Logic also commonly refers to specific instructions (in hardware and software) where "Boolean" logic is used: IF/THEN/ELSE, AND, OR, XOR (exclusive or), etc.

logic board, logic card, logic chip

The **logic board** is the *board,* or *card* (piece of plastic that holds the *chips*) that has the **logic chip** on it. The logic chip does the processing and controlling of the computer. Generally only the bigger computers have a separate logic board, since the *microprocessor* itself in the smaller computers (like our personal computers) has the equivalent of a logic chip built into it. That's why the *motherboard* in personal computers is sometimes referred to as the logic board.

log on, log off, log in

Whenever you want to use a *network* or a *bulletin board service* or an *online service* through your computer, first you have to **log on** or **log in.** You'll know you are logging on because you have to sign in, usually with a *password* of some sort. Once you log on and satisfy the computer with your proper identity, you are allowed to use the service or the network. When you want to disconnect, the process of disconnecting is called **logging off.**

When you connect to a service or network, your computer is actually calling another computer. So the process of logging on and off is similar to you calling up your friend and saying hello (logging on), chatting for a while and doing business, then saying goodbye (logging off).

look-and-feel

Each type of computer and its *operating system* has its own **look-and-feel,** resulting from its particular ways of operating and its visual *interface.* For instance, a computer that uses a *command line interface,* where you must type lines of codes to make the machine do what you want, has a very different look-and-feel from a computer with a *graphical user interface,* where you can click on pictures *(icons)* to accomplish tasks.

Whether you enjoy a particular computer's look-and-feel boils down to your overall psychological response to things like the keyboard, the mouse, the software, how it works, how it looks, etc. When you use a Macintosh you can sense the very different look-and-feel from an IBM PC, which has a different look-and-feel from an Amiga. And each software program has its own special look-and-feel. Just like people.

loop

A **loop** is a set of commands that repeat themselves. Sometimes you plan to create a loop; sometimes it is an accident. Programmers create loops on purpose all the time, but you might accidentally create a loop in your spreadsheet or in a HyperCard *stack* or in a *macro.*

Loop can be a noun, as in "I don't know how to get out of this loop," or it can be a verb, as in "Loop this until you get to the last card." What you **don't** want to create is an endless loop that just keeps going forever because you forgot to tell it when to stop, or because maybe you told it to stop somewhere but it can't find that place. For instance, maybe you set up a situation where you want the program to add 2 to the number someone entered in a field, and to stop when it reaches 100. Well, if the number in the field was originally 3, the computer will never see the number 100 and thus it will never stop.

lossless, lossy

When a graphic image is very large or complex, it often needs to be *compressed* if you want to move it from one computer to another, or if you want to save it for future use. Compressing compacts the data, or the information, that tells the computer what to display on the screen or what to print to the printer. A compressed file then takes up much less space on the disk. Usually you must uncompress an image if you want to work with it.

A compression scheme that works in such a way so as to prevent any loss of data while compressing the image, and thus preventing any degradation of the image, is called **lossless.** Some people claim there is no such thing as lossless compression.

A compressor that loses data every time an image is compressed is called **lossy.**

Lotus 1-2-3

Lotus 1-2-3 is (at least primarily) a *spreadsheet* program that was a major factor in the early success of the IBM PC. It's often just called 1-2-3, and sometimes it's even just called Lotus, although that's really the name of the company that makes it.

Spreadsheet software lets you make calculations on rows and columns of numbers, which is what much of business management and planning is all about. 1-2-3 wasn't really anything new—it was more or less an imitation of *VisiCalc,* the highly successful spreadsheet program for the Apple II. In 1978, VisiCalc was the program that started the personal computer revolution. 1-2-3 was just the first good spreadsheet to come along for the IBM PC, and the first one to be marketed well. For many years it was the best-selling business software of any kind in the IBM marketplace, and it made Lotus one of the biggest software companies. Over time, the program has been steadily enhanced, and there are now two separate versions for DOS, one for Windows, one for OS/2, and one for the Mac. But 1-2-3 no longer dominates the market for PC spreadsheets—Borland's Quatro Pro and Microsoft's Excel have taken away big hunks of the market.

1-2-3 does offer some features besides calculations: automatic charts based on the numbers in the spreadsheet, and primitive database capabilities. The clever name 1-2-3 is supposed to suggest two ideas: that the software can do three things (spreadsheets, charts, and databases), and it's as easy as . . .

lowercase

Lowercase letters are the small, uncapitalized letters, such as you are reading right now. The word lowercase is from the days when all type was created with metal letters. The tiny pieces of metal type were kept in large, flat cases with a separate cubby for each letter. There was a case for the capital letters and a separate case for the small letters. In the racks that held the cases, the case for the capital letters was above the case for the small letters. Thus the capitals became known as uppercase and the small letters as lowercase. There is an illustration of a case in the entry for *fonts*.

low-res, low resolution

If an image is displayed on your screen or printed on the page in **low-res** (short for **low resolution**), that means you are seeing a low-grade quality. Some graphics are just low-resolution to begin with, such as graphics made in the *paint file format* at 72 dots per inch. Some graphics are created as complex, *high-resolution* images, but you may choose to display them on the screen or print them in low-res just to save time, since it takes longer for a screen or a printer to create the high-res version. For instance, if you are producing a brochure on your computer and in the brochure you have several high-resolution photographs in full color, it can take a long time to turn pages or change views. So you can choose to view these images in low resolution while you are working, just so you can move around the screen faster. You can choose to print them in low res just so you get an idea of the look of the brochure without waiting to reproduce the entire high resolution images.

The lower the resolution, the less information there is in a given amount of space (in a square inch, for instance). It may mean that each pixel in that square inch of the screen is not providing enough information to resolve the image clearly, or it may mean there are less printed dots per inch on the page. Please see *resolution*.

LPT1, LPT2, LPT3

DOS uses the acronym LPT to refer to its three printer *ports:* **LPT1, LPT2,** and **LPT3.** The acronym is a contraction of **l**ine **p**rin**t**er.

When a DOS program prints, it usually passes the information to be printed to one of these three ports—LPT1 is used unless you specify a different port (LPT1 is the *default*). If you have more than one printer, LPT1 is usually the port you use for your primary printer.

The confusing thing is that the names LPT1, LPT2, and LPT3 aren't permanently linked to the actual, physical *ports* on your computer. Almost every DOS computer has at least one *parallel port*, and initially, LPT1 is identified with one of these ports. Using DOS commands, however, you can change things around so that when your program sends something to LPT1, DOS sends that information to one of your *serial ports* instead. This is called "redirection."

Why would you want to do this? Actually, it's not often very useful, especially now that the software we have is getting smarter.

But here's a possible scenario: Let's say you're using one of your serial ports to connect your computer to a *network* (a collection of computers and other devices all wired together). On your desk, you have a funky *dot matrix printer* connected to your PC's parallel port, but there's a beautiful high-resolution *laser printer* out there on the network. The dot matrix machine is OK for printing draft copies of your document. But when it's time to print the final masterpiece, you redirect LPT1 to that serial port, and now you can print on the laser printer, without changing the port your software uses (you'd still have to change the *printer driver*, and that's probably more trouble than changing the port).

LSI

LSI stands for **l**arge-**s**cale **i**ntegration, a measure of the density of *chip* technology. LSI describes the density of a single chip that holds between 100 and 5,000 transistor devices on it. If you think of those big transistors that used to be in radios and computers and imagine 5,000 of those in a space smaller than your baby fingernail, you begin to have a clue as to the scale of this technology.

VLSI stands for **v**ery **l**arge-**s**cale **i**ntegration, which can put over 100,000 transistors and related components on one chip. Current chips like the Motorola *68040* (used in the Macintosh Quadra) can hold about a million.

luggable

Luggable is an unofficial term (most of this stuff is unofficial) that refers to a computer that you can pick up and carry, but that is still too big and heavy to move around conveniently. Laptop and notebook computers like the Macintosh PowerBook have made luggables obsolete.

"Luggable" is usually used to describe computers that were designed, optimistically, to be portable—they fold up neatly into one box with a handle, but that box is about the size of a small suitcase and it weighs

25 or 30 pounds. These things had built-in, really small picture tube screens. Two of the first luggables were the Osborne and the Kaypro, and Compaq sold thousands upon thousands of IBM PC-compatible luggable machines. Even IBM made a luggable called the PC Convertible.

The self-contained Macs (the original Mac, the Plus, the Classic, or any of the SEs) could be considered luggables—they do have a handle on top, but you have to carry the keyboard separately. "Luggable" is also a fond nickname given to the Macintosh Portable; even if you had the strength to lug it all the way to the airplane, you could barely fit it on your lap. The Macintosh *PowerBook* has made the Mac Portable obsolete.

lunchbox

A **lunchbox** is a portable computer designed in the shape of a lunchbox, more or less. The screen is mounted vertically, connected to the top of the computer on a pivot or hinge, and swings out from the bottom. The keyboard folds up onto the front of the computer, and is usually attached by a cable so you can put it on your lap or wherever is comfortable. Lunchbox designs are out of date now, because you can't balance them on your lap on an airplane. Today's laptop and notebook computers have a clamshell configuration.

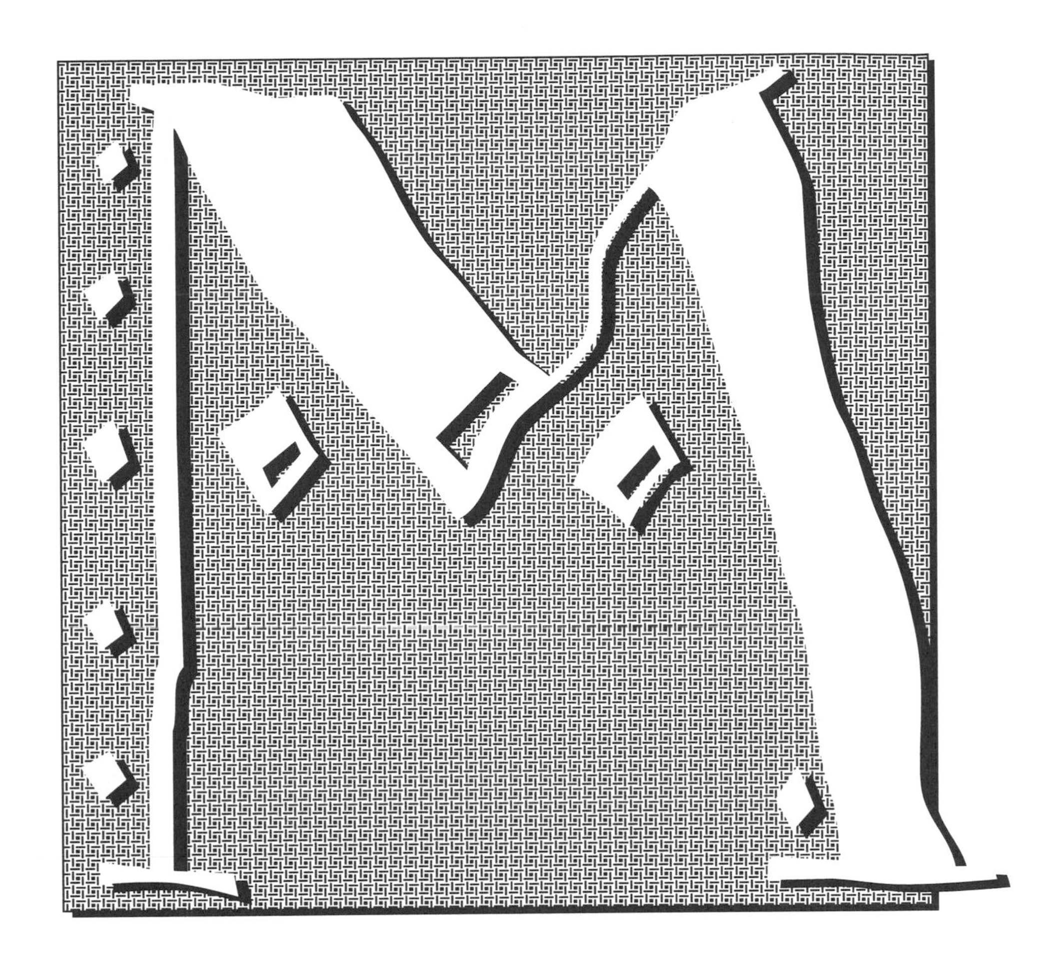

MacBinary

Macintosh files are divided into three parts: a *data fork,* a *resource fork,* and a "finder information block." If you want to transfer a Mac file to another type of computer, you have a problem: the other machine isn't prepared to recognize those three separate parts of the file, and will most likely lose one or all three. The information you're trying to send may not get there, and if you try to retrieve the file later, it won't come back in a useable form.

To solve this problem, a standard *file format* was developed that other types of computers can use to store Macintosh files. This is the MacBinary format. Almost any *telecommunications* or file transfer software on the Mac that can exchange files with other kinds of computers recognizes this format. When you send a Mac file to a foreign computer, the software smooshes the three parts of the file into one so that the whole thing gets where it's going. If you later transfer a MacBinary file back to your Macintosh, the software automatically converts the file back into its original tripartite form.

machine code, machine language

Machine code is the only language a computer understands. It's just a bunch of numbers, represented inside the computer as 1s and 0s (ones and zeros). These ones and zeros are electronic signals that tell the computer what to do. Only a very few humans write machine code (of course, they write it out in numbers, not electronic signals). Instead, humans usually write the instructions for the computer in some form of *programming language*. Then the programming language instructions go through a *compiler*, an *assembler*, or an *interpreter*, which translates the human-understandable language into machine language.

MacinTalk

MacinTalk is Apple's "speech generating *utility*" (represented by the icon you see below, left) that works with certain programs to make things talk on the Mac. There are several educational software packages where kids can type words, click a button, and the Mac "reads" and "speaks" the words, using MacinTalk. It's too much fun. I've caught my kids typing mean and sometimes naughty sentences, but I must admit it really is funny to hear a robot say, "You are a bad m—f—. I am going to whop you upside the head and stick you in the garbage can, close the lid, and roll you down the hill." The little man that speaks the voice can even tell if the sentence is a question, can pause for commas, read misspelled words, and talk backwards.

Macintosh

Macintosh is the name of a model of *personal computer* produced by *Apple Computer* in 1984. A radical departure from other personal computers, the Mac (as it is so fondly called) uses a *graphical user interface* (GUI, pronounced "gooey") with *windows* and *icons*. Along with a *mouse*, the graphical user interface lets the user *point-and-click* to open programs, choose commands, copy, delete, print, etc., without memorizing complex codes. Because this interface is part of the *operating system*, all programs on the Mac have a similar *look-and-feel*, making it easier to learn new applications.

It's been very interesting to watch over the years how the Mac was scorned as a "toy" because of its graphical user interface, but now the operating system vendors for IBMs and *compatibles* are creating GUIs, such as *Presentation Manager* for *OS/2* and *Windows* for *MS-DOS*.

Yes, the name of the edible apple is "McIntosh," but the corporate Apple had to spell the computer's name "Macintosh" because there was already a trademark called "McIntosh" for an audio equipment manufacturer.

Macintosh Bible

The Macintosh Bible is a large book, now published by *Peachpit Press,* that has an enormous amount of information in it on a wide variety of topics relating to the Macintosh.

macro

A **macro** is a programmed shortcut, programmed by *you,* to make life at your computer easier. Whenever you find yourself repeating the same sequence of steps in your software, you could set up a macro to do the whole thing for you in a single step. Typically, a macro *runs,* or does its work, when you press a certain key or combination of keys, or when you choose the macro by name from a menu.

For instance, if you are typing several pages of names and addresses into a *database,* you'll probably get tired of typing "P.O. Box." So you can set up a macro that will type it for you. If most of the addresses are in the same city, you can set up a macro to type, for example, "Santa Rosa," "California," and "95401" in each of their respective *fields.* You can use macros to automatically activate a complex series of commands in your application; you could cook up a macro that sums your deposits, opens a deposit slip, records the information, prints it out, and then quits/exits the application—all with the click of one key. Macros are really too cool.

Some applications have macro capabilities built in, but you can only use the macros within that particular application. You can also buy software whose sole purpose is to enable you to create macros for other applications. Some of these macro programs even let you string together commands from different programs and from the operating system in the same macro. This way, you can bring all the resources in your system to bear on your particular problem in an automated, customized way.

Even though they may sound intimidating at first, macros are easy to use, and the simple ones that just type text or activate a few commands are very easy to create. In fact, you can often just record what you type at the keyboard or do with the mouse, and automatically turn those steps into a macro.

MacsBug

MacsBug is a software product from Apple that is used for debugging (getting the *bugs* or *glitches* out of) software. It's available *online* for free. It's mainly useful to programmers, but there are a couple of things it can do for mere mortals—such as occasionally being able to get you back to the Finder after a *crash*. It's rather technical to use.

Mac the Knife

Mac the Knife is a column in the weekly tabloid *MacWeek*. Supposedly no one knows who writes it. The author tells secrets and leaks news and exposes embarrassing moments.

MacWorld Expo

The **MacWorld Expo,** held in Boston, San Francisco, and about 17 foreign countries, is the largest Macintosh-specific tradeshow, drawing 40,000 to 60,000 people at each American show. It's sponsored by MacWORLD magazine, hosts hundreds of vendors and a multitude of seminars on a wide variety of topics. Plus you can find some of the greatest parties in the world.

magnetic media

Magnetic media refers to any kind of disks or tapes that rely on magnetism to store information so you can use it again later. The only common alternative is optical media like *CD ROMs* that use light to store information. See *optical disc.* Data (information) on magnetic media can be erased and recorded over and over again, as opposed to the data on *laser discs* and CD ROMs, although (sigh . . .) this technology is changing rapidly.

magneto-optical

Magneto-optical is a hybrid technology for storing and retrieving information that relies on a combination of magnetism and light (in the form of a laser beam). You can erase and re-record a magneto-optical disc as many times as you like, but the information is safer than on a hard disk. See *optical disc.*

mainframe

The term **mainframe** has shifted from its original reference to the main housing, or frame, that contained the *central processing unit* (CPU) of the computer. In those days, all computers were big—like the size of a garage—and the frame for the CPU might have been as big as a walk-in closet. Now mainframe refers to the kind of large computer that runs an entire corporation. While "large" can still mean as big as a room, most of today's mainframes are much smaller, although they're still quite a bit bigger than a *personal computer* or even a *minicomputer*. A mainframe has an enormous storage space on disk and tape (like thousands of *kilobytes*, measured in *gigabtyes*), and an enormous amount of *main memory*. Theoretically, it works a lot faster than the fastest personal computer. A mainframe also costs big bucks, from half a million or so on up.

Mainframes are tended by special technicians who feed them the programs they run and who scramble around trying to fix them whenever they stop working, which is often. All mainframes are *multi-tasking, multi-user* machines, meaning they are designed so many different people can work on many different problems, all at the same time.

Mainframes serve most often as information storers and processors. An army of smaller computers is connected to the mainframe. These smaller computers are not in the same room; they may be connected through phone lines across the world. Ordinary people in the company never touch the mainframe itself. Instead, they interact with the computer using a *terminal*, which is more or less a keyboard and a monitor connected to the mainframe with wires, or by modem over the phone lines. People use the smaller computers and get information from and send information to the mainframe.

The difference between a minicomputer and a mainframe is arbitrary, and different people may use either term for the same machine. Even if you don't work for a large company, you might have contact with a mainframe: when you connect to an *online information service* or a commercial *e-mail* service from your personal computer, you are often connecting to a mainframe.

In the '60s the mainframe vendors were called "IBM and the seven dwarfs": Burroughs, Univac, NCR, Control Data, Honeywell, GE, and RCA. They turned into IBM and the BUNCH after Honeywell ate GE's computer division and Univac ate RCA's.

main memory

Main memory is the computer's primary supply of *random access memory (RAM),* the part of memory in which the *operating system* and your software runs, and where information currently in use is stored. Your computer probably has other collections of memory too, one for storing the information that appears on your screen, maybe one for a *cache*, and so on. Please see *RAM*.

male connector

A **male connector** is a plug with pins or some other protruding part that is designed to fit into a female counterpart. The female connector, of course, has openings for the pins. When you see the connectors on the end of a cable, it is not very difficult to guess which plug is the male connector and which is the female

Maltron keyboard

The **Maltron keyboard** is a keyboard with a different shape from the standard rectangular board. The Maltron keyboard is designed to prevent problems with your hands such as *carpal tunnel syndrome*. The keys are laid out in two concave areas conforming to the shape of hands, which minimizes long finger stretches and keeps the forearms aligned with the wrists. It's less tiring to use because the most-often used characters are situated under the strongest fingers, including up to eight keys for each thumb (although a Maltron keyboard can be set up in the standard *Qwerty* layout).

manual

Whenever you buy any software or hardware, it usually comes with a **manual** that tells you how to use the item. Manuals get a lot of flack, and most manuals really aren't terrific. But even the worst manual has information that can help you. Many problems could be avoided, many headaches prevented, much tension relieved, and much knowledge gained if you just read your manual first. Remember the Number One Computer Tip: RTFM (Read The Manual).

map

Map is a common word in mathematics, computer science, and programming. A map is just a means for you or your computer to find something. Conceptually, a map is a table of corresponding pairs of items, though the "table" may not have rows and columns. The user of the map can find what they want by looking up the corresponding item in the table.

For instance, your computer keeps a "memory map," which relates the operating system's method for organizing memory (the "logical" memory) to locations in the actual memory chips inside the computer (the "physical" memory). A "bitmap" matches each pixel on the screen with a memory location (in *RAM*). And a "keyboard map" tells you what characters or action will result from pressing each key. Map can be a verb, as in "the pixels in the picture are mapped to bits in memory."

margin

When you are creating a document on your computer, you can almost always choose how close to the edge of the paper the text or graphics should begin. The space between the edge of the paper and the beginning of the text is the **margin.**

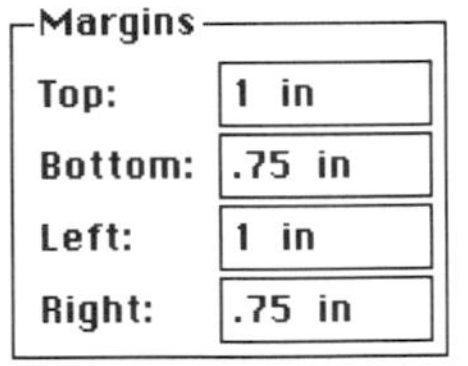

When you set the margins, you determine the space around all four sides of the text.

Now, don't confused the margin with the left and right *indents.* In most programs, if you indent from the left or the right, the indent moves in from the *margin,* not from the edge of the paper. Also see *first line indent,* a wonderful feature for eliminating the step of having to tab over to indent the first line of your paragraph.

marketing

Marketing refers to the plan for getting a product out to the people who need it, but first those people have to be convinced that they do indeed need the product the marketing person is trying to sell. Marketing includes figuring out just who it is who needs the product (or can be talked into needing it), plus advertising strategies, merchandising, pricing, distribution, etc. Marketing, rather than the quality of the product, is often the key to whether a product is successful or not.

marquee

When you press-and-drag the *mouse* to select *text,* the selected portion is highlighted, visually indicating what portion of the text has been chosen.

When you drag to select text,
it appears highlighted.

When you press-and-drag a pointer or tool around *objects* to select them, you usually see a border that looks like little marching ants. That marching border is called the **marquee** (pronounced "mar kee´"). Anything enclosed within that marquee is selected.

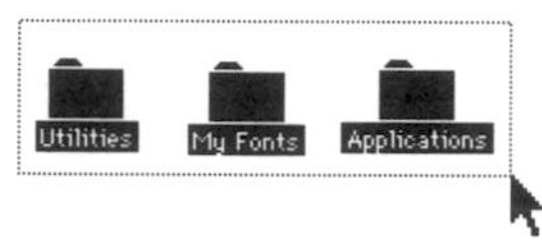

Sometimes you have to contain the entire object within the marquee; sometimes you just need to grab a portion of it.

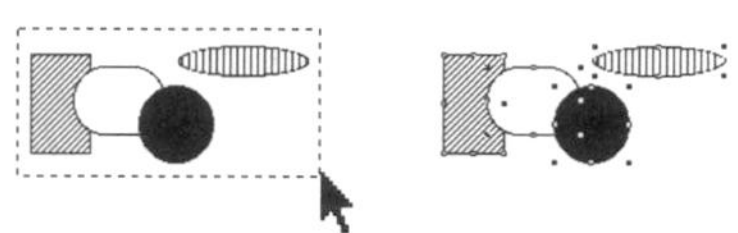

As soon as you let go of the mouse button, the marquee disappears and the objects are selected.

mask

A **mask,** in a graphic sense, is when you cover up something you don't want to see, or cover something that you don't want to be affected by another action.

The word "mask" can be a noun, as in "We need to put a mask over this area." Or it can be a verb, "Please mask the red portion so when we spray-paint it, the red won't get splotched." You can create a mask of an object and then copy the mask to do some other creative operation on it.

master, golden master

When a software product has gone through its *alpha* and its *beta testing,* it is ready to be **mastered.** The developers create one disk, a **master disk,** and set it up exactly as they want the user to find it—all the files are named properly, and everything is in the right folder or subdirectory; the software is set up so that the first time the customer uses it, the windows open to the correct position on the screen. This master gets sent to the company that duplicates disks. They duplicate the master and send it back to the software vendor for verification. This verified copy, the **golden master,** is sent back for massive duplication for the eagerly awaiting public.

master disk

A **master disk** to a *software vendor* (a company that creates and sells the programs you buy) is the disk that has the original program on it from which they make copies for distribution. See *master, golden master.*

A **master disk** for a *user* (someone who uses the product, like you) is the original disk that came in the box when you bought the software package. They always recommend that you first make a copy of that disk. Then use the copy to *load* the program onto your machine (which means to copy it onto your hard disk). Put the original master disk in a safe place. A master disk is also known as an *original disk.*

math coprocessor

A **math coprocessor** is the same thing as a *floating-point processor.* Please see the *floating-point* definition for a full explanation.

matte

A **matte** finish, or texture, is one that is not glossy, but has a tiny bit more shine than a flat finish. It may be smooth, but has no shine to it. Matte may refer to paper or to photographs or to a paint/varnish finish, or to a graphic technique that gives the illusion of a matte texture rather than a glossy texture.

MAUG

MAUG stands for **m**icronetworked **A**pple **u**sers **g**roup, a group on *CompuServe* (a major *online service*). MAUG holds at least thirteen separate *forums* dedicated to Macintosh topics and technology. MAUG holds conferences, has libraries, message boards, and other online features. It's one of the best places to mingle with other Macintosh minds and keep up on what's going on in the Mac world.

maximize icon

In Microsoft Windows, you can click the **maximize icon** to enlarge, or maximize, a *window* to fill the screen. If you click the icon (sometimes called a button) again, it will reduce the window back to the size it was before you enlarged it. This is equivalent to the *zoom* button on a Macintosh *window.*

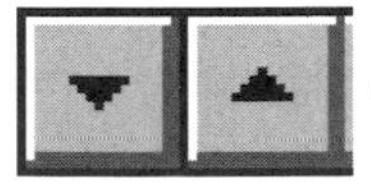

The maximize icon, or button.

Mb, MB

Once upon a time, **Mb** (with a lowercase b) was technically an abbreviation for *megabit,* as opposed to **MB** (with a capital B) which is an abbreviation for *megabyte,* but most people unfortunately use both Mb and MB to mean mega*byte.* One megabyte is 1,048,576 *bytes* (usually rounded off in speaking to one million bytes) and is a measure of the size of computer information or storage space, whether referring to document size, amount of *memory,* the storage space on a *hard disk* or *tape,* etc. Please see the definition of *byte* for more complete details.

Mega*bit* is usually used only when describing *memory chips.* A mega*bit* is 1,048,576 *bits,* which is one eighth the size of a mega*byte.* In other words, a megabit is 128K, which is a number familiar to people using the early Macintoshes, which had 128K *ROMS.*

MCI Mail

MCIMail is the largest commercial *e-mail* service on earth. It's part of the same company that runs the MCI long distance telephone service. If you sign up for MCI Mail, you can call up MCI with your modem and then electronically send memos and letters to just about anybody on the planet. Of course, you can read the electronic mail that other people have sent to you. There are a bunch of great things about MCI. For starters, it's big, so lots of other people subscribe. But with MCI, it doesn't matter if the person you want to send something to isn't a subscriber. MCI lets you send and receive messages to and from many other public and private e-mail systems, including *CompuServe.* You can also have your message delivered as a fax to anybody's fax machine. And if the person doesn't have a computer or a fax machine, MCI will convert your text into a paper letter and send it by mail. At the moment, at least, MCI is cheap—a $10-a-month flat rate lets you send something like 45 messages. The drawbacks are that it's not particularly easy to send anything other than text files (without the aid of special software like Lotus Express or Norton Commander); and the *online* editing features are terrible, so you should compose all your text before you dial.

MDI

The initials **MDI** stand for **m**ultiple **d**ocument **i**nterface, which is the term used in Microsoft Windows to describe *applications* that allow you to use more than one document (file) at the same time. Compare *MOP.*

media

Media can refer to the computer parts that store information, such as *disks* and *tapes*. Most forms of computer media are *magnetic media*. If any of these storage devices go bad on you, you have a "media failure." This is guaranteed to make you unhappy.

Media can also refer to the various data types of communication, such as video, audio, and print. When a project involves more than one type of communication media, it is considered *multimedia*. For instance, a multimedia presentation of a story (text) might include pictures on the computer screen where the reader can click a button to turn the pages or animate the scene (video), plus sound effects (audio). If the user can control some of the effects, the presentation is "interactive multimedia." (The term "multimedia" is considered by some to be redundant, since "media" is already plural for "medium.")

Media Lab (M.I.T.)

The **Media Lab** at M.I.T. (the Massachusetts Institute of Technology) is a mysterious place where they do all sorts of advanced research into interactive books, artificial intelligence, beyond-high-definition television, and so on. Stewart "Whole Earth Catalog" Brand wrote a book about them a few years ago, *The Media Lab*.

The Media Lab used to be one of the brightest spots in purely American computer research, but (surprise!) suddenly they're largely funded by Japanese companies. It's very hard to even get a tour of the place, much less get candid information, unless you're a corporate sponsor—minimum ante of $75,000. They have been described as a big, expensive, digital Disneyland for rich kids.

megabyte

A **megabyte** ("meg" for short) is a unit of measure, measuring the size of things like documents, *hard disk* storage space, *programs,* or *memory*. Thinking in terms of typed pages, a document containing 620 pages of double-spaced typing would take up about one megabyte of disk space; you could say, "This is a one-megabyte document." Thinking in terms of hard disk storage space, you could say, "Oh, a 40-megabyte hard disk is too small; I need 180 megs." You might want to read the definitions of *bit* and *byte* for an in-depth explanation. Megabyte is abbreviated as *MB* or sometimes incorrectly as *Mb*.

megaflop

A **megaflop** is one million *floating point operations* per second, which is used as a measurement of computer speed and performance. Please see the *floating point* definitions for the big picture.

Megaflop can also describe the one short marriage in my life.

megahertz

A *hertz* is a measurement of how often, or how frequently, something happens in one second. So one hertz would mean once per second. In electricity, hertz refers to the number of cycles of an electric current in one second. A *kilo*hertz means the electric current cycles through a thousand times per second.

A **megahertz** means the electric current is cycling through at a *frequency* of one million times per second. You've probably seen the abbreviation for megahertz, *MHz,* in computer ads, where it is referring to *clock speed.* The faster the clock speed, the faster the computer operates. A fast *mainframe* (giant computer) has a clock speed of about 50 MHz. A good personal computer runs at about 16 to 25 MHz, and some, with acceleration, can clock up to 50 or even 66 (although they may still not be as fast as a mainframe, because there are other factors that affect the overall speed of a computer).

memory

Memory is the temporary storage space in little *chips* in your computer, as opposed to the permanent storage space on the hard disk. If you think of your hard disk as a filing cabinet where you permanently store everything you need, you can think of *memory* as your desk where you put things *temporarily* while you are working on them. When you're done working, you return those items on your desk (the items that were in memory) back to their filing cabinet (which is your hard disk).

Memory can be either *volatile,* like *RAM (random access memory)* which means anything stored in it will disappear when the power is turned off or the computer *crashes.* Or it can be *non-volatile,* like *ROM (read-only memory),* which means the information will *not* disappear when the power is turned off or the computer crashes. A computer needs both RAM and ROM to operate, and how fast the computer can get to the memory and use what's in it is a factor in the speed of your machine.

If you're wondering about how much memory your computer has or how much you really need to get done what you do, you are most likely concerned with the *RAM*. Please read the section on *RAM* to understand it better and to know how to get more if you need it. And it wouldn't hurt to read the entry on *ROM,* also.

memory management

A computer usually has several different sorts of *memory* to deal with, and it has to have a plan for administering all the memory to get its work accomplished. It has to know how much memory is available, what type it is, where to find it, and how to organize it into usable chunks. Once you start filling the memory with programs and data, the computer needs to keep track of where all that information is located in memory. All this is called **memory management,** which is one of the most important functions of the system. There's a special little *chip* called the MMU (memory management unit) that handles the basics of managing memory.

Depending on the type of computer and memory you have, the system may also require a software "memory manager." On a PC, for instance, if you have *expanded memory,* you need to *load* an "expanded memory manager"; if you have extended memory, you need an "extended memory manager."

SOMETIMES BOB NEEDED A JUMP START.

menu

A **menu** is a displayed list of commands or options on the screen, from which you can choose and activate the command you want to use, such as to open a file, close a window, print a document, or change a typeface. Usually, choosing a menu item activates the corresponding command immediately. Sometimes a main menu will lead you to a secondary menu, or to some sort of *dialog box* that offers a whole set of expanded options, like what sort of style to apply to your text or how you want a *spreadsheet cell* to be formatted.

Generally, if you see a little arrow after the command or option on the menu, it's a visual clue that choosing that command will display a secondary menu, called a *cascading menu* (PC) or *hierarchical menu* (Mac).

If you see an *ellipsis* (three dots: ...) after the item in the menu, it is a visual clue informing you that you will get a dialog box when you choose that item.

Depending on the computer system and the *software,* you may use a *mouse* to select the commands from the menus, or you may use the keyboard to choose the menu and the command, or you may have a choice of doing either.

Press on the title in the menu bar to see the menu. On a PC you can usually just click on the word; on the Mac, the menu will appear only as long as you hold the mouse button down.

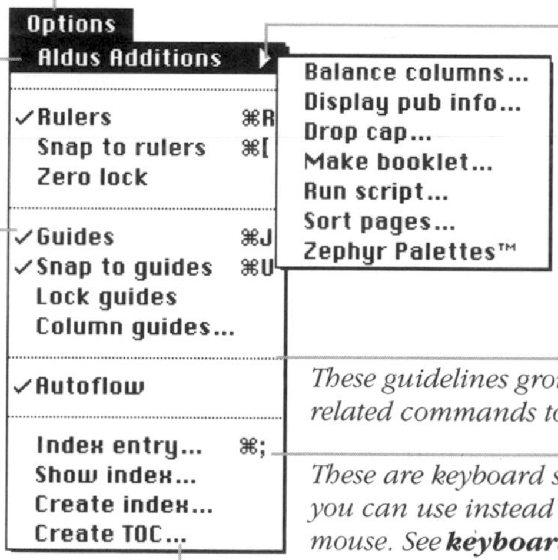

Options
Aldus Additions
✓Rulers ⌘R
Snap to rulers ⌘[
Zero lock
✓Guides ⌘J
✓Snap to guides ⌘U
Lock guides
Column guides...
✓Autoflow
Index entry... ⌘;
Show index...
Create index...
Create TOC...

Balance columns...
Display pub info...
Drop cap...
Make booklet...
Run script...
Sort pages...
Zephyr Palettes™

A menu item highlights to show you the command that is selected.

An arrowhead indicates another menu. The submenu does not always pop up on the side of the arrow! It depends on the size of your screen.

A checkmark means that item is in effect; to remove the checkmark and the effect, choose the item again. This kind of command is called a toggle switch.

These guidelines group related commands together.

These are keyboard shortcuts you can use instead of using the mouse. See ***keyboard shortcuts.***

An ellipsis indicates that you will get some sort of dialog box when you choose this item.

menu bar

Menu bar may refer to two very similar but slightly different ideas. In the Macintosh, Windows, and other graphical user interfaces, the **menu bar** is the row of labels across the top of the computer screen that displays the names of the various *menus* from which you can choose. The label in the menu bar gives you a clue as to the purpose of the list of menu commands. Of course, you will only find a menu bar in a program or computer *operating system* that uses menus.

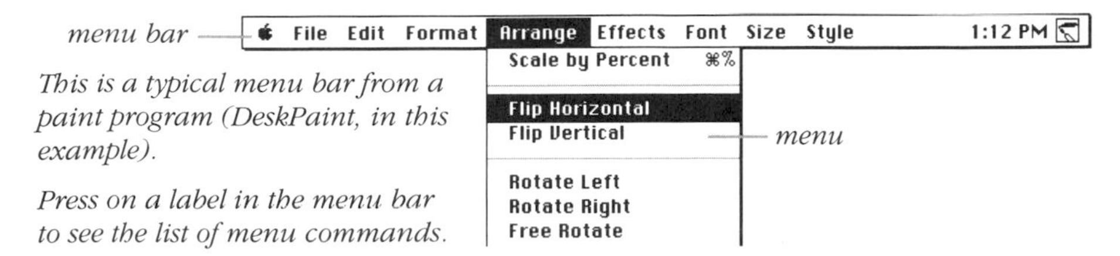

This is a typical menu bar from a paint program (DeskPaint, in this example).

Press on a label in the menu bar to see the list of menu commands.

In many PC and in some Unix programs, the menu bar is actually a menu in itself. That is, choosing one of the items on the bar may activate a command directly rather than open a menu (some items may produce secondary menus, however). This type of menu bar was first popularized by the spreadsheet program *Lotus 1-2-3.*

menu-driven

A program or computer system that uses *menus* extensively is said to be **menu-driven.** For most people, menu-driven programs are much easier and much more fun to learn and work with than "command-line" driven systems, where you send commands by typing in codes.

merge

Merge in computer jargon has the same sense as in Real Life—the act of combining two separate things into one. You might merge information from one *database* into another. You might merge information from two separate *spreadsheets* into a new, third spreadsheet.

Also see *print merge,* which is a little different—a print merge is when you want to make multiple copies of a letter, for instance, each one personalized with a name, address, and perhaps some other info from your database. So in a print merge you are merging the different names and addresses from your database into a common letter.

MFLOPS

MFLOPS stands for **m**illion **fl**oating-point **o**perations **p**er **s**econd. Please see *megaflop.*

MHz

MHz stands for *megahertz,* a measure of *frequency* (how frequently something happens in one second). Please see *megahertz.*

Micro Channel, Micro Channel architecture

The terms Micro Channel, Micro Channel architecture, or just MCA, all refer to the same thing: a kind of *expansion bus* used in PCs. An expansion bus consists of *expansion slots* inside the computer where you plug in *add-in boards,* along with the circuits the computer needs to communicate with those boards. The Micro Channel bus was introduced in 1987 with the first of IBM's new line of PS/2 computers. These days, most PS/2s still come with the Micro Channel bus, and a few other manufacturers also make PCs with Micro Channel slots. But most PCs sold today come with the *ISA* or *AT* bus, which is a modified version of the bus on the original IBM PC.

The Micro Channel has several theoretical advantages compared to the ISA bus. It carries at least twice as much data at a time (32 bits instead of only 16 or 8), it moves that data faster, and it lets the computer set up the boards for you automatically. Micro Channel PCs are particularly good as *network servers* and in other high-performance jobs. But the big drawbacks of the Micro Channel are that ISA boards, which are by far the most common, won't fit in Micro Channel slots. If you buy a Micro Channel machine, you may not be able to get the add-in board you want, and even if you can, it may cost more than the equivalent ISA board. For personal use, get a PC with the ISA bus.

microcomputer

Your *personal computer* is a **microcomputer.** Technically, a microcomputer is a computer in which the *CPU* (central processing unit, the brains of the computer) is contained on one single *chip,* a *microprocessor.* Most *workstations* are also considered microcomputers, for the same reason, although some personal computers are as fast as the fastest workstation. And a computer used by more than one person (a *multi-user* computer) is still a microcomputer as long as it has a microprocessor for its CPU.

The next step up from a microcomputer (in size, speed, capabilities, and price) is a *minicomputer.* Then a *mainframe.* Then a *supercomputer.* For a comparison of each, see *computer.*

microprocessor

We usually think of a **microprocessor** as a single *chip* that contains the entire *CPU* (central processing unit, the brains of the computer). A computer that has a microprocessor for its brain is a *microcomputer.* This is big magic—thirty years ago the equipment that was equivalent to this chip wouldn't have fit in my house. (Actually, thirty years ago the equivalent computing power didn't even exist.)

This is a silicon wafer. It has had millions of transistors etched and implanted into it. The wafer gets chopped into little chips and bonded to a package like you see on the right.

Chips off the wafer.

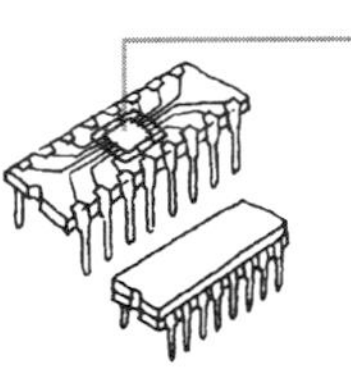

This tiny chip in the middle is bonded with tiny wires to the feet on the package, then a cap is put on to protect it all.

Microprocessors are also used on *add-in boards* that you plug into your computer (where they function as sort of accessory brains or *coprocessors*). and many printers (like LaserWriters and LaserJets) have microprocessors inside them, which convert the data from your computer into the design you see on the printed page. However it's used, a microprocessor can be *programmed,* unlike most other kinds of chips.

Microsoft Corporation

Microsoft Corporation, founded in 1975, is one of the largest and most influential corporations in the computer world. Its founder and head guy, Bill Gates, is supposed to be the richest man in the U.S. (maybe 3 or 4 times richer than Ross Perot). Microsoft's most outstanding success is the **d**isk **o**perating **s**ystem *(DOS)* that runs IBM personal computers (PC-DOS) and IBM-*compatible* computers (MS-DOS, the MS standing for Microsoft). Microsoft also makes a whole bunch of application programs for both the Mac and the PC (like Word and Excel), and the graphical user interface for the PC called *Windows.* The corporation is usually just referred to as "Microsoft."

Microsoft Windows

See *Windows* with a capital W.

MIDI

MIDI (pronounced "middy") stands for **m**usical **i**nstrument **d**igital **i**nterface. MIDI is a standard that describes a way for computers and musical instruments or *synthesizers* to interact. MIDI is both a physical, hardware thing, and a *protocol,* or set of rules for how music is encoded and transferred between different devices, like two electronic instruments or an electronic instrument and your computer. The hardware portion consists of a MIDI cable, an electronic instrument or synthesizer, and a MIDI *port* on both the computer and the electronic instrument or synthesizer. A MIDI port (like a socket for a plug) is standard on most newer Macintoshes, and on *Atari*

STs. To use MIDI with an older Macintosh or a PC, you need to buy a separate MIDI *interface*. On the Mac, the interface plugs into one of the *serial ports;* on the PC it's on an *add-in board*.

With your computer and the right music software, you can use MIDI to create, record, and play back electronic music. You can synchronize notes used on several synthesizers into one musical score, edit the music, edit only the violin score, or transpose an entire arrangement from one key to another. But you should understand that MIDI doesn't carry the actual sounds you hear—just the information about which notes are being played and how loud, which sounds in the synthesizer should be used to play those notes, and so on. To play music with your computer using MIDI, your synthesizer needs to be hooked up, turned on, and connected to some speakers.

If you hear of a keyboard being a "MIDI keyboard," that doesn't mean it is made by the MIDI company—there isn't one—it means the keyboard has the MIDI port built in so it can communicate with other MIDI devices.

minicomputer

A **minicomputer** isn't very mini. At least, not in the way most of us think of mini. You know how big your personal computer is and its related family. A *workstation* is the next step up in size, performance, and price, and is similar to a personal computer in that it is used by one individual.

Then comes the *minicomputer*. A minicomputer costs from $20,000 to $200,000 or so. It is built to perform different tasks for different people at the same time. Each person using a minicomputer has their own *terminal* attached by wires or via a modem to the computer proper. (A terminal isn't a computer—it's basically just a keyboard and a monitor; see the full definition.) The minicomputer spends a little bit of time on one person's task, then moves on to the next, and so on, juggling the work based on which jobs it thinks are most important. If you're the only one using a minicomputer, this can be one fast machine. But once many users (people) are "on" the system, the thing begins to slow down—you may type something and then wait for a minute or more before you see a response on the screen. Minicomputers used to be the only option for companies. Now, many firms are turning to networks of personal computers to accomplish the same thing faster and cheaper.

MIPS

MIPS stands for **m**illions of **i**nstructions **p**er **s**econd, one of several measurements of how fast a computer does what it's supposed to do.

A big *mainframe* performs at around 10 to 50 MIPS, while a personal computer jogs along in the approximate range of .05 to 25 MIPS. But MIPS specifications can be manipulated just like averages or miles per gallon—you have to know if they're talking city miles or freeway miles, or if they're taking an average of the highest and lowest, or the most common, or the worst-case scenario, etc.

MIPS has also been said to stand for MisInformation to Promote Sales, or Meaningless Indicator of Processing Speed.

MIS, MIS Department

MIS stands for **m**anagement **i**nformation **s**ervices. The MIS Department within a corporation takes care of processing and funneling information. An MIS manager supervises the choice of computers in a corporation, as well as the use of them.

MMU

MMU stands for **m**emory **m**anagement **u**nit, a little *chip* that controls how *memory* is managed in a computer, which includes memory protection, memory swapping, *virtual memory* (if used), and other functions. It is sometimes built right into the *CPU,* the *central processing unit.* See *memory management* for more details.

model, modeling

An electronic **model** is a mathematical or graphical representation of a prototype of something, whether that something is a business-related financial problem, like the financial planning in a spreadsheet, or a geometric model of a physical object. They are both used to better understand the projected reality of the real-world object, outside the computer.

Graphically, there are variations of **modeling,** all of which are used to create on-screen representations of physical objects, such as tools, buildings, vehicles, molecules, etc. **Wire-frame modeling** displays a three-dimensional object as if it were a model built of wire strands, including the insides of the object. **Surface modeling** usually involves *rendering* an object (adding color and substance and lighting effects) to make it appear to have a solid surface. **Solids modeling** is an advanced and mathematically complex method that represents an object as if it was solid, not only on the surface, but all the way through. A solid model can actually be tunneled through, graphically.

modem

A **modem** is the device that allows computers or other electronic devices to communicate via the telephone lines. Computers represent information *digitally*, as discrete packets of current (the current can be either "on" or "off"—see *bit* for more on this subject). Telephone lines, in contrast, carry *analog* information: continuous, fluctuating waves of current whose frequencies are in the range you can hear. A modem (the word is a contraction of **mo**dulator/**dem**odulator) translates in either direction between the different languages used by the computer and the telephone.

Think of the computer's digital information as ice cubes, which are solid and separate and countable, or *digital*. Like a hose, the phone line can't handle ice cubes, so it turns them into water, which is a free-flowing *analog* form of the same thing. For the other computer to get the information, though, the water has to be turned back into ice cubes, so the other computer must have a modem too. It is one modem's job to modulate the ice cubes into water and send the water through the hose. The modem at the other end demodulates the water back into ice cubes.

Your modem is either on a *board* inside your computer, as in most PCs, or it is a little box that sits next to your computer, as with most Macintoshes. If the modem is built inside (internal), there will be a standard phone jack on the outside of the computer where you can plug in your telephone cord. If the modem is external, the phone jack will be somewhere on the modem's case, and there will be a cord running from the modem into your computer, and the modem will have another cord from the modem to the phone or the phone jack. With a modem hooked up and with *communications software* installed, your computer can exchange information with any other computer that is also hooked up to a modem. Most modems can also dial telephone numbers for you, answer incoming calls, and perform other minor miracles.

An incredible amount of business goes on through the modems. You can send any file through the phone lines. If you live out in the boonies, you can send your desktop-created publications out for *high-resolution output* to a *service bureau* hundreds of miles away; they'll just express the finished pages back to you through the mail. If your magazine column is due by 12 noon on Friday, you can finish it at 11:45 and still have it in the editor's hands across the country by noon. You can conduct research, make plane reservations, join clubs, play long-distance chess, meet interesting people—oh, the possibilities are endless.

Do you wanna know what all the modem settings mean? Here is a brief, simple explanation of each item.

Serial port

Choose the *port* or plug, that matches the port on the back of the computer where you connected the modem cord. All Macintosh computers and a few PCs have a little picture, or *icon,* above the port. On a PC, the serial port may be labeled COM1, COM2, etc. If you have an internal modem, the kind built into your computer, you must set up the modem for a particular *serial* port even though you don't actually plug it into one. (A serial port, in general, means that data [information] flows through the port in one line, only one *bit* at at time, as if the data were marching single file. With a *parallel* port, information flows through in eight bits simultaneously, marching abreast in eight columns, and thus at a much higher speed.)

Baud rate

Sometimes referred to as *bps,* or *bits per second,* even though they are not technically quite the same thing, the *baud rate* is a measure of how quickly computers can send messages to each other. Standard baud rates are 300, 1200, 2400, 4800, 9600 bps. Less commonly, you may see 19200, 38400, and 57600 bps. Both computers that are communicating with each other must use the same baud rate. If the other person has a modem that can only send at 1200 bps and your modem can send at 2400 bps, you must choose the lower rate of 1200 if you want to connect. (Actually, the fast modem will automatically drop down to 1200 bps.)

Databits

Databits refers to whether the data you are sending is composed of seven bits per character, or eight bits. Eight bits is the standard setting for most personal computer communications. In fact, it is usually the *default* (automatic choice).

Parity

Parity is a form of error checking, interacting with the number of databits you have turned on. If you are using 7 data bits, you will probably specify some sort of parity. If you use 8 databits, you will always IGNORE parity or use NO parity.

Stop bits

A stop bit refers to the delay after one character (typically one *byte*) has been sent before the next character (byte) can be sent. It tells the receiving machine that the entire byte has been sent. Always set stop bits to 1, unless someone tells you specifically to set it to something else.

Duplex

Duplex controls what your software does with the characters you type, whether they go straight through to the other modem (FULL), whether they go through the modem plus show up on your screen (HALF), whether they show up on your screen plus echo back any characters received (ECHO), or whether the characters show up on the screen but don't get sent to the other modem (NULL), which is mainly used for testing.

If you don't see what you type, turn on HALF duplex. If you get double characters (HHEELLPP!!), turn on FULL duplex.

Handshaking (Xon/Xoff)

This is one way for the receiving machine to tell the remote to stop sending for a while because the receiving buffer is full.

Flashing lights

Ever wonder what all those little flashing lights on your modem mean? (Don't worry if your modem doesn't have all of them.)

MR Modem Ready. This light goes on when you turn on the modem.

TR Terminal Ready. This light means the computer is ready to send or receive data. It will flash on as soon as you start the communication process.

SD Send Data. This light flashes when information is sent from your computer to the modem. For instance, you'll see it flash whenever you send *e-mail*.

RD Receive Data. This light flashes when data comes into the modem and is about to appear on your computer, such as when you receive *e-mail* while you are logged on, or if you *download* a file.

OH Off Hook. This light is on when the modem has done the equivalent of taking the phone off the hook. If the volume on your modem is up, you will hear the dial tone when this light goes on. Just like you, the modem has to take the phone off the hook to dial.

CD Carrier Detect. When your modem gets a modem tone (a carrier tone) from another modem, this light goes on. It may take as long as 15 seconds after the tone actually starts for this light to go on, because the modems have to agree on a common speed and protocol.

AA	Auto Answer. This light will be on if you have set your modem to answer incoming calls automatically. It will flash off each time the phone rings.
HS	High Speed. This light is on when the modem is operating at its high speed. This light is only found on 2400/1200/300 modems.
RS, CS	Request to Send (sometimes RTS) and Clear to Send (sometimes CTS). These are sometimes used to stop and start transmission when the receive buffer in the computer is full. It's not a very useful light.
ARQ, MNP	These lights are only found on some modems and are used for testing.

modem rat

A **modem rat** is someone who spends a lot of time prowling around on the *bulletin boards* and *online services,* looking for information or a date or socializing or shopping.

modifier key

A **modifier key** is one of the keys on your keyboard that doesn't do anything all by itself, but is useful only in combination with other keys. For instance, the Shift key is a modifier key—it does nothing except modify other character keys and sometimes it modifies an action. Other modifier keys include Command, Option, Control, Alt, and Caps Lock.

Many *menu* commands can be activated by *keyboard shortcuts* instead of having to pick up the mouse and go into the menu. The shortcuts often consist of a key combination that includes one or more modifier keys. You must **hold down** the modifier key(s) while you **lightly tap** the *character* key (a letter or a number or a *function key*) associated with the shortcut. It's just like when you type a capital letter—you must hold the Shift key down while you type the letter key. Use this same procedure with any modifier key in any shortcut.

For instance, if the keyboard shortcut for pasting an item is Command V (or Ctrl V on a PC): You **hold** the Command (or Ctrl) key down, then tap the V once. If you *hold* the letter V down, you'll end up pasting in multiple copies of the item.

Many actions can be modified with these keys, also. For instance, in most applications if you hold the Shift key down while you press the arrow keys, you will *select* the text that the cursor passes over. If you are in a drawing program, you can often hold the Shift key down while you erase or draw and your tool will move in a straight line, regardless of how much you wiggle the mouse.

moiré

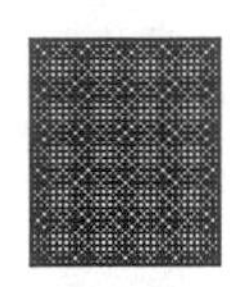

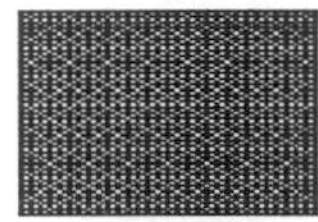

These examples are all the exact same pattern, each with a different moiré because they have been resized improperly.

Moiré (pronounced "mor ay," accent on the second syllable*) refers to a generally unwanted pattern that can appear in graphic images. It can be the result of *scanning* an image that has already been printed. (You should read about *halftones* to understand why scanning a printed image does this to you.) Sometimes it will happen if the *resolution* of your graphic image does not match the resolution of the printer. Some graphic software can adjust your image to prevent this. Sometimes a moiré will happen because the "screen angles" (the angle on which the dots of a halftone are aligned) were not applied properly.

Moiré can be either a noun, as in "If you do that, you'll get a moiré for sure," or it can be an adjective, as in "That moiré pattern makes an interesting effect on Grandma's face." Doug says you can see moiré patterns when driving by two parallel chain-link fences.

*You will see all kinds of interesting pronounciations for this word, similar to the ones you find for *beziér*. Many guides tell you to pronounce moiré as "mhwa-ray." Try saying mhwaray fast in conversation. If you just call it "mor ay," no one will ever even snicker.

monitor

Monitor is another word for the computer *screen*. But "monitor" encompasses the whole piece of equipment, rather than just the screen part that you look at. You also might hear a monitor called a *display,* as in "Oooh, I got a new two-page display," or *VDT* (video display terminal), as in newspaper journalism, or *CRT* (cathode ray tube), which is the technical term for a picture tube. However, flat panel screens like *LCDs* are not referred to as monitors, even if they're housed externally from a computer.

Some monitors are built right into the computers, like in the small Macintoshes. When you purchase a larger Macintosh or most other kinds of computers, you must buy the monitor separate from the computer itself (that's why they're called "modular"). Monitor size is measured like a television, from one corner to the diagonally opposite corner.

Some monitors are *monochrome,* meaning they can show only one color on a background, like black on white (Macs), green on black, or amber on black (PCs). *Grayscale* monitors can display different shades of gray, rather than imitating the different shades with combinations of black and white dots.

And there are many different color monitors. A color monitor can display any of several levels of resolution and can display varying numbers of colors, determined by several factors, such as amount of memory in the computer or the type of *card* that is controlling the monitor. See the section in Appendix A on how to read a computer monitor advertisement.

monochrome

If an item is **monochrome,** that means it uses only one color on a differently colored background. In a monochrome graphic, for example, the picture is created with, say, black dots (or lines) against a white background. Most printers are monochrome, meaning they only print black toner on white paper. A monochrome monitor displays only black-and-white (as on Macintoshes) or amber or green on a black background (as with most monochrome monitors for PCs). Monochrome monitors, especially black-and-white ones, are much clearer and present a sharper image than do color monitors, plus they put out less *radiation.*

A *grayscale* monitor can display a variety of gray values, not just black or white, even though there is still no color. See the information in the Symbols section on *2-bit, 8-bit (etc.), color.*

monospaced, monospacing

Text on your screen or a printed page is created with characters that are either **monospaced** or *proportional.* Monospaced characters each take up the same amount of space along a line of text; that is, the letter i takes up the same space as a capital letter W. Even a period takes up the same amount of space as a capital W. In monospaced text, the letters on different lines all line up exactly; you could actually draw lines through the columns of letters. Most typewriters print monospaced text. Many computers have monospaced fonts available for your occasional and specific use, like when you want to make your text look like it came out of a crummy old typewriter.

```
This is an example of the typeface named Courier.
Notice how the letters align in columns. The typefaces
called Monaco, Pica, and Elite are also monospaced.
```

Moof!™

Moof! is what the *dogcow* named *Clarus* would say if he could make noise. Please see *dogcow* for the whole story.

MOP

MOP stands for **m**ultiple **o**pen **p**ublications. In many software applications you can *open* more than one document at a time. For instance, in a word processing application that *supports* MOP you can open the letter you wrote to your mom yesterday, your graduate thesis that you finished last week, and a new document in which you are going to combine the two. In an *integrated software package,* the MOP concept is standard—you can have a word processed letter, a database of names, and a spreadsheet of financial obligations open on your screen at the same time, and can work back and forth between them.

mortal

If you are looking up words in this dictionary, you are probably **mortal** (meaning you will die someday). *Wizards* are not mere mortals. Most *hackers* are beyond mere mortality. Me, I'm just a *nerdette,* definitely mortal.

motherboard

The **motherboard** is the mother, the matriarch, the controlling force, the life spirit, the brains, the soul, the heart of the computer. Actually the motherboard is just a plastic *board* with *chips* on it and some *printed circuits.* But the secret to the motherboard is that she not only contains the main *memory* and support circuitry, she holds the chip with the biggest magic—the *CPU,* otherwise known as the *central processing unit,* the chip that runs the computer. If you don't know what the CPU is, you might want to check out that definition, and see the illustrations in Appendix A. The motherboard is sometimes known as the *logic board.*

There are no fatherboards nor son boards, but there are *daughterboards,* which just goes to show you who really runs the show. Or the computer.

Motorola

Motorola, Inc. is an electronics equipment manufacturer that developed the *microprocessors* (the *chips* that run the computer) in the *68000* family. These chips are used in the Apple *Macintosh,* as well as in the *NeXT,* the Commodore *Amiga,* and the *Atari* ST.

mount, dismount

On a Macintosh, when you insert a *floppy disk* into a *disk drive,* you see the icon of the disk on the screen. At that point, the disk is considered **mounted.** When you want to connect to another *hard disk* besides the one that is running your computer, you need to **mount** it. If you are using a *cartridge hard disk,* you have to insert the cartridge into the cartridge drive before it can mount, just as you insert a floppy into the floppy disk drive. A hard disk can be in its drive and up and running, but your computer cannot use the disk until it mounts, until its *icon* appears on the screen. You usually have to install certain software so the computer can recognize the other drive. Then, if everything works right, the hard disk will mount by itself once it is up and spinning.

For a variety of reasons, a hard disk may not mount. Perhaps it is the order in which you turned everything on. Perhaps it is a lack of the proper software. Perhaps it is a *SCSI* problem. Perhaps it is witchcraft. That's why *utilities* like *SCSIProbe* were invented—they can go find the stubborn disk and usually make it mount.

When you eject a floppy disk, the disk just comes popping out. You cannot eject a cartridge hard disk from its drive, however, until the disk is **dismounted,** until its icon is gone from the screen. You dismount the hard disk by either selecting the icon and using the eject command (Command Option E in System 6; Command Y in System 7), or by dragging the icon to the trash, just as you would eject a floppy. After the icon has disappeared (indicating the disk has been dismounted), you can push the buttons on the drive to pop the disk out. (Always be sure to wait until the disk stops spinning before you pop it out!)

So you "eject" floppy disks, but you "dismount" hard disks. If the hard disk is a cartridge, then you can physically eject it after it has been dismounted.

The terms mount and dismount aren't usually used in connection with a PC—at least it's not an official term in the DOS or Windows vocabulary. However, disks still must be recognized by DOS before you can use them. In most cases, this process is automatic. With certain types of hard disks, you need to add commands to your CONFIG.SYS file before the disk will be recognized. As far as floppy disks are concerned, DOS doesn't notice that you've removed one floppy and inserted another until it actually goes to use that particular floppy drive.

mouse

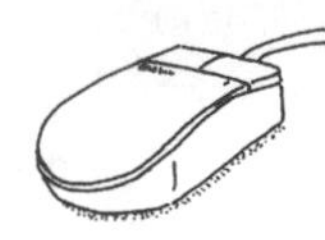

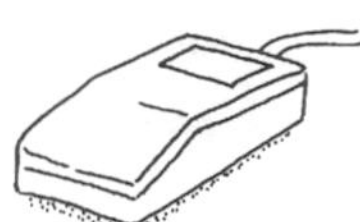

These are examples of two slightly different mice.

A **mouse** is a little *device* that you move across the top of your desk to control the pointer or cursor on the *screen*. Don't hold the mouse up in the air, or up to the screen, or in the palm of your hand. It doesn't work if you talk into it, either. The mouse needs to be resting on the desk with the little rolling ball on the underside of it in contact with the desk. Hold it so the cable points directly away from you. [Some people hold their mouse backwards, with the tail facing their body. When they move the mouse to the right, the pointer moves to the left. Personally, this makes me twitch.]

As you move the mouse to the right, the pointer moves across the screen to the right, and so forth. You tell your software what item on the screen you want to work with by *pointing* at the item—that is, by moving the mouse so the pointer is positioned over the item. (The "pointer" may not always look like an arrow; whatever form the cursor takes, the mouse can control its position on the screen.)

A mouse has a button on it, or maybe two or three buttons (PC mice commonly have more than one button). You slide the mouse around the table and when the pointer is positioned where you want it on the screen, you click or press the mouse button to make something happen: to pull down *menus,* to *drag* objects across the screen, to *select* items, to move the text cursor in any program that uses type, and many other useful activities. Some applications are completely dependent on the mouse; others don't need a mouse at all (which means you have to memorize hundreds of keyboard commands).

If you flip the mouse over, you'll usually see a little ball that barely sticks out from the bottom. Sensors inside the mouse detect how that ball rolls around as you move it across the desk and then sends the information to your computer. Some mice don't have the ball—see *optical mouse.*

A mouse isn't the only way to move the cursor on the screen and to choose menu commands—there are many other *pointing devices,* including *trackballs,* light pens, *stylus* and *tablets,* and *joysticks.*

On the PC, you can get software utilities that let you use the keyboard to move the mouse pointer and simulate the buttons. That's good for times when you don't have a mouse, but it seems odd to use the keyboard to replace a device that was supposed to replace the keyboard.

The device is called a mouse because if you pick up the cable about six inches from where it connects with the mouse itself and hold it at the end of your outstretched arm and squint your eyes and wrinkle your nose, it looks like a dead mouse.

mouse pad

A **mousepad** is a little pad on which you can roll your mouse around. There is nothing special about a mousepad—your mouse will work just fine on the plain ol' desk or on a piece of cardboard (unless you have an *optical mouse* which needs a mousepad with a reflecting grid built into it, but you would know if you did). The advantage of the mousepad is that the surface is usually designed to give a better grip on the mouse ball than you can get on your desk.

mouse port

The **mouse port** is a special port (socket, or connection) for hooking up a mouse to your computer.

Among IBM-compatible personal computers, a special mouse port first appeared in 1987 with IBM's PS/2 line. This is a small round connector with six metal pins, plus an additional plastic piece to to ensure that you plug it in right-side up. The PS/2 mouse port hasn't become universal, but some other manufacturers do include it on their IBM-compatible machines, especially on *laptop* and *notebook* PCs. Actually, the PS/2 port can be used for some other external devices such as hand *scanners*. And by the way, don't be worried if your PC doesn't have a mouse port: you can attach a mouse to the standard *serial port* on any IBM-compatible machine.

On the Mac, the mouse plugs into an *ADB port,* or Apple Desktop Bus port. ADB ports are on the back of the computer and also on both ends of most keyboards. You can plug the mouse into any ADB port you find.

ms

The abbreviation **ms** stands for **milli****s**econd, which is one-thousandth of a second. Disk *access time,* the time it takes the computer to go to the disk and get the information it's looking for, is measured in milliseconds. Twenty milliseconds is considered pretty fast.

MS-DOS

MS-DOS (pronounced "em ess doss") stands for **M**icrosoft **d**isk **o**perating **s**ystem, the most widely-used *operating system* for *IBM PC* and *compatible* computers (an *operating system* is the master control software program that runs the computer itself). This means that MS-DOS is the most widely used computer operating system, period, since there are something like 80 to 100 million PCs in the world, and most of them use MS-DOS. There are at least two other versions of DOS that are compatible with MS-DOS (meaning they work the same way and run the same programs). So please see the definition for *DOS* for the whole story.

The history of MS-DOS is a fascinating study in how business success often depends more on good timing, a nose for a good deal, and aggressive marketing, than on the technical merits of your product. When IBM first developed the IBM PC, it wanted to license a crude operating system called CP/M, which was the dominant one for personal computers at the time (ever hear of a Kaypro or an Osborne?) Apparently IBM's offer was too low for the makers of CP/M, so IBM went shopping at Microsoft. Microsoft didn't have a suitable product at the time, but the boss, Bill Gates could smell money. He bought out a little company that was making an imitation version of CP/M, and rushed out a revision that worked on IBM's machine. When the IBM PC caught on, Bill got really rich—even though MS-DOS is really not so hot.

Guy Kawasaki says that MS-DOS stands for Microsoft Seeks Domination of Society.

MTBF

This sounds kind of deprecatory, but **MTBF** stands for **m**ean **t**ime **b**etween **f**ailures, a standard for measuring how dependable a piece of *hardware* is. The MTBF is determined by counting the total number of hours the device was on and dividing that by the number of times it broke down and needed to be repaired. Usually the number is in the thousands or tens of thousands of hours. If not, forget it.

MultiFinder

MultiFinder is *System 6 or earlier** software that enables you to open more than one application at a time. Then you can bounce back and forth between the applications, and also bounce into the *Finder* (also called the *Desktop*) whenever you feel like it.

For instance, you could have your *page layout program* open and your *paint program* open at the same time. You could be creating a page, then pop over into the paint program to create a graphic, then pop back into the page layout program to place the graphic, without having to close either application. Or you could go back to the Finder, rearrange your hard disk, check your e-mail, *initialize* a new floppy, then get back to your document, which was hanging around in the ether all this time, patiently waiting for you.

Now, the concept might not seem like a big deal at first, but once you get used to MultiFinder it is hard to live without it. MultiFinder tends to confuse beginners, though, because you seem to lose menu items and windows and memory without knowing why. In System 6 or earlier, MultiFinder is optional. You can use the "Set startup…" command in the Special menu to turn it on or off (off is "Finder Only"). Anytime you switch from MultiFinder to Finder Only (also known as *UniFinder*), you need to *restart* your Mac.

To be able to use MultiFinder you need to have enough *RAM* (random access memory) to be able to have your System running **plus** a couple of applications at the same time. "Enough RAM" means at least 2.5 MB, really, and the more the better. You can't have too much money, too many fonts, or too much RAM, you know.

Under *System 7* you have no choice of whether to use MultiFinder or not; it's always on. Which means it's not really appropriate to call it *Multi*Finder, because it's just the way the basic Finder works. Under System 7, we just say it is a *multi-tasking* environment. (Just so you know, though, this is still not true multi-tasking but only "multiple program loading," because the Mac can have only one of the open applications *active* at a time. True multi-tasking can have one operation performing in the background, like a huge database sorting, while you are working in another application.)

multimedia

Multimedia is the buzzword of the '90s. When a demonstration, training or educational package, or any sort of computer presentation involves not just still images, but also moving video, animation, sound, or a combination of these, it's considered a multimedia presentation. When the presentation allows the user to control what she sees and hears by doing things like clicking buttons on the screen or typing in answers to questions, it is considered "interactive multimedia."

Since multimedia involves so much more than words to be printed on paper, it requires a powerful computer system with which to create the presentation. You need lots of *memory,* lots of *hard disk storage space,* sound capabilities, special software, etc. Many multimedia presentations need to be pressed onto *CDs* (compact discs) because of the amount of data necessary to create and run the presentation.

Multiple Masters

Multiple Masters is a font technology from Adobe Systems, Inc. Please see the section on *fonts (Macintosh)*.

multiscan

Multiscan refers to a type of computer *monitor* that automatically matches the synchronyzing signals sent from the computer's video adapter (the video circuitry). On a standard television-type monitor, the image you see is formed by a single beam of electrons scanning lickety-split across the picture tube. The beam starts at one corner, traces a narrow horizontal line, then moves down a bit and traces the next line. The speed with which the beam travels horizontally and vertically (the horizontal and vertical "scan frequencies"), must match the synchronizing signals from the computer's video circuits.

The problem is, the synch signals vary with each type of video adapter for PCs (*EGAs, VGAs, Super VGA,* and so on). Since the scan rate is fixed in an ordinary monitor, you can only use the monitor with one type of video adapter—a VGA monitor only works with a VGA adapter, and so on. By contrast, a multiscan monitor will work with many different types of adapters, within limits.

When you buy, be sure your monitor's range of scan frequencies matches all the adapters you may use it with. At a minimum, it should have a 50–75 *Hz* (*hertz,* times per second) vertical frequency and a 30–50 *kHz (kilohertz)* horizontal frequency. The vertical frequency measures how fast the entire screen is "repainted," and is also called the *refresh rate*. You should also insist on a variable frequency monitor, one that can match **any** frequency within those ranges, rather than one that simply operates at several different but fixed frequencies. And, by the way, multiscan monitors are more expensive than fixed-scan rate monitors. See also *MultiSync*.

There's much less inconsistency in the Macintosh world, so a multiscan monitor isn't so important. But many of them will work with a Mac.

MultiSync

MultiSync is the trademarked name for the original and most popular brand of *multiscan monitors,* the ones made by NEC. In the venerable tradition of co-opted brand names, some people use the term "multisync" for any multiscan monitor.

multi-tasking

Multi-tasking means that the computer can work with more than one program at a time. For instance, you could be working with information from one database on the screen analyzing data, while the computer is sorting information from another database, while a spreadsheet is performing calculations on a separate worksheet. This is different from "multiple loading" of applications, also known as "context switching" or "task switching," which is what happens when you use *MultiFinder* or *System 7* on the Macintosh or the DOS task switcher. Context switching simply allows several applications to be open, but only one is working at a time.

Actually, even in true multi-tasking, only one application is ever running at any one instant. But because the computer automatically switches from one to the next so quickly, all the programs *seem* to run simultaneously. (With context switching, *you* decide when to shift from one program to another, by hitting a key or clicking the mouse.) Each individual program runs slower, of course, since each gets only a portion of the computer's time, and since some time is lost in the process of going from program to program.

Ideally, multi-tasking capability is built into your computer's operating system. DOS has absolutely no multi-tasking features, but DESQview and Windows provide it somehow.

The Mac does allow cooperative multi-tasking in its *background* printing feature, which allows one document to print while you work on another.

The word "multi-tasking" is often written without the hyphen; I have kept the hyphen in the word specifically to make it easier to read.

multi-threading

A *multi-tasking* operating system may also permit **multi-threading,** which allows individual chores *within* an application to run at the same time, whether or not other applications are also running simultaneously. A multi-threaded graphics application might have one thread that "draws" your picture on the screen while another thread accepts new commands from the keyboard or the mouse. That way, you don't have to wait until the picture reappears each time you make a change before you use another command. Regular Windows does not allow for multi-threaded applications, but OS/2 does, and so does Windows NT.

multi-user

One definition of **multi-user** refers to a large computer system that lets more than one person use the machine at the same time. This only works if you have a multi-user *operating system,* such as *Unix.* Except for the person sitting right in front of the large computer, everyone who uses it needs their own *terminal.* A terminal is basically just a keyboard and a monitor, attached by wires or a *modem* to the large computer. The person using the terminal types commands on the keyboard and sees the results on the screen, just as if they were using a personal computer. But in a multi-user system, the computer is located somewhere else, not on or near your desk, and other people are using it at the same time. The multi-user computer keeps track of who is typing what and makes sure that they see the pertinent information on their screen.

Traditionally, most multi-user computers were *mainframes* or *mini-computers.* However, almost any computer can serve as a multi-user machine, as long as you give it the right operating system. Unix and other multi-user operating systems are available for IBM-type personal computers, and there's a Mac version of Unix called *A/UX.* A multi-user system can be cheaper than giving each person their own computer, since terminals don't cost as much as full computers. But the main reason businesses use them is to store all the company information in a central place to avoid duplication, and to keep the data secure. The same thing can be accomplished by a *network* of personal computers, with the advantage of improved performance: since each person has their own computer, they don't have to wait on the central computer to do the processing, just to supply the data.

Multi-user can also refer to a software application running on a network in which more than one person can work in a program at a time, each at a different computer, without affecting the other person's work. For instance, TouchBASE, a neat little *database* for personal contacts, is multi-user. My sister and I each have the TouchBASE program on our computers, and our computers have a little PhoneNet connector to *network* them. I set up TouchBASE to be multi-user (you can choose to have it single user). We can both use the same TouchBASE file at the same time. When Shannon adds a name and address to a TouchBASE file, it automatically appears in the file on my computer, right in front of my face, and vice versa. It's too cool.

mung

Ocassionally misspelled as "munge," **mung** stands for **m**ash **u**ntil **n**o **g**ood. It's supposed to be pronounced "muhng," but is so often pronounced "muhnj" that this alternative pronunciation has also become somewhat acceptable (ha! "acceptable pronunciation" of complete slang). Mung is a verb meaning to destroy something. Usually it is accidentally destroyed, although someone could mung something intentionally and maliciously. A definition in a *BMUG* newsletter likened munging to the feeling of stepping on a warm, sticky gob of gum. "Ooh, did I mung that file!"

Murphy's Law

Murphy's Law states, "If anything can go wrong, it will." Did you know there really was a Murphy who said it?

In 1949, the U.S. Air Force was doing those rocket-sled experiments to test human acceleration tolerances. Edward A. Murphy, Jr., was one of the engineers involved in the experiments. In one test there were 16 accelerometers that were supposed to be mounted onto different parts of Major John Paul Stapp's body. There were two ways each sensor could be glued to its mount, and someone carefully installed all 16 the wrong way around. Murphy was quoted as saying, "If there are two or more ways to do something, and one of those ways can result in a catastrophe, then someone will do it."

p.s. Here's an ironic anecdote: Each of these chapters is a separate PageMaker publication. I index each pub separately, then combine all the indices from all the pubs into one major index. Well, the index in this Section M developed a "bad record index" (called "RIX" which has nothing to do with the actual index) which corrupted the file. Although I was not able to save the file, I was able to trace the source of the bad RIX to an index entry made in this definition of "Murphy's Law."

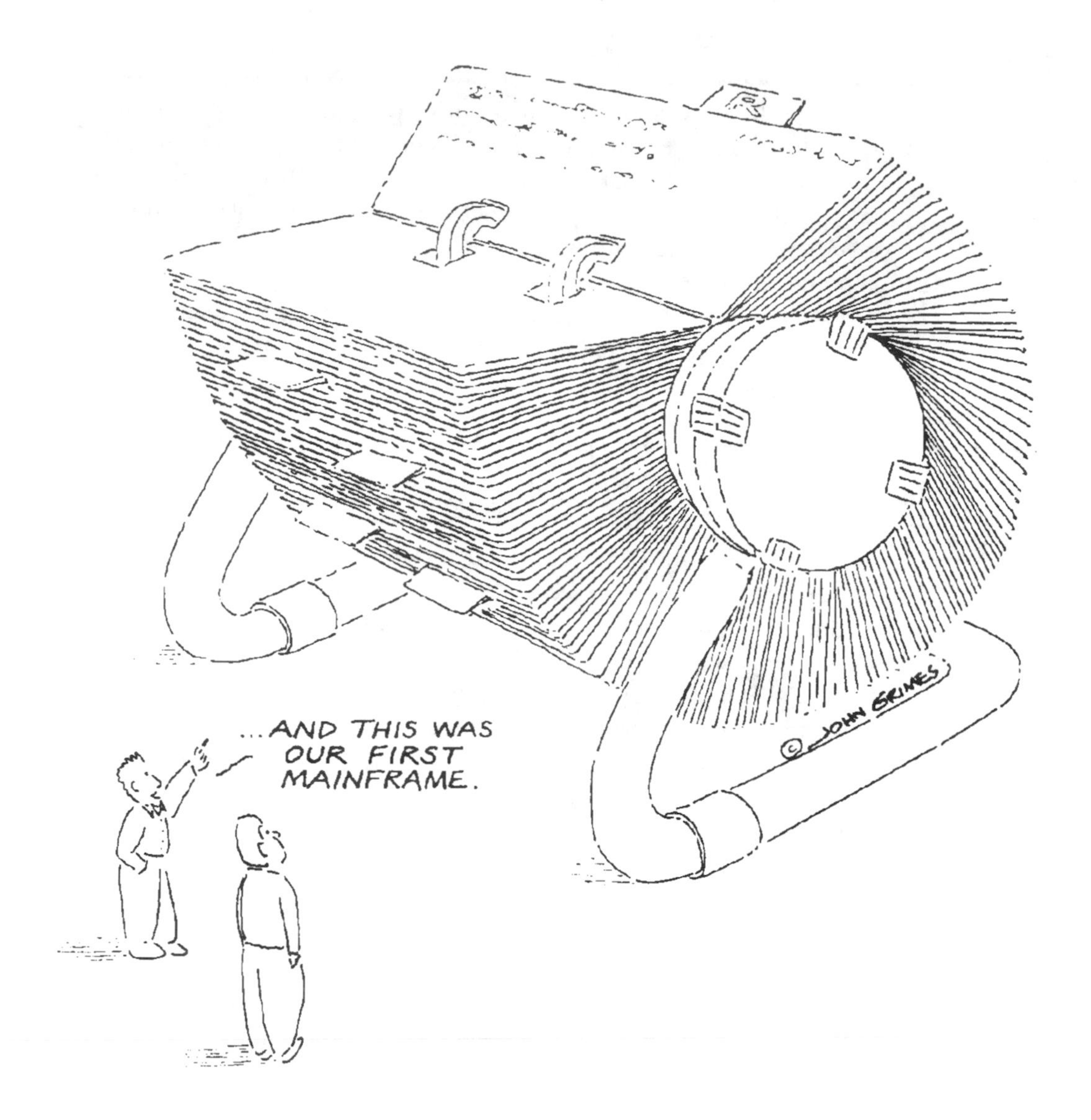
...AND THIS WAS
OUR FIRST
MAINFRAME.
© JOHN GRIMES

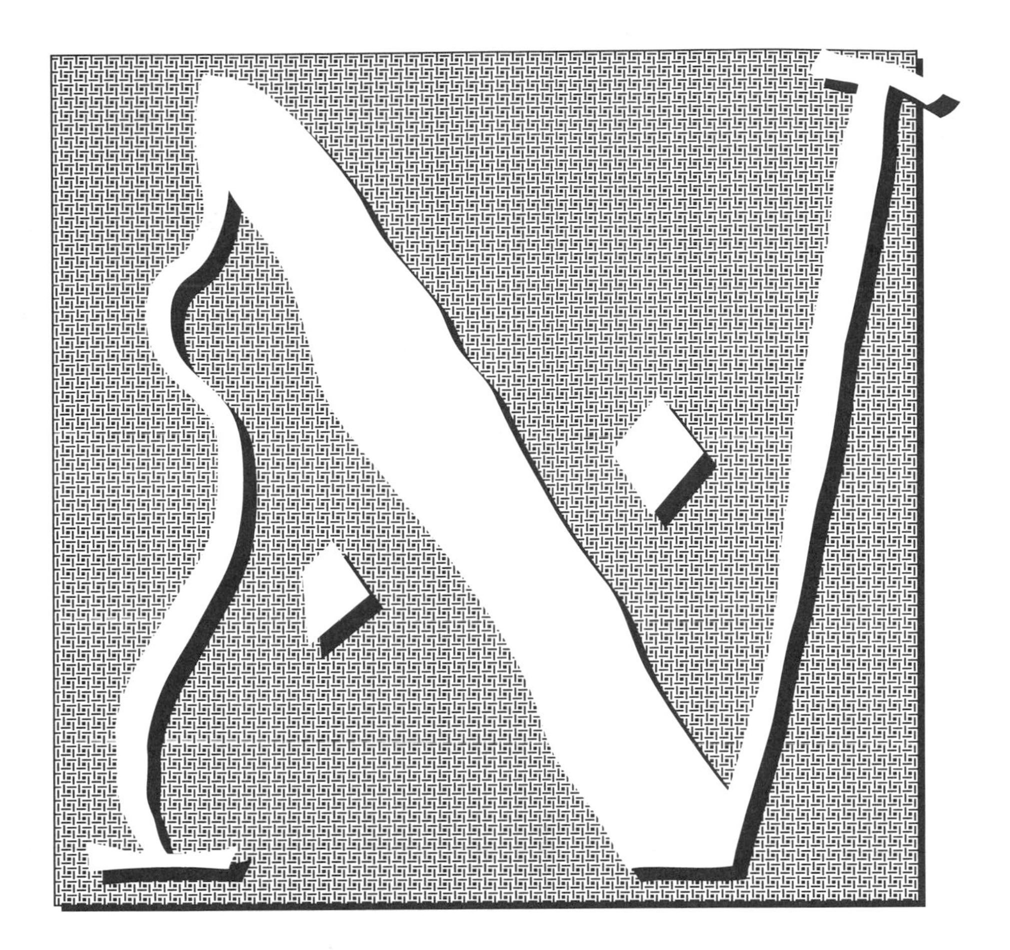

nanosecond

A **nanosecond** (abbreviated as "ns")is one billionth of a second. Light and electric current travel about one foot per nanosecond. So if all the tiny *chips* in your computer are connected with these tiny wires, the shorter the distance between the *chips* in your computer, the faster the computer can do its job.

The speed (technically, the *memory access time*) of *RAM chips* is measured in nanoseconds, such as "120ns RAM chips" (the lower the number, the faster the speed). RAM chips are the memory in your computer. The faster your *CPU* (the *central processing unit,* the chip that runs the computer), the faster your memory has to be to keep up. Which means that certain computers need faster memory (RAM) chips; for instance, Macintosh computers that use the *68030 microprocessor* need RAM chips that run at least 80ns or faster.

Whoever figured out how to measure time in nanoseconds?

natural language

A **natural language** is a language that humans use to communicate with each other, as opposed to a *programming language,* which is the way humans communicate with computers. The closer any programming language is to our natural language, the more sophisticated and *high-level* the language, and the easier it should be to use (but it may have other shortcomings). *Artificial intelligence* is the ultimate in reaching the highest level of programming known to humankind, where we will be able to talk to our computer and it will talk back to us ("which I hope will be never," says Doug).

navigate

Navigate is a verb that means to go where you want in the computer, amidst the morass of drives and directories and subdirectories and folders and *nested* folders and files. You need to navigate to find documents you want to open, to make sure you are saving a file in the right directory or folder or on the appropriate disk, or perhaps to view what is in a particular folder. It can be a tricky thing for a beginner to understand the pathways for getting from one place to another.

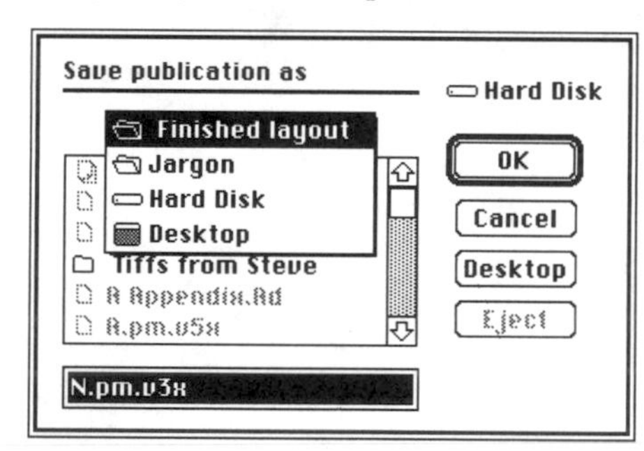

When you try to find a document you created earlier, when you look for a folder in which to save a document, when you try to find the disk you want to switch to, you are navigating.

Navigator

Navigator is the name of a software program that provides an easier way to use the *online service* called *CompuServe.* It's not very easy to bumble your way around CompuServe, so Navigator, which has a friendlier *look-and-feel,* makes it easy to figure out what you're doing. You can tell Navigator what you want, then Navigator logs on to Compuserve, does all those things you wanted it to do, like get your mail and send some mail and find something in the library and download a file, then it logs off and gives you everything.

Navigator also has cute little face files so people know what you look like. Or what you want them to think you look like.

near letter quality

See *letter quality,* and then figure out what **near letter quality** (NLQ) must be. Perhaps it means barely legible? Dot matrix printers often have near letter quality mode for better-looking (but slower) print.

near typeset quality

Text from a laser printer is considered **near typeset quality.** At 300 and especially 600 *dpi (dots per inch),* the quality of the letterforms is close to actual typesetting. This term "typesetting" refers back to traditional typesetting, when trained typesetters would take a designer's copy (words) and set the text in metal type or, later, in photographic type. Now the quality of type we can get from *imagesetters* is much better than we have ever had in history, and near typeset quality is a minimum standard we have come to expect.

need

As in, "Sure I've got a IIfx, honey, but I **need** a PowerBook!"

nerd, nerdette

Websters defines **nerd** as "A boring, dull, or unattractive person." Now, I wouldn't agree with that at all. (Of course not, I'm a **nerdette.**) Nerds just have different interests from the mainstream of yuppies and prom queens and football players. Nerds tend to be very creative in very different and interesting ways. As The Girl said, "Please, God, don't let me be normal!"

Guy defines a **nerd** as anyone who considers *CompuServe* a dating service.

Robin says, "That's right. Anyone who's really cool knows that *America Online* is the dating service."

nested

When one object is tucked inside another object, literally or figuratively, it is considered to be **nested**.

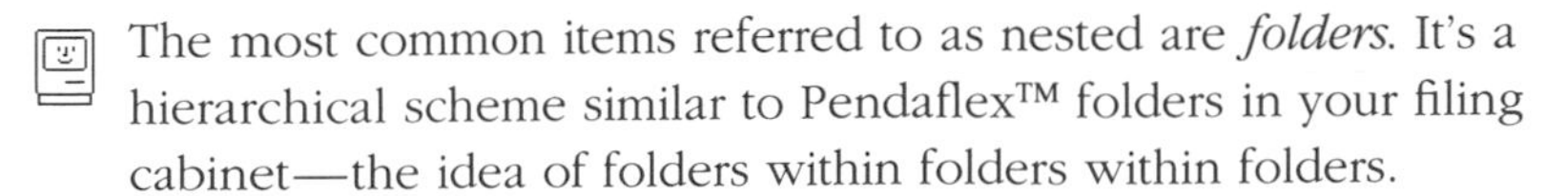

The most common items referred to as nested are *folders.* It's a hierarchical scheme similar to Pendaflex™ folders in your filing cabinet—the idea of folders within folders within folders.

Menus can be nested, so that an item on one menu brings up a subsidiary menu, and so on (these are referred to as *hierarchical menus*).

 On the PC, *directories* and their *subdirectories* can be nested: one within the next, within yet another. Menus can be nested, so that choosing an item on one menu brings up a subsidiary menu, and so on (these are usually referred to as *cascading menus*).

Dialog boxes can be nested—you click on a button in a dialog box and you get another dialog box, where you can click a button and get another dialog box.

A table can be nested within a table, or a graphic from one program can be nested inside a graphic from another program. Programmers also nest instructions in their code, something like this: "*if* a is true, then see *if* b is true, in which case *if* c is also true, then display a big watermelon on the screen"—those are nested *if* instructions.

> Paragraphs can also be referred to as nested, where one paragraph is *indented,* perhaps on both sides, like this paragraph.

network

A **network** is the communication or connection system that lets your computer talk to another computer or to a printer, hard disk, or any other sort of device. The connection system might be through cables. The connection system might be through a *modem* using the telephone lines, connecting you with an *online service, bulletin board,* or directly to another computer. A network could be a small system, such as two computers and a printer in the same room (a *local area network,* or *LAN*), or it could be a huge system connecting computers all over the world (a *wide area network,* called *WAN*).

The purpose of a network is to allow computers or their users to send and receive information electronically. It may be that several computers in one building all need to get information from one main computer or office member. Several people working on a project may need to update a common file regularly, or students in a lab all need to be able to *access* the instructor's worksheets.

NeXT

The **NeXT** is a powerful computer, billed as a *workstation,* which means it is a step beyond the *personal computer* that most of us have at home—bigger, arguably more powerful, and definitely more expensive. It is *Unix*-based, currently uses the *68040 microprocessor,* has a speed of 25 *MHz,* and can use a *high-resolution* color monitor. It uses a *graphical user* *interface,* which means it *looks and feels* very much like a Macintosh. (Instead of a trash can, it has a black hole.) The NeXT is not a wildly popular computer at the moment, which might be because it costs so much and because most people are pretty happy with their Macs and PCs. Also, perhaps, because there is not a lot of software for it yet. The NeXT is definitely the *sexiest* computer in the world.

The NeXT corporation is selling their hardware operation and will be focusing on software, namely the NeXTSTEP *object-oriented* application development. Basically, they're aiming for the *Intel*-based market, ditching the *680x0* processors.

NFNT

NFNT (pronounced "en font") stands for **n**ew **f**o**nt**. An NFNT is a *font resource* (chunk of information for a typeface) that tells the computer how to draw the characters. Every typeface (font) has its own NFNT resource and NFNT identification number that supposedly separates it from every other font. Unfortunately, it is not unusual for fonts to have similar ID (identification) numbers, in which case the computer gets confused and the fonts suffer identity crises—Korinna may appear as Gill Sans. (If you have this problem, invest in *Suitcase* or *MasterJuggler* to help you manage your fonts.)

There is also a *resource* called a FOND (**fon**t family **d**escriptor) that keeps all the members of a *font family* together, such as 10, 12, 14, 18, and 24 point Bembo Regular, Bold, Italic, and Bold Italic.

nibble

A **nibble** is a cute little term for half of a *byte,* but hardly anyone really uses the word anymore. One byte is eight *bits,* so a nibble is four bits. If you really want to know what bits and bytes are, please see their definitions under *bit.* (Nibble is sometimes spelled "nybble.")

nicad, NiCad

Nicad is a new word formed from **ni**ckel and **cad**mium, the metals used to make the most common type of rechargeable battery, and the word now applies to the batteries themselves. Nicads have been the batteries of choice for *laptop* and *notebook* computers, but they have some important draw-backs. For one thing, a nicad "remembers" how much charge was left in the battery at the time you recharge it. The battery acts as though it's empty when it runs down to this charge, even if it's still got plenty of juice. Because of this "memory" effect, you must completely exhaust a nicad before you recharge it, at least every few times—otherwise the battery will hold less and less charge, which means you won't be able to work as long without plugging in to the wall. The other big problem with nicad batteries is that they're environmentally dangerous—the cadmium is a poisonous waste, and nicads should be carefully disposed of.

Nicads are now being supplanted by a new type of battery made from "nickel-metal hydride," which holds more charge, doesn't suffer the memory effect, and is environmentally safer.

node

Any device that is connected to a *network* is considered a **node,** especially the computer itself. For instance, you may have a network in the computer lab at school that has 16 computers, 4 printers, and 1 file server connected. This network has 21 nodes.

noise

Noise in *telecommunications* is another word for the static in a phone line. In graphics, noise refers to the visual equivalent of static, such as bits of unsightly junk or dirt on an image. More generally, noise is any meaningless, undesirable information mixed into the "real" information you're working with. Noise is kind of like weeds in your lawn.

non-disclosure agreement

Anyone who is privy to secrets or information about a product before the rest of the world is informed must usually sign a **non-disclosure agreement,** in which they promise they won't tell anybody anything. It's a standard form of procedure for anyone *beta testing* a product or researching or writing a manual, etc. NDA, of course, is the TLA (three-letter acronym).

non-impact printer

A **non-impact printer** is any printer that does not use little hammers or pins to pound the ink from a ribbon onto a piece of paper. If you cannot make multiple copies through carbon paper on your printer, it is a non-impact printer. Laser printers, and inkjet printers like the HP DeskJet and the HP DeskWriter, are non-impact.

If your printer can make multiple copies through carbon paper, it is an impact printer. Any standard dot matrix printer, like an Epson or a Toshiba or the Apple ImageWriter, is an impact printer. So are daisywheel and ball (Selectric) type printers, the old *letter-quality* machines that you don't see much anymore.

non-volatile

In Life, "volatile" can mean "easily provoked," which is not appropriate to our point here, or "transient and fleeting," which *is* appropriate to our point. **Non-volatile** is an adjective that means something is *not* transient and fleeting, but will stick around.

You'll usually hear the term non-volatile in relation to *memory*. There are different ways to classify memory, but from one perspective, there are two kinds: *volatile* memory that empties its contents when the power is turned off, and non-volatile memory that does not empty its contents when the power is turned off. The most common type of non-volatile memory is *ROM* (read-only memory) in which information is permanently encoded in the *chip*. A special type of ROM, known as flash ROM, can be electronically erased and rewritten with new information. Some laptop computers can use flash ROM memory *cards* to store applications and other data, and some PostScript printers have built-in flash ROM for storing *fonts*.

However, some computers come with non-volatile *RAM* (random access memory). Most RAM is volatile, but RAM chips are available with little built-in batteries that keep alive the information stored in the chip when you turn off the computer. Or the system itself may have a battery that keeps the contents of RAM alive (non-volatile) when the power is off.

Please see the definitions for *ROM (read-only memory)* or *RAM (random access memory)* for more details.

Norton Utilities™

Norton Utilities is a particularly wonderful collection of small programs that helps you manage your computer. It is particularly useful for *data recovery,* those times when you accidentally deleted or threw away a file, or your computer seriously crashed and left no survivors. The set includes utilities for security, disk management, and assorted miscellaneous functions. The Norton Utilities are available for PCs and Macs, although the two versions don't do exactly the same things.

notebook computer

A **notebook computer** is a complete computer that looks about the same size as a big notebook. Amazing. A notebook computers covers roughly the same area as a piece of letter paper (8.5 x 11) and is no more than about 2.5 inches tall. With those dimensions, it fits easily into a briefcase. Notebooks usually weigh around six pounds. People distinguish notebook computers from *laptops,* which are a little bigger (maybe 10 x 13 x 3) and weigh more (laptops weigh up to about twelve pounds).

Note Pad

Note Pad is a *desk accessory* that comes with your Mac. You'll find it in the *Apple menu*—just press on the apple in the *menu bar,* at the far left corner of the screen, and then choose Note Pad from the menu. You can write notes in the Note Pad, just like you would write in a notepad on your desk. There are eight pages to type on. The messages stay there until you change them, even if you turn off the computer. It's a great place to store little pieces of information, or to leave messages for someone else who uses your computer. or to leave a reminder for yourself, or to send love notes to your favorite person.

Notepad is the name of the little text *editor* that comes with Windows, and is used for editing plain *ASCII text files.* It's not as automatic as the Note Pad on the Mac—you have to save your files youself—but it does hold much more text (up to 64K).

Novell

When you hear the word **Novell** it is usually in relation to a *network.* Novell, Inc. is a company that sells NetWare, the most widely used operating system software for *local area networks* (LANs). A "Novell network" is a LAN that uses NetWare to manage the network. A *local area network* means it connects computers together that are in close proximity to each other, allowing people to share information.

Novell also sells *DR DOS,* an operating system for IBM-type personal computers. DR DOS is fully compatible with *MS-DOS* which most people use instead, but DR DOS has some advantages and is cheaper. Also, DR DOS can take the credit for pushing Microsoft to make the major improvements we've seen in MS-DOS over the past couple of years.

NTSC

NTSC stands for the **n**ational **t**elevision **s**tandards **c**ommittee. They do not make sure that all television programs rise above sub-human level—no, the standards the NTSC sets pertain to the television and video playback, the *resolution* and speed and color. Clay, who works with high-definition television, says NTSC stands for Never The Same Color.

NuBus

Most Macintoshes have a *slot,* a connection point, where you can insert a *board* (also known as a *card*). A "board" is a piece of plastic with a printed circuit that adds some new function to the machine. The smaller Macs, like the SE and the SE30, have just the one slot. In the bigger machines, Mac IIs and above, Apple started adding more slots. If there is only one slot, it is usually a *PDS,* or *processor direct slot.* In a PDS, the circuit on the board talks directly to the *processor* (the *chip* that runs the machine).

You can use slots to add extra boards, or cards, to the computer system to make it work faster, to add more *memory,* to run a new monitor, or for various other reasons. As the computers grew up, Apple started adding more and smarter slots, the **NuBus** expansion slots. You can plug NuBus cards into these NuBus slots (you can *only* plug NuBus cards into a NuBus slot, and you can *only* plug a PDS card into a PDS slot).

The circuit on a NuBus does not talk directly with the processor (as does a PDS)—it passes through computer *logic* first. Because it has some logic involved with it, a NuBus card is considered more "intelligent." It "auto-configures" your machine; that is, it tells the Mac what kind of card it is and where it is and what to do with it (a PDS card can't do that—you must "configure" or set up your Mac yourself). These NuBus expansion slots were meant to help your Mac function at top speed, but because they are so intelligent they have sometimes become a bottleneck, as the information must pass through the computer logic on its way to and from the processor. Even with the occasional bottleneck, NuBus slots provide a powerful way to expand the capabilities of your Macintosh. See Appendix A for an illustration of slot and cards.

null modem

A **null modem** cable is a cable that plugs into the communication *port* (the *serial* port, the one where you would plug in the cable for a *modem*) of two computers that are pretty close to each other. Then the two computers can share information back and forth, even faster than if they were connected by a modem and phone lines. If you want to be able to connect your portable computer directly to your bigger computer without modems, you need to hook them together with a null modem cable. Or you can use a standard serial cable and get a null modem adaptor, which is just a plug with a female jack on one side and a male jack on the other.

More technical info: The reason you need a null modem cable, not a standard serial cable, is that the port where you plug in on the modem is wired as a mirror image of the port on a computer—the computer might send signals on wire 2 and receive them on wire 3, while vice is versa at the modem end. Since the ports on both computers are wired the same way, the signals get crossed if you use a regular serial cable. In a null modem cable, the wire going to pin 2 of one computer is crossed so it goes to pin 3 of the other, and so on (only 3 pairs of wires need to be reversed).

number crunching

Number crunching is an informal term referring to the process of performing lots of math on lots of numbers, as in a bank's *spreadsheet* or a complex financial report. A *math* or *floating point coprocessor* is a special *chip* that can make certain computers and programs whiz through number crunching at amazing speeds; thus this chip itself is sometimes called a number cruncher.

numeric keypad, numeric pad

The **numeric keypad** (or **numeric pad**) is the set of number keys and a few other keys arranged in a block on the right end of the keyboard. It can be used as a little calculator (if a calculator *utility* or *desk accessory* is running), as well as being a convenient way to type numbers.

Sometimes it can make a critical difference to use a number or a symbol from the keypad as opposed to the keyboard, especially in some *keyboard shortcuts* or when using *ASCII* characters on PC machines. For instance, if your application can make a character a superscript character (like the rd in 3rd) by pressing Control Shift +, the shortcut might work only with the plus sign (+) from the keypad or it might work only with the + from the key*board*. Then again, it might work with either, but if you have problems, try each one separately.

The slash (/) on the keypad is used for dividing, and the asterisk (*) is used for multiplication. In many applications you can use the slash or asterisk from either the keypad or from the alphabetic keyboard for math calculations.

If your keypad moves the cursor instead of types numbers, try pressing the Num Lock key or the Clear key (usually found above the 7). In many applications this will toggle the keypad back and forth between typing numbers and moving the cursor. See *Num Lock key.*

I TELL YA PHIL, "WORKAHOLIC AIR" IS THE <u>ONLY</u> WAY TO FLY.

Num Lock key

Many keyboards have a *numeric keypad* on the right side. Often, that keypad can either type numbers or it can move the *cursor* around on the screen. On most computers, the **Num Lock key** is a key that switches, or *toggles,* the keypad back and forth from either typing numbers *or* moving the cursor.

The Num Lock key is usually located at the top left of the *numeric keypad.* Some keyboards don't have a key specifically labeled Num Lock, but they may have a key that does the same thing. For instance, the key at the top left of the numeric keypad might be labeled "Clear," but still functions as a Num Lock key—try it to see. How do you try it? Watch your cursor as you use the keypad. If it types numbers, press the Clear key once. Type again. If it still types numbers instead of move the cursor, then the key is not functioning as a Num Lock.

Individual software packages may use other keys as a Num Lock key. For instance, in PageMaker (Mac) you can press the Caps Lock key to toggle the keypad back and forth between numbers and cursor.

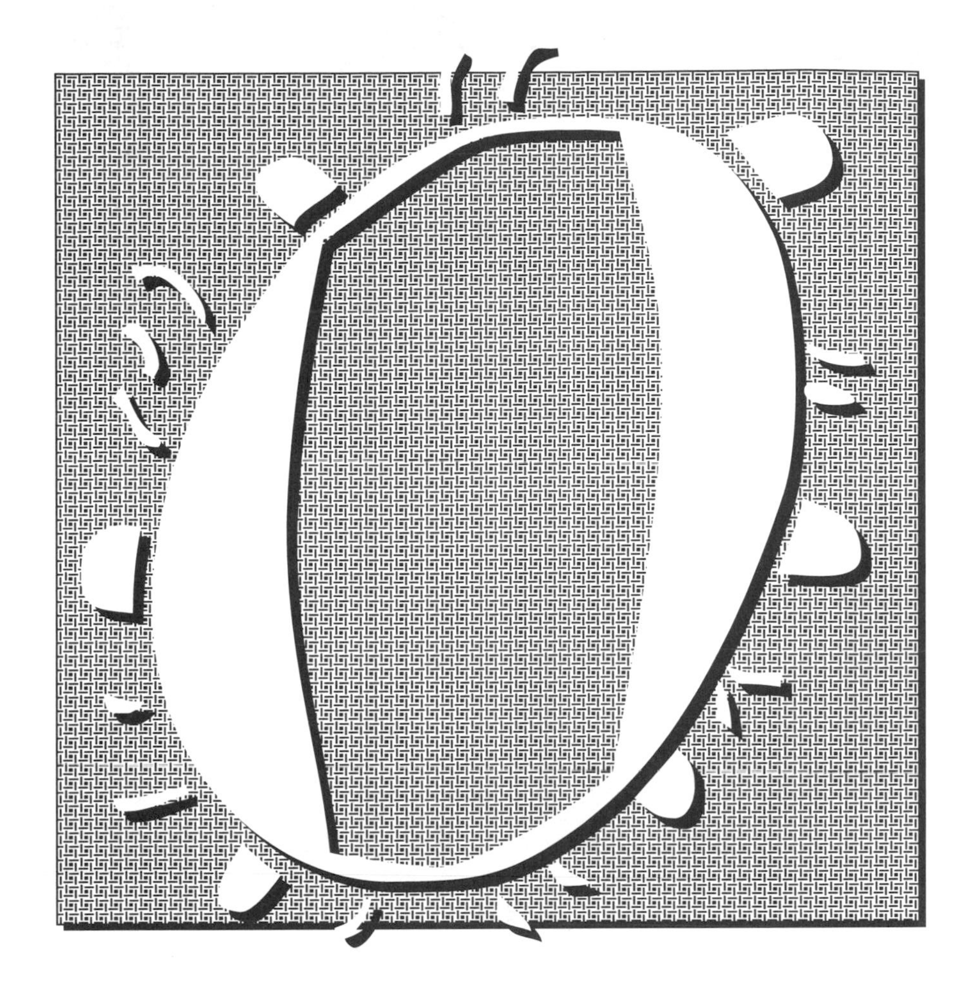

object

Used in the most general way, an object is just a thing, any thing you can think of. In computer jargon, the term **object** is often used in preference to "thing" when you're talking about something abstract, that really only exists conceptually. It's a good term for lumping together items that have characteristics in common, but aren't just alike. For instance, on your screen there may be many objects you can click on, which might include *buttons, scroll bars, icons* (little pictures), and so forth.

A bit more specifically, the term object suggests that what you're talking about has a well-defined, distinct existence (again, we're talking conceptually). In many software packages, you can place *objects* that were created in one application into a document you're working on in another application. An object might be a graphic image, a portion of a spreadsheet, text from a word processor, or even a piece of music. The object acts like a piece of wood floating in a pond—it's in the document, but it's still a separate thing (see *OLE*).

When you're talking about graphics in a *draw* or *object-oriented* graphics program, an object is any distinct shape that you can manipulate independently—a line, a circle, a triangle, an irregularly shaped blob. *Paint programs* with their *bitmapped graphics,* though, don't have objects—the shapes you see in a bitmap are just part of a collection of dots and have no independent existence.

In *object-oriented programming,* an object is the fundamental unit of a program and consists both of the data you're working with and the functions or *methods* used to manipulate that data. See *OOP*.

object-oriented

The term **object-oriented** has three different shades of meaning. One of these has to do with graphics, and is covered in the next definition. The second refers to a kind of computer programming, and that's covered in the definition for *OOP*. But in the frighteningly vague sense we'll tackle here, object-oriented is applied to almost everything these days.

Boiled down, the idea is this: something that's object-oriented is something that you can work with easily as a unit. In this sense, the Macintosh's Finder gives you an object-oriented way to work with applications and documents—they're represented as little icons, visual metaphors for objects that you can pick up and move around at your whim. If you want to get rid of a program or document file, you just put it in the trash, which is an object too. But believe me, "object-oriented" has no strict definition, and it can be tough to know what the advertisements and marketing people are talking about.

object-oriented graphics, object graphics

Also known as *vector graphics,* **object-oriented graphics** are shapes represented with mathematical formulas. (This is very different from *bitmapped graphics,* in which the image is *mapped* to the *pixels* on the screen, dot by dot. Please see that definition for a comparison.)

In a program that uses object-oriented graphics, each separate element you draw—every circle, every line, and every rectangle—is defined and stored as a separate object. Each object is defined by its vector points, or *end points.* Because each graphic object is defined mathematically, rather than as a specific set of dots, you can change its proportions, or make it larger or smaller, or resize it, stretch it, rotate it, change its pattern, etc.,

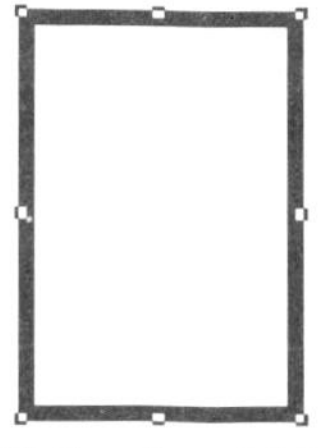

Notice the eight handles on this selected object, one at each corner and one on each side.

without distorting the line width or affecting the object's sharpness and clarity (the *resolution*). Because each object is a separate entity, you can overlap objects in any order and change that order whenever you feel like it. To *select* a graphical object in an object-oriented graphics program, you usually just *click* on the object with the pointer. When you select it, a set of *handles,* little black squares, appear on or around the object (compare *marquee*). By *dragging* the handles, you can change the size or shape of the object or the curviness of any curved lines. You can also copy or cut a selected object to the *Clipboard,* or move it around on the screen, without disturbing any other object.

The *resolution* of object-oriented graphics is *device independent*. This means that if you print a graphic image to a printer that has a resolution of 300 dots per inch, the graphic will print at 300 dots per inch. If you print the same image to an imagesetter that has a resolution of 2540 dots per inch, the graphic will print at 2540 dots per inch. (Bitmapped graphics, though, always print at the same resolution.)

Fonts, or typefaces, can also be objected-oriented, but they're not usually referred to this way—instead, such fonts are also known as *outline* fonts, *scalable* fonts, or *vector* fonts.

object-oriented programming

See *OOP.*

oblique

oblique

Oblique (pronounced "oh **bleek**") refers to slanted type (or to a slanted graphic *object*). In type, there is a definite difference between text that the computer has "obliqued" and text that has been designed as an italic font (see *italic* for an illustration of the difference between oblique and italic).

If you want a professional look, please avoid using the computer to oblique, or slant, your type.

OCR

OCR stands for **o**ptical **c**haracter **r**ecognition, a wonderful and marvelous technology. It enables you to convert previously printed text material into information your computer can understand, without having to retype it. Have you ever had a story or an article or a magazine clipping that you wanted to have in your computer, but the thought of retyping the entire thing was overwhelming? Or just boring? That's what OCR is for.

OCR requires a *scanner* and *software.* Typically when you scan anything on paper (text or graphics), to get it from a printed page into your computer, the image scans in as a picture. That is, if you scan a typewritten page, you would have a *picture* of the typewritten page, not the typed letters themselves. But with OCR software, you can *scan* the printed page and the software figures out what characters those little shapes are supposed to be, and it turns the scanned image into real text that you can put into your word processor! Since the OCR software almost always makes some mistakes, you should then pass the text through a *spelling checker;* ideally you should read it over yourself, because some mistakes still give you the wrong word, but a word that's properly spelled. Then you can edit the text as much as you like, or use it as a story in a newsletter, or add it to your thesis paper. Some people use this technology effectively right now, and some complain about the efficiency of it—it isn't really much faster than typing—but OCR software is getting better all the time.

OCR also refers to a special *font* that an OCR system can recognize easily—you know, those typefaces that *look* like a computer did it.

offline

When you are communicating through your modem, you are considered to be *online* when you are connected to the *information service* or *bulletin board* or your friend's computer at the other end of the telephone line. You are considered to be **offline** when you are no longer connected (if you were online, the phone line has been hung up, even though the communications software may still be running) the communications program. It's a common scenario to compose *e-mail* offline, go online to send it, then go offline when you're through, to put together other business.

In another sense: even though a printer or a hard disk or any other device is physically connected to a computer, the device is considered **offline** if it is not turned on or perhaps it is turned on but not ready for use. Many printers have an On Line or Ready light. When the light is off, the printer is offline. See *online.*

OLE

OLE stands for **o**bject **l**inking and **e**mbedding, a capability introduced into Microsoft *Windows* 3.1. OLE gives all Windows applications a standard way for incorporating *objects* created in one program into documents created with other programs.

An object might be a passage of formatted text, a part of a spreadsheet, some sounds, or a picture. Unlike information that you copy from one document and paste into another the standard way, a linked or embedded object retains a connection to the application that originally created it. You can return to that application to edit the object whenever you want to just by double-clicking on the object—you don't have to bother with finding the icon for the application, loading the right file, and so on. Better yet, the changes you make automatically appear in the document where you linked or embedded the object.

When you **embed** an object, you place a *copy* of the information into your document. This copy is connected to the original application, but not to a particular document in that application. The only advantage to embedding an object instead of copying the information the ordinary way is that you can edit the object more conveniently.

By contrast, when you **link** an object, you place a "reference" to a particular document from another application into the document you're working with. Let's say you have a spreadsheet that totals your third quarter sales figures. You link that spreadsheet document into a report you're preparing in your word processor. Later, when revised sales figures come in, you go back to the *spreadsheet* application and change the numbers. The next time you open the report document in your *word processor,* the new figures from the spreadsheet appear automatically in the report. This is the same idea as a "hot link," and it may help to read the generic definition for *link.*

OLE only works if both applications involved have been designed to use it, and even then it may only work in one direction (like, you can link a graphic into a text document, but not text into a graphic document). And it doesn't work exactly the same way in every application. Even so, it's easier and more consistent than the old method, called DDE.

online

Online, in general, refers to communicating to other computers or to the people sitting at other computers or to *devices* such as printers or *file servers* through your *modem* or *network.*

With a modem, you're "online" when the modem is actively connected to and communicating with another modem. For instance, you need to go online to get your *e-mail* or to send e-mail, or to use a *bulletin board service.* "Honey, could you get dinner started—I'll be online for a while."

If your computer is hooked up to a network, either throughout your office or even internationally, you are "online" when you have *logged onto* the network.

When a *device* like a printer or a hard disk is up and ready to be used and controlled by the computer, it is considered to be **online.** Your printer may have an On Line, Ready, or Select button and a corresponding light. If the light's not on, you press the button to "take" the printer online.

online help

Online help is like having a manual on your screen whenever you need it, accessible while you are actually working in the program. It's a *help* system built into the software program. Of course, as with all manuals, some programs have more useful online help than others.

To display the help info, its usually a matter of choosing Help from a menu or pressing a keyboard combination, and some programs let you use either method. If your keyboard has a key labeled Help, try pressing that. On a PC, the Help key is almost always the one labeled *F1*, though WordPerfect uses F3.

What you see when the help information pops up on screen depends on the software. If the program offers "context-sensitive" help, the helpful information you'll see should pertain directly to whatever command you're working with at the time. If the help isn't context sensitive, then you'll see a list of topics; you choose the topic you need help with, and that information appears on the screen right in front of your face.

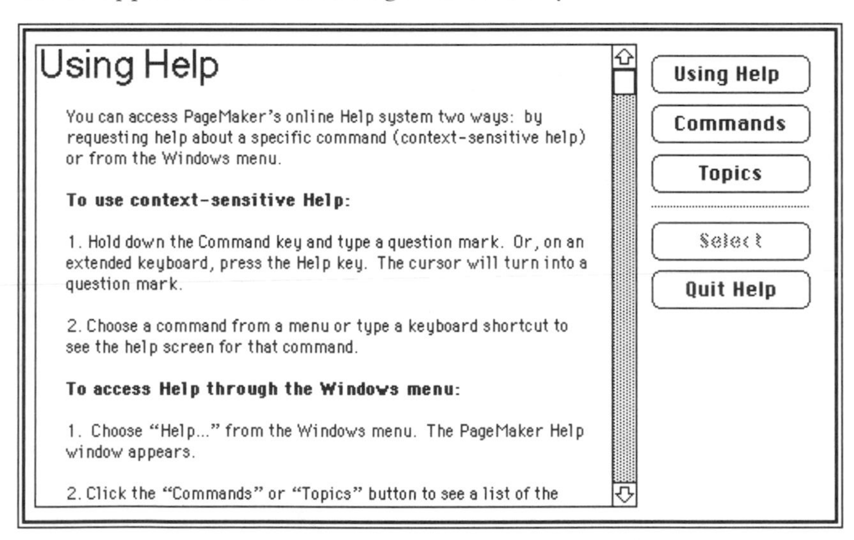

This portion of the online help system teaches you how to use it

online service

With a computer and a *modem* you can use an **online service** to research through encyclopedias and newspapers and magazines; you can make hotel and plane reservations; copy software; attend forums and conferences; make friends; send and receive *electronic mail;* get answers to questions; and oh so many other interesting things.

To use any online service, you do need to have a *modem*. The modem may be a little box that has one cord plugged into your telephone and another cord plugged into your computer (plugged into the *modem port,* of course) or the modem may be an internal *card.* You also need to have some sort of *communications software*. You can buy general communications software (sometimes called "telecommunications software") that will provide the link for general telecommunication tasks. But when you sign up for an online service, part of what you buy is the software that connects you with the service. When you use the software to go *online* (as in phone line), your computer uses the phone to call the service. In most services, you typically land in a generic sort of area from which you can choose the particular subject area you want to explore further—personals, computing, business, finance, travel, etc.

An online service is a paid service*, like any other kind of service. Sometimes you have to pay a subscription fee to join, and then a monthly fee or perhaps a fee based on how many hours you spend online. America Online, CompuServe, GEnie, and Prodigy are the names of a few of the most popular services.

*A *bulletin board service* is a little different from an online service in that a bulletin board is typically run by a not-for-profit group such as a computer user support group, or by a manufacturer such as a software company, and so the focus of interest of a bulletin board is typically narrower than with a paid online service. The BBS (bulletin board service) tells you the number for your computer to call to connect with the bulletin board and then your only fee is the phone call. You still need a modem and appropriate *communications software*. Sometimes the group that sponsors the bulletin board can provide you with the software.

online sex

Don't laugh—**online sex** is a here-and-now *virtual reality* scene that is thriving around the world, from France's Minitel to Japan's communication system to America's *bulletin boards* and *online services*. Hormones are buzzing through the wires over our heads and whizzing through the satellite channels. Computers all over the world become whorehouses at night. How could this happen, you ask? I mean, like how could you have *sex* on your computer? Its simply a matter of typing what you would be doing if you were in person. Now you think about that for a little bit.

Online sex brings up questions of changing moralities, the muddy line between fantasy and reality, the power of pretending, the significance of the need for intimate human interaction, the historical perspectives of our sex drives, and oh so many other interesting thoughts.

on the fly

When you hear of a product that performs some function **on the fly,** it means the product does its thing automatically while you work in your accustomed way—you do not have to quit whatever program you are in or shut down or reboot or do anything special to get the results.

For instance, if you are working in an application and you need other fonts, you can use a font utility program to add the fonts without closing the application. But some applications can't see the new fonts until you quit the application and open it back up again. An application that can see the new fonts instantly is said to be able to "update its font menu on the fly."

Or let's say you want to import a graphic into your document, but yesterday you *compressed* that graphic so it wouldn't take up so much space on the disk. Most programs will not let you import a compressed graphic, but the file compression utility that "decompresses a file on the fly" means it will decompress the graphic as you import it, rather than having to quit the program, decompress the graphic, then go back to your document to import it.

OOP

OOP stands for **o**bject-**o**riented **p**rogramming, a relatively recent development in programming technology. In traditional computer programs, the procedures (the programming commands) that get things done are separated from the data they work on. By contrast, object-oriented programs are put together from building blocks called *objects;* each of these self-contained software modules includes all the commands and data needed to do a given set of tasks when it receives the right "messages." Because it is "encapsulated" in this way, an object can be reused as a unit in as many programs as needed. By design, OOP makes it easy to generate new objects that automatically "inherit" the capabilities of existing objects. The programmer can then modify a function or two or add some new ones, but she doesn't have to start from scratch.

All this sounds great on paper, but designing the objects in the first place turns out to be a big job. OOP is best suited for large-scale programming projects that are likely to need modification and expansion over time.

The most widely used object-oriented languages are C++ (a modified version of C) and SmallTalk. Various software packages let ordinary people partake in some of the benefits of object-oriented programming. In *HyperCard,* for example, it's easy to copy a button you've programmed from one card to another, carrying all its programming with it. You can then customize the programming in the copied button, without having to rewrite all the programming. But HyperTalk, the language that comes with HyperCard, isn't really an object-oriented language in the way that C++ and SmallTalk are.

open, Open

The term **open,** with a lowercase o, means to make the file visible on the screen. If you want to write a letter, you must first open a word processing application. If you want to make corrections to your brochure, you must first open the brochure document.

The *menu command* called **Open,** with a capital O, lets you open a file (document) that was previously created and stored with a name. It gives you a dialog box in which you can find the name of that file you wish to open. When you see the two choices, "Open" and "New" in the menu, it's easy to confuse the terms because we think, "Well, I want to *open* a *new* one." The key thing to remember is that New starts a brand new, blank document with no name, while Open opens an existing document that someone has previously created.

Open Apple

This is the symbol on the Open Apple key.

Open Apple is what Apple II users (as opposed to Macintosh users) call the key with the little picture of the apple on it. The Open Apple is a *modifier key,* meaning it does nothing by itself, but only works in combination with other keys. Macintosh users call this key the *Command key.*

open architecture

When a manufacturer publicly publishes the specifications for their computer, the computer is said to have an **open architecture.** This allows other companies to create *add-ons* to enhance and customize the machine, and to make *peripheral devices* such as external hard disks and scanners that work properly with it. With a *closed architecture,* only the original manufacturer can make add-ons and peripherals.

A computer with an open architecture, such as the IBM PC, also means that other developers can not only make add-on parts, but can also make a similar machine, which spawned the whole world of IBM clones. The Macintosh architecture is only partway open. Apple gives other manufacturers enough information to make add-in boards and peripherals, but not enough to create a Mac clone.

operating system

The **operating system** (sometimes called the "OS") is the master control software that runs the computer itself. When you turn on your computer, the operating system is the first program that gets *loaded* into the *memory* of the machine. With the help of instructions built into the computer's *ROMs* or *BIOS,* the operating system sets up the means for your own programs to interact with the computer and its parts, such as the disk drives, your screen, and the keyboard. It organizes the computer's memory into chunks your programs can use. And when you give the command, the operating system runs your programs.

The typical software program interacts frequently with the operating system—whenever you open or save a file, for instance, or when the program needs more memory, or when it wants to print. When you *exit* or *quit* the program, the operating system handles the technicalities, deactivating the program and making the memory it used available to other programs. If you have a *multi-tasking* or *multi-user* operating system, the operating system is responsible for seeing to it that each task and each user gets their fair share of the computer's attention.

Because they depend so heavily on the operating system, all your programs must be written specific to the operating system in use on your computer. You can't use a program prepared for one operating system with another operating system. Some program *code* is *portable* from one operating system to another, but it must be compiled separately to make the actual program for each operating system.

The most common operating systems for personal computers at the moment are the *Macintosh* system, *DOS* (*PC-DOS* and *MS-DOS*), and (trailing badly) *OS/2*. Many *workstations* use the *Unix* operating system. Unix is available for PCs, the Macintosh (as A/UX), and minicomputers, also. As you can see, the same computer can run different operating systems. Other minicomputer and mainframe operating systems have names like VMS, VM, MVS, and even DOS, but these must be supplemented by "transaction processors" like CICS, TSO, and alphabet soup.

Drew Cronk sent me the following, which he downloaded from the North Coast Mac User Group bulletin board. Unfortunately, there were no credits attached to it. If you know who wrote this, let me know so I can give due thanks!

What driving to the store would be like if operating systems ran your car

MS-DOS	You get in the car and try to remember where you put your keys.
Windows	You get in the car and drive to the store very slowly, because attached to the back of the car is a freight train.
Macintosh System 7	You get in the car to go to the store, and the car drives you to church.
UNIX	You get in the car and type GREP STORE. After reaching speeds of 200 miles per hour en route, you arrive at the barber shop.
Windows NT	You get in the car and write a letter that says, "Go to the store." Then you get out of the car and mail the letter to your dashboard.
Taligent/Pink	You walk to the store with Ricardo Montalban, who tells you how wonderful it will be when he can fly you to the store in his Lear jet.
OS/2	After fueling up with 6000 gallons of gas, you get in the car and drive to the store with a motorcycle escort and a marching band in procession. Halfway there, the car blows up, killing everybody in town.
S/36 SSP (mainframe)	You get in the car and drive to the store. Halfway there you run out of gas. While walking the rest of the way, you are run over by kids on mopeds.
OS/400	An attendant locks you into the car and then drives you to the store, where you get to watch everybody else buy filet mignons.

optical disc

A CD, such as the kind you play to listen to music, is an example of an **optical disc.** So is a "video disc" (properly called a *laser disk*), such as the kind you can rent at the video store that has an entire movie on it. Optical discs for your computer can hold an incredible amount of information—up to 6,000 *megabytes* (which is 6 *gigabytes*) of data. Entire encyclopedias, Shakespeare's works, or representations of the art in the Louvre have been recorded onto optical discs.

So what makes a disc an optical disc? Just that the information is stored and read using light, rather than magnetism like a standard hard or floppy disk. (Why is it optical dis**c** but hard dis**k**? I give up.) There are three basic types of optical discs, and each one requires its own special kind of *drive* to use the disc. All three types are like floppy disks in that you pop them in and out of the drive as needed, giving you a potentially unlimited amount of storage. Here are the types:

CD-ROMS are the same size as the music CDs, and they store information in exactly the same way—in fact, you can play a music CD on your computer's CD-ROM drive. The information is permanently encoded on the disk and you cannot change it, but you can read it as many times as you like (*ROM* stands for read-only memory). CD-ROMs are great for publishing or distributing large amounts of information.

A *WORM* (**w**rite **o**nce, **r**ead **m**any) disc lets you write to (put information on) the disc yourself, but only once. After that, a WORM disc works just like a CD-ROM. Since you get to decide what information goes on the disc, WORM systems are good for keeping *archival* copies of information you have to store permanently.

Erasable optical discs can be erased and recorded over many times, a technology that is in its infancy but one that may change the way we store data. The erasable optical drives now available are a lot slower than hard disks, and they're more expensive, but the disc itself can't crash like a hard disk. So they're used mainly for backing up hard disks. Actually, the current version of this technology is more properly called "magneto-optical," since the information is still encoded on the discs magnetically. However, a laser beam assists in the process, and a laser is used to read the information from the disc.

optical mouse

An **optical mouse** is a *mouse* that must be used on a special *mouse pad* which has a reflective surface with an almost microscopic grid. The mouse keeps track of its position by shining a little beam of light onto the surface of the mouse pad. The mouse knows how far you've moved it and in which direction by counting the grid lines you've crossed. Optical mice are supposed to be more reliable than standard mechanical mice. ("The problem is," says Steve, "you have to use that special pad—since I like to sit back and run the mouse across my knee sometimes, an optical mouse just won't do.")

optimize, optimization

There are two common uses of this word. I'll explain the simple one first. If something has been **optimized,** that just means it works better or faster in a particular way. For instance, a software program can be optimized for speed, meaning that it's been designed with a priority on running fast; or it can be optimized for size, meaning that a special effort has been made to make it smaller, so that it takes less room on disk.

You may also hear of the need to *optimize* or *defragment* your hard disk. Here's why: When you save your documents, the computer tucks away that information all over the hard disk. As your hard disk fills up and you continue to write and erase files, fewer areas with large amounts of free space are left on your disk. The next time the computer has to write a long file, it finds a free space on the disk and sticks in as much of the document as will fit. If there's information left over, the computer uses the next open spot, and so on. The computer does keep track of where it put the information, but the little pieces of the document are spread out all over the disk. When you want to work on that document, the *disk heads* must go running all over the disk to gather up the parts you want. The bigger the file, the more head movement and the slower everything works. This is called *fragmentation.*

Fortunately there is software that can *defragment* your disk. It gets all the little separate parts, gathers them up, and puts all the little pieces next to each other. Now, a good software package that defragments your disk goes a step further and also **optimizes** those files—it monitors which files you use regularly and arranges them on the disk in a prioritized order. This lets the disk heads reach the files fastest that you use the most often. Amazing.

Option characters

The **Option characters** are those characters you get when you hold the Option key down and then press a key. You know what happens when you hold the Shift key down and press a key—you get a capital letter or a symbol above the numbers. Well, the Option key does the same thing but you just can't see the characters on the keyboard. Option characters include some very useful symbols, like ¢, ™, ©, and ®. It's also the place to find accent marks, such as those you need to create ñ, é, è, or ç (see the next entry). See *Key Caps* for more info on these characters, as well as the Option Shift characters.

Option key ⌥ ⌥

The **Option key** is next to the *Command key,* down near the Spacebar. The Option key, like any of the *modifier keys,* won't do anything if you press it by itself—it is always used in conjunction with another key or combination of keys, or with a mouse action. For instance, to create an accent mark over an e, (as in résumé) hold the Option key down and type the letter e. Nothing will happen. Now type an e again and the letter will show up with the accent mark above it. Or hold Option and press 8 to type a bullet (•). Many programs use the Option key in many different ways. It's sometimes symbolized in the *menu* as ⌥ or ⌥.

Option-click, Option-drag

When a direction tells you to **Option-click** or **Option-drag,** it means to hold down the Option key while you click on an item with the mouse button or while you drag the item. An Option-click or Option-drag has different results than a plain click. For instance, if you *select* a file at the Desktop (also known as the Finder) and drag it into another window or folder on the same disk, the file moves from the original location into the new window or folder. But if you hold the Option key down and drag it to another location, then you don't *move* the file, you make a *copy* of the file.

orientation

Orientation refers to whether a page is printed (or displayed) in a tall format or in a wide format. If you use a rectangular standard paper size like standard 8.5 x 11 and print on it like we usually do, it is a *tall* orientation (the paper is taller than it is wide). If you lay out your page so it is sideways, with the longer edge of the paper at the top and bottom, that is a *wide* orientation (11 x 8.5). A tall orientation is often called *portrait* or vertical, and a wide orientation is often called *landscape* or horizontal. See the illustrations under *landscape.*

In most programs you can choose your orientation when you set up your page to create the document. You may also be able to choose orientation when you go to print, or by sending special commands to your printer. When you print, though, make sure you check to see that the **printing** orientation matches the **page layout** orientation or you may get a surprise.

original disk

The **original disk** is the disk you got in the box from the software company (there may, of course, be more than one original disk; for instance, Aldus PageMaker arrives on six disks). The original disk(s) contains the software program and all the related files and documents that belong to it. An original disk is sometimes also called a *master disk.*

It's always a good idea to make backup copies of your original disks, then use the copies to transfer (or *load*) the software onto your hard disk. Keep the originals in a safe place, perhaps even in another building. The original disks are usually *locked* (the Mac term) or *write-protected* (the PC term) when they arrive. It's a good idea to leave them this way—locked; it will help prevent accidentally erasing valuable and necessary information on the disks. Lock or write-protect your backup disks too.

If you destroy or lose your original disks before or after you load the program onto your hard disk, and if your hard disk goes bad and you didn't make the back up copies, you are out of luck. You can try your sob story on the software vendor, but don't expect much. You might also want to read about the software *license*—it may surprise you.

orphan

There is and always has been and probably always will be quibbling about the exact meaning of an **orphan.** I define an orphan as the last line of a paragraph that is stranded alone at the top of the next column or page. The first line of a paragraph alone at the bottom of a column can also be called an orphan (see examples). Some people call this typographic faux pas a *widow* (which I define as something else). It doesn't matter at all what you call it—the point is never to do it. Also see *widow.*

Wants pawn term, dare
worsted ladle gull hoe lift
wetter murder inner ladle
cordage, honor itch offer
lodge, dock, florist. Disk
ladle gull orphan worry
putty ladle rat cluck wetter
ladle rat hut, an fur disk
raisin pimple colder Ladle

Rat Rotten Hut. ——— *This is an orphan.*
Wan moaning, Ladle Rat
Rotten Hut's murder colder
inset. "Ladle Rat Rotten
Hut, heresy ladle basking
winsome burden barter an
shirkle cockles. Tick disk
ladle basking tutor cordage
offer groinmurder hoe lifts

OS

OS stands for **o**perating **s**ystem, the software that controls everything in the computer. Please see *operating system.*

OS/2

OS/2 stands for **o**perating **s**ystem/**2**, an *operating system* that was developed by *Microsoft Corporation* and *IBM,* and now being improved and marketed exclusively by IBM. OS/2 runs on *286-*, *386-*, and *486*-based IBM PCs and *compatibles.* OS/2 has many of the same commands as DOS, can read DOS disks, but it is more sophisticated than DOS. In particular, OS/2 is a *multi-tasking, multi-threaded* operating system. Multi-tasking means it can run more than one application at a time; multi-threaded means that a single program can perform multiple separate tasks at the same time.

OS/2 also includes a *graphical user interface,* or *GUI,* called *Presentation Manager,* and which makes OS/2 look and work something like the Macintosh or Windows. Using Presentation Manager is optional, depending on the software you run; if you'd rather, you can type commands on a *command line* in *character mode,* just as in DOS. OS/2 also comes with *LAN* Manager software for networking (connecting several computers together so they can share information).

If you read PC-oriented magazines, you'll see lots of arguments about which is better, OS/2 or Windows. From a strictly technical standpoint, OS/2 is superior: it's theoretically less prone to crashing; it uses a better type of multi-tasking (called pre-emptive multi-tasking) that gives you more control over which application gets priority; and permits multi-threaded applications, which Windows doesn't. OS/2 runs most DOS and Windows programs, though Windows programs run more slowly than they do under Windows 3.1. Although OS/2 is really popular in corporate *MIS* circles because of its excellent mainframe connectivity and application development tools, Windows is more popular by far, and for that reason far more applications and *utilities* designed for Windows are available than for OS/2, and more kinds of hardware devices work with Windows.

outdent

An *outdent* is when the first line or first several lines of a paragraph start to the left of the rest of the text, as in this paragraph. This is the opposite of a *first-line indent.* See *first-line indent* for more examples. This is also called a *hanging indent.*

outline font

An **outline font,** now also known as a *scalable font* (and on the Mac often referred to as a *printer font*), is a font in which each character is stored as an "outline" of the character shape (the outline is represented by a mathematical formula). When it's time to display text on the screen or print it on paper, the character outlines are translated, or scaled, into visible characters at whatever size you need (the scaling process can happen in the computer or in the printer).

PostScript fonts and TrueType fonts are the two most widely used types of outline fonts, although their mathematical formulas are different. Please, I suggest you read the information under *font* for the big picture and lots of details and illustrations on all the different types and parts of fonts.

BOB LOVED HIS LAPTOP
BUT FOUND HIS DESK
OFTEN GOT IN THE WAY.

outliner

An **outliner** is an application (or a feature within another application, such as a word processor) that makes it easy to create a conventional outline, like the kind you laboriously learned to make in grade school. You know, where you have a main point with a Roman numeral, and then an indented subpoint with a capital letter, and then a further indented sub-subpoint, and so on.

Instead of having to format your outline manually by using the Tab key and indents, the outliner does everything automatically. All you have to do is tell it which are the main points and which are the subpoints, and that's pretty easy. Another feature of outliner applications is that they let you expand and collapse the outline structure so that you can choose what you want to display. For example, you can display only the main points, just by entering a command. Outliners can make your life a lot easier if, for some reason, you often need to make outlines, or you like using them. ("For those of us who thrive on chaos," says Steve, "they're threatening.")

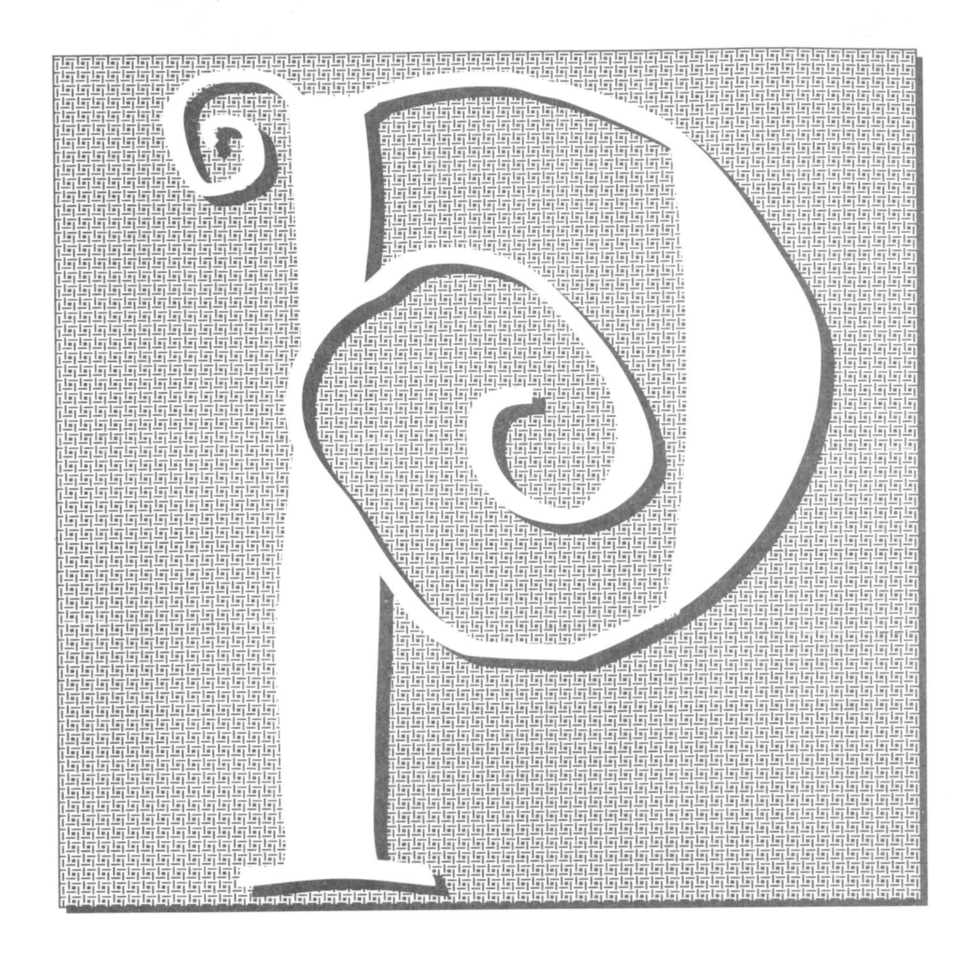

p

When you see a lowercase **p** with some numbers, such as 3p7, the **p** stands for *pica,* a typographic unit of measure. The number in front of the p indicates how many picas are in the measurement, and the number after the p indicates how many *points.* Think of picas and points as feet and inches: The measurement 4′5″ means 4 feet and 5 inches; the measurement 3p7 means 3 picas and 7 points. The number 0p6 means 6 points (0 picas, 6 points).

There are 6 picas in one inch. There are 12 points in one pica. Thus there are 72 points in one inch. (Yes, you true typographers and typesetters, traditionally 72 points did not equal exactly one inch, but only .996 of an inch. In computer type, 72 points now equals exactly one inch because the original Mac screen was 72 pixels per inch.

package

A **package** is simply a product, as in a "software package." The term suggests that the product consists of a number of components all wrapped up together, and that's what you get with software—you get the disks, the manuals, and maybe a reference card or a keyboard *template.* Usually, people use the term "package" to refer to a full-scale application program like WordPerfect or Aldus PageMaker or an operating system product like DOS or Windows, as opposed to a little utility program.

A package is an off-the-shelf product. The kind of software a specialist custom-designs for a business is not usually called a package.

page description language

See *PDL.*

page layout program

You use a *word processing program* to process words, to create pages of text. You use a *paint* or *draw* program to create graphics. You use a **page layout program** to put the text and the graphics on the page together. Often it's also called a *desktop publishing program*. Whatever you call it, this kind of package gives you much more control and flexibility in designing a page than a word processor does. You have more options as to the kinds of graphics you can use and what you can do with them, and you have much better control over your type and where you can place and arrange text. Examples of page layout programs are Aldus PageMaker, QuarkXPress, Ventura Publisher, and Frame Technology's FrameMaker.

page printer, page-oriented printer

A **page printer** forms the complete image of a page, and then prints that page all at once, more or less. Most page printers are *laser printers.* "Line printers," by contrast, print one line at a time, then pull the paper through a little and print the next line.

Page Up key, Page Down key

The **Page Up** and **Page Down** keys do various things, depending on which application you are using, but generally they move the page on the screen (particularly in word processors) up one screenful or down one screenful. On PC keyboards (and some other keyboards) the Page Up and Page Down keys are abbreviated PgUp and PgDn.

If the keyboard has a separate *numeric keypad,* PgUp and PgDn will be located on the 9 and the 3 keys of the pad. If your keys are on the numeric keypad, they only work when the Num Lock key is off.

If your Page Up and Page Down keys are located in the *edit keys* (between the characters and the numeric keypad), they always work, whether Num Lock is on or not.

paint program

A **paint program** is a software application that provides electronic versions of a paintbrush, a pouring paint can, spray paint, pencils, scissors, an eraser, rubber stamps, etc. Paint programs are always *bitmapped,* which gives the application some advantages in creative play over *object-oriented graphics.* Bitmapped graphics can have any *resolution,* but that resolution is permanent; it does not adjust to the printer (as object-oriented graphics do). Because the resolution is fixed, though, you can degrade (or improve) the quality as you resize the image (see below). Compare with *draw programs.*

This bitmapped image was created at 72 dots per inch, and it will always be 72 dots per inch when it is actual size. When the image size is reduced, those dots gets smaller and the image appears *to have better resolution; when it is enlarged, the dots are also enlarged, making it appear to have a lower resolution.*

palette

A **palette** in computer jargon is similar in concept to a paint palette in art. It is a little space, or perhaps a *window,* or perhaps a *windoid,* that holds tools or patterns or colors or styles. As with a traditional art palette, the idea is to make the tools or colors easily accessible while working. Many palettes *tear off* to make them even more accessible—just drag the pointer down through the menu and it will tear off. Wherever you let go, that's where the palette will stay. You can move the palette by dragging in the gray *title bar,* or you can put it back where it belongs by clicking in the *close box.*

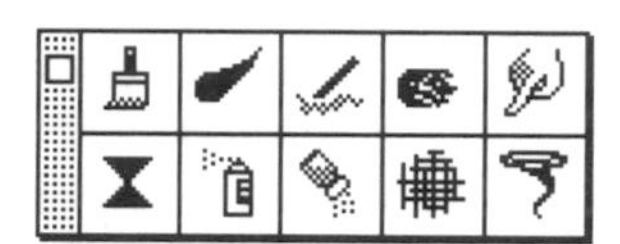

This is a palette of tools in SuperPaint.

palmtop computer

A **palmtop computer** is a very small, hand-held computer like the Sharp Wizard or the Apple Newton. Most palmtops can be connected to a larger computer for exchanging data.

panic

A **panic** is the same as a *crash,* or a *fatal error* that causes a crash. No matter what you call it, it has the potential to be a catastrophe of considerable dimension. Please see the definition under *crash.*

Pantone™ Matching System

The **Pantone Matching System** (abbreviated as PMS, really) is the name of a system for matching colors when printing on a commercial press (not for printing on your personal laser printer). Before the Pantone system, a designer might have told a printer that she wanted a lovely pale teal background color and dark steel gray-blue text. Well, of course that color definition is open to wide interpretation. With the Pantone system, the designer would pick the lovely pale teal color out of a swatch book and give the printer the number that specified the color, such as 1523C. She would also give the printer the number that corresponded to the gray-blue she had in mind. The printer could then mix the colors to match those standards, knowing exactly what the designer had in mind, and the printed job would make everybody happy.

The Pantone system has been converted to electronic media (with the label CV for computer video). Traditionally there was almost a monopoly with Pantone, but new electronic color standards are being developed, such as TrueMatch and Focoltone.

paperless office

Remember the predictions of the **paperless office**? Where all office correspondence and forms and data would be transferred electronically and need never be committed to *hard copy* (that is, paper)? We all know what a joke that is. For one thing, every office and non-profit organization in the country that never had the time or expertise or budget for a newsletter is now creating a monthly tabloid.

paradigm shift

The word **paradigm** (pronounced "pair uh dime") means an example, a pattern, a standard, an ideal, a paragon. Your own paradigm of the world is your own version of reality—those things that define your reality. What is happening in this high-tech world is that many people and corporations are going through **paradigm shifts,** where the reality of their worlds is changing. Some make the transition; some can't and are out of the race.

For instance, in the typesetting world (pre-desktop publishing computers), fonts were a high-margin, proprietary product that demanded a high retail cost. They were sold only to typesetters who in turn sold galleys of printed type to clients who pasted those pages onto boards. If there were changes in the text, the client had to take the job back to the typesetter, pay large fees, and wait one, two, or three days for the new galleys, then paste in the changes. Well, Adobe PostScript changed the paradigm for everybody in that entire chain. Designers who used to send their type to professional typesetters began setting their own type and had to raise their consciousness of typographic subtleties. Typesetters no longer had clients asking for type, and either closed up shop or made the paradigm shift to becoming *service bureaus.* Type foundries like ITC, Monotype, Linotype, and Agfa-Compugraphic had to make the paradigm shift into creating and selling PostScript fonts that were no longer proprietary and could no longer be sold at high cost with such a high margin. It's an interesting point to note that (generally speaking, *of course*) the longer a designer, a typesetter, or a type foundry and its bureaucrats have worked with a proprietary system, the more difficult it is for them to make the paradigm shift.

paragraph formatting

Paragraph formatting refers to the kind of changes you make to text that apply to the **entire paragraph**, whether you select two characters, the entire paragraph, or even if the cursor is just flashing within the paragraph. (Remember, the computer sees a new paragraph every time you hit the Return or Enter key, so even a three-line address is considered three paragraphs.)

Alignment, such as *flush left, flush right, centered,* is paragraph formatting because it's not possible to have just a few characters flush left or centered—either the entire paragraph is centered or it isn't. Tabs and indents are also paragraph formatting, and so is extra space before or after a paragraph. If a particular formatting feature applies to an entire paragraph, it is called "paragraph-specific formatting."

This is different from *character formatting,* where changing text to bold or italic affects only the **characters** you selected. Changing characters to a different point size or typeface is also character formatting.

parallel port

You have probably heard of printer ports and modem ports and perhaps ADB ports or some other kind of *port*. A port is a plug, or receptacle (known in other computer dictionaries as an input/output connector). Once you insert one end of a cable into a port, information can flow between your computer and whatever device is attached to the other end of the cable.

All ports, no matter how you connect to them, are either *serial ports* or **parallel ports,** which refers to the way the data flows through the wires (of course there are different cables for serial ports than for parallel ports). A parallel port accepts the cables that have parallel wires, so data can flow through the cable at high speed (in a serial port, data is transmitted in a single line). The special thing about a parallel port is that it can transfer a complete *byte* of information at a time. The port has eight data wires, one for each *bit* in the byte, so all eight bits can travel side-by-side and arrive at the same time. A serial port, by contrast, only lets one bit through at a time. All things being equal, a serial port is slower than a parallel port; however, a serial port can send and receive information at the same time, which many parallel ports can't do.

On PCs, parallel ports typically connect to *printers,* and serial ports to *mice* and *modems*. Here's the take-home message for PC users: if your computer and printer both have parallel ports, that's how you should connect them, unless they're farther apart than about ten feet. The longer a parallel cable gets, the greater the chance of "crosstalk," or interference, between data traveling on the two parallel wires. So if your printer and computer are fifteen or twenty feet apart, hook them up with a serial cable, or move them closer together and use parallel.

On Macintoshes, there are no Centronics parallel ports like those on PCs; modems are connected through serial ports, and printers are connected either through serial ports, LocalTalk network ports, or SCSI ports. LocalTalk and SCSI connections are usually faster than parallel printer ports, and are used for laser and color printers. Serial ports are slower and generally used to connect dot matrix printers.

parallel processing

Parallel processing is a technique that's recently come into vogue as a method for solving complex problems quickly, while using relatively inexpensive computers. A computer designed for parallel processing has more than one microprocessor—usually several. The computer's *operating system* lets all the processors work on one problem at the same time—in parallel—by breaking down the task at hand into smaller subproblems. A given processor tackles one subproblem, and when it's finished, goes on to the subproblem that's up next. As you can imagine, all this requires a sophisticated system for communication between the processors to make sure each one knows what it's supposed to do, and when. At present, parallel processing is not for home use—these computers are used by universities and large corporations to work on complicated engineering, mathematical, or scientific problems. Because they rely on standard "off-the-shelf" processor chips, they're cheaper to build than computers that rely on a single ultra-fast processor. Contrast parallel processing to *multi-tasking*, which applies to a single processor working on multiple tasks at the same time.

parameter RAM

See *PRAM* (pronounced "pee ram").

parameters

A **parameter** is some value, name, option or characteristic that you add to a command to customize it. If you don't state a parameter, then the program uses the *default,* or the automatic choice. For instance, if you want to re-open the letter you wrote to your sister which you wrote in

Microsoft Word and you called it TOSIS.DOC, you could type WORD TOSIS.DOC and the computer would open Microsoft Word and put your letter on the screen. In this case, "TOSIS.DOC" is the parameter telling the computer exactly which file you want. If you did not type that parameter, the computer would open (or *load*) Microsoft Word and display a new, blank window so you could write a new letter.

On the Mac, the concept is similar (as above) when talking about scripting or programming. As *end users,* we usually refer to parameters as *attributes.* For instance, if you are making a new field in a database, the automatic "parameter" for the field is text; that is, the field will view whatever you put in it as text and will sort it alphabetically as such. You can change the parameter *(attribute)* to numbers, for instance, if you are going to enter zip codes and want to be able to sort them numerically.

Don't confuse "parameter" with *parameter RAM.*

PARC

PARC stands for **P**alo **A**lto **R**esearch **C**enter, a research and development center established by Xerox Corporation in 1970. Macintosh's icon-oriented *interface* (and later, Windows) including the mouse, menus, and windows, was originally invented and developed at PARC.

parent directory

On a DOS or Unix system, the **parent directory** is the directory immediately above another directory in the directory heirarchy, like a parent above a child. In true hierarchical style, every directory (except the *root directory,* the one at the top of the ladder) is located beneath another directory. The one above is the parent, the one beneath is the child, otherwise called a *subdirectory.* You can identify the parent directory in the listings by the two periods (**. .**).

parity, parity bit

Parity is a form of "error checking" where the computer checks to see if all the data it was supposed to get really did come through.

You will most likely be confronted with parity when you use a *telecommunications package* to communicate through your *modem.* In fact, that's probably why you're reading this. The dialog box where you can set the *serial port settings* always wants to know the parity. The *default* setting is probably the safest thing to use if you don't know a reason to change it.

But here are a few tips, according to Scott Watson's *White Knight* manual: If two personal computers are talking to each other, both sides should use no parity and 8 bits (databits). If your personal computer is connecting to a *bbs (bulletin board service)*, use no parity and 8 bits. If you are using 8 bits, you'll always use no parity or ignore parity. Use "ignore parity" if you are sending international characters like ñ or Ç. Otherwise use no parity. If you are using 7 bits (databits), you need to use some form of parity.

A similar parity scheme is used in most PCs and many other types of computers to continuously check memory chips for errors. You can get parity errors right in the middle of your work, especially during small-town power problems. Some Macs purchased for the government have parity checking as an option; IBM is a firm believer in parity checking.

For data transmission over a serial connection, parity checking is less reliable than *XMODEM*, *Kermit*, and other *checksum*-based protocols.

park

When you use a *hard disk*, the disk spins rapidly (like 3600 revolutions per minute). The *read/write head* floats just a hair's breadth above the disk to pick up *(read)* information or to store information. Actually, it's not even a hair's breadth above—it's more like a fingerprint above. Seriously. When you turn off your computer, the read/write head **parks** itself, or locks itself in a safe position so that if you move the machine the head won't bump into the disk.

If you use a *cartridge drive*, it is particularly important to go through the steps of *dismounting* the disk (on a Mac, drag the *icon* to the *trash* or *Shut Down*), then park it before you eject the cartridge from the drive. Generally you park it by pressing the flat button in, then wait until the light stops flashing. Make sure you wait until the disk has stopped spinning and you have given the head time to park before you push the lever to eject the disk.

By the way, you should be very careful never to move or bump your computer while the hard disk is spinning because it could cause a *head crash*, where the read/write head scratches the disk, similar to a phonograph needle scratching a record. You will be seriously unhappy if this happens.

Older hard disks didn't park themselves automatically—you were supposed to run a utility program that told them to park before you shut down the computer. If you bought your computer new from a reputable dealer any time in the last five years, don't bother running a parking utility!

partition

If you have a large hard disk, like 1000 *megabytes,* some people recommend that you divide it into **partitions.** (Or, using the word as a verb, you can *partition* your hard disk.) The software for dividing your hard disk usually comes with your disk. Once you have divided it into separate partitions (also called "logical disks"), the computer thinks each partition is a completely separate storage device, just as if you had two different hard disks (or three or four or however many partitions you make). It is even possible to have a different *operating system* on different partitions of the same hard disk. In fact, in a *graphical user interface* where you can see icons for your disks, you will see a separate icon for each partition (also known as a *volume*). Having separate partitions makes it faster for the computer to find what it needs, since it is faster to search through the smaller number of files on each volume than to search through every file on the entire hard disk.

Pascal

Pascal is a *programming language.* Please see the definition of *programming language* for a brief overview of the most common languages.

password

A **password** is a special word that only one person (or a select group of people) knows that allows them *access* to a computer or to a particular program or parts of a program. When you type the password, the screen usually shows only bullets (•••••) or asterisks (******) instead of the letters so no one can look over your shoulder to discover your secret word.

A couple of cautions about a password: if you forget your password, you are in trouble. If no one else knows that word, there is usually no way to find it out. Many people have permanently lost important data by forgetting their password. Me, I lost my entire *online* account and had to re-install the whole program. Write down your passwords and find a safe place to store them. Then remember where that safe place is.

Don't use obvious words as your password, such as the names of your children or your dog, anyone's birthdate, your license number, etc. Use more than three letters. Letters combined with numbers are best. Also, it's recommended that you change your passwords regularly. (I don't change mine very often because I have a hard time remembering them.)

paste

In most applications (certainly in virtually every Macintosh application) you can *cut* or *copy* information or an image from one place and then choose to **paste** it into another place. Please see the definition of *copy* to get the whole picture regarding *cut*, *copy*, and *paste*.

path, pathname

A **path,** also known as a **pathname,** is a sequence of names that tells the computer and you where a file is located. A path is similar to your house address, and the file you want is similar to your name. For instance, to get a letter to you someone must put Your Name, Your Street, Your City, Your State, and Your Country. The Country is the *root directory*, or the bottom line. From there the path gets narrower and narrower until it reaches Your Name.

When the computer recalls a path, it does exactly the same thing (or on a DOS machine if you are creating a path, *you* must do the same thing). This file I am working in right now is called Jargon.PtoZ. It is in a folder called The Book, which is in a folder called Jargon, which is in a folder called Publications, which is on my Hard Disk. Now, don't get confused, but the computer's path appears backwards to our logical sense; the computer starts with the root first, kind of like addressing a letter starting with the country. So the path to this file is:

Hard Disk:Publications:Jargon: TheBook:Jargon.PtoZ.

On the Macintosh, names along the path are separated with a colon (which is why a colon is an *illegal character*). On DOS machines, path names are separated with a backslash (\). On Unix machines, path names are separated with a forward slash (/).

In any "Open" or "Save As" dialog boxes on the Mac you can see the path by pressing on the label above the list box. You can select any name along this path to view the contents of that folder.

In System 7 at the Finder/Desktop level, you can hold down the Command key and press on the title bar in any window to see the path.

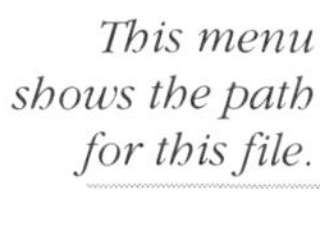

This menu shows the path for this file.

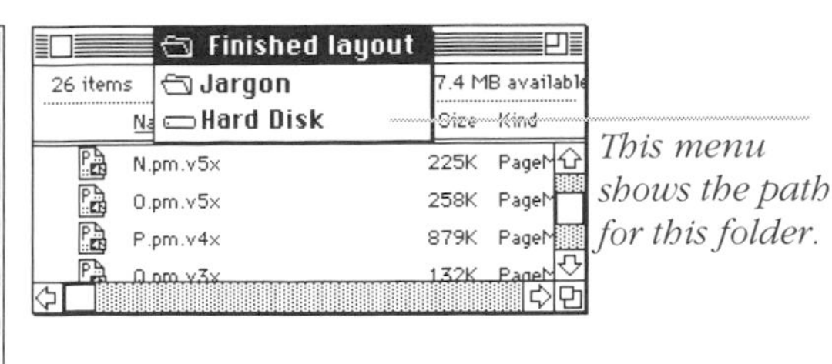

This menu shows the path for this folder.

In DOS the path starts from the top level, the disk itself, and works down to the lowest branch of the directory tree. Directory names are separated by backslash characters. Like this:

C:\FOOD\FRUITS\TROPICAL\MANGOES\

Another meaning of **path** (Mac or PC) is in some graphics programs, where you can define a curved path, or line, on which you can place text. Or you can draw a path around an object and then act on that path, like remove the object or change the color of what's enclosed within the path.

Pause key

Extended or *enhanced keyboards* have a key labeled **Pause,** but it doesn't do much. You'd think it would temporarily interrupt whatever is going on, and sometimes it does work that way (in some *telecommunications programs,* for instance, you can press the Pause key to stop information from scrolling off your screen). But most of the time it doesn't do anything—try it and see.

PC

The term **PC** stands for **p**ersonal **c**omputer, but it was also IBM's trademarked name for its original personal computer model, introduced in 1981. Although earlier machines like the Apple II and the Kaypro were fairly popular, the IBM PC really ignited the small computer revolution in business.

Since then, many other manufacturers have built computers that are compatible with the IBM PC (meaning they work just like the original, can use the same software, and they look very similar). For want of a better generic term, "PC" became the standard way to refer to any PC-compatible computer.

A *Macintosh* computer, even though it really is a personal computer, is not known as a PC, but simply as a Mac. If you mention that you have a PC, people would not think you have a Macintosh. In fact, if you tell someone you have an *Apple,* they will not think you have a Macintosh. The term "Apple," when speaking of a kind of computer refers only to the Apple II, Apple IIE, Apple IIGS, etc.

PC compatible

PC compatible refers to any computer that can *run* the same programs that run on an IBM PC.

PC-DOS

PC-DOS used to be IBM's brand-name version of *MS-DOS,* the standard *operating system* for all IBM-compatible PCs, whether made by IBM or not (see *DOS*). IBM's version of MS-DOS had a few minor extra features but was otherwise identical to the more generic versions. Nowadays, since IBM's line of personal computers are called PS/1s and PS/2s instead of PCs, IBM refers to the version of MS-DOS it sells as IBM DOS.

PCL

PCL, short for **p**rinter **c**ommand **l**anguage, consists of a large set of commands for controlling the Hewlett-Packard LaserJet and DeskJet families of printers, and compatible printers from other manufacturers. PCL commands are used to tell the printer where to place text or graphics on the page, which *font* to print, whether to print bold, italic or underlined text, and so on.

As HP has released new printers, PCL has steadily evolved, gaining new, increasingly sophisticated commands. HP recognizes several versions, or levels, of PCL. PCL Level 5, which first appeared in the LaserJet III, represented a major jump in capabilities, with the ability to print *scalable fonts* (not just *bitmapped fonts* as in earlier levels) and fancy, *PostScript*-like graphics. All PCL printers respond identically to a common core set of PCL commands, but older printers equipped with an earlier level of PCL can't understand the commands from higher levels.

You might see the term PCL or *printer command language* used more generically to describe the set of commands for controlling other types of printers. In general, each major make of printer (for example, Epson, Toshiba, and Okidata) has its its own command set, although many printers recognize the Epson commands.

PCMCIA

This long, unpronounceable acronym, **PCMCIA,** stands for **p**ersonal **c**omputer **m**emory **c**ard **i**nternational **a**ssociation. This is a group of manufacturers who got together to decide on standards for little credit card-like things you plug into special sockets in some computers. PCMCIA cards are only 2.1″ wide by 3.4″ long and typically, .2″ thick. You can buy *memory circuits, modems,* and *network adapters* on PCMCIA cards. The only problem is that the cards are so thin, there's no room to plug in a telephone cable or other connection to the outside world—instead, a little cable with the necessary receptacle hangs off the card, where it's vulnerable to breakage.

The sockets for the cards are accessible from outside of the computer, so you don't have to open up the whole case to plug in a card. Today, sockets for PCMCIA cards are found in some PC notebook computers, where size matters the most. But within a few years, PCMCIA technology may replace the conventional expansion slots in desktop computers, which would mean that desktop machines could be a lot smaller than they are now.

So far, there have been three PCMCIA standards, differing primarily in the thickness of the card: Type I is the thinnest, Type II is the most common, and Type III is the thickest. The point of having standards is that eventually, the cards should be interchangeable among different computer models. But while any PCMCIA card will fit physically into the corresponding type of PCMCIA socket, that doesn't necessarily mean it will work—you have to be sure that the card you buy is designed to function in your computer.

.PCX

.PCX (pronounced "dot p c x") is probably the most popular *graphic file format* for storing *bitmapped* graphic images on PCs (a "file format" is the particular way information is organized and stored in a computer file). The PCX format was originally developed for the program PC Paintbrush by Zsoft. PCX files are accepted by most graphics and desktop publishing applications, and are often used to store images created by *scanners*.

PDA

PDA stands for **p**ersonal **d**igital **a**ssistant, a term for a new breed of hand-held computers. PDAs are supposed to combine the power and flexibility of a real computer with the convenience of those little electronic organizers—the kind made by Sharp and Casio—that let you keep track of your names and addresses and to-do items. A PDA can run much more complex software than an electronic organizer, and you're not stuck with the built-in software—you can buy separate programs to customize the PDA to your specific tasks. With many PDAs you can use a *stylus* to write on the little screen. The PDA most people are talking about is the Newton, made by Apple, but many other companies are planning their own versions.

PDL

PDL stands for **p**age **d**escription **l**anguage, which is an actual *programming language* (such as *PostScript*) that tells a printer how to each page should look in print. A page description language has commands that give the specifications for a printed document, including the margins, the text (which typefaces at what sizes, as well as the actual text content), the graphics (dots, lines, circles, or whatever), and so on. Together, all the commands required to describe a document are considered a program—it's just that your printer runs the program, not your computer.

Although a human programmer can write a program in a page description language, computers usually do the job. When you tell your computer to print a document, it translates the document into page description language commands, and sends them to your printer.

Exactly what constitutes a PDL depends on who you talk to. Every printer needs special codes to tell it what to print and where, and these codes can be considered a programming language of sorts (see *printer command language*). But what most people mean by a PDL is a full-blown language like PostScript, which has at least a vague resemblance to English and can be written out with a *text editor* fairly readily.

PDS

PDS stands for **p**rocessor **d**irect **s**lot, which is a slot, or opening, inside the computer where you can insert accessory *cards*. It is similar to a *NuBus* slot, but a direct slot is so-called because its *pins* are connected directly to the *CPU (central processing unit,* the *chip* that runs the computer) (compare *NuBus*). Usually direct slots in one computer are not compatible with direct slots in any other model of computer, even of the same make, which makes PDS's more limiting than NuBus slots.

Peachpit Press

Peachpit Press in Berkeley, California, is very likely the greatest publishing house in the world—honest, ethical, respectful, interesting, daring, and provocative, and staffed by the nicest people. They publish computer books primarily. I'm lucky and proud to be a Peachpit author. [They suggested I delete this entry, but I have never been very good at doing what I'm told. It's my book.]

Pentium

Pentium is *Intel's* trademarked name for the latest generation *microprocessor* in the line of *chips* used in IBM PCs and compatibles. The first PC had an *8088* microprocessor; then came the *80286,* the *80386,* and finally the *80486,* and now, instead of the *80586,* the Pentium. The Pentium is much faster than its predecessors, even operating at the same *clock speed,* and it can transfer twice as much data to and from *memory* at a time than the 80386 or 80486 (64 *bits* instead of 32).

Intel broke with its numbering sequence (the Pentium should have been the 80586) for marketing reasons. Intel's competitors have been producing chips that work just like the 80386 and 80486. These companies are free to use "386" and "486" in the names of their clone chips because the courts decided that Intel's chip designations are part numbers and can't be trademarked. The Pentium name will be impervious to trademark violations, but it sounds more like a luxury car or a toothpaste than an electronics part.

Performa

The **Performa** series of Apple computers are Macintoshes in a slightly different package. They were specially created to be even more *user-friendly* to the home and educational market than the regular Macs, both in ease-of-use features and in price.

The Performas use a software *utility* called the Launcher to supposedly simplify opening applications and files, but paradoxically you have to really know what you're doing to install applications and documents into the Launcher. Every document you save ends up in a folder called Documents, whether you want it there or not (to get around this, just rename the Documents folder). One particularly great feature for beginners, though, is that while under the Launcher's control, you cannot accidentally click on the Desktop or another application and lose the application you are currently working in. This prevents beginners from leaving applications open that they didn't know were still open, which leads to "out of memory" messages.

peripheral

The term **peripheral** refers to all those items that are outside—or peripheral to—the main computer box. Peripheral *devices* are things like keyboards, monitors, modems, printers, joysticks, disk drives, mice, etc.

permanent storage

When you consciously *save* your document, you are putting it into **permanent storage** (typically onto your hard disk or a floppy disk). Until you save your work, it is held in *temporary storage*, which is the computer's *memory (RAM, random access memory)*. If there was a system failure of some sort, all your work that is in RAM would be gone. Anything you had saved to disk would be intact.

Even though it's called "permanent" storage, it isn't **really** permanent—you can erase any files you store on a disk, and they can be destroyed accidentally by magnetism, physical trauma, deterioration of the disk, children, or pets.

personal computer

A **personal computer** is a computer designed to be used by an individual person. This person, the *user,* has control over which applications to use, customization of the system, amount of time on the computer, etc. This is different from the way computers were typically used ten years ago, when people worked on data processors hooked into *mainframes* and the computers were controlled by the data processing manager. Of course, personal computer users can choose to hook into a mainframe, but the user still controls when the mainframe takes over the screen.

Some technically minded people define a personal computer by the kind of *microprocessor* it uses because the processor is an indicator of the speed and power of the computer. But the technical distinctions between a personal computer, a *workstation,* a *minicomputer,* and a *microcomputer* are becoming very blurred as the smaller computers become more and more powerful and more interconnected with the bigger computers, with networks and communications software and *cross-platform* information exchanges. Oh, it's an exciting time, isn't it?

If I had been a single mother of three children a hundred years ago, I would have taken in mending and washing to support us and be a mother to my kids at the same time. Making a living on my personal computer at home makes me feel very much a part of the history of women who have always had to find work that could be done at home in those moments between cooking and feeding and washing and caring, and during the nights when all the rest of the world, including children, have gone to sleep and we finally have quiet, uninterrupted time. s i g h . . .

PhoneNet™

PhoneNet is a *networking* system that connects many different *AppleTalk*-compatible devices, which can include Macintoshes, IBM-PCs and compatibles, Apple IIs, printers, etc. It uses ordinary telephone cabling and little connector boxes. Even I can put it together. See *LocalTalk* for an illustration.

photo manipulation program

A **photo manipulation program** (such as Adobe PhotoShop or TimeWorks Color-It) is an application that can open *scanned* files—photographs or other artwork that have been scanned into the computer. Using this application, you can then manipulate or change any aspect of the scanned photo. In fact, you can manipulate them to the point where it is becoming impossible to tell in printed work or video what was in the original photo and what has been changed. These applications are now commonly called *image editing programs.*

photo retouching

If a photograph has an unsightly blemish, or if the model has wrinkles he doesn't want showing, or if there is a potted plant drooping behind someone's head, or if the fruit isn't quite moist enough, or for any number of reasons, a **photograph** can be **retouched** to make it more desirable. This used to be done with airbrushes and paintbrushes and was a laborious, highly-skilled task. It can now be done electronically within *photo manipulation software* (also known as *image editing software*). Photo-retouching is still a laborious and high-skilled endeavor, but the possibilities for changing a photo have expanded incredibly. You can not only remove his wrinkles, you can remove his face and replace it with Godzilla's.

pica

A **pica** is a unit of measure typesetters have used for the past 100 years when working with type. Six picas equals one inch. Each pica is divided into 12 points. Thus there are 72 points per inch. Type size and *leading* is measured in *points,* and horizontal measurements like line length are usually measured in picas. For instance, you might create a brochure with columns having type at 10 point, leading at 12 point, and 24 picas wide in line length, (traditionally written as 10/12 x 24).

For purists who know that traditionally 72 points does not equal exactly one inch, but only .996 of an inch, you will be interested to know that on the Macintosh, 72 points really does equal exactly one inch. In fact, each pixel on the original screen is one point (72 pixels per inch), now embedded in stone due to *PostScript.*

A **pica** is also the term for a certain kind of generic type that has ten characters per inch, as opposed to "elite" type that has 12 characters per inch. The terms "pica" and "elite" are not really referring to a particular typeface, but to the *pitch,* or the number of characters in a given measurement. Both pica and elite typefaces are *monospaced.*

picklist

A **picklist** is a list of items on your screen to choose from. To make a choice using the keyboard, you use the arrow keys to move a highlight through the list until it rests on the item you want. Then you just press the Enter key to pick the item. If you're using a mouse, all you have to do to pick an item from the list is click, or double-click, on the item. In the really old days of PC computing, many programs made you type in the full name of whatever you wanted to choose, like a command or a file. Windows and the Macintosh make extensive use of picklists, but they're not called that any more. Instead, you have "drop-down lists" and "list boxes."

PICT, PICT2

PICT is a *graphic file format,* which means it is one of the ways a graphic image can be created in the computer. PICT is Apple's proprietary format (meaning they invented and own it) and uses Apple's *QuickDraw* language to display on the screen. PICTs can be either *bitmapped* or *object-oriented.* **Pict2** files technically have more capabilities, including high-resolution color, than PICT files, but you will rarely see files labeled differently.

Unfortunately, PICT files are not very reliable. They cannot use a line width less than one point (so they change it on you). If it's an object-oriented PICT, they don't like to include bitmapped items (so they change 'em on you). They don't like to remember fonts, especially *downloadable* fonts (the ones that are not built into your printer), partially because PICTs do not speak the *PostScript language* of the typefaces. PICTs will often just decide to change their mind and give you a different font. Sometimes a PICT will decide to create a typo where there wasn't one before, or remove the last few words in a table cell. And you never know when it is going to change its mind. It might be while on the way to the imagesetter. It might be in the middle of the night, after you finished proofing the job and had it approved. I'm serious. You never know about those creatures.

In *graphic design,* avoid using PICTs if at all possible. Sometimes you have no choice, but if you do, use something else. Most graphic programs and *scanners* give you an option as to what kind of file format to save an image in. And keep in mind that when you copy items out of the Apple *Scrapbook* they always turn into PICTs, no matter what they were to begin with.

In *color imaging* work, however, PICT files are much smaller in size than the same *TIFF* file. If you will be scanning a color image into your computer, manipulating it on the screen, then outputting it to a slide recorder or similar device, a PICT will work fine because it is created with the language that is meant for the screen. PICTs use a language that is not

PostScript, so as long as you do not try to output it to a PostScript device, it will usually behave nicely.

Now, of course, there will be people who will write and tell me all about what good luck they have had printing PICT files to PostScript printers. But I can tell you ten horror stories per one good luck story. Ask your service bureau. And I've even had trouble with PICTs in an animated presentation that I show directly from the computer.

PIF

A **PIF** is a **p**rogram **i**nformation **f**ile, a small file that Microsoft Windows uses to run DOS (non-Windows) programs. A PIF tells Windows how much memory the program needs, whether the program runs in text or graphics mode, and whether you want the program to fill the whole screen or run in a window, among other details. Each program can have its very own PIF, which you make with the PIFEDIT program that comes with Windows. If a program doesn't have its own PIF, Windows uses the *default* (generic) one.

PIM

PIM stands for **p**ersonal **i**nformation **m**anager, software that keeps track of information you need at your fingertips to conduct day-to-day personal business. You use a PIM to store things like the names, phone numbers, and addresses of your clients and customers; your calendar of meetings, due dates, blind dates, and travel plans; your to-do lists; your expense records and billing logs; random notes to yourself; and whatever else fits. The definition of this software category is really quite vague, and the specific capabilities of the different PIMs vary a whole lot. One common PIM feature (but it's by no means universal) is the ability to look at the same information from different perspectives. You might browse through a list of the people in your company to find out which projects each person is working on, and then switch to a list of projects and see which people are assigned to each project. PIMs are currently one of the hottest new areas of software development.

Lotus' Agenda for PCs more or less defined this category, but Agenda is a complicated program that never became tremendously popular. You can create your own PIM in Apple's *HyperCard* for the Macintosh.

For most people, a simpler product or a combination of several basic programs satisfy; for instance, the combination of TouchBASE and DateBook is my favorite PIM.

PIN

PIN stands for **p**ersonal **i**dentification **n**umber, which is your own personal password (pass-number, really), just like the one you use with your automatic teller card. With computers, you might need a PIN to log on to a particular computer system or *online service.*

pin

A **pin** is a slender metal prong that you find in the plugs, or connectors, at the end of some cables. A connector with pins is a *male connector.* (The connector with the holes that accept the pins is the *female connector.*) The number of pins and their arrangement is often the identifying feature of a connector. For instance, I needed a cable for a new *video card* that I put in my computer. I didn't know what kind of cable I needed, so I counted the little pin holes and told the guy I needed a 15-pin connector and he thought I knew what I was talking about.

These little prongs are called pins.

Sometimes the little legs on the *chips* that plug into a *board* in your computer are also called **pins.**

pin fed, pin feed

You've probably seen the fan-fold paper that has those tear-off strips with the little holes along either side. Those holes fit into round plastic pins in printers with a **pin-feed** mechanism, which pulls the paper through the printer. The pin-feed mechanism is also referred to as a *tractor feed,* so you can refer to these printers as pin-feed or tractor-feed printers. Many such printers have a switch so you can use either the "pin feed," which takes the stack of folded paper, or "friction feed" that can pull in one piece of regular paper at a time.

Pink operating system

The **Pink operating system** is an *operating system* (software that runs the computer) of the future (like 1995). It is a combined effort between Apple and IBM through the infamous Apple-IBM alliance, which spawned the company called Taligent. Pink was Apple's code name for the project; the operating system's official name is now Taligent.

Do you wonder why the operating system is called Pink? Not because the red Apple and the blue IBM make Pink. But because when the programming team at Apple was creating *System 7,* they made two lists for the features of System 7. On blue index cards they listed the "We gotta have this" features. On pink index cards they listed "Wow, it would be cool to have this" features. (Well, they didn't really have those titles, but that's what they meant.) For the operating system of the future, Apple and IBM set the programmers of Taligent to work on the Pink list, which they are doing at this very moment.

pirate

A **pirate** is a person who makes illegal copies of the software that most people pay a lot of money for.

pitch

Pitch describes the horizontal width of *monospaced* characters. A 12-pitch typeface can fit twelve characters in one inch and is called *elite*. A 10-pitch typeface can fit ten characters in one inch and is called *pica*. The pitch actually measures both the character and the space around it, which is what makes monospaced fonts monospaced—every character takes up the same amount of space. The letter "i" in 10-pitch takes up one/tenth of an inch, which is the same amount of space as the letter "m," which is the same as a period. (Don't get this "pica" mixed up with the measurement system of *points* and *picas*.)

Pitch only refers to monospaced type. In proportionally spaced type, where the width of each character is different, we talk about characters per *pica* (pica as in the typographic measurement unit), not per inch, and the number of characters per pica (the "cpp") is always an average, not a fixed number as in the pitch.

Pitch also refers to the space between *pixels* on a monitor. See *dot pitch* for a complete explanation of this term.

pixel

Pixel is short for **pic**ture **el**ement. The computer screen is composed of hundreds of thousands of tiny little spots of light. Each one of these spots is a pixel and is the smallest element that a screen can display. Everything you see on your monitor (screen) is created using those tiny spots. On a color monitor, each pixel is actually made up of a varying number of triads, three separate dots of color—one red, one green, and one blue—placed immediately adjacent to one another. The color you see from a given pixel is a blend of the light from those three dots.

Each pixel is controlled by information stored in your computer. If a pixel is controlled by just a single on-or-off unit of information—one *bit*—it can only display one of two "colors": it can either be on (white, green or amber, depending on what kind of monitor you have) or off (black). If a pixel is controlled by two bits, the pixel can appear as any of four colors or shades of gray. With four bits you can see sixteen colors or shades of gray. And so on. If you want more details, see the Symbols and Numbers section (before A) and look up 24-bit color.

PKZIP, PKUNZIP, PKSFX

In the PC universe, **PKZIP** and **PKUNZIP** are the most widely used utilities for *compressing* or *archiving* files for storage or modem transmission (PKZIP), and then uncompressing (un-archiving) the files when someone needs to use them again (PKUNZIP).

PKSFX can compress a set of files and store them as a "self-extracting archive" *(.sea)*. As a self-extracting archive, the person you give the archive file to doesn't need a copy of PKUNZIP; the file will uncompress automatically as they open the file.

All of these programs are distributed as *shareware,* so if you use them you should pay for them. The PK is for Phillip Katz, author of the programs.

platform

A **platform** is a somewhat vague term that can be used in slightly different ways. Most broadly, it's simply just a snooty way to say "computer," or to refer to a particular type of computer (the Macintosh is one platform, the PC is another platform.) The term is a little more useful when it refers to a combination of a particular type of computer running a particular type of operating system software. For instance, an IBM PC running only DOS as the *operating system* is one platform; the same computer running DOS with Windows is another platform; and the same computer running OS/2 or Unix is yet another platform.

If you hear a particular product described as "cross platform," it means that the product will work on several platforms, or that it's something that's applicable to more than one platform.

This is different from a product that has been *ported* to another platform—to port a product means to rewrite it so it will work on another platform, not that the same product can function on both.

plotter

A **plotter** is a graphics printer that literally uses ink pens to draw the images. The pens move around on the surface of the paper like something out of *The Sorcerer's Apprentice.* Plotters can only draw data in vector graphics format, graphics that are made of straight lines (the curved forms are actually drawn with many tiny straight lines). There are flatbed plotters where the pen moves across the page in the x and y axes. There are drum plotters where the drum moves along one axis while the pen moves along the other axis and the paper rolls through. There are pinch-roller plotters where the pen moves along one axis and the paper moves back and forth on small rollers. Some plotters draw the image with electrostatically charged dots, then apply the toner and fuse it, similar to laser printers. Plotters are very interesting and often fun to watch, especially the fast ones.

PMMU

PMMU stands for **p**aged **m**emory **m**anagement **u**nit, which is a *chip* or part of a chip that manages the *memory* tasks, such as controlling the amount of memory used by applications, making *virtual memory* available, or swapping data between the *hard disk* and memory. High-end *processors* (chips) have a PMMU or the equivalent built right in (like the Macs with a *68030* or *68040* processor, or PCs that use an *80386* or *80486* processor). Without a PMMU, a computer cannot implement *virtual memory*. Some computers, like the Mac II, have a special spot on the *motherboard* where you can insert a PMMU chip, but other computers may need an *accelerator board* installed before you can add a PMMU.

PMS

PMS stands for the Pantone Matching System, a method that standardizes printing ink color. Please see *Pantone Matching System*.

point (noun)

A **point** is a unit of measure, equal to 1/72 of an inch (on a Macintosh, it's exactly 1/72 of an inch; a typesetter's point is very slightly different). Type sizes are measured in points, as you've probably noticed. *Leading,* the space between the lines of type, is also measured in points. Please see *pica* for a more complete definition in relation to picas and inches.

point (verb)

When a direction tells you to **point** at something on the screen—something like a menu choice, an icon, or a button—it doesn't mean to point your finger. The direction to point means to use the mouse or perhaps the arrow keys on the keyboard to position a *pointer* (or perhaps a *cursor* or a *highlight*) so that it's over the item you want. A software manual might tell you, "Point at the OK button, then click the mouse." See *point-and-click* and *click*.

This arrow is pointing at the OK button.

pointer

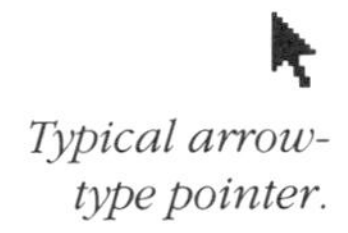

Typical arrow-type pointer.

A **pointer** is a marker on the screen that you move around by moving the mouse or trackball or some other kind of *pointing device*. You can tell your computer which item on the screen you want to work with by moving the pointer over that item. The most common type of pointer looks like a little arrowhead.

The pencil.

In many applications the pointer shape changes depending on what function you're using. For instance, if you choose to use the pencil tool in a paint program, the pointer turns into a little pencil. Most of the time, though, when the pointer changes shape, we call it by another name (like "pencil"). And a pointer can be considered one type of *cursor*.

In most DOS *character-mode* programs that use a mouse, the pointer is just a highlighted block the size of one character. It may be called the *mouse cursor*, but it's different from the text cursor.

point-and-click

The most basic technique for working with a mouse is to *point* at something on the screen using the *pointer*, and then click a mouse button. When you read the directions, you probably won't see the term "point-and-click" per se, but you'll probably be told something like, "Point to the large button marked Help! and then click the mouse button." Or you may just be told to click on something on the screen, which is a shorthand way to say "point and click."

Sometimes you may be led astray with directions that tell you to point-and-*click,* when actually you're supposed to point-and-*press*. If you click, as the directions say, and what is supposed to happen doesn't, or perhaps you see something flash in front of your face and then disappear, then try *pressing:* hold the mouse button down. The item (a menu, for instance) will be visible **as long as the button is held down.**

The term point-and-click can also be an adjective describing a product, as in describing the incredible *interactive* computerized story called "The Manhole": "It's an exciting, point-and-click storybook."

point-and-shoot

Point-and-shoot is a descriptive term for software you can operate by picking choices from menus rather than by typing commands, but it's usually used to refer to programs controlled by the keyboard rather than the mouse. The "point" part of the term means to move a highlight on the screen so that the menu choice you want is highlighted. Then you "shoot" by pressing the Enter key. Compare *point-and-click.*

pointing device

A **pointing device** is anything you use to control the movement of a *pointer* or *cursor* on the screen. A pointing device always has at least one button that you press, or some equivalent method of telling the computer when you want the pointer to do something.

The *mouse* is the original and most common pointing device. The mouse is a small plastic item with one or two buttons on the top. It's connected to your computer and sits on your desk. Most mice have a ball on the bottom that rolls when you move the mouse. The movement of this ball is tracked by tiny rollers inside the mouse, so when you move the mouse, the pointer moves around on your screen in a corresponding direction. You press the button on the mouse when you want to do something to the item you're pointing at.

There are many other pointing devices. The most common mouse substitute is the *trackball,* which is like a mouse flipped upside down: the ball is on top where you roll it directly with your fingers, while the device itself sits stationary on your desk. There is a *graphics tablet;* a flat plate where you draw with a plastic stylus, and the pointer on the screen follows that movement. Some graphics tablets, combined with certain software, are *pressure-sensitive,* where you can get different effects from your drawing tools with different pressures of the stylus. There is also something that looks like a fat ballpoint pen, and that's the way you use it, as if you were writing at your desk.

pop-up menu

The typical menu, like the one that comes from the menu bar across the top of the screen, is considered a *pull-down menu* because it slides downward onto the screen. A **pop-up menu,** by contrast, pops up from wherever you press the mouse button, which may be in a dialog box, or in a *palette* floating at the bottom of a window, or in the middle of the screen.

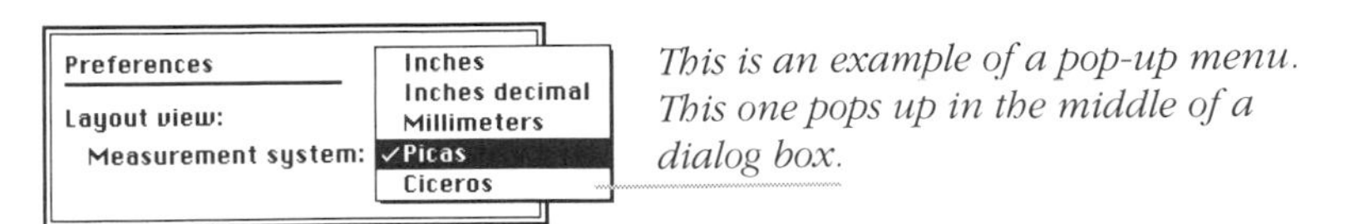

This is an example of a pop-up menu. This one pops up in the middle of a dialog box.

port (noun)

A **port** is a socket, a kind of receptacle (also called an "input/output connector"), usually found on the back of hardware like the computer, the monitor, or any other peripheral device. The port is where you connect the *cable.* Ports come in many different sizes and shapes: long and narrow, short and narrow, tiny and round. The computer can send and receive information in both directions through a port (as opposed to a "plug," where power goes in only one direction).

There are parallel ports and serial ports, SCSI ports and ADB ports, video, audio, and MIDI ports, and who knows how many other types. Fortunately, most computers give you visual clues to tell you which port matches which cable. Besides the port's size and shape, which you can match to the corresponding *connector* on the end of the cable, there's the matter of gender: female ports have holes, male ports have little pins sticking out, and you must always mate male with female. All Macintosh computers and some PCs have a little picture above the port describing what the port is used for; for instance, the port for the printer cable has a little picture of a printer, and the port for the modem has a little telephone.

port, port over (verb)

Port is also a verb referring to the process of rewriting software that originally worked on one *platform* (computer hardware and software system) so the program will work on another platform. For instance, Aldus PageMaker was originally written for the Macintosh; eventually it was ported (or "ported over") to the PC. You may hear complaints about programs that were ported too faithfully (especially from PC to Mac), so that they work on the new platform almost exactly like they did on the old one. That can be a problem if there are conflicts with the way people expect things to work on the new platform. So look for programs that have been modified during the porting process to respect the new platform's "customs."

portable

The term **portable** usually refers to a computer that is small enough to be carried. As a noun or adjective, "portable" of course refers to **any** piece of hardware small and light enough to be carried from one place to the next, at least from time to time. There are portable hard disks, portable tape drives, and so on, but unless the context tells you otherwise, what the term usually refers to is a portable computer. A portable can refer to any computer that can be carried, from the tiniest palmtop on up, but it's usually applied to "luggable" computers, meaning only that you don't need a wagon to move it from one place to another. A portable computer doesn't necessarily run on batteries. In that sense, a Macintosh SE is a portable. Portable with a capital P usually refers to the bulky Macintosh Portable, which wasn't very portable at all.

Portable can also refer to software or files that will work on different *platforms* (different computer hardware and software systems). A portable application is a program that can run on two or more platforms, but only if it is *compiled* separately for each of those platforms. In other words, the programming *code* is what's portable—if you have a version of the finished program that runs on your Mac, for instance, that program file won't work on a PC if you just copy the program file to the other computer. On the other hand, if a document file is truly portable, it should work fine on another platform without any changes.

portrait

When you set up your document or go to print, you usually have a choice of whether you want the page **portrait** or *landscape.* Portrait is synonymous with **tall,** in that the page will be set up or printed taller than it is wide. Portrait is what we usually think of as a "normal" orientation. The other choice, landscape, or wide, is what we think of as "sideways."

This is a portrait, or tall orientation.

This is a landscape, or wide orientation.

posterization

Posterization is a special effects technique that can be applied to scanned photographic images. Posterizing reduces the number of different shades of gray or color in the image. Instead of a natural-looking continuous variation in the shades, you wind up with something that looks like a silkscreened poster, with obvious contrasts between a few different shades. Many applications have a menu command or choice that will just apply a posterized look to an image, which you can then adjust.

This is the normal image.

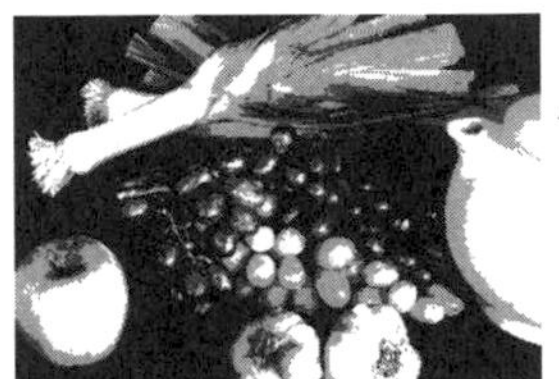

This is the posterized image.

PostScript

Created by Adobe Systems, **PostScript** is a programming language designed specifically for defining how documents look—primarily printed documents—in precise detail, one page at a time. But you don't need to know anything about programming to use PostScript. What you do have to have is a printer or other device that understands the PostScript commands (please also see *PostScript printer*).

PostScript has a wealth of commands that can be used to specify anything that might appear on a printed page: the placement of a line, how long it is, how thick it is, and exactly how it bends and curves; the contour, size, and color of a shape such as a triangle or star; the look of the characters in a font, their size, and what the text says.

The set of PostScript commands needed to define a document or the characters in a font is referred to as a "PostScript program." People can write PostScript programs from scratch, but usually the necessary commands are generated for you by your software when you tell it to print.

PostScript defines the elements of a document, text as well as graphics, by describing the *outline* of each shape, representing these outlines with mathematical formulas called *Bézier curves*. These PostScript commands are sent to the printer, which contains a special computer that understands the commands (see *RIP*, or *raster image processor*). This computer handles the details of actually "drawing" the shapes. Thus PostScript is called "device-independent," because any make or model of printer that understands the PostScript language can produce the correct image in print.

That also means that the images will always be created at the maximum *resolution* of whatever printer you're using: for a laser printer, this might be 300 or 600 *dots per inch;* on a PostScript *imagesetter* (a very expensive, professional-level printer), it's typically 1270 or 2540 dots per inch.

On most computers, including PCs and Macs, PostScript is used only for printing—the version of the document you see on your screen is generated within the computer using some other method. However, the Next computer uses a variant of PostScript called *Display PostScript* to display the document on the screen as well.

PostScript Level 2

Adobe has rewritten PostScript from the ground up, and the new version is called **PostScript Level 2.** It sports many improvements over the original PostScript design (now called Level 1) though most software still doesn't take full advantage of PostScript Level 2. Improvements in color halftone screening and color matching won't really affect people printing on black-and-white laser printers. A new *memory management* scheme, however, helps print very complex jobs more reliably. PostScript Level 2 can also handle the composite character sets required for printing Japanese Kanji and other Asian-lanquage fonts. It also has built-in image file decompression so that large image files can be compressed and sent to the printer more quickly. As a practical matter, the move to PostScript Level 2 has been gradual and will probably remain so. Software developers, in particular, haven't been in any great hurry to built Level 2 support into their products. If you're doing fine with your Level 1-based printer, there's not much point going to the expense of acquiring a Level 2-based printer.

PostScript clone

A **PostScript clone** is a printer or software that uses a work-alike version of the *page description language* called *PostScript* instead of the true Adobe PostScript. This kind of clone is also known as "PostScript compatible." These printers are usually significantly less expensive and can print most of what a true PostScript printer can print.

Lots of people are very happy with the clones, especially since, in many cases, they provide features not available with genuine Adobe PostScript–based machines. If you can find and Adobe-based PostScript machine with the features you want, at the price you can afford, then you might be better off with it, but clones are getting a lot of commercial use these days. Currently, all 600 *dpi* tabloid-sized PostScript printers are clones, and these are cranking out commercial jobs, including magazines and newspapers, around the world.

PostScript printer

A **PostScript printer** is a printer that interprets the *page description language* called *PostScript* (see the definition above). Many of the sophisticated computer printers (and all PostScript printers) are actually special-purpose computers themselves, containing their own *microprocessor* (electronic brain, CPU) and *memory (RAM)*. Most PostScript printers are *laser printers,* but a few use other technologies, such as inked ribbons, to put images onto paper.

PostScript printers are capable of printing complex graphics and patterns and typefaces that you just can't produce otherwise (because many complex graphics and typefaces are PostScript-based). However, with the increasing sophistication of graphics and page layout software, and with *ATM* and *TrueType* to generate high-quality fonts at any size, you can now do some of the fancy things on cheaper, non-PostScript machines. These days, only graphic-oriented users really need a PostScript machine.

Is your printer a PostScript printer? If you bought it, you probably already know, because PostScript printers are more expensive. These are examples of printers that are not PostScript: the HP LaserJet, DeskWriter, or DeskJet; the Apple StyleWriter, Personal LaserWriter, LaserWriter SC, and Personal LaserWriter LS.

PowerBook

The **PowerBook** is Apple's *laptop* Macintosh. It weighs about seven pounds and comes in a variety of combinations of RAM, hard disk space, and now even color monitors. It's the coolest thing. For workaholics (who, me?), a PowerBook can change your life.

power down, power up

Power up means to turn on the power to the computer, otherwise known as a *cold boot.* This is different from merely *restarting,* which is when you start over again after the power is already turned on. To **power down** means to turn off the power switch.

Power-on key

The **Power-on key** is only found on the newer Macintosh keyboards (you won't find it on the keyboard for a 512 or a Plus). It's the key either in the top on the center of the keyboard, or in the top right corner. It has an indented triangle on it, pointing left. This is the key you use to turn on the Mac, rather than having to stretch around behind the computer, as on

and Option keys are down, which will remove any notes you might have put in the *Get Info* boxes.

To zap your PRAM in System 6

Just to be on the safe side, close any open applications first. You don't have to restart for this. While you're looking at the Desktop, hold down the Command and Option keys. While they are down, choose "Control Panel" from the Apple menu. You'll get a message explaining that you are about to zap your PRAM. Just click OK.

After you zap your PRAM, you need to **reset the Chooser** before you try to print or you just won't be able to print.

On a PC, the corresponding chunk of memory is referred to as the CMOS.

predator

A **predator** is a person, usually male, prowling the *online services.* He seeks out a companion, acting as a dear, caring friend, alluding to future meetings and committments, but he is in reality often married, carrying on with several people at the same time, and eventually causing Big Trouble. Sometimes he's more than just a heartbreaker. One guy I heard about would carry on a relationship with a woman until she agreed to meet him in person, then he would wrangle a significant amount of money from her, dump her, and move on to the next. He hurt five women this way, all of whom he met online. I suppose this is no different than what people do offline anyway—it's just another method for the madness.

Preferences, preferences file, Prefs

Many applications have an auxiliary file called the **preferences file,** or simply **Prefs.** This file contains the specifications for the program, like what colors to use in displaying the window, the placement of the ruler and tool palettes, etc. Don't toss out those Prefs files or put them in other folders or directories—some programs will just recreate a new preferences file, but most programs need the file before they can open.

On the Mac, in System 7, the preferences files for almost all software programs are stored in the Preferences folder within the System Folder.

On the PC, preferences may be stored in a file named with an extension like .PRF or .CFG (for *configuration*). Windows programs, however, usually store your preference settings either in their own .INI file (for *initialization*) or in the main Windows WIN.INI file.

pre-formatted

When you buy a disk, whether it is internal or external, a hard disk, a cartridge, or a floppy, you can buy a "raw" (completely blank) disk or one that has been already prepared to accept information, **pre-formatted,** either at the factory or by your dealer. Every disk must be *formatted* and initialized before it can accept data, so it saves you time and bother if the vendor does it for you. Pre-formatted floppy disks cost quite a bit more than the unformatted kind, but your time may well be worth the difference.

pre-press

When you take your pages down to the commercial press to be reproduced, whether in black-and-white or in full color, you take *camera-ready* pages. That means you have done the layout, and perhaps the paste-up, so your pages are ready to be shot by the printer's cameras and have printing plates created. Well, traditionally, after you took the press person your pages, they had other work to do to actually get those pages on the press. If it was a *four-color* job, the *separations* had to be made. *Halftones* had to be created and "stripped" in. Two- or three-color work needed separate "plates." This work is called **pre-press.**

Now, as we gain more and more control over our pages in the computer, we are trying to do as much of the pre-press work electronically as we can. We are pushing the edge of the technology and technology is trying hard to keep up. Some systems can already take a camera-ready page straight from the computer to the printer's plates.

presentation graphics

When you take plain ol' charts and graphs, such as bar graphs, pie charts, text charts, line charts, etc., and make them visually interesting, they are then considered **presentation graphics,** sometimes known as *business graphics*. Presentation graphics are often shown in slide form.

presentation graphics program

A **presentation graphics program** is an application that makes it easy to create slide presentations. The slides can include titles and text, and add graphics and *clip art*. A full-featured program will even let you make fancy charts and graphs with your data. You can then show your slide show straight from the computer, or *output* it to paper, or output it as slides.

Presentation Manager

Presentation Manager (PM) is the *graphical user interface (GUI)* in *OS/2*, IBM's operating system for PCs. Presentation Manager looks and works very similar to *Windows,* with *windows, pull-down menus, on-screen typefaces,* and *icons,* though there are some differences.

pressure-sensitive tablet

A **pressure-sensitive tablet** (also known as a *graphics pad*) is a thin, flat pad that sits on your disk. You use a *stylus* (shaped like a pen, but it doesn't have any ink) and draw directly on the tablet. The tablet has an intricate grid that tracks the movement of the stylus and sends the information to the computer screen. Not all graphics pads are pressure sensitive, but if the pad is, and if the software can take advantage of it, you can mimic brush strokes and paper absorbency with a tablet and stylus. When you press harder with the stylus you create a thicker line on the screen, as if the paint was flowing off the brush and spreading on the paper.

print buffer

Most *dot matrix printers* have a **print buffer,** a special chunk of memory for temporary storage. When you print a document, the computer sends out the necessary information faster than the printer can print it. If the printer has a buffer, the information the printer can't deal with immediately goes into the buffer, where it waits until the printer is ready for it. As soon as the entire document is in the buffer, you can use your computer again for other work, while the printer takes its time to print the "buffered" information. The larger the print buffer, the more it can hold and the faster you get your screen back.

Print buffers are similar in concept to print *spoolers,* but a spooler works by parking the extra information to be printed on your *hard disk,* rather than inside the printer.

Laser printers have *memory,* but there's no separate section set aside as a buffer—any memory not used for the page currently being printed (or for fonts), is available for temporary storage.

printed circuit board

A **printed circuit board** (sometimes called a PCB) is a flat piece of plastic or fiberglass that has metal pathways printed onto it. *Chips* and other electronic components are mounted onto the board. The little metal legs of the components are soldered onto the metal pathways to create a connection through which electricity can flow. The *adapters, cards,* and *boards* you buy to enhance your computer are built on printed circuit boards.

printer

A **printer** is that device that takes the text and images sent from the computer and puts them on a piece of paper that we can then curl up in bed with, send through the mail, save for posterity, post on walls, or use as incriminating evidence. There are *laser printers, line printers, ink-jet printers, dot matrix printers, thermal transfer printers, imagesetters,* and probably a few others. There are *impact printers* and *non-impact printers.* Print quality and speed vary. The price range is wide.

Printer can also refer to the person who runs the commercial press where you have your job reproduced—that person is a printer. These two meanings can cause confusion, such as when I warn my graphic design students, "Before you start designing any job, first talk to your printer." No, I don't mean talk to your laser printer.

printer driver

The **printer driver** is a software file. Your applications use this file to figure out what kind of printer is attached to the computer so documents can print properly.

On PCs, each application has to have its own print drivers for each of the many many printers that it may print to. These drivers will only work with the application they come with. Fortunately, all applications that work with Windows can use the same driver.

On Macintosh computers, the printer driver is part of the *operating system.* This means that individual applications don't need their own drivers (although they can, like versions of PageMaker before 5.0). Any program on the Mac can print to any printer it can hook up to. See *LaserPrep.*

printer memory

Page-oriented printers such as laser printers need a healthy supply of **printer memory** in the form of their own *RAM chips.* The printer uses some of this memory to store the instructions it receives from your computer (if you have a *PostScript printer* these instructions are in the form of PostScript language commands). Based on these instructions, the printer creates the image of each entire page in *bitmapped* form, storing that image data in another area of printer memory until the page is actually printed. (See the entries for *laser printer, PostScript printer,* and *page description language.*)

Any printer that can accept *downloadable fonts* has some memory where it stores the font information. This might also be called printer memory. Many simple *dot matrix printers* come with a supply of memory to use as a *print buffer* which stores incoming data and commands until the printer is ready to process them. People usually don't refer to buffer memory as printer memory.

printhead

In printers that print one character or one line at a time, the **printhead** is the part of the printer that moves back and forth across the page to create the text or graphics. It does this either by striking the page through a ribbon or by spraying ink out of tiny jets onto the page. Laserjet printers don't have printheads, but *dot matrix, daisy wheel,* and *inkjet* printers do.

Print Manager

Windows comes with a print *spooler* called **Print Manager.** If you want to use it to lessen the time you spend twiddling your thumbs waiting for your printer to finish, be sure the box labeled "Use Print Manager" is checked in the Printers dialog box in the Control Panel. See print *spooler.*

print merge

A **print merge** is when you take information from two or more places and merge them (combine them) into one printed document. The information is not merged into another file—it is merged just on the printed page.

This is a basic print merge: You have a *database* of some sort with a list of the names and addresses of people to whom you want to send letters. You've typed a letter that you want to send to all these people. In the places where you want to personalize the letter, as in the address, the salutation, and perhaps in the line about how much money they owe you, you have a *placeholder.* Each of these placeholders has been linked to a particular *field* in the database, like to the first name, the last name, the street address, the amount owed, etc. When you tell the application to print those letters, it prints the first letter and fills in the name and address and all that other information from the first *record* in the database. When the application finishes printing that first letter, it prints the second letter and fills in the placeholders with the information from the second record. And on and on until all the letters are printed, each one "personalized." This kind of print merge is also known as a "mail merge."

Print Monitor

The **Print Monitor** in *System 7* and under *MultiFinder* in *System 6* is a print *spooler,* which means it is a *utility* that takes care of the printing for you so you can get back to work. Without a print spooler of some sort, you have to wait until the printer is finished printing your job before you have control of your screen again. Print Monitor will take all the printing information, hang on to it, and send it to the printer as necessary. You can send several files at once (just select the documents and print as usual) and the Print Monitor will line them up and send them off while you do something else. Also see *spooler.*

printout

Printout is another one of those computer contractions, this time a convergence of **print**er and **out**put (since it comes out of the printer). It means the printed paper, the same as *hard copy.*

Print Screen key, PRT SCR

PC keyboards have a key labeled **PRT SCR** or **Print Screen.** Depending on the type of PC you have, you may find the Print Screen key in one of several places. If you have an early PC, Print Screen is usually the shifted version of the asterisk (*) key, the asterisk on the *numeric keypad* on the right of your keyboard. In other words, you have to press Shift and the asterisk key at the same time to get the Print Screen key to work. More commonly, the Print Screen key is a separate key at the top of the numeric pad. You may still have to press Shift and Print Screen together to activate it, though. If you have a laptop or notebook PC, all bets are off, but generally, Print Screen is to the upper right of the keyboard (you may have to press a special *function key* to activate it).

In DOS, pressing the Print Screen key sends whatever text or image is on your screen to your printer. Ideally, you end up with a printed copy of what was on this screen. Whether or not this works depends on two things: the type of printer you have (it won't work with a PostScript printer), and whether you've set up DOS for that type of printer (use the DOS command GRAPHICS to do it).

In Windows, the Print Screen key has an altogether different function. You can use it to copy the image on the screen—all the windows, the menus, everything you can see—to the Windows *Clipboard.* From the Clipboard you can paste the screen image into a graphics program, edit

privileges

Privileges in computer jargon is similar to privileges in Life. If you want to go through the door marked "Employees Only," you need to be privileged to do so. If you want to get into information that is locked up within a computer, you need *access* privileges, which usually involves a *password*. "I tried to get that information for you, but I don't have access privileges on that computer."

process colors

Process colors are the four colors that a commercial press uses to print *full-color* photographs and illustrations. The process colors are cyan (a light blue), magenta (a kind of red), yellow, and black. You'll see the process colors abbreviated as CMYK in any program that deals with full-color photographs or illustrations. By using these four colors in a color *halftone* technique, a printing press can simulate a wide range of colors, enough to reproduce color photos and illustrations.

To employ process-color printing, a computer program is used to separate a scanned image into its cyan, magenta, yellow, and black components. Separate printing plates are then prepared, one for each of the four color components. The finished page is printed in four passes, once for each color component, using the corresponding cyan, magenta, yellow, or black ink. If everything goes right, small dots of each color cluster tightly together or even overlap to create the impression of many different colors and shades. Take a look through a magnifying glass at a color photo in a magazine or newspaper and you will be able to see the dots of the four colors. Also see *CMYK*, *four-color separation,* and *full color.*

processor

A **processor** consists of a set of electronic circuits that perform computations. The term is often used as short for *microprocessor,* which is a single *chip* that contains an entire processor, also known as the *CPU (central processing unit),* which is the processor that runs an entire computer. (A CPU can be a microprocessor, but in some computers the CPU is a whole set of chips.) Please see either *microprocessor* or *CPU.*

Prodigy

Prodigy is an *online service* developed by IBM and Sears. It has all the services that most any other online service has, like shopping and travel arrangements and information (but you can't *download* anything) and *electronic mail*, but you also pay for a constant stream of commercials running in front of your face. Personally, I find paying for commercials offensive.

product manager

Every product has a **product manager.** This is the person who identifies the market, coordinates between the engineers and the quality control, and between the sales people and the press. This person often also identifies the functionality of the product and is responsible for its content. In the company, the product manager is the person most in touch with and most knowledgeable about the product. So if you ever get to see a demo by the product manager, you are getting the best information.

program

A **program** is basically a coded set of instructions, written by humans, that tells the computer what to do. The program can be stored on disk (in which case the program is *software*) or in a chip (which is *firmware*). Either way, the instructions have to be transferred into the computer's *RAM (random access memory)* before the program will operate.

Some programs tell the system as a whole what to do; this is the *operating system*. Some programs are little *utilities* that we can use to make our computing lives easier, like making a clock show up on our screen, or managing our fonts. The programs we are most familiar with are the *applications,* the ones we use to do something productive—a spreadsheet program, a word processor, and a page layout program are all applications. The term "application" is sometimes used synonymously with "program," but technically an application is just one type of program.

If a program is finely crafted, clearly structured, and executes smoothly, it is said to be "elegant." If it is haphazard, patched together, and poorly designed, it is said to be a *kluge*.

Program Manager

Program Manager is the "homebase" software in *Windows* 3.0 and 3.1; it's the place from where other programs run. Program Manager displays your programs as little pictures, or *icons*, allowing you to start a program by double-clicking on its icon with the mouse. You can organize your programs into separate "groups," each in its own window. You could have one group for word processing programs, one for *utilities*, one for financial software, and so on.

This is what the Program Manager looks like.

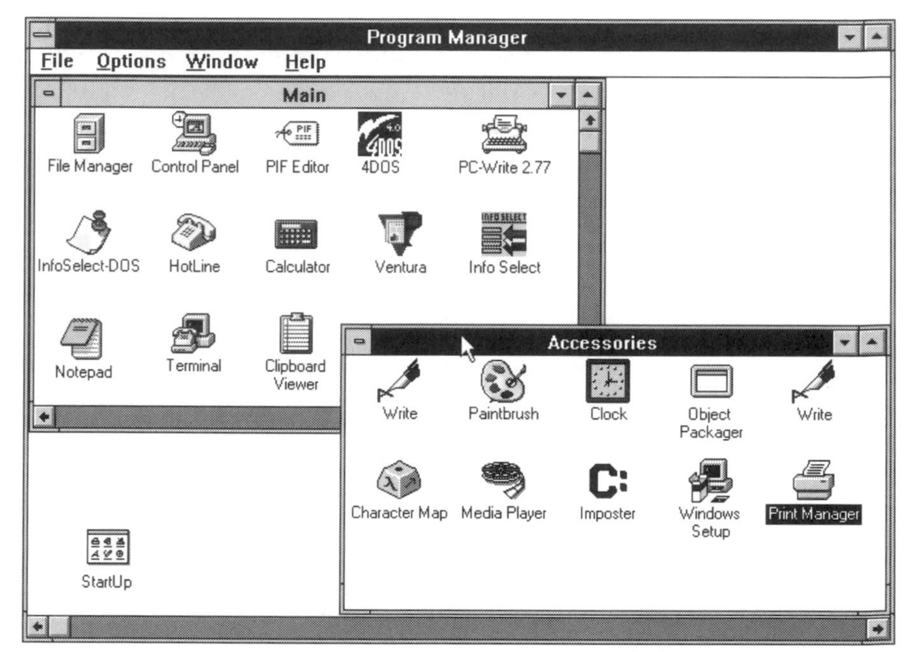

Starting programs with Program Manager is much easier than with *DOS*, where you have to type out the program name yourself. But Program Manager has its problems. One of them is that it only shows programs, not your documents (files). To open a specific document, you have to start the program first, then open the document (in *File Manager*, you can do the whole thing in one step by clicking on the document's name, but you don't get to see those nice icons or organize your work into logical groups). Macintosh advocates justly criticize Windows for these deficiencies (the Mac doesn't suffer from them), but you can buy Windows *shells* like the Norton Desktop that combine the functions of Program Manager and File Manager in one.

programmer

A **programmer** is the magician who writes the *programs,* the coded instructions that tell the computer what to do. Some programmers might have degrees in computer science, but lots of them have taught themselves. You too can learn to program, by reading books, asking questions, and most of all, lots of experimenting. What, you say you don't have the slightest interest in programming? No problem—you can use a computer to its full potential without knowing the first thing about programming.

programmer's switch, reset switch, interrupt switch

There is a little piece of plastic that comes with every Macintosh (except the Classic, LC, IIsi, and the PowerBook) called the **programmer's switch.** Your manual might tell you not to install it, but do. It has two little tabs to depress; one is the **reset switch** (usually has a little triangle on it) and one is the **interrupt switch,** also called the **debugger switch** (it usually has a broken circle on it). If the tabs are on the side of the computer, the one closest to you is the reset switch. If the tabs are installed on the front of the computer (usually under the Apple logo), the one on the left is the reset switch.

Have you ever crashed? Yes? Or has your screen ever *frozen?* And did you have to turn off the computer from the power on/off button? Well, the **reset switch** does the same thing as if you restarted without turning off the power so it's easier on the computer.

The other switch, or button, is the **interrupt switch.** If you want to try to get back to the Finder (the Desktop) without restarting the computer or turning the power off, you can try pressing the interrupt switch when the screen freezes. This will usually make a box appear on your screen with this symbol in it: > . If you see that box, type SM 0 A9F4 (make sure you type a zero and not the letter O, and make sure you put a space before and after the zero). Then press Return and type G 0 (that's a G, space and a zero). Press Return again. If this doesn't take you back to the Finder, then push the reset switch.

In *System 7* if the screen *freezes* on you, try pressing Command Option esc (escape). About 75 percent of the time this will return you to the Finder. If not, try the interrupt switch.

No matter how you get your screen back, go into any applications that may still be open, save your document, quit, and restart the computer.

programming

Programming is the creation of a computer software program, the coded instructions that tell the computer what to do. See the entry below for the kinds of instructions, or languages, that have been developed.

programming language

A **programming language** is a written language that human beings use to control computers. Programming languages truly are languages, in that they communicate meaning with organized sequences of words and symbols. No programming language is as complex as a human "natural language," but each does have a vocabulary, a grammar, and a syntax—and a few quirky exceptions to the rules.

You can classify programming languages in several ways. For one, there are low-level languages and high-level languages. A low-level language is one that directly controls the *processor*, the chip that runs the computer. Therefore, a program you write in a low-level language will only run on computers with that processor. The lowest of the low is *machine language*. Machine language consists solely of code numbers, each representing one of the individual operations, or "instructions," the processor can perform. *Assembly language* uses words instead of numbers to represent those same codes, so it's easier to work with than machine language, but it's still low-level. Since each type of processor understands a different set of codes, there are many different machine and assembly languages, one of each for each processor.

High-level languages like BASIC, COBOL, Pascal, or FORTH sound more like English (or some other natural language). In a high-level language, a single statement or command, such as "PRINT" or "WAIT," can take the place of many low-level language statements. And high-level statements don't correspond directly to particular processor instructions. For this reason, a program written in a high-level language often can run on different types of computers with relatively minor modifications. (You must *recompile* it first. And actually, whether a given program is *portable* from one computer to another is a complicated matter, but this is one of the factors.) The language C is often described as an intermediate-level language, since you must define what you want the computer to do in fairly small steps, and C allows you to control the processor directly in some ways if you want.

Another way to classify programming languages is to look at how "structured" they are. A structured language has well-defined rules about what goes where in the program, so that the programmer can put each function the program does into a separate, self-contained module. In an unstructured language like BASIC, you can tell the program to jump from here to there at your pleasure. That gives you more freedom, and it's okay for short programs, but in the long run structured programs are better.

Finally, there's a big face-off these days between procedural languages and object-oriented languages. Procedural languages include BASIC, Pascal, C, FORTRAN, and COBOL. A procedural language divides a program into two broad compartments: the information the program works with, and the procedures, which contain the programming commands that manipulate that information. An object-oriented language structures things differently, allowing you to assemble programs out of distinct, building block-like modules called objects. Each object contains both a set of programming commands and the particular data those commands act on. The distinction between procedural and object-oriented languages may not sound like a big deal, but some people think it is (for the reasons why, see the entry on *object-oriented programming*). The most popular object-oriented languages are C++, FORTH, SmallTalk, Prolog, and ADA.

Here is a brief rundown of most of the high-level programming languages you're likely to hear about:

COBOL (pronounced "co ball"), short for **co**mmon **b**usiness **o**riented **l**anguage, is still the most widely used programming language in business applications.

FORTRAN is a dominant language for solving engineering and scientific problems.

Pascal was originally invented to teach computer science students how to program, but it soon left the classroom and became one of the leading languages for home and business programming. The original version of the language has been extensively modified. These days, Pascal sees its heaviest use on PCs in various permutations sold by Borland. Pascal used to be a popular language for Macintosh programming.

Modula-2 is based on Pascal, but is supposed to be a better language because, for one reason, you can set aside separate program modules. Be that as it may, it has only a small following.

LISP is often used in *artificial intelligence* research because you can write programs that modify themselves.

FORTH is often used to control other hardware devices, such as robots, arcade games, and musical devices.

Prolog is used a great deal in developing *expert systems* in research.

C, C++ is a programming language developed at Bell Labs (the renowned research center of AT&T) in the 1970s. C quickly became extremely popular and to this day remains the language of choice for developers of commercial software. C++, developed at Bell Labs in the 1980s, is an *object-oriented* version of the more traditional C.

Smalltalk is the programming environment that inspired *HyperTalk,* the programming language in *HyperCard*. The purpose of SmallTalk (and HyperTalk) is to make programming accessible to *mortals* (you and me). It's used in research and development.

ADA was developed by the U.S. Department of Defense in the late '70s in response to an overwhelming proliferation of incompatible language systems. ADA is the language required for all military programming applications. It's named after Lady Augusta Ada Byron, Countess of Lovelace (1815–1852), daughter of the romantic English poet, Lord Byron. The Countess was the first female computer scientist and is acknowledged as the first computer programmer in the world.

Logo is a variation of LISP that was specially designed as a language for teaching children the concepts of programming. Logo uses "turtle graphics," where a child can write a program that controls the movements of a little turtle on the screen, and the turtle can draw geometric diagrams.

progress indicator

A **progress indicator** is that little box that tells you what's going on, like who is in line to print or how many of the files have been copied or how long it will take your file to *download*.

project management software

Project management software helps you organize and schedule the multitude of tasks that must be accomplished in order to complete a big project, such as building a shopping mall or getting a new vegetable slicer-dicer onto the store shelves. Each of the tasks is a separate chunk that needs to be scheduled properly in the context of the overall project. Project management software helps keep track of what resources (like people, machines, and money) are available and helps keep the schedule coordinated so the entire project can be completed on time. Project management software lets you set up schedules in the form of charts (Gantt charts, PERT charts, time charts, and so on) so you can get a grasp on the overall project visually.

PROM

PROM (pronounced "prom") stands for **p**rogrammable **r**ead-**o**nly **m**emory. PROM is a *read-only memory chip.* Typically a read-only memory chip has information permanently embedded into it when manufactured and cannot be changed—like information that helps run the system. A PROM can be programmed with a device called a "PROM programmer." PROMs are typically used while making a prototype of a product (software or hardware) because the proms can be created and discarded until the product is perfected.

An **EPROM** (pronounced "ee prom") stands for **e**rasable **p**rogrammable **r**ead-**o**nly **m**emory. It's the same as a PROM but it can be erased by exposing the chip to ultraviolet light, and then it can be re-programmed.

An **EEPROM** (pronounced "ee ee prom") is an **e**lectrically **e**rasable **p**rogrammable **r**ead-**o**nly **m**emory chip. An EEPROM can be erased with a tiny electric signal applied to one or more of the pins, and can be reprogrammed while it is still attached to the circuit board.

prompt

A **prompt** is a symbol or perhaps a question on the screen that "prompts" you (asks or tell you) to tell the computer what to do next. Different programs may each have their own prompt. For instance, when you're using the computer system in the library, you may see a message (a prompt) like, "Type G to go to the next page of listings." A *DOS* prompt may be C>, which tells you the computer is waiting for you to tell it what to do next. See *DOS prompt.*

propeller head

Propeller head is similar to computer *geek,* except it refers to people who wear beanie caps with little propellers on the top. *The New Hacker's Dictionary,* which is the comprehensive compendium of *hacker* lore and wisdom, says that nobody actually wears propeller beanies except as a joke. The hackers have obviously never met my brother.

My big brother, Jeffrey Williams.

proportional spacing

Proportional spacing means that the letters on the page each take up a relative, or proportional, amount of space according to how wide they are. For instance, the letter "w" takes up significantly more room than the letter "i" or a period. Now, you may think this is only logical. But on the typewriter that most of us grew up with, proportional spacing was just about non-existent—typewriters use *monospacing*. On a typewriter and on some primitive computer screens, the letter "i" takes up as much space as the letter "w." If you drew vertical lines between the letters of a monospaced paragraph, all the letters would line up in columns (see the example). The fonts Courier, Monaco, Pica, and Elite are examples of monospaced fonts.

```
This is the font Courier. Notice it is monospaced. The
rest of this dictionary uses a proportional typeface.
```

Most computers at this point in history are capable of producing text with proportional spacing, which looks much more professional and is much more *readable*.

protocol

You know how protocol works in diplomacy or in the military—it refers to the rules for how things are suppposed to be done. In the realm of computers, a protocol is a particular set of standards or rules usually having to do with communications between two computers. If the two computers trying to exchange data use different protocols, they won't be able to talk to each other.

There are *modem* protocols that specify the particular warbling tones two modems must use to communicate with one another (see *v.32*). There are software *communications* protocols such as *XMODEM* and *Kermit* that specify the messages your telecommunications software must send and receive in the process of transferring files.

As Scott Watson explains in his great manual for White Knight (a *telecommunications program*), "Consider a telephone conversation between people. Both persons must agree to speak in a language both participants understand, and a certain degree of etiquette (or protocol) must be followed throughout the conversation (such as not hollering wildly at the same time). Using computers and modems is very much the same."

prototype

A **prototype** is a working example or model of a computer system or software program. A prototype is used for testing and refining, both for hardware, programs, and entire systems of information management.

PS/1, PS/2

In 1987, IBM tried to re-exert its leadership of the PC marketplace with a new **PS/2** line of computers (PS/2 stands for Personal System/2). Partly this was just a marketing ploy, because "PC," which originally was the trademark name for an IBM computer, had become such a generic term. But most of the PS/2s were equipped with a new *Micro Channel expansion bus* (the expansion bus consists of the slots inside the computer where you insert add-in boards and the related circuits). The Micro Channel bus won't accept cards that work with the IBM PC, PC/AT, or compatible computers, but it has technical advantages over the expansion bus in those computers (see Micro Channel and ISA). IBM continues to refer to its line of business personal computers as PS/2s; a few of the current models still have the old-style slots. The company has also come out with a **PS/1** series intended for home users.

PS/2 port

See *mouse port.*

public

If a corporation is not public, that means it is privately owned and the owners own all of the stock. When they go **public,** the general population has the opportunity to buy shares of stock. Going public can be an indication of the strength and stability of a corporation.

public domain software

Public domain software (sometimes called PDS) means that someone has taken the time and trouble to create a piece of software and then has offered it up to the world for free, out of the kindness of their heart. Public domain software is technically a little different from *freeware.* When the author places a program in the public domain, she gives up all the rights, which means anybody can modify the program, or even sell it for a profit if they can find a gullible buyer. With freeware, the author retains the copyright and can place restrictions on how the program is distributed.

Publish and Subscribe

Publish and Subscribe is a technology that Apple built into *System 7,* the operating system for the Macintosh. It allows you, from within any application that supports this technology, to *publish* (save) a document or a part of a document (text or graphics) as a separate file, called an *edition.* Other people on the *network* using other programs and documents can then *subscribe* to this edition file; they can incorporate it into their own documents. For instance, you can publish an edition file of a report from Microsoft Word. Then in another Word document or even in another application, someone else can subscribe to that edition and it will appear in their document. In fact, any number of people can subscribe to it.

The key benefit is this: if you update the original published edition file, then all the editions in the subscribers' documents will be updated automatically. It's great in theory, and eventually it will be a wonderful tool.

pull-down menu

A **pull-down menu** specifically refers to the standard kind of *menu* that pulls down from the top, as opposed to a *pop-up menu* that pops up from the bottom, or a *hierarchical menu* that pops out to the side of another menu. (Each of the definitions in italics has an illustration.)

Put Away

Put Away is a command in the File menu at the *Finder* (the *Desktop*). If you have moved a file from its folder to the Desktop or the *trash,* it's easy to forget where it came from. If you change your mind about throwing away the file, or if you want to tidy up the Desktop, you can use the Put Away command to send the file back where it came from. Select the file (click once on it), and choose the Put Away command; back it goes to its own folder.

If you are using System 7, you can use Put Away to eject a disk. The keyboard shortcut is Command Y, so just select the disk (click once on it or use one of the keyboard shortcuts), and press Command Y to eject it. This does not leave the disk in *memory,* meaning you will not be left with a gray icon of the disk on the screen.

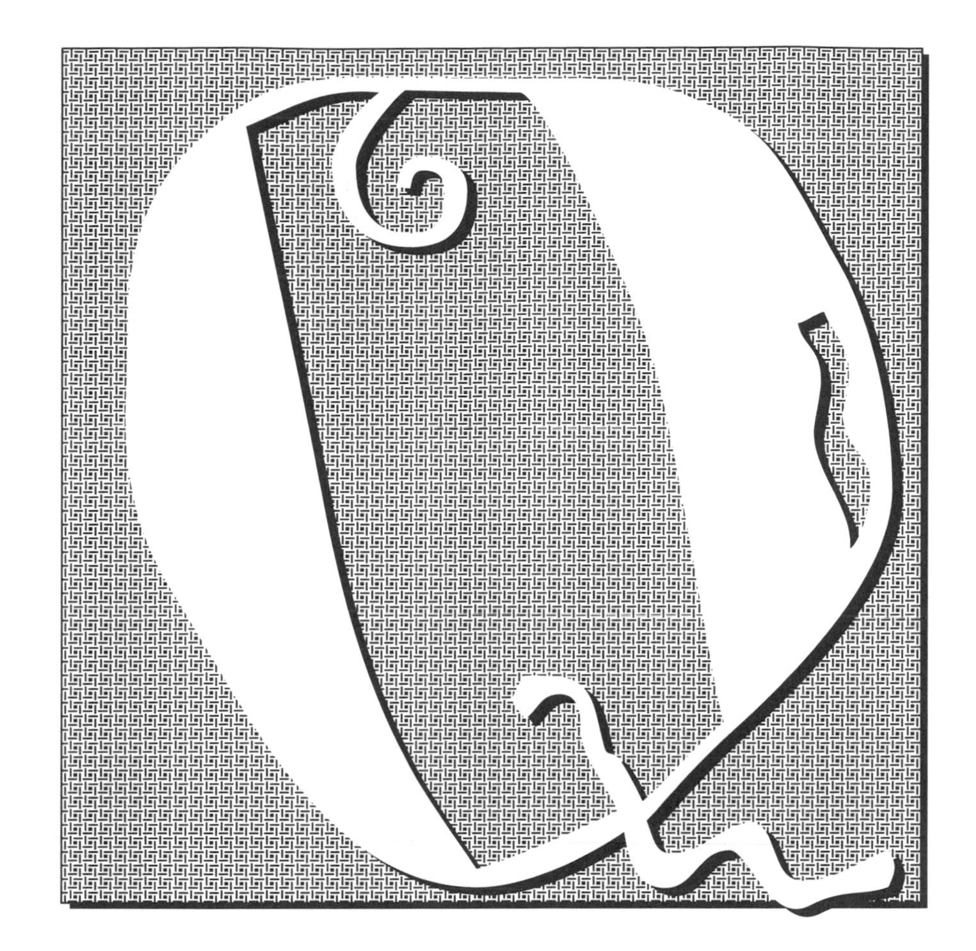

Quadra

The computers in the **Quadra** line are the fastest and most powerful Macintoshes that Apple makes at the moment. These machines encroach on the territory of workstations made by the likes of Sun and HP. At the time of this printing, the Quadra 900 models run at a 25 *MHz clock speed,* and the 950 at 33 MHz (by comparison, the SE/30 trots along at just 16 MHz). Quadras currently use the *68040 microprocessor* (that's the *CPU, the central processing unit,* the *chip* that runs the computer). The Quadra is a *tower,* meaning it is so big and tall that it looks like a tower. It's supposed to sit on the ground, not on your desktop.

queue

A **queue** (pronounced "cue"), as any good Englishperson knows, is an orderly line, such as a queue of people waiting patiently to get on a bus. When more than one person on a *network* prints to one printer, or if you use a *print spooler* (*background printing* on the Mac), the print jobs queue up, or get in line, and the printer takes them in order, first come, first served. Your software may show you a listing of the print queue on screen, so you can see which job is printing and which is next.

QuickDraw

QuickDraw is Apple's computer language that tells the Macintosh how to draw text and images on the screen. QuickDraw is built into the *ROM (read-only memory)* of every Macintosh, and is responsible for the graphic display of windows, icons, the menu bar, etc., as well as the *PICT* graphic images.

QuickDraw printer

There really isn't a type of printer called a **QuickDraw printer—**the term just refers to most printers that are not *PostScript*. See, for most programs, what you see on the screen is displayed using the *QuickDraw* language, even if the document is ultimately headed for a *PostScript printer.* If the document does go to a PostScript printer, the LaserWriter *driver* converts the QuickDraw information to the PostScript *page description language* (unless the document is already coded in PostScript, as with Adobe Illustrator or Aldus FreeHand). The Mac then sends the PostScript information to the printer (a PostScript printer actually has an entire computer within it). With PostScript, the printer is able to create smooth, detailed graphics and text. But when you print to a printer that is *not* PostScript, all that printer can do is recreate what is on the screen. Since what is on the screen is done in QuickDraw, a printer that prints what is on the screen is called a QuickDraw printer.

Adobe Type Manager (ATM) takes the PostScript information for creating smooth fonts and makes those same fonts appear much smoother on the screen (see *ATM* for details). It will also send that information for creating smooth character shapes straight to a QuickDraw printer. So if you use ATM and *Type 1 PostScript fonts,* your type will look as beautiful from a QuickDraw printer as it does from a PostScript laser printer. (ATM does not help print any PostScript graphics, though.)

TrueType fonts use a technology built into System 7 (you can also use it in System 6; see *TrueType* for details) to create smooth on-screen type. If you don't have ATM but you use TrueType, your type will print beautifully on a QuickDraw printer.

QuickDraw GX

QuickDraw GX is an "imaging model" software *extension,* available in 1993, that works alongside the normal *QuickDraw* that is built into the Mac's *ROM chips.* It is designed to make the Macintosh the first truly international computer, as well as the *platform* for a new color publishing standard.

QuickDraw GX automates consistent and predictable color input, display, and output with ColorSync color management technology—colors in a scanned image reproduce automatically and accurately on a display, and these colors reproduce accurately in print.

The printing process is simplified and more powerful, and you can display and control selected printers via icons on the Desktop.

QuickDraw GX allows you to modify any shape (which means any graphic image, including text): you can rotate, skew, clip (use the shape as a window that shows a graphic in its shape), change the line thickness and joins, add patterns, and apply other tricks.

Aaack! Another font technology—GX fonts, both TrueType and Type 1—that work with the Line Layout routines. The concept is pretty amazing—refined, professional typography that happens automatically in any application. This includes automatic substitution of fractions, *ligatures,* and decorative letters where appropriate; optical scaling; *kerning; tracking;* optical alignment; *hanging punctuation;* arcing, skewing, rotating, and adding perspective to editable text; using text as a window for a graphic image; setting different reading directions on the same line; the potential for 65,000 characters; and more.

QuickDraw GX will work on any color-capable Mac that has System 7.1 installed, a hard disk drive, and 5 megs of *RAM,* minimum.

QuicKeys™

QuicKeys is a very popular *macro* program. A macro is a series of steps you set up to perform automatically, with just the touch of a keystroke. Anytime you find yourself repeating a process, you can (should) set up a macro to do it for you. Do you ever tranpsose letters when you type? You can set up a macro to fix 'em. Do you switch your monitor from grayscale to color every now and then? Set up a macro to do it. Macros can get quite complex, and QuicKeys is many people's favorite utility for creating and organizing them.

QuickTime™

QuickTime (often abbreviated as QT) is a software product from Apple. It's actually just an *extension* that you drop into your System Folder. You don't really *do* anything with the QuickTime software—it's meant for synchronizing the recording and playback of images and sounds (little movies) that you have created in other applications. It also supports various types of *compression,* which is a big deal—movies can take up an enormous amount of disk space.

This is the QuickTime extension icon.

The QuickTime software is about 440 *kilobytes* and is available free from any Apple dealer, *user group,* or *online service.* If you are running System 7, look in the Extension folder within the System Folder—it may already be there.

For an interesting *Easter Egg,* turn on *Balloon Help* from the Help menu at the Desktop. Then go into the System Folder, open the Extensions folder, and point to the QuickTime icon.

If you have a QuickTime movie, which you can get from many sources, you can name it Startup Movie (two words), drop it into your Startup Items folder in the System Folder, and that movie will play every time you turn on your Mac. There are entire CDs of movie clips available.

QWERTY keyboard

The **QWERTY keyboard** is probably the one you have. It is the standard sort of keyboard whose first six letters, reading from the left, are q, w, e, r, t, y. In the 1870s, C. L. Sholes intentionally arranged the keys in an inefficient pattern to slow down typists because the early typewriters jammed if the keys were pressed too quickly. Well, we no longer have the problem of keys jamming (gosh, some of you are probably too young to even have a concept of typewriter keys jamming—that's a weary thought). An alternative to the QWERTY layout is the *Dvorak* layout, designed to take advantage of our most nimble fingers.

R&D

R&D stands for Research and Development, the process of dreaming up new products, figuring out how to make them, and testing prototypes. Large corporations often have an R&D department. Some business parks set aside a section of the buildings for R&D, for office sorts of work—meaning it is not to be used for any warehousing, manufacturing, heavy equipment, etc.

radiation

Radiation is the process in which energy is emitted as particles (minute pieces of atoms) or radio waves. Human bodies absorb radiation, which can cause a number of physical problems, from headache or naseau to various forms of cancer.

All monitors emit some amount of radiation in various forms. Color monitors emit more than monochrome monitors. The concern is not so much with x-ray radiation, but with *VLF* (very low frequency) and *ELF* (extremely low frequency) electrical radiations, and especially with ELF magnetic radiation. It is the *cathode ray tubes*, also known as *CRTs* (the same as in your television) that release the radiation. The sides and back of CRTs often radiate stronger magnetic fields than the front: don't sit closer than four or five feet from the side or back of a CRT, and try to sit at least an arm's length away from the front of your monitor.

The most common anti-radiation devices sold are filters that fit over the front of the screen that can block out electrical radiation. There is no transparent material yet available that can block much of the magnetic radiation (even lead doesn't have much effect at stopping magnetic waves). Newer models of monitors are being made with higher standards of radiation control, but nobody really knows what the standards should be. Several foreign countries have more stringent guidelines than the U.S., so you may hear of monitors that "meet Norwegian radiation levels."

radio button

There are several kinds of *buttons* in computer applications, buttons that you click on with the mouse pointer or that you activate with a key stroke. Buttons provide a way to tell the computer what to do. The buttons with the little round dots are called **radio buttons,** and they always come in groups of two or more. A set of radio buttons is a visual clue that you only have a choice of *one* of those options. Choosing another button automatically turns off the button that was currently selected, just like pressing a button on your car radio switches you from one station to the next.

Like other buttons, radio buttons are mostly found in a *graphical user interface* like the Macintosh or Windows or Presentation Manager, but they're popping up in *text-mode* programs, too. Compare radio buttons with *checkbox buttons,* where you can choose as many of the checkboxes as you like.

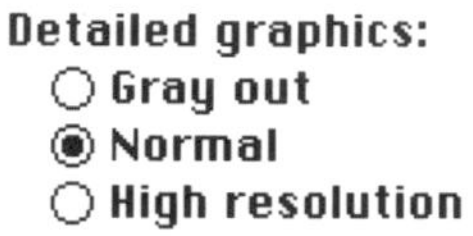

Detailed graphics:
○ Gray out
◉ Normal
○ High resolution

Radio buttons are a visual clue that you can choose only ***one*** *of the options.*

Type style: ☒ Normal ☐ Italic ☐ Outline
☐ Bold ☐ Underline ☐ Shadow

Checkbox buttons are a visual clue that you can choose any number of the options; in fact, you can choose all of them or none of them.

ragged right

Ragged right is a form of text alignment where the text is all lined up on the left side, but the lines on the right side end wherever they end, creating a ragged, or unsmooth edge, as in this paragraph. A ragged right alignment is easy to read because it has consistent spacing between letters and between words, as opposed to *justified* text where space is forced between words and letters to make every line the same length.

If you are going to leave the right margin ragged, try to keep it as smooth as possible. That is, try to avoid having some long lines and some short lines; adjust the lines so they are all close to the same length.

RAM

RAM stands for **r**andom **a**ccess **m**emory. Be prepared, because this definition is pretty long (but it's important!). First we'll tell you what RAM in general is, then we'll go over how RAM works in Macintoshes and PCs. Also, there are different kinds of RAM, including VRAM, PRAM, DRAM, SRAM, and LA RAM. See the entry for each separate one.

Technically, RAM (memory) is nothing more than electronic circuits (little *chips*) in your computer and in other kinds of equipment (like printers and *add-in boards*). How much RAM a computer has in measured in *megabytes,* and that is often what confuses people: you might know that you have an 80-megabyte hard disk and 8 megs of RAM, but what's the difference?

You probably have at least one *hard disk,* and you definitely have a collection of floppy disks, filled with all kinds of applications and documents. Disks are for storing large amounts information (your applications and your files) over the long term. The hard disk is known as *permanent,* mass storage.

A hard disk in your computer is like a filing cabinet in your office. Your computer, though, can't actually work from the hard disk; it would be like you climbing inside your filing cabinet to work. You probably take things out of the filing cabinet and put them on your desk, right? Well, the computer does a similar thing. When you ask it to open a particular application, the computer goes to the hard disk (the filing cabinet), finds the application, and puts a copy of that application into RAM, into memory. RAM, to the computer, is like your desk is to you. Likewise, when you start a new document, that document lives in RAM, just like a document would stay on your desk while you're working on it.

When you give that document a name and *save* it, the computer puts a copy of the document onto the hard disk, just as if you would put a copy into your filing cabinet for safekeeping. Whenever you save changes to that document, the version on your hard disk (in your filing cabinet) is updated with those changes.

RAM is *volatile,* meaning that everything in it will disappear as soon as the power is interrupted, either because you turn off the computer or there is a power outage, no matter how brief (even if there is just enough interruption to cause your lights to flicker). Anything that you changed in the document but did not yet save to the disk will just disappear, never to be seen again. For instance, if you saved your document at 2:10 in the afternoon, then worked for another hour and the power went out at 3:10, you will have lost all the work you did since 2:10. *Everything you stored on your hard disk is permanent, though, and is always saved intact.*

There is a finite amount of space in RAM. Some computers have as little as 640K (less than one megabyte), or even less. New Macs usually have at least 5 megs, and these days, most PCs come with at least 2 megs. An Apple Quadra can have up to 256 megabytes! The more RAM you have (the bigger your desk), the more stuff you can work on at one time. You can add more memory to most computers. It comes in the form of little *chips* you stick inside your machine. These days most computers require *SIMMs* (single inline memory modules), which are little pieces of plastic (*cards*) with several memory chips attached and you (or someone) stick them inside your computer. But there is always a limit to the amount of RAM your computer can actually use, or *address* (see *32-bit addressing* for more info).

Sometimes you might get a message telling you that you can't open an application (run a program) or complete a process because the computer is out of memory. Have you seen that message? You may have said to yourself, "How can I be out of memory? I have 8 megs left on my hard disk," but remember: *Memory isn't the same as hard disk storage space.* Some systems will just *crash* if RAM gets too full.

In some situations, it may be that your computer just doesn't have enough memory installed to do what you want it to. If your machine does not have a lot of RAM, you have to work smarter. Larger documents may not fit into the RAM you have, but you can sometimes split up the document into small ones that will. Save every few minutes, not only so you don't lose any work, but because the longer you work without saving, the fuller RAM gets. So you see, the more RAM you have (the bigger the desk), the more projects or the bigger the project you can work on at one time. And the bigger your hard disk (your filing cabinet), the more stuff you can file away in it.

If you find you are running out of memory often, there may be some things you can do to help yourself, short of buying and installing more SIMMs (memory chips).

Macintosh RAM

A Macintosh puts a lot more information into RAM than other personal computers, which is what makes the Mac a little slower than a PC running DOS, and also what makes the Mac much easier to work with. The data to run the *Finder,* to display the *icons,* to use the *fonts,* the *desk accessories,* and the *extensions,* etc., all loads into RAM on startup. All this information is waiting and accessible to you so you can just click buttons and press on menus to make things happen, instead of memorizing codes to tell the computer what to do. In fact, many of the items in your System Folder get loaded into RAM, so be careful what you put in there. Don't put anything in the System Folder that doesn't *have* to be there.

If you need to free up some of your existing memory, you can remove any *extensions,* desk accessories, or fonts that you don't use regularly (have your *power user* friend help you). If you have a lot of fonts, they shouldn't be in the System Folder anyway—buy and use a *font management utility,* either Suitcase or MasterJuggler.

If you run out of memory consistently while working in a particular application, such as PageMaker, try allocating more memory to the application itself. See the entry for *application heap* for step-by-step directions on how to do it.

Otherwise, the most common reason for running out of memory is that applications are left open when you thought you closed them. This is not a problem in System 6 (because you can't have more than one application open at a time), unless you are using *MultiFinder.* It is a constant problem in System 7. It is very easy to pop in and out of various open applications without even knowing it (as a beginner) simply by clicking in the wrong place. There are two things to remember and do:

> Whenever you stop working in an application, **quit** the application, don't just close the window you see on your screen. When you open that *application,* remember, the computer puts a copy of the app into memory. If you open a word processor, this is similar to taking a typewriter out of your filing cabinet and putting it on the desk. When you work on a *document,* that's like typing onto a piece of paper with the typewriter. When you close the *document,* that is like taking the piece of paper out of the typewriter and filing it away in the cabinet, *but the typewriter (the application) is still on the "desk," in RAM.* If you open another application, such as a spreadsheet, that is

like taking a large calculator out of the filing cabinet and putting it on the desk, along with the typewriter. If you did not actually **quit** from the File menu or by pressing Command Q, every application you open is still sitting in RAM. RAM, your "desk," is getting full. So when you are finished with the application, **quit**—don't just close the window.

Any applications you see listed here are still open and taking up RAM.

Also, regularly check the Application menu (shown to the left), which is the little icon in the top far right of the menu bar. The icon there changes depending on which application is currently *active.* Press on that menu and check the list at the bottom. If you see the name of any application other than "Finder," that application is open and is in RAM. Choose its name, and *even if you don't see the application on your screen,* press Command Q to quit. Trust me. Go back to the Application menu and do the same thing to any other application you don't need to use.

If you're not sure **how much RAM is in your Mac,** go to the Finder (the Desktop). From the Apple menu, choose "About this Macintosh...." You will get this dialog box:

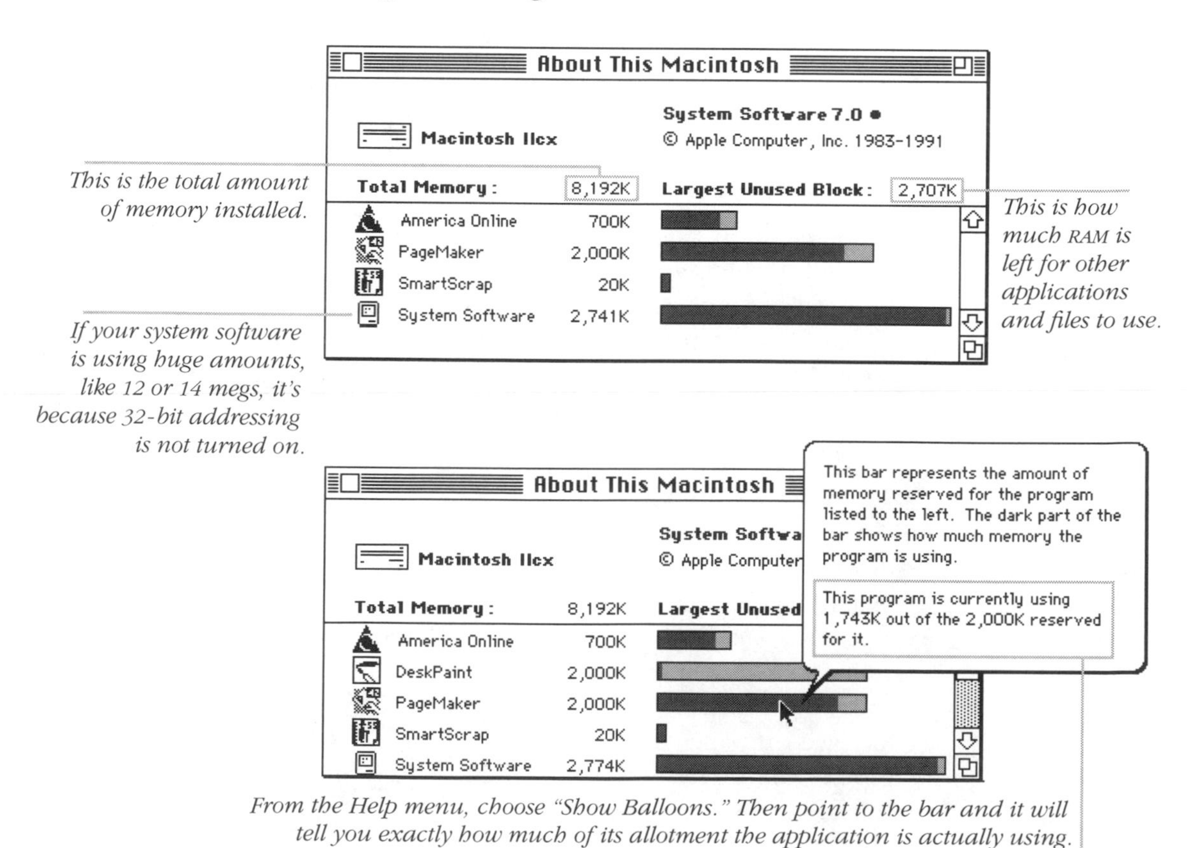

This is the total amount of memory installed.

This is how much RAM is left for other applications and files to use.

If your system software is using huge amounts, like 12 or 14 megs, it's because 32-bit addressing is not turned on.

From the Help menu, choose "Show Balloons." Then point to the bar and it will tell you exactly how much of its allotment the application is actually using.

PC RAM

Understanding RAM on a PC gets pretty complicated, especially if you're using DOS instead of Windows. If your system is working okay already, don't bother with this information because it's pretty technical. But if you're running out of memory—if certain programs won't run or you can't create large files—you need some background knowledge before you march down and buy more memory. You may already have enough!

> If you are running out of memory on a PC, you may have too many *memory resident programs (TSRs)* loaded. *Unload* the ones you don't absolutely need, or "load them high." Otherwise, your computer may not be able to get to the memory you do have. Be sure you run the right "extended memory manager" or "expanded memory manager" (later in this definition). If you are working in *Windows* on a PC, you may have too many applications open—close something.

Here's the scoop: You can think of your computer's memory as a series of mailboxes, each containing a single scrap of information. Mailboxes have addresses, and so does each location in memory. This concept is important because the address of a given memory location determines what your computer can do with it. In DOS PCs, memory is divided into four regions based on address, and two of the four overlap. I warned you this was going to be technical:

> *Conventional* memory, the first 640K of RAM
>
> *Upper memory*, the memory between the 640K and 1024K (1 megabyte)
>
> *HMA* memory, the 64K of memory starting at 1024K
>
> *Extended memory*, everything above 1024K (including HMA)
>
> In addition, there's a fifth type of memory that can't be pinned down to a specific region. *Expanded memory* is actually located at addresses outside the first 1024K, but your computer thinks it is inside that first megabyte.

Conventional memory, the first 640K of memory addresses, is where the PC ordinarily keeps DOS itself and the programs and files you're actively working with. Because large programs and files won't fit, the last address in conventional memory is known as the "640K barrier."

Upper memory, the region between 640K and 1024K, contains memory addresses set aside for the system's own use, including the information representing what you see on the screen and the instructions in the *ROM BIOS* chips that tell the machine how to operate. However, every PC has lots of unused addresses in upper memory. With the DOS commands DEVICEHIGH and LOADHIGH you may be able to rig things to make the upper memory addresses available for your *TSRs* and *device drivers*, freeing up space in conventional memory for larger programs and files (see your DOS manual).

Extended memory, which begins at 1024K, is found only on PCs with *80286* or newer *microprocessors*. Although the typical PC now comes with 2 megabytes of RAM, DOS still can't use extended memory except the first 64K HMA (see below) memory. But some *programs* do. These programs have the ability to suspend DOS and switch the microprocessor into another mode that recognizes extended memory. This is how Windows works, and this technique is also used by programs like Lotus 1-2-3 Release 3, and Paradox. The key is that unless you have a program that knows how to use extended memory, it won't help at all to buy more, and adding more extended memory will only benefit that particular program.

HMA memory, the first 64K of extended memory, can be used as conventional memory by DOS and some programs (HMA stands for High Memory Area). To make this memory available, you must run a special piece of software called HIMEM.SYS that comes with DOS and Windows, or another "extended memory manager." Even then, a program can only use the HMA if is specifically developed to do so. If you have DOS 5 or DOS 6, the best way to use HMA is for DOS itself with the DOS=HIGH command.

Expanded memory, or EMS, was developed as a way to get around the 640K barrier in the old days, when PCs with *8088* processors were common and extended memory wasn't available. It works by setting aside a section of upper memory which becomes a sort of "window" through which your programs can access all the expanded memory in your computer. When a program wants to store information that doesn't fit in conventional memory, it sends the data to an address in this expanded memory window. Special software, an "expanded memory manager," steps in, copying the information into an area of expanded memory. When that part of expanded memory fills up, the memory manager "moves the window" to a different area of expanded memory. Your software keeps sending information to the window's address, while the memory manager makes sure that the information goes to an unused spot in expanded memory.

The only reason to use expanded memory is if you have a program that requires it—extended memory is faster and easier to work with. Some applications use expanded memory to store large files. Some TSRs can load themselves into expanded memory, which saves space in conventional and upper memory. Expanded memory isn't built into PCs, so if you want it, you have to add it:

If your PC has an 8088 or 80286 processor, you add expanded memory to your system on a *board.*

If you have an 80386 or newer processor, you can run a special kind of expanded memory manager to convert extended memory into expanded (DOS and Windows come with one called EMM386.EXE).

If you run Windows in 386 Enhanced mode, you don't need to do anything, because Windows automatically supplies expanded memory to any application that wants it, even a DOS one.

Clear as mud?

RAM cache

Ram cache (pronounced "ram cash") is the less technical term for a "processor cache." See *cache.*

RAM disk

A **RAM disk** is not really a disk at all—it is *memory* (*RAM,* in the form of memory *chips*) that has been set aside **pretending** to be a disk. As long as the power is on, you can use a RAM disk just like a real disk drive—you can copy files to and from the RAM disk, display the RAM disk's directory or folder on your screen, and run any programs you've stored there. A RAM disk may be an external *SCSI device* that looks similar to a regular external hard disk (but is very expensive and is slowed down by the limitations of the SCSI connection), or it may be a lot of extra memory that you set aside with special RAM-disk software, or it may be a *memory card* that you add to your computer.

The big advantage of a RAM disk is that because it's purely electronic (rather than a physical platter), it's much, much faster than any mechnical disk—hard, floppy, or optical. Normally, if a large application or program can't fit into regular memory all at once, the computer has to go to the hard disk from time to time and retrieve what it needs to continue working, which slows down the entire process. But if you copy the application to your RAM disk and run it from there, you'll barely notice when the program fetches more instructions from the RAM disk.

If you ever have to copy the same set of files to a bunch of floppy disks, a RAM disk will speed things up dramatically. First copy the files to the RAM disk, and from there to the floppies. On the other hand, it's not such a good idea to use a RAM disk for document files that you plan to work on and change. Remember, RAM is *volatile,* meaning any information that is in

RAM disappears when the power goes out or the system *crashes.* If you make changes in a document file parked on a RAM disk, you still must be careful to save the file to a real disk. Unless you're willing to go long periods without saving "permanently" and risk losing a lot of work, putting your documents on a RAM disk will actually slow you down.

RAM disks have another disadvantage: the memory they use is unavailable for other programs. Unless you always use the same programs, your system will run faster, on the average, if you use extra memory for a "disk cache" (see *cache*) instead of a RAM disk.

RAMDRIVE.SYS

Recent versions of MS-DOS and Windows come with **RAMDRIVE.SYS,** a software *device driver* that will turn part of your PC's memory into a RAM disk. If you decide you want to set up a RAM disk, place the line DEVICE=RAMDRIVE.SYS in your *CONFIG.SYS* file (if the RAMDRIVE.SYS file is not in the root directory of your boot drive, you need to put the *path* name right after the equals sign). See *RAM disk.*

range

A **range** is a collection of *contiguous* (adjacent, connected) items. For instance, in a spreadsheet you can select a range of cells to add to a formula or to format in a certain way. A spreadsheet range is often written as the top left cell name, a colon (or some other symbol), and the bottom right cell name. For instance, A1:B5 (or A1..B5) would include all the cells from A1 to A5 and from B1 to B5.

A range of text refers to a selected group of adjacent characters. For instance, you might want to do "range kerning," where you select a range of text and apply the same *kerning* value between each of the letters.

raster

A **raster** is another word for a *bitmapped* graphic. The term comes from television technology, where a raster is the pattern of horizontal lines traced by the scanning of the picture tube's electron beam. See *rasterize.*

raster graphics

Raster graphics are the same as *bitmapped graphics,* which means the graphic image is composed of many tiny dots, as opposed to *object-oriented* or *vector graphics,* where the image is composed of collections of lines. Please see *bitmapped, raster,* and *rasterized.*

rasterize, rasterizing

To understand what rasterizing does, first you need to know a little about the images in the computer: *Bitmapped (raster)* graphics and *fonts* are created with tiny little dots. *Object-oriented (vector)* graphics and fonts are created with *outlines. Output* devices, like printers (except for some *plotters*) and monitors can only print or display images using dots, not outlines. This means that when an object-oriented graphic or font is output to a printer that prints in *dots per inch* (as most of them do) or to a monitor that displays in *pixels* (as most of them do), the outlines must be turned into dots. This process of turning the outlines of the objects into dots is called **rasterizing.** Everything you see on your monitor has been **rasterized.** Everything you print has been rasterized.

When you print object-oriented and *PostScript* images and fonts to an *imagesetter,* the information about building those straight lines goes through a **RIP (raster image processor),** a piece of hardware that stands between the computer and the imagesetter (printer). The software in the RIP turns the straight lines into the dots that the imagesetter will print, in *resolutions* like 1270 or 2540 dots per inch.

When you output (print) to a laser printer that understands *PostScript,* the computer chip inside the PostScript printer rasterizes the images so they can be printed in dots, usually with a resolution of 300 to 600 dots per inch.

When the image is displayed (output) on the monitor, it has actually been rasterized so that it could be created out of the *pixels* on the screen.

And if you have a *non-Postscript printer,* you should read the definition for *Adobe Type Manager* to better understand how that software rasterizes your fonts, both to the monitor and to the printer.

ray trace

Ray tracing is an incredibly complex method of producing shadows, reflections, and refractions in high-quality, three-dimensionally simulated computer graphics. Ray tracing calculates the brightness, the reflectivity, and the transparency level of every object in the image. And it does this backwards. That is, it traces the rays of light back from the viewer's eye to the object from which the light was bounced off from the original light source, taking into consideration along the way any other objects the light was bounced off or refracted through or absorbed by. whew.

RDBMS

RDBMS stands for **r**elational **d**ata**b**ase **m**anagement **s**ystem. Please see *relational database* and *DBMS.*

read

If a computer can **read** the disk or tape or file, it means it knows the file exists and can open it for you. When a computer reads an item (as we would read a disk or play a tape), the disk, tape, or file is not changed in any way—just the information is absorbed and perhaps acted upon. Compare with *read-only* and *read/write.*

readable, readability

Readability is different from *legibility.* Readability refers to how easily you can read a large amount of text (legibility refers to how easily a short burst of text, such as a headline or sign, can be read). Readability depends on the design of the typeface, on how long the lines are, how much space is between the lines, the type style (bold, italic, etc.), and other factors. The most readable typefaces are those that are "invisible"—they have no distinguishing features that make your eye stumble over the words. Extensive studies repeatedly show that *serif typefaces* are more readable than *sans serif.*

THIS IS NOT A VERY READABLE TYPEFACE.

Readability can also be a measure of the grammatic complexity of a passage of text, and how many big words it has. In this sense, readability is often stated in terms of grade level.

Read Me, README, ReadMeFirst, ReadMe file

Whenever you buy or *download* a new software program, the first thing you should do is check out the Read Me file. Almost every software program for the PC comes with a file named READ.ME, README, README.TXT or something similar. On the Mac, it will usually be called **ReadMeFirst** or **Read Me.** If the software came on disks, you'll usually find the Read Me file on the first disk.

Read Me files are the software version of those Errata pages they put into good (but slightly defective) books. The Read Me file contains corrections and additions to the printed manuals—information that the software company discovered too late to print in the manual. Perhaps it tells you of some *bugs* that didn't get fixed in time, or known incompatibilities with other software or certain hardware, or new features that were added at the last minute. Often the information is critical to running the program properly or even to getting it installed right in the first place. It's always a good idea to read these files.

Occasionally someone thinks I am so smart for knowing something, when all I did was read the ReadMeFirst file. Did you have trouble with your fonts in PageMaker 4.2? Read the ReadMeFirst file. Wanna *really* impress people? *RTFM.*

Most Read Me files are *text-only (ASCII)* files, so you can open them with any word processor or file viewing program.

On a Mac, you will often find Apple's tiny little word processing program, *TeachText,* included on the disk; if you double-click on the ReadMeFirst file, it will often open itself in TeachText.

With some Windows programs, the Read Me files are in Write format (Write is the word processor that comes with Windows).

read-only

If a **disk** is **read-only,** that means you can look at what's on the disk, you can open any files on the disk, you can print information that's stored on the disk, but you cannot make changes to anything on the disk (you cannot "write" to it). On some read-only disks you may *think* you can make changes—you can open the document, you can use the tools to edit the text, but you'll find you will not be able to save those changes to the original file.

A disk may be read-only because that's the way it was created, such as a *CD-ROM*. A disk or a **file** may be **read-only** because you *locked* it or limited the *file-sharing* options.

Read-only also applies to *ROM* (which stands for **read-only memory**). ROM is memory that does not lose its information when the computer is turned off or crashes (as opposed to *RAM,* which *does* lose everything in it). ROM contains items that make the computer function, and neither you nor the computer itself can erase or change what is in it; the computer can only "read" and use what is in it.

read/write

If a **disk** is **read/write,** it means you can *access* and view (read) any of the information on the disk (open any files and look at them), and you can also *write* to the disk, meaning you can make changes to any of the files on the disk. Most disks are read/write because that's the point, right? But if you *lock* a disk, it becomes *read-only.* CD-ROM disks are read-only. There is also a disk technology that is *write once, read many (WORM)* on *floptical* disks.

Read/write also applies to **memory,** such as *RAM (random access memory).* The computer can "read" information that has been stored in RAM, and the computer can "write" to RAM, meaning it can put things into RAM (as opposed to *ROM,* which is *read-only memory.*)

And **read/write** can apply to individual **files,** because you can lock or limit the access and use of any file.

real time

Real time (sometimes spelled "realtime" or "real-time") means the computer can work fast enough to fool you into thinking that the activity is taking the same amount of time it would take in Real Life, according to your perception of time. For instance, a real-time flight simulator would seem to take as much time to take off from the ground as a jet would. Or when you are sitting at your computer chatting with someone *online* (you are at your computer typing and she is at her computer typing, both through your *modems*), then you are chatting in real time, as opposed to writing a letter to her *e-mail* box which she will pick up the next day.

real-time clock

A **real-time clock** is the clock inside virtually all computers that keeps the time (time as in time of day). It is almost always battery-powered so it keeps time even when the computer is turned off. The *system* uses this clock to "time stamp" when you created your files, and certain programs allow you to automatically insert the date and time into documents.

This real-time clock is not to be confused with the *clock* run by the *CPU (central processing unit,* the *chip* that runs the computer). The CPU clock determines the *clock speed* of the machine, which is a major factor in how fast the computer works.

Real World

The **Real World** is a vague place where they tell us Life happens. It's where teachers and parents always threaten us of things that are different and someday we will find out and be taught Great Lessons. Where is the Real World, I wonder? I'm hoping to avoid it.

reboot

Reboot means to start your computer over again. You don't need to *power down* (turn off the *power*) to reboot—you just tell the computer to *restart*. Ideally, all applications should be closed. On a Mac you choose "Restart" from the Special menu. To reboot on a PC, press the Ctrl, Alt, and Delete keys simultaneously.

rebuild the desktop

On a Macintosh there is an invisible file that keeps track of every disk you ever stick in your computer and every file you ever take a look at. This file is called the *Desktop file*. Every hard disk and floppy disk has one. It is possible, occasionally, for this file to become damaged or infected by the *WDEF virus*. And it gets bigger and bigger until you notice your hard disk slows down. As a preventitive measure, it has long been recommended that you **rebuild the desktop** regularly—like every couple of months. Do this by holding down the Command and Option keys when you start, restart, or insert a floppy disk. You will get a message asking if you really want to rebuild the desktop. Click OK. Doing this will destroy the WDEF virus, usually repair any damage to the Desktop File, and your disk will run faster. Unfortunately, it will also eliminate any messages you may have typed into any file's *Get Info* window.

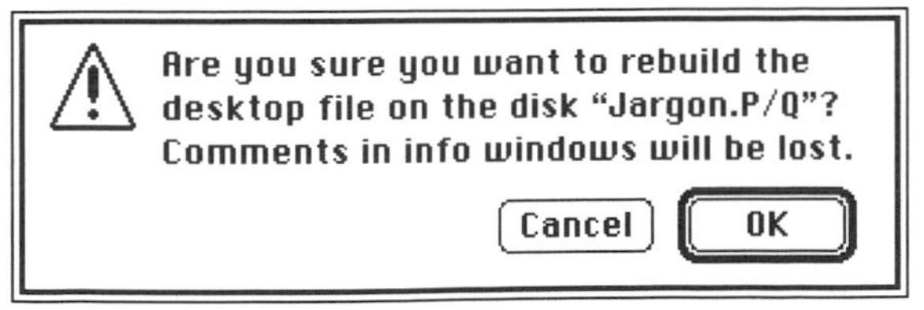

This is the alert box you will get when you rebuild the desktop.

record

In a *database,* you keep track of information. It's kind of like what you used to do on index cards, writing someone's name and address, perhaps, and the names of their kids and their birthdays, where they like to eat dinner, and if they like to eat broccoli. The equivalent of the whole index card, with all the details about one person, is called a **record.**

If you had a database of your video cassettes or your books, each record would hold the information about one item. Within each record, the information is broken down into separate *fields*—one field for the author's last names, one for the first names, one for the publisher, perhaps one for the date of publication, and a field that contains information or comments about the book. You can have the database software find all the records that have matching entries in certain fields, like all the books written by a certain author.

This entire "card" of information is one record.

The entire collection of all these "cards" is a table.

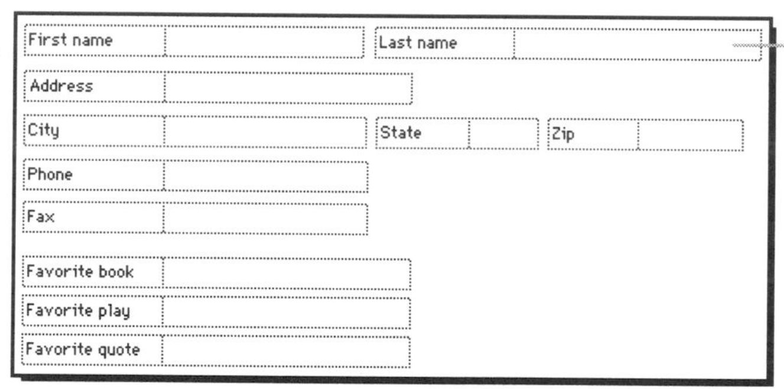

Each one of these spaces is a field. The field title always remains the same; the information, or data, in the fields will be different for each "card," or record.

recover data

When you have a serious *crash* that destroys some of your files, or if you accidentally reformat your hard disk, you would probably like to **recover data,** meaning you would like to get back those files you lost. Sometimes it's possible, sometimes it's not. It's actually possible more often than you might think. You can try to recover the data with a *data recovery* program, which everyone should have on their shelves. If you can't do it, call the local computer stores or hard disk outlets and ask around for someone who specializes in data recovery. Sometimes it takes special hardware to find information that is no longer in sight.

recursion, recursive

Did your spreadsheet yell at you because you did something **recursive?** That usually means you tried to create a formula that used itself to solve it. It's kind of like using a word in its own definition.

redraw

Whenever you change the screen image, turn a page, move an object, edit text, etc., the screen **redraws** the text and images in the next position. For simple text and images this doesn't take long at all. For text with a great deal of fine formatting, like intricate word spacing and tracking and superscript repositioning, it can take longer. For color graphics, it can take a seriously long time. For *high-resolution* color photo images it can take what seems like hours. Most graphic applications, especially those that work in color, offer you the option of displaying graphics in high-resolution or not. The lower the resolution you choose, the faster the screen will redraw (but the image will still print in high resolution).

Sometimes you may want to force the redraw of the screen. There are various ways to do this, depending on your program, but if you are in a window you can usually click in the little box in the upper right corner. This may resize your screen, also, in which case you can click the button again to return to the previous size.

reformat

When referring to disks, **reformat** means to prepare a previously used disk for reuse. This will effectively erase any existing data on the disk, so make sure that's what you really want to do.

Reformat also means to format (or apply attributes to) something over again. It may be a redesign of the layout of your database or the appearance of your text. Perhaps the text on the page was aligned in the center and you decided to reformat it to align right.

refresh rate

The computer is constantly renewing the screen, something like 50 to 75 times a second. How fast it does this is called the **refresh rate.** It's called "refresh" because the electron beam inside the computer is re-energizing the phosphors on the screen.

You'll see the refresh rate listed in ads as a *frequency* in *hertz (Hz)*, meaning how frequently the image gets renewed in one second. A good monitor has a 70–75 Hz refresh rate, or better. Less than 65 Hz will often show an annoying flicker.

When you buy a monitor, be sure the refersh rate matches the output of your video circuits (or get a *multiscan* monitor that matches many different frequencies).

registered user

You become a **registered user** of a software package when you buy the product ***and*** send in the *registration card* to the company that makes the product. Once you are in their files, you are a bona fide registered user. Usually you have to be a registered user before you can ask for help with the product over the telephone. Sometimes the vendor sends newsletters or product update information about new or updated versions to registered users. Or you'll at least get pitches to buy the upgrades, and occasionally even free software *upgrades* without even having to scream about it (sometimes you just get sold to a mailing list). If you are not a registered user, you cannot take advantage of any discounts on upgrades or *side grades.*

Notice that you are not a registered "owner" of the software. You are a registered owner of your car, which then belongs to you and you can sell it to whoever you like, but you never really own your software. See the definition for *license* to see what you actually get when you buy software.

registration card

When you open a new box of software or hardware, there is almost always a **registration card** enclosed, a little postcard that you fill out and send in. Once you send it in you are a *registered user* (see above).

registration number

When you buy a software program it usually comes with a serial number identifying your particular copy of the program. This number is sometimes called the **registration number** since the company uses it to keep track of you when you register the software. You can find the number stamped onto or printed on a sticker glued to the box somewhere, or perhaps on the first page of the manual, or maybe it's printed on one of the floppy disks. You should always send in your *registration card* with this registration number on it. The software company will keep you informed of updates and *bugs* (ha) and other interesting information. (Supposedly, anyway.) I think many of the vendors I have bought from have just sold my name to mailing lists.

relational database

A *database* is just a collection of information stored on a computer. In the simplest type of database, the *flat file* kind, all the information is organized in one way. It's like filling in a set of pre-printed index cards, all of which have the same blanks for name, address, phone number, and so on. The entire collection of "index cards" is called a *table;* each card in the collection is a *record*. Each blank space on the card where you fill in certain information, like a name or a zip code, is a *field*. A flat-file database consists of a single table. (See the illustration in the definition for *record* a few pages back.)

A **relational database** goes beyond a simple flat-file database that stores information for you and enables you to make selective lists and reports. A relational database can share information across several tables, meaning you don't have to retype similar data into related files—you enter the data into one table, and that information automatically gets entered into any other table that needs that particular data ("automatically" meaning that someone previously *programmed* it to do that).

A relational database can organize information much more efficiently than a flat-file database in many situations. Imagine computerizing your mail-order knick-knack business so you can keep tabs on which customers like little ceramic figurines and which ones like fancy marble eggs. Of course, you need a database table for the customers themselves, with their names and addresses and so on, along with their customer numbers. But it's not practical to keep track of who bought what in that same table—each customer will make a different number of purchases, so you can't plan a preset number of fields to record them.

The solution is to have a separate table for sales. Here, each entry represents the sale of an item to a particular customer, with one field for the customer number, one for the item, and one for how many of each item the customer bought. Whenever you want to know the buying patterns of each customer, you can have your database software make a report that combines information from both tables, using the customer number field to link them together.

A relational database management system (RDBMS) is the software that lets you create relational databases and extract information from them. Relational database software is larger, more expensive, and more difficult to learn than flat-file database programs, but you need a relational program to do major business-oriented work.

removable hard disk

Removable hard disk is another name for *cartridge hard disk,* a hard disk in cartridge form that you can remove from its drive. It's essentially like having a giant floppy disk.

You need a removable hard disk *drive* (around $400 to $800) to put the cartridge hard disk into, then you can use the removable cartridges. Each cartridge is 44 or 88 megabytes of hard disk space and is about the size of a thick pancake (they cost about $45 to $80 for a 44MB cartridge).

So the concept works like this: You have your internal hard disk with all the files you need regularly stored on it. You have an external removable hard disk with a stack of cartridges. When you finish a major project, you can use the removable hard disk to store the entire thing, rather than a whole stack of floppies. You can store an entire book and all its graphics on one disk. You can take the entire publication, all the linked graphic files, and all the fonts to the service bureau for output. If you work with large photographic files, a cartridge hard disk is essential. These cartridges are so cool—they make life so much easier.

You may hear people refer to any removable cartridge as a "SyQuest." SyQuest Technology, Inc., does manufacture drives and disks, but so do several other companies. All "SyQuest" drives, no matter who makes them, use a SyQuest mechanism inside. Bernoulli also makes a removable disk drive and cartridges, but until recently they were so much more expensive, (although more dependable), that they just weren't as common. SyQuests and Bernoullis cannot interchange disks. Also see *Bernoulli box, Ricoh,* and *SyQuest.*

render, rendering

Rendering is the process of applying color, shading and shadows to an image to make it appear realistic. Much of this can be done automatically on the computer, converting outline drawings to fully formed and solid objects.

required space

Some people and some programs use the term **required space** to mean the same thing as a *hard space,* or *non-breaking space.* See *hard space* for details.

ResEdit

ResEdit (pronounced "rez edit") is a software *utility* that gives you, as a *mortal,* the power to modify many parts of the Macintosh, including menus, icons, desk accessories, parts of the System, dialog boxes, etc. ResEdit stands for Resource Editor, which is what you are doing—editing *resources.*

Everything you ever read about ResEdit includes the caveat/warning/caution/careful/dangerAhead message to work only on a *copy* of the resource you want to edit. It is very easy to cause permanent damage to your entire System. If you want to play with ResEdit, which is a sign that you have advanced to the *power user* level, get *The BMUG guide to ResEdit, Zen and the Art of Resource Editing,* published by Peachpit Press. You will be inspired and instilled with a powerful sense of control over your Macintosh.

reserved character

Most *operating systems* and some software programs won't allow you to use certain **reserved characters** when you name a *file* or create a *macro.* Generally, the reserved characters are just ordinary punctuation marks (sometimes letters or numbers). They are also known as *illegal characters.* You're not allowed to use these characters because the *operating system* or your software has already claimed them for its own special use.

On a PC, for instance, DOS won't let you use any of the following characters in a file name:

^ * + = [ ] ; : " \ / , ? > <

Spaces can't be used either. You *can* use a period, but only to separate the main part of the file name from the three-character extension. Other punctuation marks are okay to use, including ! @ # $ % & ( ) - _ ' ` and ~.

On a Macintosh you should never *start* a file name with a period (.), even though the Mac will let you do so. This is a special message to the printer, and its existence at the beginning of a file name can corrupt the file (how do you think I discovered this fact?). And don't use an exclamation point (!) either. It's perfectly safe to use a period at any other point in the name.

It's safe to use a blank space at the beginning of a name; in fact, if you start a file name with a space, it will always be at the top of the list, and two spaces will come before one space.

reset switch, reset button

For information about the **reset switch** or **button,** see *programmer's switch.* A reset switch is not the same as the programmer's switch, but they are both defined in that entry because they are sort of connected.

resident

An item that is **resident** is one that lives in your computer or some other device on a permanent or temporary basis. For instance, resident *fonts* are the fonts built into the printer—they reside there (they live in the *ROMs,* the *read-only memory). TSR* programs *(terminate and stay resident)* live and work in the computer's memory, even when you're not actively using them.

resize box

The **resize box** is the little box in the **bottom** right corner of a Macintosh window. If you position the mouse in that spot and press-and-drag, you can resize the window to any size. The resize box is different from the *zoom box,* which is in the **upper** right of the window.

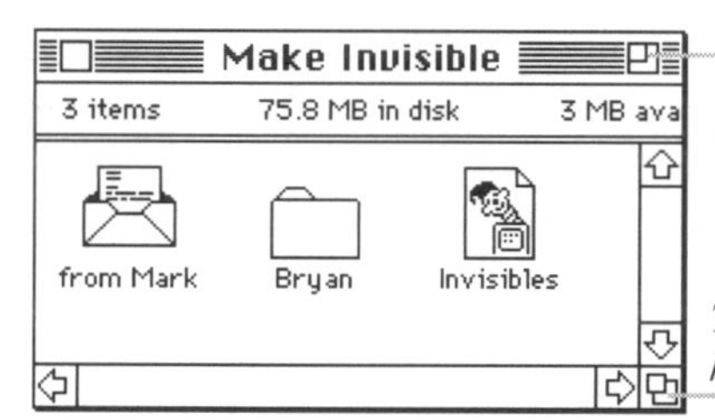

This is the zoom box. Click once in it to instantly zoom the window larger or smaller.

This is the resize box. Press-and-drag here to resize the window.

resolution

In a general sense, the term **resolution** pertains to how sharp and clear an image looks on the screen or in print, and how much detail you can see; it refers to how finely resolved an image appears to be. In a low resolution image, fine lines look coarse and overlap one another, and curves are obviously jaggedy. The higher the resolution, the less you notice these defects. If you're being quantitative, you can express the resolution of a particular screen, printer, or *bitmap graphic* in numbers (see the next two definitions).

resolution (monitor)

The **resolution** of a **monitor** is not actually perceived through the number of pixels per inch. How clearly an image is "resolved" on the screen is determined by how easily the computer can fool our eyes. Our eyes get fooled when there are so many colors (or shades of gray) that we can't easily differentiate where one color blends into another. When there are only two colors on the screen, black and white, an image is not resolved very well because our eye can clearly see the separation between black and white. When there are shades of gray, an image appears to be more resolved, because our eyes can't tell as easily where one gray shade

Black and white.

4 levels of gray.

16 levels of gray.

256 levels of gray.

blends into the next. When there are 16 million colors, our eyes perceive that the image is highly resolved, or looks very real. So the same monitor with the same number of pixels per inch can actually show varying "resolutions," depending on how many shades of gray or colors are displayed. See the screen shot examples to the left that show the same image on the same monitor with varying levels of gray values; as the numbers of grays increase, the "resolution" increases.

Early PC monitors could display only a limited number of colors, and Macs with built-in monitors only display black and white, with nothing in-between. However, Mac color and grayscale monitors and newer PC monitors (beginning with VGA models), can display an incredible range of colors or gray shades. In this case, it's not your monitor that determines how many colors or gray shades you see on the screen—every one of these *monitors* is capable of displaying images in high resolution. The number of colors or grays you see depends on the kind of *video card* you use, how much *memory* it has, how much *RAM* the computer has, and whether or not the *operating system* can handle large numbers of grays or colors.

Other factors that affect the clarity or resolution of an image on the screen are the *dot pitch* of the monitor, as well as how tightly focused the beams are that come from the three separate electronic color "guns."

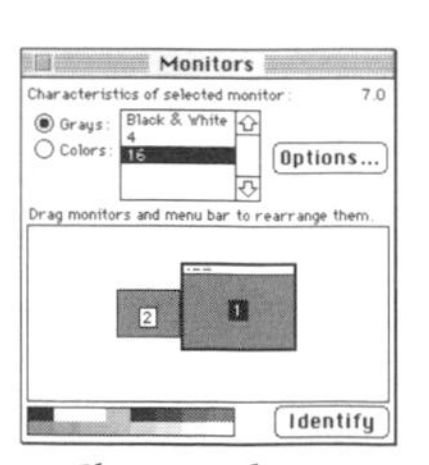

Change the gray or color level, then click on the image to view the changes. It's really quite remarkable.

If you have a color or grayscale monitor you can experiment with the varying "resolutions" on your monitor. Open a color photograph or painting in an application like SuperPaint or Photoshop or Color-It (you can usually find a photo in the folder that contains the sample images that came with the application). While the image is on your screen, open the Monitor Control Panel. Notice how clearly the image resolves or doesn't resolve depending on the number of colors or levels of gray you choose to use. If you don't even have the option to display "millions" of colors, it's not because of your monitor—it's because you don't have enough memory in your computer, or perhaps the video card in your computer is not capable of working with that number of colors, or your operating system cannot work with *24-* or *32-bit color*. (You can only get a maximum of 256 gray levels, which is the maximum number a human eye can distinguish anyway.)

Have you ever wondered why the ruler on your screen is not exactly one inch? Or do you sometimes think what prints out looks smaller or larger than it does on the screen? The original Macintosh screen, such as on a MacPlus, has 72 pixels per inch. The Apple High-Resolution 13-inch monitor has 69 pixels per inch. My 21-inch grayscale monitor has 76 pixels per inch. Well, software applications are written based on the original 72-pixels-per-inch size. When your word processor or page layout application opens up, it says, "Okay, I'm supposed to put the zero in the ruler right here at this *x/y coordinate*. Then I'm supposed to count over 72 pixels and put the one-inch mark there." On a screen with 72 pixels per inch (ppi), the ruler measures exactly one inch. But on a 69-ppi monitor, the one-inch mark is still sitting on the 72nd pixel; this means one inch on the screen is actually ***larger*** than one real inch. On my 76-ppi monitor, the one-inch mark is still sitting on the 72nd pixel; this means one inch on my screen is actually ***smaller*** than one real inch.

So don't be fooled into thinking that a greater number of pixels on the monitor means higher resolution—monitor resolution is not like printer resolution. For one thing, *of course* a 21-inch monitor has more pixels than a 13-inch monitor! It is possible to get a monitor that can display several different pixel-per-inch standards, such as 64 ppi, 72 ppi, and 90 ppi. Assuming you don't change the magnification or zoom level, what you will see at 64 ppi is a closer view of the same image. It will *appear* to be in lower resolution because the pixels are larger and you can see them more easily; you eye is not fooled into "resolving" the image clearly. What you will see at 90 ppi is a bird's-eye view, as if you had stepped back a few feet. It will *appear* to be in higher resolution because you are looking at it from farther away. It will also be much smaller. Your menu bar, also, will be in a bird's-eye view—the letters will be tiny, because they are still being told where to position themselves according to the 72-pixel grid.

The resolution quoted for a PC monitor tells you how many pixels there are on the entire screen, as in "640 by 480 pixels." The first number is always how many pixels fit across the screen, horizontally; the second number is how many pixels fit vertically, down the screen. In *graphics mode,* the resolution of a *CGA* monitor is 640 by 200, while a *VGA* monitor in graphics mode is 640 by 480. With more than twice as many dots on the screen, the VGA image will obviously look a lot sharper (if the two monitors are the same size). Also see the monitor section in Appendix A, How to Read a Computer Ad.

resolution (printer)

When you're talking about printers, **resolution** (used to describe sharpness and clarity), is measured in *dots per inch (dpi).* The measurement of dots per inch refers to how many dots are printed horizontally in one inch. For instance, in 72 dots per inch (dpi) there are 72 rows of dots in one inch. This output resolution ranges from low resolutions such as 74 dpi on an Apple ImageWriter, to 360 dpi on an HP DeskWriter, to 2540 dpi on a

Linotronic *imagesetter*. The more dots per inch, the more clearly the image is resolved. Also see *lines per inch,* which affects printed resolution.

Jimmy photo and text:

300 dots per inch

53 *lines per inch*

Jimmy photo and text:

1270 dots per inch

133 *lines per inch*

resource

In software, a **resource** is a well-defined collection of information used by a program for some specific purpose. In *graphical user interfaces* like the Macintosh or Windows, a program may display many different little pictures, or icons; the data that defines each icon is a resource. Likewise, when a program lets you pick commands from a series of menus, the contents of each menu is a resource. Each dialog box you see is also a resource. There are many other types of resources, including fonts, cursors (mouse pointers), and sounds.

Applications have their own resources that they alone use and apply just while that application is running (see *resource fork*). But some resources are system-wide, meaning they are or should be available no matter what application is running. On the Mac, these include *fonts, desk accessories, alert sounds,* and *FKEYs.*

Apple puts out a free *utility* called *ResEdit* (short for Resource Editor) which allows you to edit parts of most Macintosh resources. You can change icons, change file types, change menus, etc. In fact, it is very easy to destroy your entire System. Be very careful with ResEdit. Check that definition for further info.

Utilities similar to Apple's ResEdit (above) are available for Windows, but you have to buy them separately.

Another definition of the word **resource** has to do with hardware, and refers simply to the parts of your computer system: the screen, the keyboard, the printer, the memory, even the *processor (CPU)* itself. You're most likely to come across the term used this way if you work with Windows.

RGB

The acronym **RGB** stands for **r**ed, **g**reen, and **b**lue. All color monitors use red, green, and blue input signals to create the colors on the screen. There are several different ways to get these signals into the monitor. In the method called RGB, the three signals are sent separately (as opposed to combining them into one "composite video" signal, the kind used by television sets, VCRs, and *Amigas*).

An RGB monitor (any color monitor, actually) has three dots in each *pixel:* a red dot, a green dot, and a blue dot. Inside the monitor are three electronic "guns"; each gun fires electrons at one of the colored dots. The more electrons that fire away at a particular color dot, the brighter that dot glows. Combining varying intensities of the three colors creates the different colors we perceive.

If all three color guns hit the pixel with full intensity of color, the pixel appears white. If none of the guns hit the pixel at all, the pixel appears black. I know, that's different than what we are used to doing with our crayons, but monitors use "transmitted" light instead of "reflected" light. That is, the color is not being reflected off of an object, like off of an apple, into our eyes. Rather, the colors are being sent (transmitted) through the monitor straight to our eyes. With transmitted light, combining equal amounts of red light, green light, and blue light creates white light. The "additive primary colors" are red, green, and blue. We grew up with "subtractive primary colors" of red, yellow, and blue. If you are going to work with color images on a monitor, you will have more control over your work if you learn how to use the additive primaries.

Any color monitor is capable of displaying millions of colors. It's the computer and its *video card,* not the monitor, that determines whether you see millions of colors, or only hundreds of colors, or only eight colors. Add more *memory* or switch to a better video card and your monitor will be able to display more colors, which in turn will create the illusion of higher resolution (see *resolution [monitor]).*

ribbon

The **ribbon** is what Microsoft Word calls the little display at the top of the ruler that allows you to click on icons to change font, font size and, in some styles, to create sub- and superscripts, columns, and a few other items. Its purpose is to make life easier so you don't have to pull down the menus all the time. You can customize it with the features you want.

This is the ribbon. You can just click on icons to format the text or drawings.

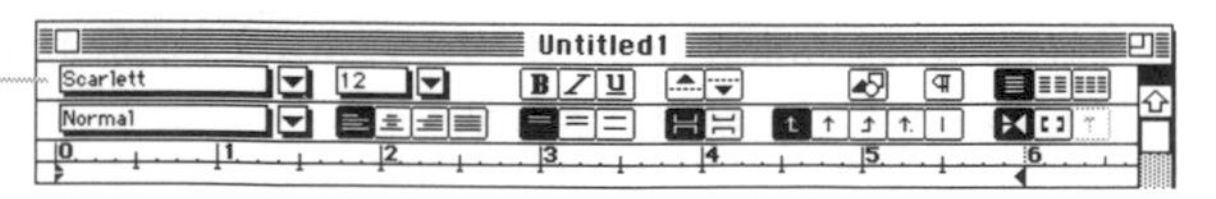

Rich Text Format

Rich Text Format (abbreviated as RTF) is an optional *file format* for text files. Microsoft developed the Rich Text Format as a universal means for exchanging text between different software programs, and even different types of computers. RTF files preserve the appearance of the text—characteristics such as fonts, boldface, italics, and colors—as well as its content. They even retain information about the layout of the paragraphs, such as the margins, the line spacing, and the *style* assigned to each paragraph. But unlike most word processor file formats, the RTF format stores all this information in a readable, if terse, coding system (\ul means underlined text, \i means italics). Even if you don't understand the codes, at least there are no weird symbols to deal with.

Despite Microsoft's intentions, the RTF is by not yet universal. Since so many text-oriented programs understand the Microsoft Word and WordPerfect formats, they're usually just as good for exchanging text. But if you do want to save a file in the Rich Text Format, you can, as long as your word processing software will let you (use the Save As or Export command).

Ricoh drive

A **Ricoh drive** is a *removable cartridge hard disk drive* that competes with *Bernoulli* in claiming to be faster and more dependable than the more popular *SyQuest drives*. They *are* faster and more dependable; they also cost much more, which is one big reason they aren't as popular. Part of the reason for owning a cartridge drive is to be able to transport large files to and from a service bureau (or even send them through the mail), or to take a multimedia presentation on the road. But if the service bureau or place of presentation does not have a Ricoh, you must also cart along the entire disk drive box. At the moment, SyQuest drives are much more common and so they are much more convenient to use. It's a circle.

right justify

In a **right justified** paragraph the text lines up on the right edge, like this one, and each line starts at a different place on the left (the text is *ragged* on the left). This can be an interesting effect for short pieces, but it's not good for long passages of text because your eye has to find the first character in each new line in a different place.

RIP

RIP (pronounced "rip") stands for *raster image processor,* which is a piece of hardware and/or software that translates *object-oriented* graphics and fonts into printable dots (remember, you always hear people talk about the *dots per inch* of the printer). Please see *raster* and/or *rasterized.*

RISC architecture, RISC chip

A computer with **risc architecture** is one that uses **risc chips** (pronounced "risk"), an IBM technology that stands for **r**educed **i**nstruction **s**et **c**omputing. The idea is that most computers use only a few instructions most of the time, and these few instructions are relatively simple. If those basic instructions could be done faster, then the overall performance would be increased most of the time.

A general purpose *CISC chip* (pronounced "sisk," **c**omplex **i**nstruction **s**et **c**omputing) such as the *Intel 80386* or the *680x0* chips, can recognize a far greater number of instructions than can a RISC chip, but generally, the more instructions a chip can handle, the slower the overall performance.

So a RISC chip can only process a limited number of instructions, but these instructions are optimized so they execute very fast. When it comes to complex tasks, though, the instructions must be broken down into lots of machine code before the RISC chip can deal with them. If the RISC chip runs fast enough, though, it can still execute these complex tasks faster than a traditional CISC chip.

RISC *processors* (chips) are less complex so they are faster and cheaper to design and debug. Sun Microsystem has a RISC chip called SPARC, Motorola has the 88000, and Intel has the i860.

RISC and CISC chips are *microprocessors,* the *CPU (central processing unit),* the chip that runs the entire computer. There is much debate over the relative advantages of the two systems. Apple's union with IBM involves a cooperative development project based on RISC technology, and it is projected that the next generation of personal computers will be based on this technology also.

river

When you *justify* type (align it on both sides of the column) on a line that is too short for the size of the font, the computer has to force extra space between words to make the lines reach out to the edge. When several of these wide word spaces stock up near each other in a paragraph, it creates what appears to be a **river** of white space flowing through the text. Turn your text upside down, angle it slightly away from you, and look at it with your eyes squinted. You will be able to see any rivers clearly.

Notice the rivers of white space running through the text.

Rivers distract the reader from the message, so you should avoid them. This might mean rewriting copy, adjusting the margins of the page, adjusting the point size of the font slightly, or a variety of other techniques, depending on the capabilities of your software.

robotics

Robotics is a term coined by Isaac Asimov. It refers to the use of computer-controlled robots to perform repetitive and manipulative tasks, as on an assembly line. Sometimes the term is applied to the particular area in *artificial intelligence* that relates to robots. The word "robot" was first used by Korel Capek in his play, "R.U.R." (Rossum's Universal Robots).

rotfl, rotflol, ROTFL, ROTFLOL

When you're *online* "talking" with other people via your keyboard and your modem and somebody says something that strikes you as funny, tell them you're **lol**—laughing out loud. After all, they can't hear you laughing, they can't see you—so how else are they gonna know? If something really cracks you up, you might want to tell your online friends you are **rotfl,** which means you are rolling on the floor laughing. Or you may be **rotflol,** which is, of course, rolling on the floor laughing out loud.

Keep in mind that typing with all capital letters online (like HELLO THERE) indicates you are shouting, which can really bother other people in the room. They will tell you to stop shouting. But if you've seriously fallen out of your chair with the hilarious comment someone made, you just may really be **ROTFLOL.**

These acronyms are part of an *online* shorthand language. More acronyms are defined under *baudy language*.

ROM

ROM (rhymes with "tom") stands for *read-only memory*. The computer can only work with information that is held in *memory,* the little *chips* where information is stored. Most of the memory in a computer is *RAM,* which is *random-access memory*. When you turn the computer off, the information in RAM vanishes (RAM is referred to as *volatile* because the contents disappear when this happens).

By contrast, information stored in ROM remains intact; it does not disappear even when the power is turned off. The information in ROM is actually programmed into the chip at the factory and can't be altered. That's why it's called "read-only." The information can only be read (acted upon), it cannot be changed.

ROM is almost always used to store programs for a computer or some other device (programs stored in ROM are called *firmware*). The ROM chips that come built into your computer hold instructions that the computer needs to do essential tasks. In a Macintosh, the ROM chips contain part of the *operating system*, the part the Mac needs to get itself started up. The corresponding allotment of ROM in a PC is called the *BIOS.*

Other devices in your system have ROM, too. The typical *add-in board* has ROM chips that instruct the board how to operate. In a printer, ROM tells the machine how to handle the data and commands it receives from your computer, and how to turn that information into a printed page. If your printer has built-in or cartridge fonts, those fonts are stored in ROM, too.

root

Root is short for *root directory*. See the next definition.

root directory

In computer systems which keep track of files in a hierarchical system, a disk's **root directory** is the first level in the heirarchy. It's the level which contains everything else, and from which everything else branches out. Every disk has its own root directory. The root directory of any disk is simply the letter name of that disk drive: A: or B: or C: or whatever. In DOS commands, you indicate the root directory by a backslash, \, without any characters after.

RS-232, RS-232-C, RS-422, RS-423, RS-449

These spoonfuls of alphabet-number soup designate different standards for connecting *serial* devices (like modems, mice, and printers) to the computer by plugging their cables into *serial ports*. Through a serial port, the computer exchanges information with the device back and forth "serially," or one bit at a time.

The **RS** stands for **r**ecommended **s**tandard as defined by the EIA (Electronics Industry Association). People mostly use names like RS-232 and RS-422 to identify the type of *serial ports* on a computer, or the cables that connect them. "Ah, you've got an RS-232 port—you need an RS-232 cable." But technically, each standard defines a particular electronic system for passing information through a serial port—what voltages to use, that sort of thing—not just the physical connections between your computer and a cable.

RS-232 is the most common type of serial port, the kind used on modems and the vast majority of PCs. It's simple and cheap, but unreliable at distances over 50 feet. The C in **RS-232-C** means it's the third version of the RS-232. This version is the most common. RS-422 permits faster communications over longer distances. RS-422 and 423 are both subsets of **RS-499,** which specifies which pins in the cable connector are supposed to do what.

But actually, these so-called standards aren't all that standardized. For example, the Macintosh serial port can serve as either an RS-422 (for *AppleTalk* networks) or an RS-232 (for modems, and for serial, non-AppleTalk printers). However, the Mac uses incomplete versions of both standards, so you can't necessarily hook it up to a true-blue RS-422 or RS-232 serial device. This has become a particular problem with some of the newer high-speed modems. Anyway, you can look in a technical book if you want the details. In fact, you can ignore this RS business altogether—when you buy a serial device, just make sure it's designed to work with your type of computer.

RSN

RSN means **r**eal **s**oon **n**ow, a comment often heard when asked about the shipping date of a *vaporware* product. My kids heard this phrase quite a bit this year, whenever they would beg to know when this stupid dictionary was gonna be finished.

RTFM

RTFM stands for the best and most reliable user tip in the world:
Read The Manual.

ruler

A **ruler** on the screen is just like the one in your desk drawer—it's for measuring. Many graphics packages and *page layout applications* like *paint* or *draw programs* have optional horizontal and vertical rulers so you can measure the size and placement of objects. But remember this: an inch on a screen ruler indicates an inch in real life—on the printed page—not an inch on the actual screen. As you change the magnification of your document on the screen—the ruler magnifies also.

Most Mac and Windows *word processing programs* and *page layout programs* have another type of ruler you can use to format the text. You can set left and right indents, the first-line indent, and tabs in the ruler. Each paragraph, which you create when you press the Return or Enter key, can have its own ruler settings. The rulers sometimes have tiny icons that represent different text alignments, such as *centered, flush left* or *right,* and *justified.* Below is a typical ruler (this one is from Microsoft Word).

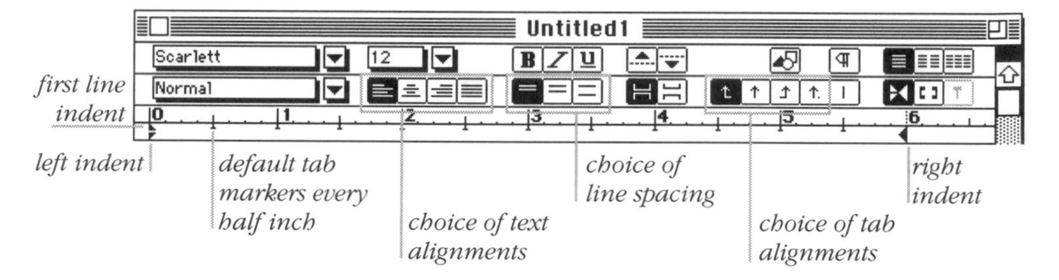

run, running

To **run** a program means to start it up and use it on the computer. A program that is **running** is one that is currently operating, whether or not you're actively using it at the moment. But you might also have a program running in the *background.*

run-around

In a **run-around,** text flows around the edges of a graphic or a shape, as in this paragraph. This is also known as *text wrap.* Not all applications can create a text wrap. Typically you will only find a run-around feature in a *page layout program.*

runs on

If someone says that a software program **runs on** a particular sort of computer, it means the program will work on that computer—you will be able to start it up and make it do whatever it's supposed to do. Depending on the context of the sentence, the person may be implying that the program *does not* run on any other kind of computer.

runs under

If a software program **runs under** DOS, Windows, System 7, or what have you, it means that the program works when the other software is in control of your machine. This "other software" may be an *operating system* (like DOS or System 7), or it may be something that enhances the way the operating system works (such as Windows does to DOS, or MultiFinder did to System 6). Someone might use this term simply to tell people which operating system (or other software) the program works with. They might also use it to distinguish between programs that work and those that don't with a particular version of, or enhancement to, an operating system. For instance, you may hear that a certain program will **not run under** MultiFinder, even though it works fine otherwise on the Macintosh.

run-time version

When you get a software product that tells you it uses a **run-time version** of a software program, that means you do not need to own or possess or even have a copy of the ***original*** program, yet you can open the *document* that was ***created*** in that original program. A run-time version of a software product does only one thing: it runs programs or displays documents that were created with the full version of the software.

For instance, TMR Multimedia transferred my presentation on type into a beautiful, animated, MacroMind Director presentation. I don't have MacroMind Director yet, but Todd set this up so I just double-click the icon and the presentation opens in a special run-time version of Director (Cache's MacroMind Player), even though MacroMind Director itself is nowhere to be found on my hard disk.

In the business world, you're likely to come across a run-time version when you're working with a customized database program, the kind someone has written just for your company using a database programming language like *dBase* or "PAL" (the Paradox Application Language). You can run this type of program using the complete version of the matching database software (dBase, Paradox, or whatever). But if you don't need the complete software, the programmer can give you a run-time version of say, dBase, whose only job is to run dBase programs. The run-time version is much cheaper than the complete version—in fact, it's often free. And since you can't do anything with it except run your program, you don't have to worry about getting confused by all the options and commands you don't need.

On the PC, any program designed for Windows will only run if you have Windows running first. In the days before Windows became really popular, it used to be common for software companies who made Windows programs to throw in a run-time version of Windows. Aldus PageMaker, for instance, came with a Windows run-time which was only good for running PageMaker. Since Windows 3.0 came out, however, Microsoft hasn't allowed software companies to give away run-time Windows.

Sad Mac

When you turn on your Mac, you should see a little picture of a Macintosh smiling at you—that's the Happy Mac (which is the little Mac icon in these definitions, shown to the left). If you see a flashing question mark instead, that means the computer cannot find a *System* with which to *boot* (turn itself on). And if you see the **Sad Mac,** (instead of the Happy Mac, the computer found a System, but it is incompatible in some way with the computer (maybe one of your kids took the Finder icon out of the System Folder). Or maybe there is a *SCSI* conflict or your *SIMMS* are loose or you have a bad *memory chip.*

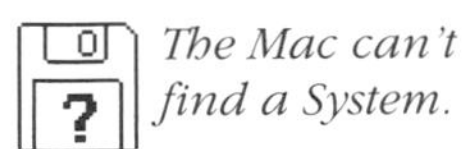

The Mac can't find a System.

The Mac found a System, but the System is damaged in some way. Or this may indicate a hardware problem. Turn the page for a suggestion on troubleshooting.

If you get a Sad Mac, you can try *rebooting* the computer with a floppy disk in the drive—a floppy disk that has a System file and a Finder file on it. If you get a Happy Mac with the floppy disk, your problem is probably that the System files on your hard disk are bad or missing. You may be able to fix the problem by replacing the System on the hard disk (just copy the System files from the working floppy to the System folder on your hard disk). But I strongly suggest you call your *power user* friend before you try something like replacing the System.

SANE

SANE stands for **s**tandard **A**pple **n**umeric **e**nvironment. Apple built SANE routines into every Macintosh to do all the common kinds of math with very high precision. So when you ask a progam to do some sort of calculation, the program doesn't have to go find out how to do it all by itself—it just calls on the SANE routines. To get these extremely precise answers, though, takes a relatively long time. If you do a lot of *number-crunching,* you might want to check into one of the software packages that are much faster than SANE but not quite as precise (depending on whether you want speed or precision, keeping in mind that both are relative).

sans serif

Sans serif (pronounced "san SAIR iff," *not* "sans sair EEF") is a category of type that does not have *serifs* ("sans" is French for "without," as in sans souci). Serifs are the tiny crossbars on the ends of the strokes on letters in some type designs. Sans serif type, in general, is considered more *legible* than serif type, but less *readable.* Fascinating studies. This paragraph is set in a sans serif font; the other definitions are in a serif font.

This is sans serif type. Notice the ends of the letter strokes just end bluntly.

Serif

This is serif type. Notice the ends of the letter strokes have tiny crossbars.

save, save to disk

While you are working on a file, all your work is kept in *memory (temporary storage)* because that's where the computer needs it. When you **save,** the computer puts a copy of your entire file on the *disk (hard* or *floppy)* for *permanent storage.* As you continue to work, the changes you make are kept in memory. When you save, again those changes are added to the file on the disk. Remember, anything left in *memory* disappears when you turn the computer off, or when it *crashes* (stops working against your will, due to a hardware or software problem), or when there is the least little power failure. So until you save, any work you do is liable to disappear without notice. Rule Number One on any computer is "Save Often, Sweetie" (S.O.S.) Like every three minutes.

The phrase **save to disk** is the same as *save.* It's just a more specific expression. "Did you save it to disk?"

scalable font

A **scalable font** is a font that can be resized on the screen or when you print to any point size without becoming distorted. *TrueType* and *PostScript* are both scalable font formats. Please see the definition under *font* for more details and comparisons.

scan

When you have an image outside the computer, like a photograph or a drawing on paper, and you want to use it inside your computer, you need to **scan** the image. A *scanner* is a device that takes a picture of the image, *digitizes* it (breaks it up into dots that can be recreated on the computer screen with electronic signals), and sends this *digital* information to the computer. Once the scanned image is in the computer, you can view it in different sorts of applications and change it.

You can also use a "video scanner" and its appropriate input software to scan three-dimensional objects, such as your body parts, into the computer.

Scan can be a verb, as in "I need to scan this photo," meaning you need to get an image of the photo into your computer. It can be a noun, as in "I need a scan of this photo." It can also be an adjective, as in "This is a scanned image."

scanner

A **scanner** is a device that takes a picture of an image that exists outside the computer, such as a photograph or a drawing on paper. As the scanner takes the picture, it *digitizes* the image (breaks it up into dots that can be recreated on the computer screen with electronic signals), and send this *digital* information to the computer as a file. Then you can take this file of the scanned image and use it in your work.

Scanners come in several varieties. There are "hand-held" scanners that you hold in your hand and roll over a flat image. These are relatively inexpensive (around $200) and are surprisingly good. There are "flatbed" scanners that look kind of like small copy machines, where you lay the artwork on the glass. There are "video scanners" that use a video camera to capture an image, which means you can *input* three-dimensional objects, including your children.

A scanner also needs *software* to get the job done. Various software can give you different kinds of options, such as *dithering* patterns, *resolution, file formats,* etc. Although software is always supplied with the scanner, you can buy better scanning software, such as Ofoto, to use with almost any scanner.

scrambled

When something in your computer (such as its *memory* or a *file*) gets **scrambled,** its contents have been goofed up, altered so that it no longer works right. For example, if you scramble some of the computer's memory, your data may turn into meaningless gibberish, or your programs may go haywire. If you scramble a file, you'll get a bunch of nonsense when you try to read it. And if you scramble the special part of a disk that tells your computer where to find the files (the FAT, or *file allocation table*), then the disk becomes unusable. A scrambled disk can sometimes be restored using a *disk repair utility.*

Scrapbook

The **Scrapbook** is a *desk accessory* that comes with your Macintosh. You can find the Scrapbook under the *Apple menu,* where all desk accessories are found. The Scrapbook is a place to store graphic images or text or parts of spreadsheets, etc., that you may want to use again, either in the same document or in another document or program altogether. For instance, you can create your business logo in MacDraw and paste it into the Scrapbook. Whenever you want to use that logo, you can just copy it out of the Scrapbook and paste it into your document.

You can copy text into the Scrapbook and when you paste it somewhere else, it appears as text again. In System 7 you can put sounds into the Scrapbook, then click the button to hear them. When you copy graphic images into the Scrapbook, though, it turns all images into the *graphic file format* called *PICT,* which can be placed into just about any Macintosh program, including word processing programs. If you don't want your file to be a PICT, then don't stick it in the Scrapbook.

To put an item into the Scrapbook, you need to *select* the item and *copy* it. Then choose the Scrapbook from the Apple menu (since it is a desk accessory you can open it while working in any other program). When the Scapbook is visible, choose "Paste" from the Edit menu. The existing pictures will move over and the copied item will appear in the window. Close the Scrapbook (click in the little close box in the upper left corner; in System 7 you can press Command W to close any desk accessory).

To take an item out of the Scrapbook, open the Scrapbook from the Apple menu. *Scroll,* if necessary, to find the image you want. While that image is visible, from the Edit menu choose "Copy." Close the Scrapbook (click in the little close box in the upper left corner). While your document is open, from the Edit menu choose "Paste." The item will paste in and **replace** whatever is currently selected on the screen, or if there is one cell or an insertion point flashing, it will appear at that spot. Some applications cannot accept graphic images of any sort.

There is a similar Windows program called "Scrapbook+," but you have to buy it.

scratch disk, scratch file

A **scratch file** (also known simply as the "scratch") is a portion of the *memory* or *hard disk* that the application you're working in creates for itself and maintains without you even knowing it. It's a temporary storage place the program uses like you would use a scratch pad. When you quit the program, the scratch file gets emptied automatically.

Adobe Photoshop users may come across messages about the scratch disk, usually telling you it's full. If you have several hard disks attached to your computer, Photoshop uses the hard disk that holds the Preferences file as its virtual or scratch disk, and it uses hard disk space as *virtual memory,* unless you tell it otherwise (which is why you should not use System 7's virtual memory while working with Photoshop).

script

A **script** is a series of commands that can be executed without you, the user, interfering. Scripts are similar to *macros* and *batch files*. A "scripting language" is a simple programming language you can use to create your own scripts. You can use a scripting language to control *HyperCard*, for instance, and you can use scripts to automate procedures in Aldus PageMaker. With the software package called *Frontier*, you can write scripts to automate almost anything in a Macintosh.

Script also refers to the rehearsed spiel that salespeople use when showing their products, or that trainers use in instructive videos, or that presenters use in lectures.

scroll, scrolling

Often there is so much information on a computer screen you can't see it all at once. Most programs have a way of smoothly and gradually moving information across the screen, as if you were reading through a document rolled up on a scroll (as opposed to flipping through index cards). The **scrolling** action may be horizontal or vertical. In some applications, you **scroll** by pressing the *arrow keys* on the keyboard. In some applications, you scroll with the mouse using the *scroll bars*.

scroll arrow

At the ends of the *scroll bar* in a *window*, there are **scroll arrows.** You can *scroll* the information in the window by positioning the mouse pointer on one of the scroll arrows and pressing the mouse button. Please see *scroll bars* and *window* (with a lowercase w) for illustrations.

scroll bar

In a *window* there are **scroll bars,** usually both across the bottom and along the right side. If a scroll bar is **white,** it is a visual clue that you are viewing everything that can be seen in that direction (vertically or horizontally). If a scroll bar is **gray,** it is a visual clue that there are too many items in a window to view all at once, or that the program can display more cells or pages than will fit on the screen.

When the scroll bar is gray, you'll see *scroll arrows* at either end, and a *scroll box* within the scroll bar. You can press (using your *mouse* button) on either arrow to scroll the contents of the window past your vision, like the scenery flowing past the train window. Or you can press in the scroll box and drag it anywhere along the gray bar to pop to that position instantly. Or you can click in the gray area, which usually moves the contents one screen full at a time. (The scrolling action seems to go backwards. You get used to it after a while.) See *window* for more details.

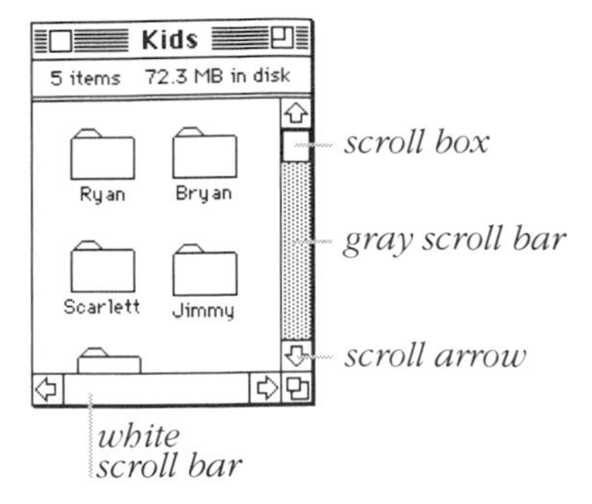

scroll box

When the *scroll bar* of a *window* is gray you'll see a little **scroll box** (see the illustration on the previous page). You can press on this scroll box to move the contents of the window up, down, or sideways. Some people call the scroll box an *elevator,* a *thumb,* or a *widget.* On a *grayscale* or color *monitor,* the scroll box has little ridges on it so you don't slip off. See *window* for details.

Scroll Lock key, SCR LCK key

PC keyboards and Mac *enhanced* (extended) keyboards have a key labeled **Scroll Lock** or **Scr Lck,** but it doesn't do anything—it doesn't lock the scroll, whatever that would mean. So don't think you're a bone-head if you can't figure out what it's for. Actually, there are a few applications that make use of the Scroll Lock key, but you'll have to rely on their manuals to tell you what happens when you press it.

SCSI

SCSI (pronounced "scuzzy," not "sexy") stands for **s**mall **c**omputer **s**ystems **i**nterface. SCSI is a standard for *interfacing,* or connecting, personal computers to *peripheral devices* (like scanners, hard disks, or CD-ROM players) and having them send information to each other.

The interface works through a *SCSI cable* that you connect to a *SCSI port* (which simply means that you connect a cord with a particular shape into the socket, or port, that matches). The computer itself usually has only one SCSI port, but you can attach up to seven other devices together in a chain of cables (see *SCSI-to-SCSI cable*). Then your computer can send information to and receive information from these devices.

All Macintoshes (except the very earliest models, the 128 and 512) have a built-in SCSI interface. Only a very few PCs have SCSI interfaces.

SCSI-2

SCSI-2 is a new standard for the small computer systems interface (see *SCSI*). It can transfer data faster, up to 10MB per second, whereas the first SCSI specifications were limited to about 4MB per second. SCSI-2 is sometimes called "wide SCSI" because the interface can send 16 *bits* or 32 bits of information at a time. This means the speed could increase up to 20 and 40MB per second, as soon as the computers and printers all get the matching interfaces to take advantage of the new standard.

SCSI address

Computers that have a *SCSI port* (all Macs [except the ancient 128 and 512] and some PCs) can have up to seven devices attached to the computer, such as an external hard disk, a scanner, a CD-ROM player, etc. Since information travels through the cables to these separate devices, each one must have a different **SCSI address** so the information gets to the right place. A SCSI address is also called a SCSI ID.

A SCSI address is a number from 0 to 7. When you add a new device to your system, you determine the number yourself, either using software provided with the device or by flipping a little switch somewhere on the hardware (the manual will explain how; yes, you probably should read that manual). You can pick any number you choose for any peripheral device, just make sure the number for each address on the chain is different. The last device on the chain must be *terminated*. On a Mac, the computer itself always has a SCSI address of 7; the internal hard disk always has a SCSI address of 0 (zero). All other devices must have a number from 1 to 6. (My friend Clay is superstitious and always connects his SCSI devices in ascending numerical order. See *termination* for some very important information about connecting SCSIs.)

Because the computer always has a SCSI address of 7 and there cannot be more than one device with the same address, you cannot connect two computers to a common device at the same time, such as an external hard disk or a scanner or a CD-ROM player. You can't change the address of the computer.

Don't confuse a SCSI address with any other kind of address, like a *memory* address or *32-bit addressing*.

SCSI bus

The **SCSI bus** refers to the electronic connections between hardware that connect *SCSI devices* to each other—the *connectors* and the *cables*. Information (data) travels through the SCSI cables and the SCSI *ports* (which is where SCSI cables connect to the computer). They say this data travels on a *bus*. There are also other kinds of buses. See the definition of *bus* for detailed information.

SCSI cable

A **SCSI cable** is the thick cable with the large, rectangular connectors on either end; this cable is used to connect *SCSI devices* together. For instance, if you want to connect a CD-ROM player or an external hard disk or a scanner to your computer, you need to hook it up with a SCSI cable.

SCSI chain

A **SCCI chain** is when several *SCSI devices* are linked together with *SCSI cables*. Because there is usually only one *SCSI port* on the back of the computer, all the other devices you want to attach must be linked one to another in a chain. The first and last device on the chain must be *terminated*. Please see *SCSI* and *SCSI-to-SCSI cable*.

SCSI device

A **SCSI device** is an external piece of hardware that connects to the computer via a *SCSI cable* (those big fat cables with the big rectangular connectors on the ends). Scanners, CD-ROM players, and external hard disks (both regular and removable cartridge) are examples of SCSI devices. Mice, modems, keyboards, and most printers are not SCSI devices; they are considered *serial* devices and they connect to the computer through the *serial* or *LocalTalk ports* (those unintimidating little outlets, sometimes round and sometimes just like the outlets for our telephones).

SCSI port

A **SCSI port** is a plug-like socket on the back of a computer or on any SCSI *peripheral device* (such as a scanner or external hard disk) where you connect a *SCSI cable*. A SCSI port is a high-speed *parallel port*. It's standard on all but the oldest Macintoshes, on some IBM PS/2s (model 65 and above), and on the IBM RS/6000. You can install a port on IBM PCs and compatibles as an *expansion card* (SCSI ports are different from standard PC *parallel* or *printer* ports). Even though there is only one port, you can attach up to seven devices through this one port, connected in a chain. See *SCSI-to-SCSI cable*.

SCSIProbe

SCSIProbe is a *utility* from SyQuest Technology, Inc. that will find and mount an ornery SCSI device. That is, if you insert a cartridge hard disk into the removable cartridge drive and its icon doesn't appear on your Desktop (it doesn't *mount*), you can call up SCSIProbe and ask it to mount the disk (which make its icon appear). There are other utilities that do the same thing.

SCSI-to-SCSI cable

A **SCSI-to-SCSI cable** is a cable that connects directly from one *SCSI device* to another SCSI device (rather than from the computer to the device). For instance, you may want cables to connect from the *scanner* to the *cartridge disk drive,* then from the cartridge disk drive to the *CD-ROM player*. Through the SCSI port you can have up to seven SCSI devices communicating with your computer, each one hooked to the next with a SCSI-to-SCSI cable. You have to arrange the hardware like this because the computer has only one SCSI port, so it cannot connect directly to each device.

When you connect several SCSI devices to each other, you have a "SCSI chain," which is one form of a *daisy chain*. These devices are then said to be connected to the *SCSI bus*.

scuzzy

Scuzzy is the proper pronunciation for SCSI. Interesting term, huh? Please see *SCSI*.

.sea

This little *file extension,* **.sea** (pronounced "dot see") stands for **s**elf-**e**xtracting **a**rchive. A file with this extension has been *compressed,* usually for *archival* purposes (which means for storing) or so it could be transferred over the *modem* (the smaller a file, the less phone time it takes to transfer it). Sometimes you might hear someone call these "ess ee ay" (s-e-a) files, as in, "Perhaps this is an s-e-a file."

A .sea file has been compressed, but it is "self-extracting," meaning you do not have to have the *application* in which the file was originally compressed in order to uncompress (extract) it (see *run-time*). A file that is not self-extracting can only be uncompressed by the same or a compatible file compression *utility*.

Files may be self-extracting even if they don't have this file name extension. If your compression utility (like StuffIt, DiskDoubler, PackIt, PKZip, or an ARC program) can make a self-extracting archive, you will find directions in the manual. In DiskDoubler it is as easy as selecting the file, holding down the Shift key, and choosing "Create SEA..." from the DD menu.

search

To **search** means have the computer look for something, either a particular file or some text or data within a file. Different programs have different search capabilities. Sometimes *operating systems* let you can search for files just by the names of files (such as any file with the word "résumé" in the name). Sometimes you can add other search criteria, such as the date a file was last saved or the size of the file, to narrow the number of matching items. For instance, you might want to search for all the files with "Résumé" in the title that were created since last week.

Word processing applications let you search for certain words or phrases within a document. There are *utilities* you can add to your system that even let you search through closed files for certain words or phrases.

search-and-replace

Usually found in a good word processing application or page layout program, a **search-and-replace** feature lets you tell the computer to look (search) for certain words or phrases and replace them with other words or phrases. A really good search-and-replace feature will let you look for words with certain attributes, such as bold or italic or small caps, as well as search for special characters like Returns or Tabs or *em spaces*.

Say you wrote a novel and the main character's name is John, which is the name of your boyfriend. Well, let's say just before the book was going to press you realized you didn't like John anymore and you want to change John's name in the book to Sam. A search-and-replace feature will find all the instances of his name and replace them with Sam. You can also tell it to find only the words "John" that have a capital letter so no one in the book ends up going to the sam.

You can use search-and-replace to replace all the double spaces after periods with single spaces, or to replace two hyphens with an em dash, or to replace two Returns with just one. You can type a shortcut for a long phrase (for instance, type "sc") throughout a document, then search-

and-replace the "sc" with "Sonoma County Office of Education." You can search for all the underlined words and tell the program to make those words a different font, and bold, and **not** underlined.

Search-and-replace is sometimes called "Find-and-Change" or "Find-and-Replace" or sometimes just "Replace" or "Change."

sector

Sectors of the disk.

A disk has two sides (a top and a bottom). Each side of the disk has tracks (concentric rings) on the surface. Each ring is divided into arc-shaped **sectors,** little units of storage space on the disk, usually 512 *bytes* on a floppy disk and up to several thousand bytes on a hard disk. Whatever the size, a sector is the smallest unit the computer can *read* or *write* at a time; it cannot deal with portions of a sector. For instance, if the sectors on your hard disk can hold 1 *kilobyte* each (1,024 bytes) and you have a graphic that is 3.5 kilobytes, the computer will use 4 sectors to store that graphic, even though the fourth one is only half-full. Even though this wastes some space, it is exceedingly faster than if the computer had to write to each byte one at a time. The computer knows where the data is stored because it keeps track of exactly the track and exactly the sector within that track where data has been put, and keeps this information in a little directory on the edge of the disk.

If one of these sectors is physically damaged or flawed, like from dust or dirt, it is considered a *bad sector* and cannot be used. If there was already data in that sector when it got damaged, chances are slim that you can recover it, unless you have the specialized hardware and software necessary for this sort of operation.

seek time

Somewhat simply stated, **seek time** refers to how long it takes the *read/write head* on a hard disk to move from one track to another. Technically, seek time is only one factor affecting "average" *access time,* and is not the same as access time. Because it sounds faster, certain disk vendors may try to woo you by quoting seek time specifications for their disks, but you should really base your comparisons on average access time.

select

Before you can apply a change to something on your screen, you first need to **select** the item (the graphic, the text, the object, the region) you wish to change. That is, you need to identify the item as the one you want to do something with. For most actions on the computer, the rule is "First select it, then do it to it." For instance, if you want to change some text to bold, you must first select the text and then apply the command to change it to bold.

You know you've selected the item because of the way it looks on your screen (see *selection*). If you ***don't*** select the item, one of two things will ocur: either ***nothing,*** because the computer doesn't know what to apply the command to, or the change will apply to ***everything*** you do after that point because you have set a *default* (an automatic choice).

Selecting might be done with the *pointer tool* or with the *arrow keys* or with a combination of keys. Depending on the application you are in and the effect you want to apply, you may need to select with any number of *tools* that are available.

selection

When you *select* (choose) an item (see the previous entry), the chosen item is called a **selection.** A selection is indicated in several different ways, depending on the computer and the application. The item may be *highlighted,* meaning it appears white on black, or there may be a colored bar over the text. An object may show handles around its perimeter, or a graphic area may shimmer or may have a rectangle of marching ants surrounding it (the *marquee*).

You can't let the seeds stop you from enjoyin' the watermelon.

The text with the dark bar behind it is selected.

Sometimes you can just click on an object to select it. This graphic is showing its eight handles.

The shimmering dotted line, the marquee, selects what's inside the line.

Once an item is selected, then you can do things to it, like change the style of the type or resize or color the graphic image or delete the objects, etc.

semiconductor

A **semiconductor** is a material that is not really a conductor of electricity (like copper), nor is it really an insulator that resists electricity (like rubber). When a semiconductor is charged with light or electricity, it changes its state from conductive to non-conductive, or vice versa. A transistor is a semiconductor that acts like an on/off switch, letting electric current pass through it or not.

When *silicon* is chemically combined with other elements, it acts like a semiconductor. Computer *chips* are made of semiconductors, mainly silicon and germanium. The use of semiconductors made it possible to miniaturize electronic components. Not only are the components smaller, but they are faster and more energy-efficient.

The term semiconductor is also loosely applied to electronic components that are made from semiconductor materials.

send-receive fax modem

A **send-receive fax modem** is a *modem* that can both send and receive faxes from your computer to someone's fax machine or computer or vice versa. If you have the capability to send a fax straight through your modem, you don't need to print up a *hard copy* of it. Some fax modems are "send-only," meaning they can't receive fax transmissions, so be sure to check before you buy. And both your computer and your modem must be turned on to receive a fax through your modem, of course, which is something else you need to take into consideration.

SeniorNet

SeniorNet is a non-profit organization for older adults interested in using computers. SeniorNet members are 55 or older and have an interest in learning about and working with computers. Members who reside near a SeniorNet Learning Center can take beginning courses designed to introduce them to computers, with classes in word processing, spreadsheets, telecommunications, and other topics. Others can learn more about computers by reading SeniorNet's newsletter and other publications, participating in SeniorNet Online, or by attending national SeniorNet conferences.

SeniorNet has more than 5000 members across the United States. It publishes a variety of instructional materials and has learning centers throughout the country. SeniorNet offers discounts on software and computer publications, operates SeniorNet Online on America Online, and holds a national conference where members can learn new skills, meet computer experts, and share experiences with other members.

Members learn to use computers to write or publish anything, from a newsletter to an autobiography; to manage personal or business records; to communicate with other members across the country; and to serve their communities. SeniorNet members share a desire to continue learning and a willingness to contribute their knowledge to others.

Members receive "SeniorNet Newsline," a quarterly newsletter that contains reviews and articles about software and hardware and step-by-step computer projects for specific software packages. Periodically SeniorNet also makes other computer publications available to members, such as its "How to Buy a Computer" series, and the "SeniorNet Source Book," which is a compilation of computer projects designed by and for seniors, such as "Computers for Kids Over Sixty" and "Portraits of Computer-Using Seniors."

SeniorNet grew out of a research project begun in 1986 by Mary Furlong, Ed.D., Professor of Education at the University of San Francisco. The project sought to determine if computers and telecommunication could enhance the lives of older adults. Now a non-profit organization based in San Francisco, Seniornet receives its support from foundations, corporate sponsors, and individual members. The annual membership fee is $25; couples' membership is $35.

SeniorNet
399 Arguello Blvd
San Francisco, CA 94118
(415) 750-5030
(415) 750-5045 FAX
AOL: SENIORNET

sequence, sequencer

A **sequencer** is the electronic equivalent of a combined tape recorder/ player piano. For our purposes, a sequencer usually consists of software you run on your computer, although you can also buy self-contained sequencers that are separate devices. To record or play back your musical compositions, or **sequences,** you hook up your computer (or the sequencer device) to *MIDI* devices such as keyboards, *synthesizers,* or samplers.

If you own a MIDI keyboard or other controller (a device that generates MIDI signals), the sequencer can record the notes you play and then play them back later. But it doesn't record the sound of the notes, just which keys you pressed and when you pressed them. You can also create or edit sequences by writing music on the computer screen. When you play back the sequence, the sequencer sends the note information to a MIDI instrument, which generates the music all over again (sort of like the way a player piano plays each time you start it up). For this reason, a sequence can sound different every time you play it—you can change the sounds on the MIDI instrument, or have a different instrument play it back.

serial

Serial means one by one, one at a time. You'll most often hear "serial" being used in reference to data transmission, or sending information over wires, such as to a *printer* or through a *modem.* When data goes through a serial interface, the *bits* go through the wires one at a time. But there are two wires, so information can be sent and received simultaneously. Compare with *parallel,* and see the definition for *serial port,* below.

serial device

A **serial device** is a piece of hardware that connects to the computer through a serial cable and *serial port,* which is a particular sort of cable and socket for transmitting data serially (see the definition above for *serial*). Serial cables are usually not very thick and they have little, unintimidating connectors on the ends. Modems, mice, and keyboards are examples of serial devices.

serial number

When you buy computer hardware, just as when you buy any electronic equipment, it has a **serial number** unique to that particular item imprinted on it somewhere. Be sure to keep track of your serial numbers just as you would a serial number on your stereo or television. When you send in your *registration card,* there is always a line for you to write in the serial number.

Most software you buy has a personal *registration number* for your particular copy of the program. Sometimes this *registration number* is referred to as a serial number. You can usually find it on the outside of the box on a little sticky tag, or perhaps right inside the manual, or printed on the disk label.

serial port

A **serial port** is the socket (also known as an "input/output connector") where you plug in the cables to attach to a *serial device,* such as a printer or modem.

Data is transmitted through a serial port in a single file, one *bit* at a time, but there are two data wires so a serial port can send and receive information simultaneously. *Parallel ports* connect with cables that have parallel wires so they can transmit data faster; they accommodate eight bits of data at a time (on a PC, however, parallel data can only flow in one direction at a time).

Most Macintosh printers are *serial printers,* though the bulk of those use the Mac's *LocalTalk network* port. While technically a serial port, LocalTalk is much faster and more complex than the serial port used for modems and other serial printers like the *ImageWriter*. LocalTalk is generally faster than the parallel ports on PCs, though the standard Mac serial ports are slower. The slower rate of the standard serial ports is not really a problem since non-LocalTalk serial printers take much longer to process the information anyway than the computer takes to send it.

Because the two *platforms* (Macintosh computer systems and IBM computer systems) each use a different sort of port to connect to a printer, they can't share the same printer without some sort of adaptation (or unless the printer is designed to support both platforms simultaneously).

serial printer

A **serial printer** is a printer that is connected to the computer through a *serial port*. Some of the low-speed printers for the Macintosh are serial printers (not including printers that connect via *LocalTalk*). Most printers for IBM PCs are *parallel printers*. Also see *serial port*.

serif

A **serif** (pronounced "SAIR iff," *not* "sair EEF") is the little crossbar on the end of the stroke of some printed letters. Some *fonts* have serifs, some don't. Studies indicate that serif typefaces are generally more *readable* than type faces without serifs *(sans serif),* but serif type is not quite as *legible* as sans serif.

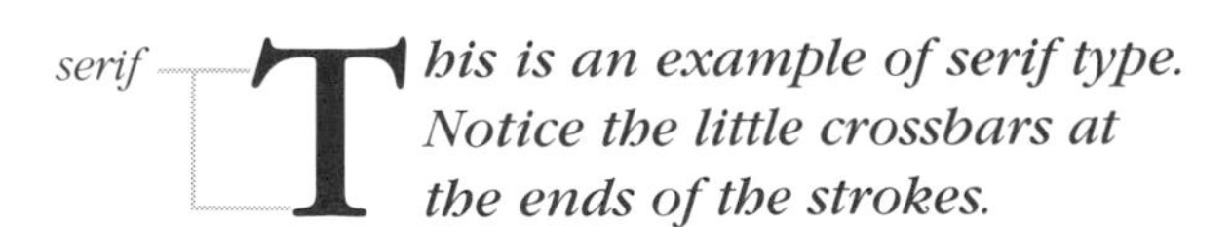

server

In a *local area network* of computers connected together (which means in a room where there are several computers, or even in a large office building where computers are connected between offices), there is usually at least one computer with a large hard disk that "serves" (provides information to) the other computers, and so it is called the **server.** Anyone using another computer on this network can use the server's hard disk as if it were their own. A "file server" just keeps the files that everyone uses in a central location, dishing them to individual users on request. A "database server" not only stores the company database files, but also the main database software for retrieving specific information from the database. "Database server" can also refer to this kind of centralized database software. File serving is a very common arrangement in school computer labs and in many offices.

sexy

When a product excites you, when you can't wait to get home to it, when you dream up work to do just so you can use it, when it makes you weak in the knees just to look at it, that product is **sexy.** Now, it's all relative, of course. To me, a PC is not sexy. A Mac is exciting and great fun and addictive, but not really sexy. The NeXT is sexy.

Seybold

Seybold is the name of a trade show billed as the Event of the Year for Electronic Publishers, started by Jonathan Seybold. It's held each year, once on the east coast and once on the west. It covers electronic publishing on all *platforms* (computer systems) and holds a very high-end and very expensive conference along with the show. The Seybold group also sponsors a number of other conferences on high-end electronic topics, such as digital media.

shade

Shade usually means the inside fill pattern of an object, the color or texture that is inside. For instance, the shade of the box shown below is stripes.

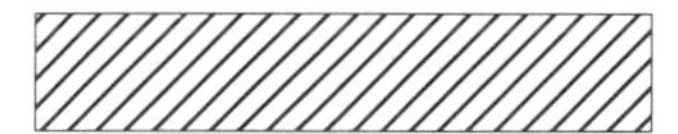

shareware

Shareware is try-before-you-buy software. It is created by nice people who put it out into the world for you to try, with a note attached (actually, it's a file) asking you to please send the author a small fee if you use the program. This registration fee can run from $5 to over $100 (and usually less than $20), but it's almost always less than you'd pay for a comparable product at a computer store. If you do register the program, the author may send you the latest version or maybe even a printed manual. It is so nice of these people to do this that we really should support them by sending their minimal fees. Compare with *freeware* and *public domain software.*

Shareware is distributed electronically via *bulletin board services* and on disks at *user groups* and by special mail-order companies. It's legal for you to make copies of shareware, so you can also get and give these programs to your friends. It doesn't cost you any money to download a shareware program from a bulletin board service, except for any charges the bulletin board makes for using the service in general. If you buy the shareware on disks, you can expect to pay $3 to $6 per disk for the disk itself, the cost of copying the programs, plus the distributor's profit.

You can get almost every kind of software imaginable in a shareware version, from simple *utilities* to complex word processors, spreadsheets, and database programs. *Fonts* and *clip art* are also available as shareware. Huge numbers of shareware products for both PCs and Macs are out there.

shell

A **shell** is the program that accepts the commands from the user and runs other programs. On PCs, *COMMAND.COM* qualifies technically as a simple shell, since it displays the *DOS prompt* where you type commands. However, most people think of a shell as easier to use than the DOS prompt. DOS 5 and DOS 6 come with a nice shell program that lets you use menus to work with files on a screen list. You can buy even better shell programs like the ones that come with XTree, the Norton Desktop, or PC-Tools.

In Windows, the term "shell" is often applied to programs that help you manage your applications and your files, like Program Manager, File Manager, and the Norton Desktop. Technically, though, the Windows shell is whatever program you specify as the shell in your *WIN.INI* file. When you exit this shell program, Windows itself shuts down.

The software application "At Ease" from Apple is the closest things to a shell in the Mac world, and it's meant primarily to prevent young or multiple users (as in a school lab) from getting to or trashing things they shouldn't.

Shift key

The **Shift key** is one of the *modifier keys* on the keyboard, meaning it is a key that doesn't do anything all by itself, but only works in combination with other keys. It works the same on the computer keyboard as it did on the typewriter keyboard to make capital letters—you hold the Shift key down while you type the letters you want capitalized.

There is no Shift Lock key on the computer keyboard, as you might have become accustomed to on a typewriter. The equivalent is *Caps Lock*. See that definition for the major difference between Shift Lock and Caps Lock.

Shift-click

When you read a direction that tells you to **Shift-click,** it means to hold down the Shift key while you click the mouse button. This does different things depending on which program you are in and even which *tool* you are using in the program. For instance, Shift-clicking might allow you to *select* more than one item, or to *deselect* one item from a group. It might restrain a tool to drawing only straight lines or squares and circles, rather than angled lines or rectangles and ovals.

shout, shouting

When you type *online,* particularly when you are in a "room" or *forum* with other people online, it is considered rude to type in all capital letters (LIKE THIS). All caps is the online equivalent of **shouting,** and people will politely ask you to refrain from doing so.

Shut Down

Shut Down is a command in the Special menu when you are at the *Desktop.* When you choose this command, the Mac ties up all the loose ends inside and prepares to *power down* (have the power turned off). You should always choose Shut Down before you turn off the computer or you run the risk of damaging data. On the larger Macintoshes, when you choose Shut Down the computer turns itself off. On the smaller Macs, when you choose Shut Down you get a reassuring message telling you it is okay now to turn off the computer.

side grade

When software developers create new and improved versions of their software, they offer the improved version at a discounted price to those people who are *registered users* of the previous version. This new and improved version of the same product is called an *upgrade.*

Sometimes a software developer may make an entirely new product that is closely related to the first product. In that case, knowing that you liked the first product (because you are a registered user), they offer you a significant discount on this second product. This is called a **side grade.** For instance, Salient Software made a product called DiskDoubler. Then they made another product, similar but different, called AutoDoubler. They offered a side grade to all registered owners of DiskDoubler.

If you are not a registered user, you cannot legally take advantage of any discounts on upgrades or side grades.

SideKick

In the early years of the PC, Borland's **SideKick** was a hugely popular piece of software, the quintessential pop-up utility. SideKick is a *TSR,* or *memory*-resident utility that combines a multitude of helpful functions, including an electronic notepad, a calculator, and a calendar. After you load it into memory, SideKick pops up on your screen when you press a certain key combination, no matter what other program you're working with at the time. You can check your calendar, make a calculation, take notes on an unrelated matter, and then put SideKick away and go back to your program right where you left off.

SideKick had its quirks—it didn't cooperate well with other TSRs—but many people swore by it. Over the years, Borland released two or three fancier versions of SideKick, and even sold a Mac version (a *desk accessory*) for awhile, but none of these caught on like the original.

SIG

Computer *user groups* (support groups for people with similar computers) have several different levels within the group. You can participate at any level. Generally there is a main membership group that has meetings dedicated to general topics. But when a smaller number of people want to get together to discuss a particular topic that may not be of interest to everyone, they usually form a **sig,** or **s**pecial **i**nterest **g**roup. For instance, there are SIGs for multimedia, desktop publishing, education, architecture, etc. I run a SIG in Santa Rosa for new Macintosh users. Well, you don't really have to be a new user to join my group—it's for anyone who has been working with their Mac for any length of time but who still prefer to hear presentations without technobabble, or who are intimidated by a roomful of *power users.*

See the definition for *BCS,* the Boston Computer Society, to get an idea of the kinds of SIGs that can be built from a user group.

signature

A **signature** can refer to anything that is used for identification. On a computer, when sending *e-mail* or a fax through your *modem,* a signature is not always your handwritten name, obviously. So a signature may be a code word or symbol that indicates who you are.

Signature is also a traditional printing and binding term that you may run across if you are *desktop publishing* booklets or books. You know when you look at the binding of a book you can see the little groups of pages glued to the binding? Each one of those is a signature. Each signature starts out as one large piece of paper. A common signature size for a book consists of eight or sixteen pages. The printer takes the text for each page and lays it out on the flat signature (the large flat blank paper); some pages are upside down and they certainly are not in numerical order. The press prints those pages, then the back side of those pages are printed on the other side of the signature, and the printer makes darn sure the right pages end up on the backs of each other upside right. The signatures get folded up and glued into the binding. Then the edges of the folded paper get chopped off so all the pages are lined up neatly. (Long ago, books were sold before they chopped the folded edges off and a person had to read with a knife in hand to slice open each page.) It's because of the signatures that you often see blank pages at the end of a book, where the actual text was not enough to fill up the signature. I always ask the press what signature size they are using and plan the final page count around that, adding a half title page or condensing the index or expanding the table of contents to make the numbers of pages come out to an even multiple of 8 or 16. It's a puzzle.

Sometimes for a small job, like a booklet printed on half of an 8.5 x 11, you need to figure out the signature yourself. That is, you need to figure out that page 16 should go next to page 1 on the *camera-ready* piece, and page 15 goes next to page 2, etc. The trick to getting this right is to make a "dummy." Get some paper and fold up all the pages into a mock booklet. Write the page numbers on every page. Then take the booklet apart and see which page ends up next to which other page. Logic makes very little sense when making a paging dummy, except to remember that when you add the two page numbers together, they should equal one more than the total number of pages.

For an eight-page half-size booklet, these are the signatures. Notice each set of pages equals 9 (one more than the 8 pages).

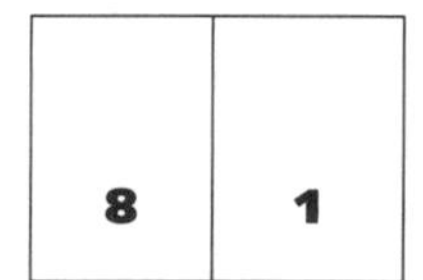

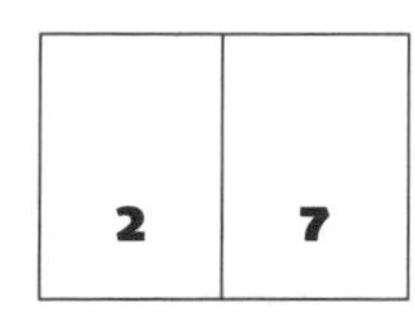

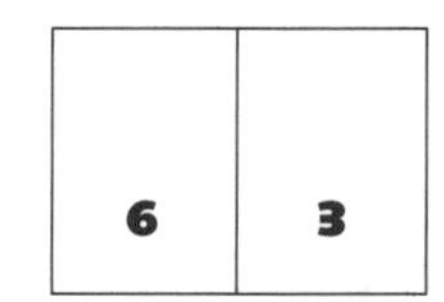

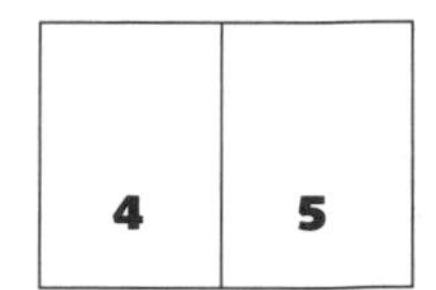

These two pages will be printed back to back, so 2 will be on the other side of 1, and page 7 will be on the other side of page 8.

These two pages will be printed back to back, so 4 will be on the other side of 3, and page 6 will be on the other side of page 5.

silicon

Silicon (pronoucned "sill i cahn," not "sill i cone") in its pure form is a lightweight metal that resembles aluminum. It's found in most of the rocks and sand on earth. In fact, silicon is the second most abundant element in nature, next to oxygen. See *semiconductor.*

Silicone (pronounced "sill i cone"), on the other hand, is a fluid, rubbery polymer. Silicon is in computers; silicone used to be implanted in breasts.

Silicon Forest

The Seattle area in Washington state is becoming known as **Silicon Forest,** with the Aldus Corporation and Microsoft Corporation as seeds.

Silicon Prairie

Austin, Texas, is becoming known as **Silicon Prairie** because of its growing number of high-tech companies. Dr. Macintosh himself (Bob LeVitus) lives in Austin.

Silicon Valley

Silicon Valley is the area in California around San Jose (where I grew up), south of San Francisco. It includes the towns of Palo Alto, Los Altos, Mountain View, Sunnyvale, Cupertino, Menlo Park, Fremont, and Santa Clara, among others. Silicon Valley has the largest concentration of computer-related and innovative high-technology companies in the world. There are actually freeway signs in this area directing travelers to Silicon Valley. It will be interesting to see what will happen when the earthquake wipes out Silicon Valley. (I now live in New Mexico.)

Don't pronounce it Sili**cone** Valley or people will laugh at you. Sili**con** is used in computers. Sili**cone** was used in breasts.

SIMM

SIMM stands for **s**ingle **i**nline **m**emory **m**odule. Before you try to understand what a SIMM is, you should read and understand *RAM* and *memory.*

A SIMM is a little flat piece of plastic, usually about as long as your finger, that holds a number of *memory chips.* This plastic card is actually a small *printed circuit board.* When you decide it is time to add more *RAM (random access memory)* to your computer, then you buy SIMMs. Make sure you buy the SIMMs that fit your computer, both physically and with the right amount and size of memory chips. In most computers you can open the box and install SIMMs yourself (first call your *power user* friend). They just pop into the little slots that are made for them. There is an illustration of a SIMM in Appendix A in the section on memory.

simulation

Simulation is when a computer program (using a lot of math) imitates a real life object or a process and then displays what happens to that object or process when conditions change. For instance, physical models of airplanes are created on the computer and then put through simulations of turbulence and storms to see how the planes react. The computer can do what wind tunnels used to do. Larger computers can simulate stars and molecules so scientists can study certain aspects of them. The game Microsoft Flight Simulator so realistically simulates flying a plane that it's often used in professional flight instruction.

single-sided disk

A **single-sided disk** is a *floppy disk* created at the factory so it can only store information on one side of itself, as opposed to a *double-sided disk* that can store information on both sides. A double-sided disk, of course, can hold twice as much information. Single-sided disks were the only kind available in the early days of personal computing. You may occasionally run across them even today, but no one sells them anymore.

On the Macintosh, when you put a brand new floppy disk into the computer or if you re-initialize a disk, it asks you to decide whether to *initialize* the disk single- or double-sided. Some people initialize their old single-sided disks as double-sided and many times this works just fine. You do run the risk, though, of losing all the data the computer stored on that second side. Few things are worth that risk.

.sit

This little term, **.sit,** is called "dot sit." When you see it at the end of a file name, it means the file has been *compressed* so it doesn't take up as much space on the disk. Most .sit files must be uncompressed before you can use them. You usually have to use the same program that compressed the file to uncompress it. (Notice it's called *un*compressing, not *de*compressing.) Also see *compress, archive,* and *.sea*.

site license

Technically, you (or a business or a school) must buy a full software package for each computer on which the product is used, even if the two or more computers sit right next to each other. For fonts, you need to buy a copy of the font for each printer they will be printed to. As this can get prohibitively expensive for institutions like schools or even for large businesses, most software companies offer **site licenses.** This doesn't mean you can just go and copy the one program you paid for onto every computer; it usually means you can buy multiple copies at a reduced rate.

sleep

Some battery-powered *laptop* and *notebook* computers let you put them to **sleep** to preserve the battery charge while you're not actively working (like when you're on the plane and you need to use the restroom). Sleep is a temporary state where the computer uses a very small amount of the charge because it's not doing anything. The screen usually goes black and the hard disk stops spinning, but the power is still on. To wake the computer up, you simply press any key.

To preserve the battery while using a Macintosh PowerBook, you can put it into **sleep** mode. Choose "Sleep" from the Special menu, or click on the battery icon on the little Battery *desk accessory* from the Apple menu. The PowerBook will automatically put itself to sleep if you don't touch it for a little while. Press any key to wake it up. Also see *battery.*

slot

A **slot** is a special place in which to insert something. On the outside of your computer you have a slot where you can insert floppy disks. Inside your computer there are slots where you can add *cards* (printed *circuit boards*) to enhance your computer.

small caps

You know what lowercase letters are. And you know what capital, or uppercase letters are. **Small caps** are capital letters that are about the size of lowercase letters, LIKE THIS. They are designed to allow you to set text in all caps, as some things must be (like all these darn *TLAS*), yet not let that word overwhelm the sentence. When a word is set in all full-sized caps, it appears so much larger than the rest of the sentence that it calls much too much importance to itself. The small caps, because they are closer in size to the lowercase letters, let an all-caps word blend more harmoniously into the page.

There is a design problem with using small caps, though. If you have ever used them, you may have noticed that the regular capital letters appear much bolder than the corresponding small caps. That's because to make the small caps the computer just reduces the size of the letters. The thickness of the letters automatically gets thinner as the size gets smaller, and the result is visually inconsistent letterforms. If you find that you use small caps regularly, invest in one of the font *expert sets* that has a specially redesigned set of small caps that are in proportion to the capitals.

REGULAR SMALL CAPS

EXPERT SET SMALL CAPS

Notice how the stroke thickness in the caps and small caps is consistent in the expert set.

smart quotes, smart apostrophes

True typographic quotation marks and apostrophes look like this:
“ ” and ‘ ’ as opposed to typewriter marks: " and '.

Notice the typographic marks turn in towards the sentence, both at the beginning and the end. Using typewriter apostrophes and quotation marks in your computer-generated material is the single most visible sign of unprofessional type. Because typewriter quotes are so tacky, many software packages have a feature called **smart quotes** that will automatically insert a real apostrophe or quote mark when you press the typewriter quotation mark key. The real marks themselves are sometimes called “smart quotes.”

If you don't have a “smart quotes” feature in your application (it may be called “Use typographers' quotes” or something similar), these are the keystrokes for typing them manually. If this is the first you have heard of smart quotes, check out *The Mac is not a typewriter.*

“ Option [

” Option Shift [

‘ Option]

’ Option Shift]

On a PC, it depends on your *configuration* of monitor, software, printer, and fonts.

In Windows, you can type smart quotes by holding down the Alt key and typing 0147 for " or 0148 for ".

In PageMaker, type Control [for ", and Control Shift [for ". Type Control] for ', and Control Shift] for '.

With other software, consult your manual. If this is the first you have heard of smart quotes, check out *The PC is not a typewriter.*

snailmail

What with overnight mail and modems and satellite communication, the U.S. Post Office has been dubbed **snailmail.**

snd

The abbreviation **snd** either in a file's name or in the information pertaining to a file indicates that the file is a *sound resource.* See *sound resource.*

Sneaker net

There is overnight mail and modems and networks and Internet and UserNet and wide area networks and local area networks and satellites and all manner of ways to get information from one place to another—but sometimes the best and fastest and most reliable is the **Sneaker net,** where you pick up the information in your hands and walk down the hall and deliver it. Wearing sneakers, of course.

snooze

There are many programs that allow you to set up alarms for yourself, where at a designated time a message will appear on your screen telling you it's time to go somewhere or do something. Some of these programs let you **snooze** the alarm, just like on your alarm clock that wakes you up in the morning. You can snooze the computer alarm for a certain number of minutes. When that time has elapsed, the message will appear again.

snow

In the early days of the PC, IBM made a graphics *board* called the *CGA*, or Color Graphics Adapter. Despite its name, a CGA card can display text as well as graphics. And among its many deficiencies, IBM's CGA card caused an annoying sparkling and flickering whenever you scrolled text on the screen. This effect is called **snow.** Many text-based PC programs have a setting that lets you control for snow at the cost of slowing down the program a bit. But even if you're stuck with a CGA system, you may not need the snow control—the majority of CGA-compatible cards weren't made by IBM, and most of these don't have the snow problem.

soft font

A **soft font** is a *font* (specifications for a typeface) that is stored on a disk that you have to *download* (send) to your printer in order to get it to work. This is in contrast to *cartridge* fonts that are stored in cartridges that plug into some printers, and built-in *(resident)* fonts that are stored in *firmware* inside the printer itself. The term soft font means the same thing as *downloadable font,* except that it is usually applied exclusively to the downloadable fonts that work with the Hewlett Packard LaserJet printers. See *font.*

soft Return

When you press the Return key on most computer programs, you get a hard *line break* and a new paragraph. But sometimes you want to break, or end, a line before the computer does it for you automatically. If your program has this feature, you can enter a soft Return (often some keyboard combination like Alt Return, Shift Return, or Option Return). You will still get a hard line break, but the advantage is that you do not get a new paragraph with all its formatting, such as a first-line *indent,* space before and after the paragraph, etc. Any new lines created with a soft Return will be considered part of the same paragraph. A soft Return is sometimes called a "line break," or "forced line break."

I must tell you, though, that the above description is the "popular" definition of a soft Return. Technically, a soft Return is what you know as *word wrap,* where the words just bump into the right margin and bounce back to the left margin (see the illustration in *word wrap*). A soft Return is really when the **computer** decides where to end the line, which then is "soft" and flexible and will be changed automatically as the text is edited.

software

Software is instructions, stored on a disk. The software determines what you can do with your machine. When you buy a computer application such as Aldus PageMaker, you get a box with several thick *manuals* (also known as *documentation*), registration papers, and several *floppy disks* or *diskettes* (however you prefer to call them). On the disk itself, encoded into the magnetic layer, is the software, the magic that makes the computer do what you tell it.

Lotus 1-2-3 or Microsoft Excel is software that lets you work with numbers *(spreadsheets)*. Word processing software like Claris MacWrite or WordPerfect gives you the capability to type multitudes of words and edit and rearrange them. Other software gives you the tools to design pages or to balance your checkbook. Without software the computer itself is an expensive doorstop.

If you can bump into it, it's *hardware*. If it's some mystical stuff on a disk that you cannot see or touch, it's *software*. Most computer *programs (applications)* are software, meaning that they come on disks. Some programs are stored in *chips* inside your computer; this is *firmware*.

Clay says hardware is what you can kick; software is what you would *like* to kick. And *liveware* (sometimes calles "wetware") is us, the live humans who use all this other-ware.

software library

A **software library** is your collection of *software*—the applications and programs that turn the computer into more than a large paperweight. I don't know why they call it a library if you are not allowed to loan them out.

solution

A **solution** is a product, any product, hardware or software. At least that's the way marketing people in the computer industry use this term. It solves a problem, doesn't it? That's the idea.

sort

Sort means about the same thing in computer jargon as it does in regular English: to organize items in some kind of order. If you're working with words, sorting them means to alphabetize them, either from A-to-Z or from Z-to-A. If you're working with numbers, sorting them means to organize them from the lowest to the highest, or the other way around. Information in a *database* is commonly sorted according to some particular detail, such as zip code or date.

sort within a sort

Sometimes you want to arrange the information in your *database* alphabetically according to a certain category of information, such as by city. But maybe you also want the zip codes within each city *sorted* (arranged in some logical order) as well. Sorting one category within another sorted category is a **sort within a sort.** If you sort by state, zip code, city, and last name, you'd have a sort within a sort, within a sort, within a sort, and each seperate category would be in order.

Many database applications allow you to sort according to several categories of information in this way. You choose the "sort order"—which type of information is sorted first, which next—based on what you need to know. For instance, if you mainly want to track month-to-month sales, you'd sort primarily by date, and then secondarily by customer. If you were mainly interested in knowing how much each customer had spent, you'd sort by customer, then by amount of purchase.

Do a sort-within-a-sort by starting with the smallest and working your way up to the biggest. For instance, to do the sort mentioned in the first paragraph, first sort by the last name. After these are all alphabetized, then sort by city. As the database sorts by city, it will keep those last names in alphabetical order. Then sort by zip code. Then choose to sort by state.

sound resource

A **sound resource** is what you call the little piece of programming that makes a sound. For every sound you hear on your computer, there is a sound resource somewhere. With the proper software you can record your own sounds, then install the sound resource into your System.

source code

Source code is the programming instruction code that the programmer (a human) originally wrote. The computer, however, cannot understand source code, because the computer is a machine. The source code has to be converted to a language the machine can read, using some sort of special conversion program (an interpreter, an assembler, or a compiler).

Spacebar

The **Spacebar** is the long, blank bar at the bottom of the keyboard that you press while typing to make a blank space. It's sometimes used in combination with a *modifier key* in keyboard shortcuts. Don't use the Spacebar to align text in columns, like you used to do on a typewriter. Always use *tabs* and *indents*.

SPARC™, SPARCstations

SPARC stands for **s**calar **p**rocessor **arc**hitecture, which is a *chip (CPU) architecture*. Sun Microsystems builds these chips into *workstations* called SPARCstations. These fast and powerful workstations are used for scientific and engineering purposes and sometimes for high-end publishing.

spelling checker

A **spelling checker** is *software* that checks your spelling for you (do you think that's why it's called a spelling checker?). Some spelling checkers are independent applications, but most often this a feature within a word processing or page layout program. When you *run* a spelling checker, it reads through your document and selects the misspelled words—"misspelled" means the word is not in the dictionary it uses. "Misspelled" includes many foreign words, technical terms, proper nouns, and sometimes words with capital letters in the middle, such as PageMaker. Most spelling checkers let you add on to the base dictionary to include words you use often.

It's important not to rely totally on the spelling checker because there are so many words in our language that sound the same and are spelled differently. For instance, a spelling checker would not care if you wrote, "He wore a blue pear of pants," because the word "pear" is in the dictionary.

This term "spelling checker" has been shortened and contorted into a verb, as in "Did you spell check the document?" And also into a noun, as in "I'd better run a spell check on this."

spike

A **spike** is a pulse of extra voltage on the power line that only lasts a fraction of a second. A *surge* is an excess of voltage that lasts longer, like several seconds. Both spikes and surges can damage your computer and destroy data. Both spikes and surges happen regularly, as you turn electrical appliances on and off in your home or office, or as the electrical company boosts the power in the late afternoon so everyone can cook dinner at the same time. Some computers have a *surge protector* of some sort built into them to protect against these spikes and surges, but you usually need a more powerful unit to truly protect your machine. See *surge protector.*

split bar

A **split bar** is a little narrow black bar at one or both ends of a *scroll bar* that you can drag with the mouse, dividing the window into two independent sections (see *split window,* below). To put the split screen back into one piece, drag the split bar back to where it came from.

split screen, split window

The size of the computer screen, or the window within the screen, is often smaller than the document you are working on, and you can only see a section of the document at any one time. For this reason, many applications offer a way to display a **split screen** or **split window,** with the screen (the window, actually) divided into two or more parts. Each part can show a different portion of the same document. This is particularly handy in databases and spreadsheets, where you may want to view the information in the first two columns at the same time as the information in the last column which is 16 inches across the worksheet. By splitting the screen you can view the first two columns in the **left**-hand portion and the last column in the **right**-hand portion of the screen.

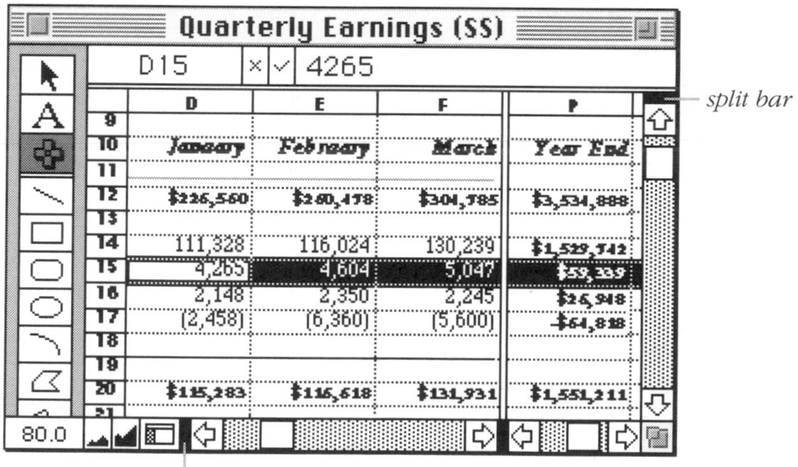

Quarterly Earnings (SS)

D15 4265

	D	E	F	P
9				
10	January	February	March	Year End
11				
12	$226,560	$260,478	$304,785	$3,534,888
13				
14	111,328	116,024	130,239	$1,529,742
15	4,265	4,604	5,047	$59,339
16	2,148	2,350	2,245	$26,948
17	(2,458)	(6,360)	(5,600)	-$64,818
18				
19				
20	$115,283	$116,618	$131,931	$1,551,211

80.0

split bar

This is the split bar. Notice there is one at the top of the vertical scroll bar as well. Each portion of the split window now has its own independent scroll bar.

spool, spooler

When you choose to print a document, the computer sends the document information to the printer very quickly, but the printer can't accept it at the same rate. The printer can only handle a chunk of information at a time, and it pauses to process and print that chunk before it's ready for more. Meanwhile, you have to wait until the printer has accepted the whole document, piece by piece, before you can use your computer again because the computer has to hang around and feed the information through. That's why you need a print **spooler—**software that reduces the amount of time during which you can't work while you wait for a job to print.

A spooler works by intercepting the information going to the printer, parking it temporarily on disk or in memory. The computer can send the document information to the spooler at full speed, then immediately return control of the screen to you. The spooler, meanwhile, hangs onto the information and feeds it to the printer at the slow speed the printer needs to get it. So if your computer can **spool,** you can work while a document is being printed.

You will notice during spooling, though, that your work gets slightly interrupted for a few seconds here and there because the computer cannot *really* do more than one thing at a time, meaning it can't keep the spooler running and your monitor running at the same time. So if your cursor doesn't move or the letters type sporadically here and there, don't worry. Total control will be returned to you when the printer is done.

Sometimes spooling software is built into the system software, as in "Background printing" in System 7 or the "Backgrounder" file on System 6 on the Mac, or the Print Manager in Windows. Sometimes you buy extra software that allows you to spool.

spot color

Spot color refers to colors on a page that are solid colors or perhaps a tint of a color. If you put the headlines in your newsletter in a different color, that is spot color. If you have the heavy lines at the top and bottom of a page and perhaps a graphic in a different color, that is spot color. But color photographs and multicolored artwork are **not** spot color—they need more complicated techniques *(four-color separations)* to print them on the page, rather than simply printing a color where the spot color is designated. Spot colors might touch each other or they might not—that's entirely up to the designer. Whether they do touch or not has nothing to do with calling them spot color.

Spot color doesn't mean there is only one other color on the page besides the basic black—there can be a page with six or eight or any number of spot colors. But when you go beyond four separate spot colors, it often behooves you financially to use another process, like the *four-color process* where any number of colors are built out of the four basic process colors: cyan, magenta, yellow, and black. See *CMYK* and *four color process*.

spreadsheet program

A **spreadsheet program** is a *software package* used for financial or other number-related information processing, like in budgeting, cash flow analysis, forecasting, etc. One great thing about a spreadsheet is the "what-if" factor you can play with—how much will my mortgage payment be ***if*** the interest rate drops a quarter percent; ***if*** I charge 10 cents more for each widget, how will that effect the overall profits of my business?

A spreadsheet is composed of rows and columns of *cells* with letter names for the columns and numbers for the rows. Each cell can hold information as numbers, text, or formulas. Formulas can be set up to get information from other cells and to act on that information (as in the calculation (B1+C3)*12). There are usually millions of cells to work with, although the typical spreadsheet document (or "worksheet," as a spreadsheet document is often called) uses a tiny fraction of the available cells. With some spreadsheet programs, you can have several related worksheets in one project, all interconnected.

Most spreadsheets can create a number of different charts and graphs for you automatically, using the data you choose.

Quarterly Earnings (SS)

M4

First Quarter 1991

	C	D	E	F	G	H	I
10		January	February	March	April	May	June
12	NET SALES	$226,560	$260,478	$304,785	$304,785	$304,785	$304,785
14	Cost of Goods Sold	111,328	116,024	130,239	130,239	130,239	130,239
15	Shipping	4,265	4,604	5,047	5,047	5,047	5,047
16	Depreciation	2,148	2,350	2,245	2,245	2,245	2,245
17	Interest Expense	(2,458)	(6,360)	(5,600)	(5,600)	(5,600)	(5,600)
20	TOTAL EXPENSES	$115,283	$116,618	$131,931	$131,931	$131,931	$131,931
23	Earnings Before Taxes	111,277	143,860	172,854	172,854	172,854	172,854
24	Income Taxes	44,511	57,544	69,142	69,142	69,142	69,142
27	NET EARNINGS	$66,766	$86,316	$103,712	$103,712	$103,712	$103,712

80.0

This is a typical, simple spreadsheet, created in ClarisWorks.

SQL

SQL (pronounced as the letters "s q l," or sometimes as "sequel") stands for **s**tructured **q**uery **l**anguage. It's a language created specifically for searching for and otherwise managing information through large *relational databases.* You can also use SQL commands to add to or change the information in the database.

SQL is more like English than many other database programming languages, which makes it somewhat easier to understand and use. Supposedly, non-technical people, as well as techies, can work with SQL. But you still have to use it very precisely—no sloppy grammar allowed.

SRAM

SRAM (pronounced "ess ram") stands for **s**tatic **r**andom **a**ccess **m**emory. Most *RAM* in computers is *DRAM,* or dynamic RAM, which loses its charge and must be constantly recharged, or "refreshed" (like hundreds of times per second). Static RAM (SRAM) does not need to be constantly refreshed like DRAM—it just needs to have power flowing through it. This is why it's so very fast—*access times* of 10 to 30 *nanoseconds* are typical, compared to 50 to 150 *ns* (nanoseconds) for DRAM. Wouldn't you know it, SRAM is much more expensive to make than DRAM. And SRAM chips can only store about one-fourth the amount of data as the same size DRAM chips. SRAM is often used in *portables* since it requires less energy to stay active, and it is also used in memory *caching* because it is so fast.

stack

A **stack** is what Apple calls the documents created in *HyperCard.* A stack, though, almost never comes out of the computer. It's not like a word processing document or a page layout publication that you print out—most HyperCard stacks are meant to be used only while the computer is on.

A stack may be a project in itself, like a directory for a shopping mall, or it may be the *front end,* the part that runs the show, like the part that guides people through the art of the Louvre on screen.

This is the icon for a HyperCard document, called a stack.

stand-alone

Stand-alone means that a software product is separate from any other software product. Most major packages, of course, like spreadsheets and databases, are stand-alone, so this term usually is applied to products that you might commonly find included in a larger package. For instance, a stand-alone *spelling checker* is an application that stands alone, or is separate from, any word processing program—although its purpose is to check documents that were created in a word processor.

standards

Standards are technical, detailed guidelines created to establish a uniformity in a field so people can have products that work with a variety of other products. The lack of *printer driver* standards in the IBM world led to an abundance of different printers, all of which need special software to use.

Official standards are set by committees like *ANSI,* the American National Standards Institute. But many procedures and uniform techniques become standards simply because they are so common already, a company would be foolish to develop a product not using that standard. These are called *de facto* standards (*de facto* literally meaning "from the fact," as opposed to *de jure,* "from the law"). The *page description language* called *PostScript* is a de facto standard, as is the *Hayes modem protocol.*

star-dot-star (*.*)

The symbol *.* looks like "asterisk-period-asterisk" but it's pronounced **star-dot-star** (thank goodness). Star-dot-star is a shorthand way to designate any and *all* files in a directory of DOS or OS/2 computers. You might type **DEL *.*** (meaning "delete all the files in this directory") or **COPY *.* B:** ("copy all the files to disk B").

The star (*) is a *wildcard* and can stand for any word sequence of characters of any length, up to the maximum of eight for the main part of a filename or three for the *filename extension.* So *.* means any filename and any extension, which includes all the files

The question mark symbol (?) is also a wild card, but it refers to only one character, not to any length word. You could use it in the same way.

startup disk, startup volume

The **startup disk** or **startup volume** refers to the disk (also known as a *volume*) that is used to start up your computer. It can be either a floppy disk or a hard disk.

On a PC, neither DOS nor Windows gives you a visual clue as to which disk is the startup disk—you just have to remember which one it is.

On a Mac, the icon for the startup disk (also known as the *boot disk*) always appears in the top right corner of the screen. If you have more than one hard disk attached to your computer, you can choose which one of these you want to boot (start) your computer with; use the *Control Panel* called **Startup Disk.** The startup disk, of course, must have a System Folder on it.

StartUp group

If you're using Windows 3.1, you have a *Program Manager* group called **StartUp.** Every time you start Windows, Windows automatically runs all the programs you put in the Startup group. So if you want a program to start automatically whenever you start Windows, put it in the StartUp group. It's generally best to create a new program item for the StartUp group or copy (rather than move) an existing one there. That way, when you delete the program from the StartUp group, you can still find the original in its old location.

Startup Items

In System 7, there is a folder within your System Folder called **Startup Items.** Anything you put in this folder will automatically open when you start up (turn on) your Mac. Actually, what you really should put in the Startup Items folder are *aliases,* not the original files. Make an alias of the folder or program or document you want to open and put that alias in this folder (select the item, then from the File menu choose "Make Alias"). That way your original item stays in its filed spot where you can always find it and where it belongs. When you no longer want that item to start up, trash the alias or remove it from the Startup Items folder.

For instance, you might add an alias of a folder that you always want open when you arrive at the Finder. You might add an alias of the big project you are working on at the moment. For a while I was opening about five Word files and a Pagemaker file for this book, so I put aliases of them in my Startup Items folder; when I turned on my Mac it opened the applications Word and PageMaker and gives me those six files on my screen.

You can store sounds in this folder and they will play when the Mac starts; they play after the Desktop appears. I showed Uncle Floyd how to use the microphone that came with his Mac to create a sound (very easy—just open the Sounds Control Panel and click Add). He yelled into the mike, "Get back in the kitchen, Woman, and fix me some dinner!" He moved the sound file into the Startup Items folder in Auntie Jeannie's Mac. Don't worry—I showed her how to get even.

After you make the sound, it will appear in your System file. Drag it out of there and into the StartUp Items folder, **or** *hold down the Option key and drag a copy into the folder. You can't use an alias for this.*

startup screen, StartupScreen, Startup Movie

The **startup screen** is whatever you see on your screen when you first turn on your computer. In particular, Windows and the Mac both display pictures on their startup screens. You can substitute pictures or messages of your own.

On the Mac, a **StartupScreen** is a file (MacPaint file in System 6; System document in System 7) in the System Folder. When you turn on your Mac, this file appears on your screen for several seconds, right after the *Happy Mac* icon. To make a StartupScreen, use an application like SuperPaint; you can also capture an image with one of the several *screen capture utilities.* In SuperPaint, create an image. It won't do you any good to fill the entire 8 x 10 space because the StartupScreen will only use the space that you would see in a small screen Mac. Use clip art or draw something yourself or just type a message. Save the document in the StartupScreen format. Name the file **StartupScreen.** Capitals letters don't matter, but it must be one word.

Then take that file and put it in the System Folder. It doesn't belong inside any other folder within the System Folder—it has to be floating around loose. The next time you start your Mac, this picture will appear on your screen. Oh all right—I won't tell you all the mean rotten tricks you can do with a StartupScreen.

You can also have a **Startup Movie** (must be two words). If you have a *QuickTime* movie clip, name it Startup Movie (include the space between the words) and place it in your System Folder. When you turn on your Mac, the movie will play. Too cool.

state-of-the-art

If a product is **state-of-the-art,** it means it is the latest innovation in its field. In the computer industry, most things are state-of-the-art for about ten minutes.

stop bit

In *telecommunications,* like when you are sending information over a *modem,* a **stop bit** refers to the delay after one character (typically one *byte*) has been sent and before the next character (byte) can be sent. It tells the receiving machine that the entire byte has been sent.

storage media

Storage media refers to anything you *save* or *store* your files on, such as a hard disk, a cartridge hard disk, a floppy disk, a tape, etc.

store

To **store** something means to *save* it on a disk. See *save-to-disk.*

string

A **string,** or **character string,** is a series of *alphanumeric* characters (meaning there can be numbers and punctuation marks as well as letters). A string can be one character or it can be pages worth of characters. A string is data, or information, that the computer does something with. For instance, when you search for a particular word or phrase in your word processor, the computer is searching for the "string" of characters you typed into the little box. See *character string.*

stuff, stuffed

If you hear someone say that a file is **stuffed,** it means the file has been *compressed.* The term comes from the popular *compression utility* called *StuffIt* (see the next definition), but "stuffed" is often applied to a file that has been compressed by any means. **Stuff** can be a verb, as in "I need to stuff this file before I send it."

StuffIt

StuffIt is the name of a *utility* that *compresses* files (makes files take up less disk space without losing any data). Sometimes you want to compress a file so it will fit on a disk for transportation or storage; sometimes you want to compress it so you can send it over the *modem* faster (since it is smaller, it takes less time to send all the information). Usually you must uncompress the file before you can use it. If StuffIt was used to compress the file, usually it must be used to uncompress, also.

With StuffIt you create an "archive," or holding place, where you can put as many files as you like in one stuffed bundle. Then you can store or send this entire archive at once.

Compressed files are often indicated by an *extension* (a one- to three-letter identification tag at the end of a file name) of **.sit** (pronounced "dot sit). For instance, if I see a file named "Ryan.sit," I know it is a compressed file. Also see *.sea.*

A compression utility that competes with StuffIt is *DiskDoubler.*

style

The term **style** has two definitions. One refers to the *character formatting* of text, such as whether the text is bold or italic or reverse. Typically a style is only applied to the characters that were *selected,* or *highlighted* before the command was chosen.

Most applications apply styles with a *toggle* command. That is, if you select text and press a keyboard shortcut to apply the bold style, you can ***remove*** the bold style by selecting the text again and repeating the same keyboard shortcut.

The other definition for **style** is a particular set of formatting characteristics that you can apply as a group to an entire paragraph or a selected block of text. With one command, you can set a paragraph style that formats the typeface and size for the text, the margins, any *indents,* whether the paragraph is *justified* or *ragged* or centered, the spacing between the lines, and so on. In this sense, styles are what make up a *style sheet*— please see the next definition.

style sheet

A **style sheet** is a collection of the *styles,* or formatting, you have applied or will apply to certain paragraphs of text (remember, on the computer you create a new "paragraph" every time you hit the Return/Enter key).

The style sheet might include one set of formatting for headlines, another for body text, another for captions, another for pull quotes, etc. Having a group of these sets of styles makes it incredibly easy and quick to format an entire document.

In a style sheet for a headline, for instance, you can include the typeface, the size, the style (bold, italic, etc.), the amount of space you want above the headline, the space below the headline, the alignment, whether it is to be included in the table of contents, whether it should start a new page or a new column, any tab settings, and other details. Then you simply click in the headline, choose that style from the style sheet, and all that complex formatting is automatically applied to the headline on the page.

The best news is, once you have applied a style from a style sheet to the text, you can make *global* changes (changes throughout the entire document) with just the click of a button or two. Let's say you have a 16-page newsletter with 12 different stories in it. Each story has a headline using the typeface Futura and body copy using New Century Schoolbook. There are also subheads and captions throughout the 16 pages. Well, you can imagine the trouble it would take if your boss decided (after you finished the newsletter) that she wanted all the headlines in Antique Olive instead of Futura, and of course all the smaller-sized subheads have to match, and all the body copy should be Bembo instead of New Century Schoolbook, and it should be *justified,* not *flush left.* If you had applied style sheets to the text, you would simply go to the style sheet definition, click a couple of buttons to change the body copy to justified Bembo, and everything in those 16 pages that had the style "Body Copy" applied to it would now be Bembo and it would be justified. You could change the definition for the headlines to Antique Olive. If the subhead definition had been "based on" the headline definition, the small-sized subheads would automatically change font to match the heads. Oh, it's too cool.

Style sheets have got to be the most important time-saving and frustration-reducing invention since the computer itself. Most good word processors and all page layout programs have style sheets.

stylus

A **stylus** is a pen with no ink. It is not meant for writing on paper, but for writing on some sort of electronic grid. The grid, whether it is on an *LCD* panel or on an electronic *tablet,* translates the movements of the stylus as you write on it into digital information that the computer can use. Some combinations of stylus and tablet can translate handwriting into text or can mimic the action of paint on canvas or ink on absorbent paper.

SuperATM

SuperATM is technically version 3.5 of *ATM, Adobe Type Manager* (see *font* for details about how and why ATM works). But SuperATM has a special purpose in life. Have you ever opened one of your documents on another computer and discovered that the formatting had completely transmogrified into a horrible mess because the fonts you had used were not in this computer? SuperATM uses Multiple Master technology to solve this problem. If SuperATM is installed in the other computer, your document will open with fonts substituted by SuperATM that exactly match the *metrics* (the width, spacing, thickness of strokes, etc.) of the missing fonts. It won't be the same font, but the look will be similar, the lines will break at the same place, the headlines will look like your headlines, the layout will stay the same. You can edit this document with the substituted fonts; when you take it back to your machine, your original fonts will be automatically replaced.

This isn't a big deal for typographers or designers who are fussy about their type, but it will be a wonderful tool for the millions of people who need to pass electronic documents around. No longer will a document appear on your screen or print in the dreaded Courier.

supercomputer

Supercomputers are the biggest, fastest, and most expensive computers on earth. The cheapest supercomputer costs well over $1 million. They're used for scientific simulations and research such as weather forecasting, meteorology, nuclear energy research, physics and chemistry, as well as for extremely complex animated graphics.

Besides raw speed, one big difference between a supercomputer and a *mainframe* is that a mainframe serves many people at once or runs several programs concurrently, whereas a supercomputer funnels its power into executing a few programs at high speeds. Mainframes are mostly used for large data storage and manipulation tasks, not for computationally-intensive tasks.

SuperDrive

A **SuperDrive** is the same thing as an FDHD, the **f**loppy **d**isk/**h**igh **d**ensity drive in the newer Macintosh models. This drive can read *high density disks* (disks that hold 1.2 *megabytes* of data) and 3.5″ DOS-formatted disks, as well as Apple II and OS/2 disks.

Superfloppy

Superfloppy refers to *high density* 3.5″ floppy disks, the ones that can hold 1.2 *megabytes* of information. You can easily recognize a superfloppy disk because it has two holes in it, one in each corner. One of the holes has a little tab so you can lock the disk; the other hole has no locking tab. Also see *high-density disk.*

Super VGA

Super VGA is the generic term for PC video circuits *(adapters)* that display a *resolution* of 800 x 600 pixels on the screen. Images produced by a Super VGA adapter are noticeably sharper than those from a standard VGA, which has a resolution of 640 x 480. To use a Super VGA, you need a *multiscan* monitor; a standard VGA monitor won't work.

support

Support has two definitions. For one, it refers to the help you get for your software and hardware from the people who sold you the product—either support in getting it up and running, or perhaps support in exchanging a defective product. Sometimes you get no support from the place where you bought it.

When you call the software or hardware vendor directly (rather than the store where you bought it) for help in actually using the product, it's called **technical support** (**tech support**). Some companies have free 800 numbers for technical support, while others actually charge you for their time (plus you pay for the phone call). Either way, it's usually quicker and cheaper to find the information you need in the manual before you call *(RTFM!).* But don't be bashful about calling tech support even for "minor" questions if they're important to you (yes, "tech support" has been turned into a noun). Vendors are always hoping that customers will only call for help with questions that aren't answered in the manual. In reality, they'll spend a lot of time teaching people the basics of their products until the products themselves and the manuals become really easy to use (says Steve; Robin says, "No matter how wonderful the manual is, most people won't read it anyway; that's why I have a job").

It works best if you call the technical support number while you're sitting in front of your computer so the technician can ask you questions about what you see on the screen and can have you try out advice on the spot. You should have your software *registration number* handy when you call, because you may not get your questions answered without it.

I once called tech support for help formatting a new cartridge hard disk. It turned out I had their old software (I don't know why they hadn't sent me the new software with the new hard disk). So the tech support guy walked me through the process of initializing and formatting my cartridge disk. Unfortunately, he mistakenly walked me through initializing and formatting my entire 80 meg internal hard disk instead, one week from the PageMaker book deadline. Yes, it was a catastrophe of considerable dimension. I was not very happy. I have since heard that this company has done this same thing to many others.

The second definition of **support** is truly jargonesque and gets used with imprecise abandon by marketing people trying to sound impressive. It can mean "to work with properly," or "to permit the use of," referring to other products, as in "PopPrinter software supports PostScript and LaserJet printers" (whatever PopPrinter does, it works with these two kinds of printers). It can mean "to adhere to the standards of" as in "MightyModem supports the AT command set and MNP 42.*bis.*" And it can simply mean "to have," referring to features or capabilities, as in "MagicWerd supports multicolored drop-down menus."

surface-mount

Did someone tell you, "Be sure to buy RAM in surface-mount SIMMs"? **Surface-mount** technology (abbreviated SMT) is a method of making *printed circuit boards,* such as *SIMMs,* by mounting the chip directly to the surface of the board instead of soldering the chip into pre-drilled holes. A surface-mount makes the board more compact, more resistant to vibration, and more chips can fit on each board.

surge

A **surge** is an extra burst of power that lasts for several seconds (a *spike* is a very short burst, lasting less than a second). A surge or a spike can destroy data and even damage your equipment. They happen all the time. Most computers have some sort of minor surge protection built in, but every computer should have a *surge protector* between the wall outlet and the computer's power plug.

surge protector, surge suppressor

A **surge protector** (also known as a **surge suppressor**) is an electrical device that goes between the wall outlet and the plug from your computer. It diffuses any *surges* or *spikes* (extra bursts of power that come through the power lines) before they can blast into your computer and harm things. Both spikes and surges happen regularly as you turn electrical appliances on and off in your home or office, or as the electrical company boosts the power in the late afternoon so everyone can cook dinner at the same time. All computers have a *surge protector* of some sort built into them, but you usually need to add a more powerful unit to truly protect your machine.

Most surge protectors look like those outlet bars that let you plug several plugs into one socket. But some of those outlet bars suppress surges and some of them don't—make sure the package actually mentions surge suppressing. You can get inexpensive surge suppressors in household hardware stores for around $15, and more expensive versions ($200) in electronic appliance stores.

swapping

Swapping on the computer is similar to swapping in Life—one thing is exchanged for another. You may have played the disk swapping game, where the computer spit out your disk, asked for another, then spit that one out and asked for the other one back. That happens when you copy information from one floppy disk to another floppy disk and you only have one drive; the computer has to take information from one disk, store it in memory, copy it onto the other disk, go back to the first one to get more information, then copy that new information onto the disk. It eventually ends.

Using *virtual memory* involves a lot of swapping—moving information from real memory into the virtual memory to store it for a while, then back into the real memory so the computer can actually work on it. This is also called "paging," and data is "swapped in" and "swapped out."

swap file

A **swap file** is a file that Windows creates on your hard disk where it temporarily deposits parts of programs and data files when they're not actively in use. In this way a swap file increases the apparent amount of *memory* in your system. When actual *RAM* gets full, Windows can move pieces of programs or data files into the swap file for temporary storage, freeing up space in memory for new files and programs. Windows still considers the information stored in the swap file active in memory, so when you call on a program that is in the swap file, Windows just reloads it back into RAM, "swapping out" enough material to make room for it in the swap file. The swap file is Windows' version of *virtual memory.*

When you run Windows in *386* enhanced mode (your computer has to have an *80386, 80486,* or Pentium *processor*) you can have either a temporary or a permanent swap file. If you use a temporary swap file, then Windows creates it for you out of whatever space is available on your hard disk. You don't have to do anything special to arrange it, but it is slower than a permanent swap file. To use a permanent swap file, you have to set aside an area on a hard disk that doesn't have any files stored on it. Use the 386 Enhanced option in Control Panel to change your swap file settings.

Windows uses another type of swap file when it runs in standard mode. This swap file is created when you run *DOS* sessions within Windows. Windows creates a temporary file on the disk that stores the programs and information you are working with in the DOS session. Then if you switch back to the DOS session later, the contents of the swap file are returned to actual RAM so you can go back to working with whatever program you were using before—exactly where you left off.

SYLK

SYLK (pronounced "silk") is short for **sy**mbolic **lin**k, which is a *file format* for *spreadsheet* information, devised by Microsoft Corporation. It's used primarily for transferring information from one spreadsheet to another; you would rarely save your spreadsheet as a SYLK file for daily use. Instead, you'd save in the "native" file format of your program ("native" meaning the format it automatically saves in). If you need to transfer your work to someone who uses a different spreadsheet program, you can save the file as SYLK, send it off to them, they open it in their program, and save the file into their native file format. But actually, many spreadsheet programs understand one another's native formats, so you may not need to use SYLK at all.

synthesizer

A **synthesizer** is an electronic device that makes sounds. It may be a *stand-alone* system, or a *peripheral device,* or sometimes a *chip* that can create sound from digitized instructions (instructions from the computer) instead of from a musical instrument or recording device. With a *MIDI* interface, a computer can control one or more synthesizers.

SyQuest disk, SyQuest drive

SyQuest is the brand name of a *removable hard disk drive* and *cartridge hard disk.* Often drives and disks by other manufacturers are called "SyQuest," partly because most removable disk drives use the SyQuest mechanism inside, and partly because SyQuest has become a *de facto* standard and we tend to call everything similar by the same name (sort of like "Kleenex" and "Xerox"). Also see *removable cartridge, Bernoulli box,* and *Ricoh.*

sysop

Sysop (pronounced "siss op") is short for **sys**tems **op**erator, a person who runs or manages a computer *bulletin board* or other *online service.* A sysop may also act as a facilitator for online conferences or *forums.* A "Wizop" is a wizardly sysop, even higher in the online chain of command.

System

The **System**, with a capital S, refers to the Mac's *operating system,* the software that actually runs the computer. The System puts the computer into gear, like you put your car into gear. Without an operating system, you won't get past a gray screen; without putting your car into gear, you won't get out of the driveway. (Generally speaking, Mac users call it the System; PC users call it the *operating system,* or OS, which is typically *DOS.*)

You don't use the System itself directly—you use *applications* that use the System to get everything input and created and output. You manipulate everything in the application program, which works with the System to talk to the *disk drives* and the *SCSI ports* and the *fonts* and the graphics.

System software gets updated just like any other software. It's important to use the current version of the operating system to ensure that any new program you buy will also work, since new software is always written for the latest version of the system.

This is the Macintosh System file.

The System on a Macintosh is represented by the *System file icon.* In System 7 you can open this file and view fonts and sounds, but don't think that just becuse that's all you see in there that that's all there is! There is an extraordinary amount of invisible data in this file that runs the whole show. Also see *operating system.* If someone asks, "What system are you running?" they want to know what *operating system* you are using on your computer.

system

With a lowercase s, the word **system** usually refers to the kind of computer setup you have—what type of computer and monitor, how much hard disk space, how much *memory,* or what software package and *peripherals* you use. Someone might ask what kind of accounting system you use or what database management system you recommend. Also see the definition above for *System* with a capital S.

system disk, System disk

A **system disk** refers to a disk that can be used to start, or boot, the computer. Typically this would be your internal hard disk, but the term can refer to any disk, hard or floppy, that has the system files necessary to start your machine. (A disk that does not contain a System on it is a *data disk.*)

On the Macintosh, a System disk has the System Folder on it (since the System and Finder files are stored in the System Folder). The icon for the system disk, or *boot disk,* always appears in the upper right corner of your screen.

Even if you start your computer with a hard disk, it pays to keep a floppy system disk handy with at least a copy of the *System file* and the *Finder* file on it. You can use this System disk in an emergency (like when you *crash* and can't open to the Desktop anymore) to boot your computer and start *troubleshooting* from there.

The system files necessary to start your PC are the hidden DOS files, plus COMMAND.COM (see *DOS*). To make a PC system disk (let's say it's disk A) you either format the disk with the command FORMAT A: /S (/S for system), or add the system files to an existing disk with the command SYS A: .

System file

The term **System file** refers specifically to the file in the Macintosh's System Folder that represents the *operating system.* Also see *System.*

This is the Macintosh System file.

In System 7 you can double-click on the System file icon to view the fonts and sounds that are stored within this suitcase. You can double-click on a font icon to see the font; double-click on a sound icon to hear the sound. There is also an incredible amount of invisible data that you don't see, data that is running the computer. Also see *System.* Don't confuse the System file with the *System Folder* (shown below).

System Folder

The **System Folder** is the one folder, the magic folder, that contains the *System file* and the *Finder file.* If these two items are in a folder, that folder is "blessed" (two syllables).

This is the blessed System Folder.

The System Folder is an extraordinarily important folder. Everything inside of it helps to run the computer. Applications look in the System Folder for vital data and resources they need to function. You never want to put anything into the System Folder that doesn't belong there! Never store any documents in it and never install any applications in it. If a utility or some other form of software is supposed to be in the System Folder, the documentation will tell you so.

System heap

The **System heap** is a section of *RAM (random access memory)* that the System sets aside for itself to store such items as *drivers* and *extensions.* System 7 watches over this heap and controls its contents and size so there are no problems. In System 6, however, it is possible to wind up with an overcrowded System heap that causes such catastrophes as system *crashes.* If you find you are crashing regularly and you can't pinpoint it to anything else (like to a particular extension), you might want to check out the *utilities* Bootman or Heap Fixer and increase the size of the heap.

SYSTEM.INI

Windows uses the settings specified in the **SYSTEM.INI** file to configure itself properly for the kinds of equipment (the keyboard, mouse, screen, and so on) in your system. Although SYSTEM.INI is an ordinary text file and you can edit it with almost any word processor, the settings are very technical and you should leave them alone unless you're a real whiz, in which case you are probably not reading this book anyway. See also *WIN.INI.*

System 6

System 6 (or some variation thereof; see the next entry) was the Macintosh *operating system* version just before System 7 arrived (the operating system is what runs the computer and all the applications you use). Because System 7 required major changes for many people, not everyone has switched to it. So for over a year now, and probably for another year to come, everything you read refers to how things work in System 6 as opposed to how they work in System 7, so no one feels left out. See *System 7* if you're not quite sure whether you're currently *running* 6 or 7.

System 6.0.x

System 6.0.x (pronounced "six point oh point x") refers to any version of *System 6,* where **x** refers to the updated version number. For instance, 6.0.x could refer to either 6.0.1, 6.0.3, 6.0.4, or 6.0.8, etc. You may occasionally see it abbreviated as 6.x ("six point x"), but it refers to the same thing.

System 7

System 7 (or some variation thereof, such as 7.1 or 7.1.2, etc.) is the latest version of the Macintosh *operating system.* It was a major upgrade from *System 6*—so major that many people needed to upgrade their computers,

hard disks, memory, and all their software packages to take advantage of it. There are some strong innovations like *Interapplication Communication* and *virtual memory* access that work quietly behind the scenes. There are a good number of wonderful little tools that make any user's life much easier, such as *aliases, Find,* easy *file sharing,* and an *Apple menu* that can be customized. There are also some great shortcuts for many daily tasks which (once you get used to using them) make it difficult to go back to System 6. Eventually all software will be written to work with System 7 and everyone will have to upgrade.

If you see this in your menu bar, you are running System 7.

How do you know if you're on System 7? Look in the upper right corner of the menu bar. If you see a little question mark inside a balloon-like thing, you are running System 7.

If you are new to System 7, you might want to invest in the third edition of *The Little Mac Book,* which is full of hundreds of tips and tricks that take advantage of System 7. (Make sure you get *The Little Mac Book* written by Robin Williams, not the imposter version with the same title that came out almost a year later from Que Corporation.)

System 7 compatible, friendly, savvy

If a program or utility is **System 7 compatible,** it can *run under* System 7 and it probably won't crash or act weird.

If a program or utility is **System 7 friendly,** it runs perfectly under System 7, but it doesn't take advantage of System 7's new features.

If a program or utility is **System 7 savvy,** it not only runs perfectly, but it takes advantage of new features such as *Balloon Help, Apple Events,* or *publish-and-sbuscribe.*

system software

System software is another way to refer to the *operating system* plus all the various and sundry files, supplied by the vendor, that come with the system to make the computer work or make it more fun and easier to work on.

On a Macintosh, system software includes such items as the standard *desk accessories, printer drivers,* the word processer called *TeachText,* and a limited selection of *fonts. HyperCard* used to be system software, but unfortunately it's no longer included.

RISCY BUSINESS
WILL DO MULTIMEDIA FOR FOOD
BIZ CARDS
JOHN GRIMES

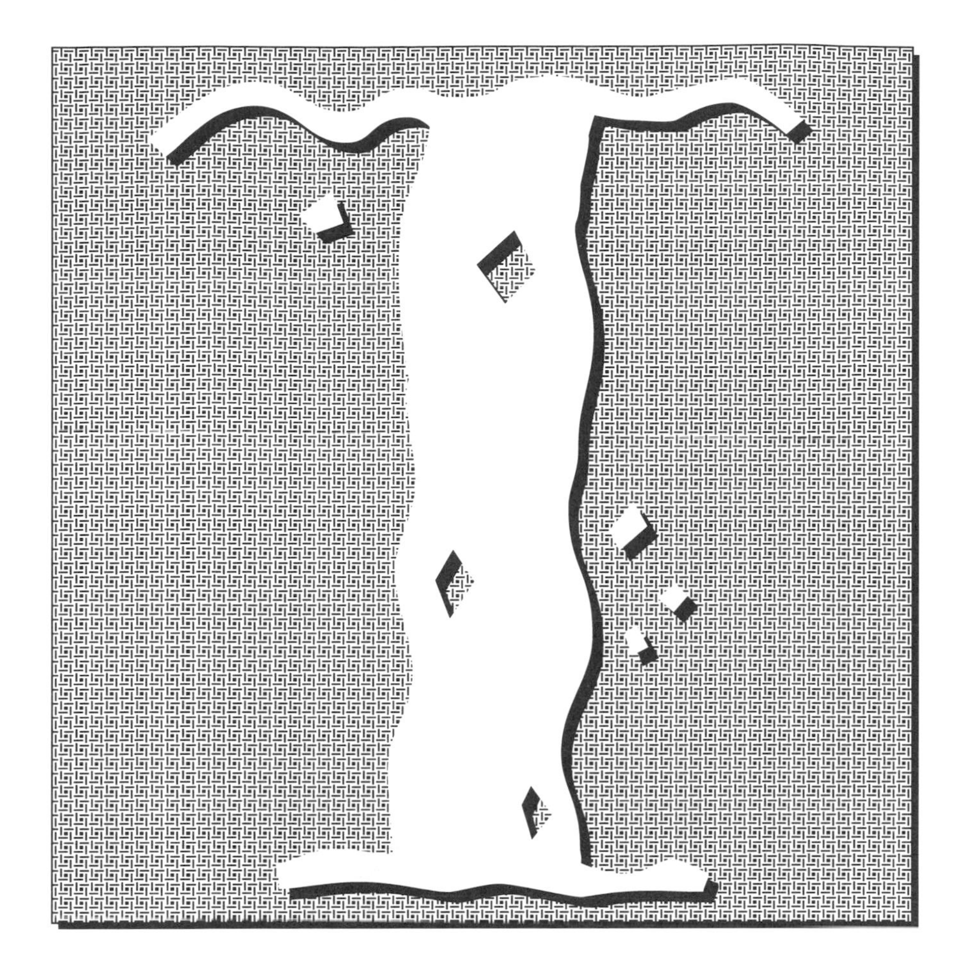

tab character

A Tab character usually looks like some kind of arrow, if you can see it.

When you set a *tab stop* within text and press the Tab key to send the cursor over to that tab stop, you actually type a **tab character.** It's an *invisible,* or *non-printing* character because it doesn't show up on the page and it doesn't print. But the computer thinks it is a character. You can select it just as you can select any other character, and you can delete it. Many applications have a command that makes the tab character visible, in which case it usually appears as a right-pointing arrow (although it still won't print). The command may say something like "Show Invisibles" or "Show ¶."

tab-delimited

If text information in a file is **tab-delimited,** it means there is a tab character between each separate piece of data in the file. For instance, in a spreadsheet or database you can move to the next cell or field to the right by pressing the Tab key. Even if you didn't press the Tab key to get to the next cell, like maybe you just clicked in it, the application considers the data to be tab-delimited.

In a word processor or page layout program you can create tab-delimited data simply by pressing the Tab key between each piece of information. You do this anyway when you make columns of information, such as in a price list.

If you're working in a *table,* the table is probably tab-delimited between the columns.

Typically, the **rows** in a database or spreadsheet are "Return-delimited," meaning there's a carriage Return character between each row because you pressed the Return key.

It's also possible to have a "comma-delimited" file, and sometimes you can specify your own *delimiter.* Please see *delimiter* for a further explanation of what delimiters are and why it is important to know how your file is delimited before you try to import or export information.

Tab key

The **Tab key** is on the upper left of the keyboard, near the number 1 (one). When you press the Tab key, it moves the cursor (and any text following the cursor) over to the next *tab stop.* Pressing Tab inserts a *tab character* in your text, but this character doesn't print, and you can't see it unless your program lets you display special, non-printing characters.

The Tab key is useful in other applications where there are no tabs stops. For instance, in many spreadsheets or databases you can press the Tab key to select the next *cell* or horizontal *field* to the right (Shift Tab will select one horizontal field to the left).

In Windows dialog boxes, pressing Tab moves you to the next item, no matter if it's a *checkbox,* a *radio button,* or an *edit box* (where you fill in text or numbers). You can also select a specific item directly by typing the underlined letter in the item label (you may have to press Alt at the same time).

Notice the underlined character in each button.

In dialog boxes on the Macintosh, pressing the Tab key moves the selection from one *edit box* to the next edit box. In the Save As or Open dialog boxes (in System 7 only), pressing the Tab key selects either the *list box* or the edit box where you type in the name of the file to save.

tab, tab marker, tab stop, and indents

A **tab stop,** or just plain **tab,** is the spot in the line of text where the *cursor* will be positioned when you hit the *Tab key.*

You use tab stops to set up positions along a horizontal line of text for alignment, as in columns of numbers or a list of items and prices. When you press the Tab key, the cursor and all the text to the right of it jump over to the next tab stop.

If your application displays a *ruler* (which all Macintosh text applications do), you can either just click in the ruler to create a tab stop, indicated by a little **tab marker,** or drag a little tab icon into the ruler (and just drag the marker off the ruler to get rid of it). Otherwise, you may have to type in the location of the tab stop (in inches or spaces).

Example of tab markers.

Most applications in which you can type and format text give you several types of tabs to choose from. The following are the most common, but you may find other sorts of tabs available. **Tip:** Always keep your text in a left alignment or justified when using tabs; never center it or set it with a right alignment or your text will not line up with the tab stops.

123.45
6789
10023.5

Left-aligned: The standard type of tab stop is left-aligned. You press the Tab key, the cursor moves to the tab stop. The text types out to the right and is lined up on the **left.**

123.45
6789
10023.5

Centered: You press the Tab key and the cursor moves to the centered tab marker. As you type, the text types out in both directions, on either side of the tab marker.

123.45
6789
10023.5
1.007

Right aligned: You press the Tab key and the cursor moves to the right-aligned tab marker. As you type, the text types backwards to the left, which looks really weird, and aligns on the **right.** The cursor stays on the right side of the text, though. As soon as you press the Tab key again to go to the next tab stop, or as soon as you press Return to get a new line, the text starts typing in the regular direction.

123.45
6789
10023.5
1.007

Decimal: This is for columns of numbers. You press the Tab key and the cursor moves to the decimal tab. As you type, the text types out to the left (backwards) until you type a period (decimal). The decimal lands right on the tab stop, and the rest of the text (until you type another Tab or hit Return) types out to the right.

A 12,345
B 6,789
C 100,235
D 1,007

Leader tab: "Leaders" are the little dots or dashes or underline characters that lead your eye over to the next column of information, like you see in a table of contents. Many applications let you apply leaders to any tab stop, although the most practical arrangement is leaders with a right-aligned or decimal tab.

You should never use the spacebar to align text—always use a *tab stop*. It is the only way to ensure your text will really be aligned when it is printed (unless you're using a monospaced font like Courier, or if you're in *text mode* on the PC).

Indents set up the outer limits of the text area, indented from the left and right *margins*. See *indent* and *first-line indent*.

Commercial: If tabs and indents on the Macintosh confuse you, you might want to check out my tiny book called *Tabs and Indents on the Macintosh*. It includes a disk with files that you use to follow the step-by-step directions in the book.

table

Table has different meanings in a word processor and in a database.

Some applications are specifically designed for creating tables—the kind you see in reports—and some word processors and page layout applications have a feature for making tables. You can set up a grid-like structure with rows and columns, forming "cells" where you type in text. The program automatically lines up the cells, and each cell will expand as you type so you can have adjacent columns of text in varying amounts. This kind of table-making feature makes it practical to set up "side-by-side paragraphs," as shown below.

On Government	On Politicians	On Diplomacy
It is dangerous to be right when the government is wrong.	Ninety percent of the politicians give the other ten percent a bad reputation.	Diplomacy is the art of saying "Nice doggie" until you can find a rock.
Voltaire	*Henry Kissinger*	*Will Rogers*

The term **table** can also refer to the information stored in a conventional *database*, the kind with a repeating structure of *records* and *fields*. That's because you can represent the information as a grid with rows for the records and columns for the fields. In a *flat-file database*, the entire database occupies one table; a *relational database* may have many tables. Also see *record* for another illustration.

This table shows a list of the records. Each record (each row, in this example) has separate fields for each piece of data.

Viewing Who table: Record 1 of 10 Main

Who	Employee ID #	Last	First	[illegible]	[illegible]
1	67543	Mordocs	Arthur	8/13/88	T
2	65437	Smith	Wendy	12/21/89	F
3	43546	Stolper	Paul	6/02/88	T
4	32355	Martin	Bonnie	12/07/06	T
5	87548	Crow	Janet	7/03/88	T
6	43357	Clarke	Henry	12/09/84	F
7	11241	Aubry	Cynthia	6/23/81	T
8	78950	Kaplan	Barbara	2/02/85	T
9	66357	Hunt	Steven	10/20/87	F
10	19941	Donovan	Joan	2/14/90	F

tablet

A **tablet,** known as "digitizer tablet" or "graphics tablet," is a flat, rectangular, electronic pad whose underlying grid corresponds to the computer screen. Some tablets are "touch pads" and you use your finger to select items on the screen, draw images and, pull down menus. But most graphics tablets are used with a *stylus,* which is a pen without the ink (although some styluses do have ink so you can draw on a piece of paper placed over the graphics tablet and see the image appear both on the paper and on the screen).

"Pressure sensitive" graphics use a tablet and a stylus that, combined with appropriate software, allow you to get a variable brush stroke, depending on the amount of pressure you apply. That is, if you brush the stylus quickly and lightly over the tablet, you apply a thin, light brush stroke to the screen. If you press harder and slower, you draw a heavy stroke. Some software can make a slow stroke appear to apply more ink, as if you drew on a piece of absorbent paper.

Although a stylus is the most popular device to use with a tablet, a "puck" is also useful for many applications. A puck looks like a mouse, but it has an extra little attachment with crosshairs for precision alignment. One advantage to the puck is that you don't have to keep picking it up—it stays where you leave it.

Tall Adjusted

In the Print dialog box, the **Tall Adjusted** option is specifically for printing to Apple ImageWriter printers. Since the Macintosh screen is about 72 pixels per inch square and the ImageWriter prints at about 72 x 80 dots per inch, round images print oval, square images print rectangular, and type looks squished. Clicking Tall Adjusted will make shapes print out in truer forms, characters have truer shapes, and the letterspacing more closely approximate the spacing you would get on a laser printer.

tape, tape backup, tape drive

A **tape backup** system is designed to *backup,* or make copies of, all the important data on your hard disk. It backs up the data onto magnetic **tape** at high speed. The tape, similar to an audio-cassette tape, can usually hold anywhere from 40 *megabytes* to 5 *gigabytes* of data, and costs about $20 for a 150-megabyte tape. There are also 8mm tapes that hold 2 to 4.5 gigabytes and cost less than (wow!).

Tape backup is one of the least expensive ways of storing large amounts of data for backups, but the drawback is that you can't use the tape for anything else. That is, if you chose to backup onto *removable cartridges,* such as *Bernoulli* or *SyQuest,* then you could also use those hard disks just like any other hard disk, and you could certainly use the drive itself.

target

When something is called a **target,** it is the object to which you are directing information. For instance, a "target printer" is the printer you will send the document to for printing. A "target disk" is the one you are copying information onto (you copy it *from* the "source disk"). The "target computer" is the one which will run the program, as opposed to the computer the program was created on.

task switcher, task switching

A **task switcher** is a special piece of software that lets you keep more than one program active at the same time so you can switch between them whenever you like. That way, you don't have to shut down one program and then start the next one every time you want to switch. Instead, you just press a special key or pick the program you want to use from a special menu. The task switcher freezes the program you *were* using in a kind of suspended animation, and then takes you to the new program. When you switch back to the original program later, the task switcher revives it exactly as it was when you left—the screen looks the same, the cursor is in the same place, and you can pick up right where you left off. Trés cool, and very convenient.

Task switching is not the same as *multi-tasking.* In task switching, only the program you're actively using is running—every other program is dormant, doing absolutely nothing until you switch to it. In multi-tasking, all the programs you've loaded are actually running at the same time.

System 7 on the Mac and DOS 5 and DOS 6 come with built-in task switchers. But if you want to task switch on a PC, you should jettison the DOS switcher and get something better. Software Carousel and Back-and-Forth have lots of features, but the one I like best is called Multiple Choice—although it limits the size of the programs you can run, it switches between programs much faster than the others and works fine even if you don't have a hard disk. Multiple Choice is available free from the Public Software Library.

TeachText

TeachText is a tiny word processing application from Apple that you often see on a disk when you buy software. You will also see a file called "ReadMe" or "Read Me First" on that disk (which you should always read). Because the information in the ReadMe file is important and because the software vendor has no idea what word processing application you use, they create the document in TeachText and then send along the TeachText utility so you can read it.

The application icon.

You only need one copy of TeachText in the Mac. I have seen up to 16 copies of this utility in computers because people keep copying everything on the new floppy disks onto their hard disk. Many times when you install a large application, TeachText is installed along with the rest of the program. Throw away all the excess TeachTexts except the latest *version*. (Check the *Get Info* box for the version number.)

The icon for a document created in TeachText.

You can also use TeachText to create ReadMe files to send messages to people when you send floppy disks. TeachText is free so you can give it to anyone you want, although you can rest assured that they already have at least six copies hanging around in their Macintosh already.

Older versions of TeachText only read *ASCII* files (*text-only*), but version 7.0 and above can also view PICT graphic images and can even play *QuickTime* movies. In fact, there is a very mean yet very harmless trick you can pull on someone while they go get a cup of coffee . . . nahh, I shouldn't tell.

tear-off menu, tear-off palette

If a *menu* or a *palette* is **tear-off,** that means you can drag your mouse down through the menu or palette, keep on dragging onto the screen, and the menu will tear off from its place and float around. Wherever you let go of the mouse, that's where the menu or palette will stay.

After you tear off a menu or palette, it becomes a little *windoid*, but it's usually called a palette (see the illustrations in *windoid* and *palette*). You can move it around again by dragging in the gray bar area, comparable to a window's *title bar*, which may be along the top or perhaps along the left side. Position the menu/palette near where you're working, and you can choose commands or tool or colors quickly and easily. To put a tear-off palette or menu away, click in the close box (on the Mac) or double-click the Control-menu box (in Windows) in the top left corner of the palette.

GEOWorks is a PC product that's similar to but better in some ways than Windows; one of its advantages is tear-off menus.

technobabble

Technobabble is what I have tried to avoid in this dictionary—those definitions and terms that are incomprehensible to plain ol' folks like us who just want to use the computer to get our work done. This is an example of a technobabble definition, taken from a book whose name I will not mention:

> **single in-line memory module (SIMM):** A memory module that contains the chips needed to add 256K or 1M of random-access memory to your computer. A SIMM plugs into a motherboard or a logic board.

See, if I knew what all those other words and numbers meant, I probably would not have to look up "SIMM" in the first place (besides, it wasn't even defined under "SIMM"). Here's another example:

> **byte:** A grouping of adjacent binary digits operated on by the computer as a unit. The most common size byte contains 8 binary digits.
>
> Yeah, right—now I get it.

Technobabble is often one of those signs of the "priesthood." To protect their place in the priesthood, some people find it necessary to intimidate others with their superior knowledge and masterful understanding of the technical world by tossing around technobabble, knowing full well no one else understands what they are talking about. Sometimes, I have discovered, even *they* don't know what they're talking about.

technoweenie

A **technoweenie** is similar to a *nerd,* but tends to be more technically-oriented. Janet says technoweenie is a sexist term. She says we are *technokitties.* Think about that one. Me, I'm just a nerdette.

telecom

Telecom is short for *telecommunications;* see below.

telecommunications

Telecommunications refers to the communications we carry on through our computer and the telephone lines, using a *modem* or a *fax machine.* It refers to using a *bulletin board service,* an *online service,* or just sending files directly to someone else. To use the modem you do need to own some sort of *telecommunications software.*

telecommunications software

When you use a *modem,* you also need software to let your modem and your computer talk to each other and to let the modem know how to send and receive the messages. This particular kind of software is **tele-communications software,** or sometimes it's just "communications software."

When you buy a modem it may come with general purpose telecommunication software which you can use to communicate with your friends and business associates, or to access any *bulletin board service.* If you subscribe to a commercial *online service,* they will give or sell you the software to access their service.

telecommute

I **telecommute.** That is, I work at home and commute to my various jobs through my telephone/modem/computer/fax machine. Plus overnight mail, of course. Telecommuting is becoming a viable way of life for many people. You can live in a meadow or a tree house or on a tropical island and take care of all kinds of business electronically. Also see *personal computer.*

template

A **template** is an original document all set up with specifications and formatting, from which you can pattern other documents. For instance, you may have a monthly newsletter to do. You can make a template of the first newsletter, including the text and graphics in it as placeholders. Each month you open a *copy* of the template and replace the existing text and graphics with the new material. The pages are already laid out and set up, and the text is already formatted, so you hardly have any work to do. The advantage of a template over using the same document repeatedly is that you start with a fresh original each time, rather than a mutating document.

On the Macintosh in System 7 you can make any *document* a template. At the Finder (Desktop), click once on the document icon. From the File menu choose "Get Info." Click in the checkbox at the bottom labeled "Stationery." Close the Get Info window (press Command W). That document is now a template (or "stationery pad" in System 7 lingo). When you open that stationery pad, you will actually open an untitled *copy* of the document. The original will stay untouched. Just uncheck the box if you don't want the file as stationery.

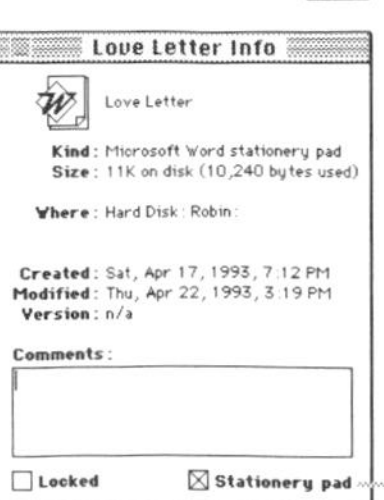

Check this box. If you don't see this box, the file you have selected is not a document.

temporary storage

People are always referring to *memory* as **temporary storage.** This is because anything held in memory is only held there while the computer needs it (the computer can *only* work on something if it is in memory). When you open an application, the computer actually puts a copy of that application in memory. When you close that application and open another one, the computer forgets the first one and puts the second one into memory. If the computer happens to crash, everything in memory, because it is only temporary storage, is gone, including any work you have not *saved to disk.*

terminal

Fundamentally, a **terminal** is simply a screen and a keyboard, along with the circuits necessary to connect it to a main computer at another location. You might mistake a terminal for a personal computer at first glance, because they both have the same sort of monitor and keyboard. But a terminal isn't a computer—it is just the means for you to communicate with a computer somewhere else. The terminal sends the commands you type on its keyboard to the computer, which processes them and sends back the results to the terminal's screen.

Terminals are always used with multi-user computers, such as a mainframe or a minicomputer, or even a PC equipped with a multi-user *operating system* like *Unix.* Each user of the computer has her own terminal.

The distinction between a PC and a terminal gets a little blurry when you talk about "intelligent terminals." A "dumb terminal" only has a keyboard and a monitor and can't do anything on its own. An intelligent terminal, by contrast, has some computing capabilities built in. This way the terminal can handle some tasks that the main computer would otherwise have to do, which speeds things up.

Traditionally, terminals could only display text, but over the last few years new "graphics terminals" have appeared; of course, they can display graphics as well as text. This allows people to use *graphical user interfaces* such as *X/Windows.*

terminal emulation, terminal emulator

Before *personal computers* people communicated with the very large computers like *mainframes* through a *terminal,* a device that was used mainly to input information and to get information out of the *mainframe.* Now, with a *modem* and *telecommunications* software (or with a special *add-in board* you install inside your computer and with the special software to go with it) you can use your personal computer to exchange information with mainframes. But since mainframes and other large computers are accustomed to communicating with terminals, the communication software has to pretend your computer is a terminal so the mainframe will talk to you. This is called **terminal emulation.** Depending on the communications software, you may someday find yourself checking in to the **terminal emulator** *online.* (I once had such a bad experience with a terminal emulator that I have forever dubbed them "terminator emulators.")

A "dumb terminal," which originally didn't even have a screen (it was bascially a typewriter with a modem), could only do the minimum necessary to input and output data. This was also known as a "teletypewriter," called a *TTY* for short.

terminate, termination, terminator

Termination has to do with *SCSI* devices, such as hard disks and scanners, hooked up one to the next in a "chain"; the devices at either end of the chain must be **terminated** for the system to work properly. Don't worry if you've been having difficulty trying to figure out the logic behind termination: be comforted in the fact that termination is known to be black magic, even among the Most High Gurus. You find what works and you don't ask why.

To understand the concept of **termination,** imagine this: Inside your computer is a hard disk. Outside your computer is another hard disk, perhaps a *removable cartridge drive.* You also have a *scanner.* And you have a *CD-ROM* player. Each of these *peripheral devices* (devices that are outside the computer) is a *SCSI* device, which means they are connected with SCSI cables and they communicate with each other through the SCSI standard of data communication. These devices are *daisy-chained* together; that is, there is a SCSI cable from the Mac to one device, from that device to the next, and from that device to the next.

As the information goes coursing through these cables from the computer to the devices, it gets to the last device and tries to continue. When they get to the end, the data signals try to go back the other way. They bump into the other data that is already coursing through the lines. Things get all mixed up, data gets confused, and none of the devices can function—including your computer.

To prevent this scenario, the first device and the last device in the chain must be **terminated.** Termination absorbs the signals that might otherwise bounce back in the wrong direction. The internal SCSI hard disk in your computer is always terminated, so you don't need to worry about that one. The problem arises when you need to decide **which** peripheral device to terminate.

As I just mentioned, the last device in the chain must be terminated. This means the last one physically in line. That seems easy, except for two things. Some devices arrive on your doorstep with *internal* termination, which means the manufacturers have taken it upon themselves to terminate the device for you. The only way to know if a device is internally terminated is to read the manual. If it is, the manual should explain how to take needle-nosed pliers and remove a couple little parts from inside the device to remove the termination. (Personally, this job does not appeal to me. Fortunately, fewer devices are being made with internal termination.) If you know that the device is internally terminated and it is the *only* one internally terminated, you can place it at the end of the chain.

To externally terminate the last device in the chain, you need a **terminator.** It looks just like the SCSI plug on the cable, except it has no cable. You insert it into the extra connector on the back of the device (every SCSI device has two connectors, one for plugging into, and one for daisy-chaining to the next device or for terminating).

The other thing about termination—the one that puts it in the category of black magic, is that even if you have all of the devices chained together properly, with internal termination on your internal hard disk, external termination on the last device, and nothing else along the way terminated, that does not mean it will work. If you turn your computer on and nothing happens, it's black magic. Turn it off and rearrange the devices. That is, put the last one second and the second one third, etc. Terminate the last device, and try again. If that doesn't work, rearrange them again. If it still doesn't work, try new cables. And make sure to use the shortest cables you can—the data can get confused if it has to travel very far, and the signals start bumping into each other (called "cross-talk").

A "self-terminating" device is one that decides for itself when it should be terminated or not. I wish they would all do that.

Remember, never connect or disconnect anything from the computer or any other device while there is any power going to any device. *Turn everything off!* I turn off my power source completely.

Some people warn that if you have devices terminated improperly you can damage hardware or software when you turn on the power. Others say the only way to figure out how to terminate properly is to turn on the power and see if it works. I am not recommending anything—I'm only telling you what they told me. I will say that I have turned on the power, discovered it didn't work, rearranged things, and everything worked fine. If you don't turn it on, how will you ever know?

If you have only two devices in the chain—your computer and one other peripheral—even that one peripheral device must be terminated (remember, the first and the last). Many SCSI devices will come with an external terminator in the box, or you can get one at any computer store.

Most printers do not have to be terminated because most printers are not SCSI devices. (Macintosh printers plug into either the *serial port* or *LocalTalk ports,* with those cute little unintimidating plugs. Thank gawd.) But if the printer *is* a SCSCI device, such as the Kodak XL-7700, then you need to work it into your SCSI chain properly.

test drive kit

A **test drive kit** is slang for the *demo disk* that vendors often give away at trade shows. The demo disk may just run a demonstration of the program. Sometimes the disk contains the actual application, like a word processor, but it always has some disabling feature, like you can't save a document or print a document or it prints with a big "DEMO VERSION" diagonally across the entire page, right over your text.

text editor

A **text editor** is similar to a *word processor* in that it is an application you use to create and edit text. But a text editor concentrates on the *content* of the text, rather than its appearance. Text editors produce plain text (ASCII) files, without any formatting codes in them. They give you only rudimentary control over the layout of the text—you can set *tab stops,* but you can't change the spacing between lines or paragraphs, nor can you use character formatting like bold or italics. Text editors are often set up so that you have to press the Return or Enter key at the end of the line, as you would on a typewriter, although many text editors offer a *word wrap* mode so that the text automatically drops to the next line after it reaches the right margin.

Text editors are good for preparing files you'll be importing into a page layout program, where you will format it, or for files that you'll be sending as *e-mail,* since many e-mail systems only accept plain ASCII files. If you use DOS, OS/2, or Windows, you would use a text editor to make changes to files like AUTOEXEC.BAT, CONFIG.SYS, or WIN.INI. And *programmers* use text editors to write the instructions, or the *code,* for their programs—

a computer doesn't care what the program code looks like, just what programming commands it contains.

DOS has always come with a text editor. Prior to DOS 5 you were stuck with an absolutely awful program called EDLIN, but the MS-DOS Editor that comes with DOS 5 and DOS 6 is pretty good. Ignore anything you read in old books about EDLIN and just throw that program out. If you're still using an older version of DOS, you can get many good editors from a *shareware* distributor (one of the best, VDE, is free for the asking).

text-only, text file

A **text-only** file is a document that consists entirely of unformatted text (meaning no bold or italic or different typefaces). This allows almost any program on any computer to *read* the file.

Technically speaking, a text-only file has been saved in *ASCII* format. The ASCII (pronounced "askee") format is a standard way of encoding characters, with a number standing for each character. For instance, the letter M is coded as 77. What you see on the screen looks like the letters you typed, but inside the computer these letters are represented by ASCII code numbers.

Some applications let you save or *export* a formatted document as text-only, but when you do, the file you save is stripped of all formatting—it loses any bold or italic formats, any special letter spacing you may have applied, any typeface changes, or any other sort of fancy attributes. A text-only file *can* and usually does contain tabs and Returns and spaces, because those "characters" also have an ASCII code.

Ladle Rat Rotten Hut

part three

"Armor goring tumor groin-murder's" reprisal ladle gull. "Grammar's seeking bet. Armor ticking arson burden barter an shirker cockles."

"O hoe! Heifer gnats woke," setter wicket woof, butter taught tomb shelf, "Oil tickle shirt court tutor cordage offer groin-murder. Oil ketchup wetter letter, and den—O bore!"

This is the formatted text.

```
Ladle Rat Rotten Hut
part three
"Armor goring tumor groin-
murder's" reprisal ladle
gull. "Grammar's seeking
bet. Armor ticking arson
burden barter an shirker
cockles."
"O hoe! Heifer gnats woke,"
setter wicket woof, butter
taught tomb shelf, "Oil
tickle shirt court tutor
cordage offer groin-murder.
Oil ketchup wetter letter,
and den—O bore!"
```

This is the same text in the ASCII format, or text-only.

text wrap

When text flows around the edges of a graphic or of any other particular shape, it is called a **text wrap.** The text in this paragraph is wrapping around the large initial cap.

Text wrap is different from *word wrap,* which is when the typed words bump into the right margin and automatically jump to the next line as you type. See *word wrap* for examples.

thesaurus

You know what a **thesaurus** is in book form—it's kind of like a dictionary, except that instead of looking up a word to find the definition, you look up a word to find another word that means the same thing (a synonym) or the opposite (an antonym). Well, you can buy an electronic version of the same thing for your computer—in fact, several word processing programs come with a thesaurus built right in.

A thesaurus is a wonderful tool for anyone who writes. It broadens your vocabulary and can make your writing more interesting. Say you're working on a story about a mountain bike race and you want a few more colorful alternatives for the word "race." You can *select* the word "race," call up the thesaurus, and it automatically displays a list like contest, river, mankind, speed. "Contest" is the sense you had in mind; you want to look further into the "contest" synonym, so you click that button and you get competition, event, meet. Choosing "competition" gives you tournament, contest, match, meeting. Choosing "event" gives you match, bout, game, sport, and contest. And so on. Somewhere amongst all these words is just the one you need.

third-party

You are the first party, the manufacturer of your computer is the second party, and the other vendor who makes hardware or software that works with your computer is the **third party.** For instance, some people replace the internal hard disk in their Macintosh with a larger, **third-party** hard disk, because even though Apple makes larger hard disks for the Mac, it is usually cheaper to get one from a third party.

Third-party vendors build computer accessories and *peripherals* either because the major manufacturer (like IBM or Apple or Sun) does not make that device or because the third party can offer them at cheaper prices.

thrashing

When you use *virtual memory,* where your computer is pretending your hard disk space is *memory,* the computer has to swap information back and forth between the disk and the memory (please see the definition for *virtual memory*). Too much swapping is called **thrashing.** You can tell when this happens because the computer slows down—a lot.

Three Rules of Life

Guy's **Three Rules of Life** are: Back up your hard disk; rebuild your desktop; and send in your registration card.

Robin's **Three Rules of Life** are: Your attitude is your life; maximize your options; and never take anything too seriously.

throughput

Throughput is a measure of a computer's overall performance. The advertising for a computer will tell you all these fast speeds for certain portions of the product, like the *clock speed* of the *microprocessor* (for instance, an 80386 microprocessor running at 25*MHz*). But they don't tell you that if the *memory chips* are slow and the *hard disk* is slow, that 25MHz doesn't mean diddly. In reality it can run much slower. It's the throughput of a computer *configuration,* or setup (including hardware and software), that matters. Unfortunately, though, there's no standard rating system for throughput because the performance you get also depends partly on what you do with your system.

thumb

Some people call the little white box within the *scroll bar* a **thumb.** Some people call it an *elevator*. I call it a *scroll box*. No matter what you call it, if you press on it with the pointer and drag it along the scroll bar, it moves the contents of the *window*. See *window* for an illustration of the scroll box.

thumbnail

A **thumbnail** is a miniaturized version of a graphic image or a page of a document. Graphic designers have always created thumbnails with pen and paper to generate a lot of ideas in a short amount of time. Hand-drawn on paper (or napkins), thumbnails were the preliminary sketch stage. Now that we can generate ideas electronically, thumbnails usually serve as a representation of a multitude of fairly finished ideas; most design applications can produce printed or on-screen thumbnails of your project. Ideally, the thumbnail displays enough content so you can identify the original image it represents. Thumbnails are useful in cataloging images or in seeing an overview of an entire multiple-page project.

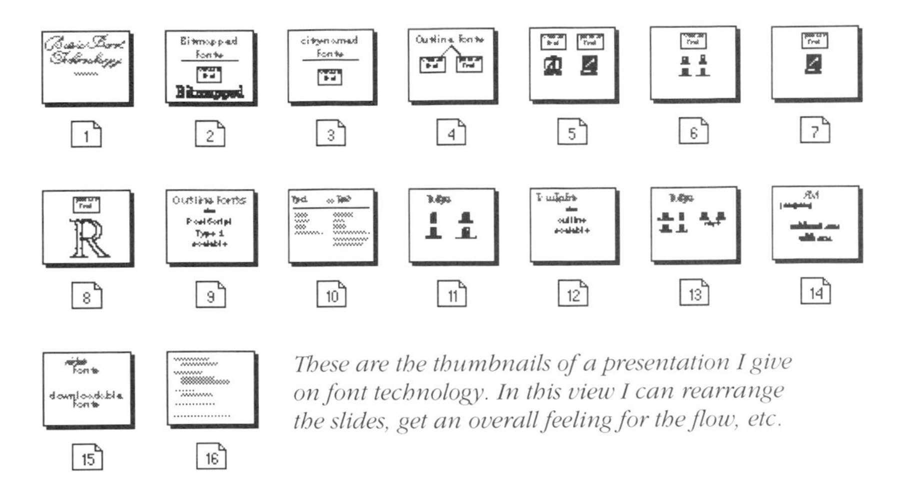

These are the thumbnails of a presentation I give on font technology. In this view I can rearrange the slides, get an overall feeling for the flow, etc.

TIFF

TIFF (pronounced "tif") stands for **t**agged **i**mage **f**ile **f**ormat. It's a *graphic file format* developed by Aldus and Microsoft, in combination with leading scanner vendors, specifically for capturing scanned images. When you scan an image, you have a choice of saving that scan in one of several different *file formats*. TIFF is always one of the choices, and is usually the best choice since it was developed for scanning; unless you know why you need to save it in a different format, save the scanned image as a TIFF. Several graphic applications also offer the option to save files you create directly on the computer as TIFFS.

A TIFF is always a *bitmapped graphic,* also called a *raster graphic,* but it is not limited to the 72-pixel-per-inch bitmap of a *MacPaint-type* image. A TIFF can be any *resolution;* it can be *bilevel* black-and-white; it can be *grayscale* with up to 256 levels of gray; it can be *24-bit* full color.

Another strong point for TIFFS is that they are the most *platform-independent* graphic; that is, TIFFS created on Macintosh machines can be used on DOS machines, and vice versa. They can be also used on NeXT machines, and Sun and Silicon Graphics workstations. The Windows version of my 800-page PageMaker book has hundreds of TIFF screen shots from a PC that I dropped right into the Mac layout. I used the *Apple File Exchange* utility to copy the files from the DOS disks to the Mac, then placed them directly into PageMaker on the Mac. Too cool.

tilde

The **tilde** (pronounced "till duh") is the accent mark you see in some Spanish words over the letter **n**, like this: **ñ**. Sometimes the tilde is used in text to indicate the omission of a word or syllable. On the keyboard, the tilde key is usually on the upper left, sometimes the lower left.

Occasionally the **tilde** key is used in combination with *modifier keys* in keyboard shortcuts. For instance, in HyperCard, pressing the tilde key will take you back one card. If you are in painting mode, it will undo the last action. In some applications, the tilde key acts like the *escape* (esc) key.

In text, however, the tilde key on your keyboard just types a tilde by itself—it doesn't put the accent mark over the **n**. If you want to type ñ, do this:

On the Macintosh, type the word up to the **n**. Hold down the Option key and type the **n**. Nothing will happen. Then type the **n** again, without holding the Option key down. The letter **ñ** will appear. See *Key Caps* for more info on creating accented characters.

On a PC, you must check your manuals to see if your software and printer permit the use of accented characters. But generally, in Windows programs, you can hold down the Alt key and type 0241 on the *numeric pad* (you may need to turn *Num Lock* on) to get a lower case ñ. For a capital Ñ, type Alt 0209. In DOS programs, you produce an ñ by typing Alt164 and an Ñ by typing Alt165 (remember, use the numeric keypad).

tile, tiling

Whether you're talking about windows on the screen or printed pages, **tiling** means arranging them like floor tile, just touching on all four sides. If two windows are arranged side by side, vertically and horizontally, so that they fill up the available space on the screen, they are **tiled.** Some applications have a menu command called "Tile." If you have more than one window open when you choose this command, all the open windows will make themselves equal size and position themselves next to each other. This is a great way to see two different views of the same thing, such as the working version and the preview version. You can do it yourself, of course—the menu command is just faster.

When you ***print*** pages of an oversized document (like 11 x 17) on regular-sized paper (8.5 x 11), you must **tile** the pages. When you print in tiles, the computer breaks up the large page into chunks just big enough to print on one piece of paper. Depending on how you arrange it, an 11 x 17 might be printed on two pages side by side, or they may be drawn and quartered and printed onto four pages that you overlap and tape together.

(Actually, when you print tiled pages there is a little overlap on each so you can easily align the separate pages).

time

Time is what keeps everything from happening all at once. And then there is *QuickTime, real time, access time, seek time, clock time,* and probably others. Check their respective definitions.

title bar

In a *window,* the **title bar** is the narrow, horizontal rectangle across the top where you see the name of the application or document or folder or disk that the window belongs to. Also see *window.*

In Windows, the title bar of the *active* window is a different color than that of unactive windows.

When a Macintosh window is *active* there are horizontal lines in the title bar, while unactive windows have plain white title bars. When you're at the Desktop in *System 7,* you can hold down the Command key and press on the name in the title bar to see the *path* to where this particular folder is stored.

This is the title bar.

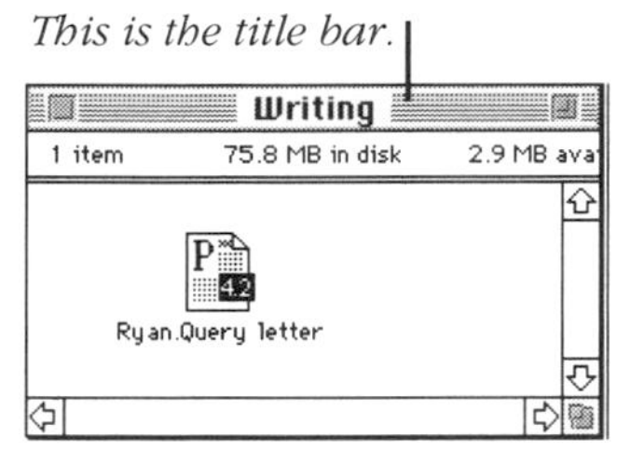

Command-press to see where this folder comes from. Choose any level to open that folder.

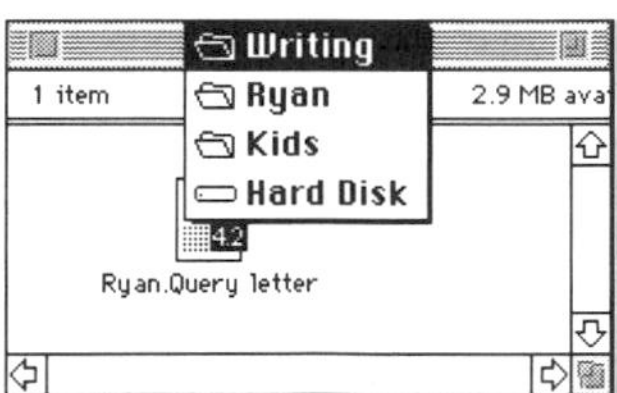

tits on a keyboard

Don't yell at *me*—I didn't make this up. **Tits on a keyboard** are those little bumps on the letters on the keyboard that let you know if your fingers are in the right place without having to look. On my keyboard they are on the D, K, and on the keypad 5. Sometimes they are on the F and the J. Steve says, "I lose out. My keyboard doesn't have tits, but little raised bars instead."

TLA

TLA is a three-letter acronym that stands for **t**hree **l**etter **a**cronym. The TLAM (pronounced "tee lamb") is the specialized, high-technology machine (the **t**hree-**l**etter **a**cronym **m**achine) that produces TLAs. The TLAM tends to be a little more popular than the FLAM ("eff lamb," **f**our-**l**etter **a**cronym **m**achine), although they both get a great deal of use. TLAMs and FLAMs are religiously maintained by those in the *priesthood,* who depend on TLAs and FLAs to keep themselves installed in their revered positions and the rest of us in our places.

toggle

If a command is a **toggle,** it means this one command turns a feature on, and the same command turns the feature off. For instance, you can press Command R to make the *ruler* visible in many word processors, and if you press Command R again, the ruler disappears.

Token Ring Network

With capital letters, **Token Ring Network** is an IBM *local area network* that uses "token passing." With lowercase letters, **token ring network** refers to any network that uses the token passing system.

Whether capitalized or not, the basic principle is this: When several different computers are connected with cables so they can either talk to each other or send information to the same printer or all access a common *file server,* they are on a *network.* Eventually more than one computer wants to print or to access the server simultaneously. This can result in conflicts called "collisions," because only one action can take place at a time.

In a token ring network, all the computers and servers and printers are connected in a physical ring. "Tokens," which can be thought of as envelopes, are circulated around the ring through the cables to the different computers. If a computer has nothing to send, it lets the token pass on by. When the computer wants to print, it grabs the token, turns on the "Occupied" sign, puts its information in the envelope, and sends it off to the printer. As the token circulates through the ring to its destination, no other computer can use that token until it has finished its job, returned to the computer, and the computer turns on the token's "Vacant" sign.

TokenTalk

TokenTalk is software from Apple that lets Macintoshes operate, or connect to a *Token Ring Network.*

toner, toner cartridge

Toner is the black powder, made of carbon and a certain form of plastic, that serves as a laser printer's "ink." In the most popular laser printers, the toner comes in a sealed **cartridge,** but with some printers you have to pour it into a refillable little container. See *laser printer* if you want to know how the toner gets fused onto the paper.

Toner cartridges cost about $100 new. There are companies that recharge used cartridges (check computer magazines). A recharged cartridge can cost as little as half the price of a new one, often give you blacker blacks and more prints per cartridge, and is more environmentally conscious.

toolbox, toolkit, Toolbox (for programmers)

A **toolbox** or **toolkit,** with a lowercase t, is a set of predefined software routines and programs that a *programmer* can use when creating a new program for a particular computer. For instance, the programmer can use a toolbox to draw a dialog box or display a menu, without having to write the required code from scratch.

The Programmer's **Toolbox,** with a capital T, refers specifically to Apple's User Interface Toolbox for programmers who are creating applications for the Macintosh. It gives them the tools they need, in the form of program modules, to support the *graphical user interface* of the Mac.

toolbox, tools (for us)

Many applications, especially painting, drawing, and page layout applications, have a **toolbox,** an area on the screen that displays a variety of **tools,** such as a brush, a paint can, an airbrush, scissors, a pencil, an eraser, etc. You click on a tool to select it, then press-and-drag the tool around (usually) in the document to use it. Don't try to drag the tool onto the screen from the toolbox—just click the tool, let go, move to the document, and then start dragging. Also see *palette* and *tear-off menu.*

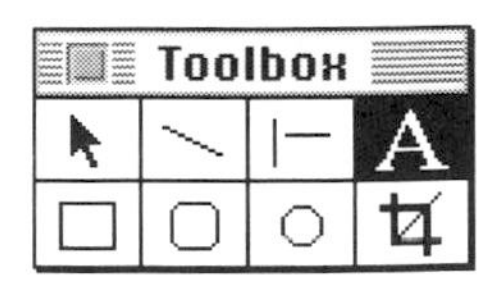

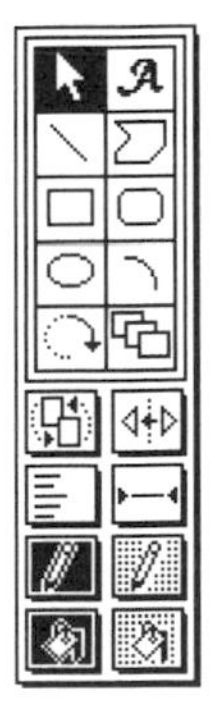

These are variations of toolboxes.

TOPS

TOPS, an acronym for **t**ranscendental **op**erating **s**ystem, is a *local area network*, meaning it is used to connect computers that are in close proximity to each other, like in one room. It can connect Macintoshes with PCS and compatibles. TOPS works with *LocalTalk/AppleTalk* and *EtherNet* networks.

TOPS has no dedicated *file server,* but provides "peer-to-peer" file transfer, where each user has access to any file any other user has publicly posted. TOPS works in such a way that Mac users don't have to figure out DOS, and DOS users don't have to figure out Macs—when a Mac file is sent to a PC, the file shows up in the directory; when a PC file is sent to a Mac, the file shows up as an icon.

For Mac-to-Mac file sharing, TOPS is being replaced with *AppleShare* and System 7's file sharing.

touch screen

Also known as "touch sensitive display," a **touch screen** is a pressure-sensitive panel that is mounted in front of a computer screen. You use your fingers to push the buttons. You've probably seen these in airports or museums or in the wedding gift section of big department stores. Now you can buy touch screens that work with personal computers.

tower

Tower refers to a computer whose box containing the disk drives and the *CPU* is so tall it looks like a tower, such as the Apple Quadra 950. They're so big they don't even fit on the desktop but have to sit next to you on the floor.

track

On disks, the tracks are in concentric bands.

A **track** is a thin channel for storing information. On a floppy disk or a hard disk, the tracks are concentric circular bands (shown left). On *CDs* and *video discs,* tracks are spiral. On *tapes,* tracks are parallel lines. See *sector* for detailed information on how the tracks are used on a disk.

tracking

Tracking is a typographical adjustment of the space between letters.

If you type 12-point text on the page, then enlarge it to 48-point, the computer enlarges the characters four times, and it also enlarges the space between the characters four times—but the larger the type, the *less* space there should be between the letters. If you reduce that text to 6-point, the characters are reduced by half, and the amount of letterspacing is reduced by half—but the smaller the type, the *more* letterspacing there should be.

Technically, when you apply tracking to text to adjust the letterspacing, the tracking value should be point-size dependent, meaning the amount of letterspacing that is adjusted depends on the point size of the type.

Unfortunately, most page layout programs call it "tracking" when you *kern* over a range of text, which is actually just "range kerning." Range kerning is simply removing or adding the same amount of space from between every letter. Tracking, however, adjusts the letterspace based on the point size of the type and the *auto kern pairs*. PageMaker has the closest thing to real tracking (plus range kerning); QuarkXPress, although it has a feature it calls tracking, has range kerning.

trackball

A **trackball** is an alternative to a mouse or a stylus. It looks kind of like a mouse upside-down, and you use it by rolling the ball around with your fingers. It has one or more buttons to click, just like a mouse.

Trackballs are great when you don't have enough space for a mouse on your desk—or when you don't have a desk at all, like when you're using a *laptop* or *notebook* computer. These days, many little computers, including the Macintosh PowerBook, come with built-in trackballs. Trackballs are also a viable alternative for people who need to use the mouse backwards (with the tail facing them).

transfer rate

The **transfer rate,** or "data rate" refers to the speed with which data is transferred from its source to its destination, such as from one computer to another over a network, or from the computer to the disk drive. It's measured like we would measure any rate of speed—in units of information per the unit of time, like miles per hour. On a computer you may hear transfer rates expressed in terms like *bits* per second, *megabytes* per second, or *characters* per second.

transparent

Transparent, of course, means you can see through it. A graphic image may have a transparent shading inside of it, which may look exactly like the *opaque* shading in another image if both images are on a plain white background. But if you put another object behind each of the images, the difference between transparent and opaque becomes very clear.

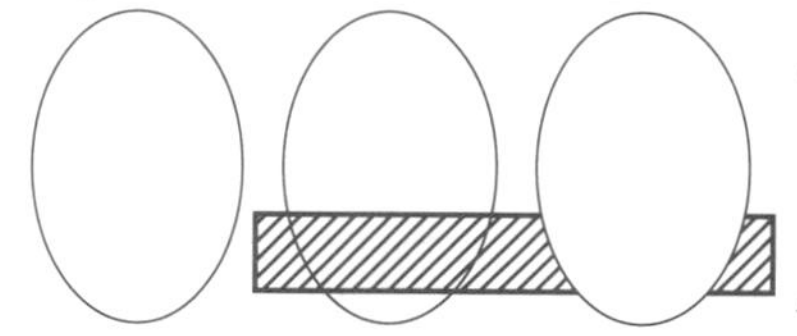

Is the first oval transparent or opaque? You usually can't tell until you place it over another image. You can also try picking it up—you can grab an opaque object in the white area; you can't grab a transparent one unless you press on the solid line.

Transparent is also an adjective describing hidden computer operations that the user doesn't have to deal with. Transparency is usually considered a positive feature because it means the user is protected from the complexity of the operations. But the term transparent can be a *technobabble* adjective, most commonly used by marketing people trying to make their products sound sophisticated. Transparent is not a transparent word.

trap, trapping

A **trap,** in printing, refers to a process that must be applied when printing two or three different colors on a press. If two colors bump against each other, they must overlap a tiny bit to allow for the paper stretching or shifting on the press. If there is no overlap, or trap, allowed, then if the paper shifts even a hair's width you will see the white of the paper show through between the two colors.

If the two colors do not overlap a tiny bit, you will often see a sliver of white. This sliver is undesirable in fine printing (très gauche, or very uncool).

Some people swear that every time two colors are next to each other they must be trapped. But a high-quality print shop using certain presses and paper can *kiss-fit* colors beautifully, using no traps at all. See more details in *kiss-fit*.

trash can

The **trash can** is an *icon* (picture) of a trash can on the Desktop. And yes, you throw things away in it. You can throw away any file on your computer simply by dragging the icon of the file over to the can. When the ***tip of the pointer*** touches the can (it doesn't matter if the shadow of the file touches the can—the tip of the pointer is the key element), the

The famous trash can icon.

can turns black. You let go and your file is in the trash. (On the NeXT machine, the equivalent to the trash can is a "black hole.")

In System 6 the trash is automatically emptied (meaning that file is gone and you will never ever see it again) whenever you copy an item, open an application, quit, or if you are running out of memory. Or you can choose to empty the trash yourself by choosing "Empty Trash" from the Special menu.

When the can turns black, you have opened the lid.

In System 7, the trash never gets emptied until you choose the "Empty Trash" command from the Special menu. Your files stay there even if you crash or quit. You can **turn off that irritating warning** that appears every time you try to empty the trash (but then you must accept the responsibility): Click once on the trash can. Press Command I to get the Get Info window. Click in the little box "Warn before emptying" so there is no checkmark in it. Press Command W to close the Get Info window.

When you have successfully trashed something, the can bulges.

To **throw away locked items,** hold down the Option key as you toss it or as you choose "Empty Trash" from the Special menu.

Besides throwing things away, you can also **eject your floppy disk** (or *dismount* a hard disk) by dragging the disk icon down to the trash. Don't squeal—it's perfectly safe; the trash can does not erase or destroy any information on a disk. In System 7 you can also eject a disk by clicking once on it and pressing Command Y. You'll see the disk icon slide down and into the trash can as it pops out of your computer.

tree, TREE

In DOS and OS/2, the term **tree** refers to the hierarchical structure of directories and subdirectories on a disk. No matter how many branches in the tree, there is always only one *path* from the root directory up to a particular file.

TREE is an external DOS or OS/2 command that shows you the directory hierarchy, or the tree structure, in diagram form (you don't actually have to type the command in uppercase letters).

Trinitron

Trinitron is a special type of *CRT (cathode ray tube,* or picture tube) made by Sony that produces an especially bright, sharp picture. Your monitor, a television, a video monitor—they are all CRT displays. Most CRTs use tiny phosphor dots, but Trinitron tubes have vertical phosphor stripes and a special wire grid that prevents distortion. There is a Trinitron tube in Apple's 13″ color monitor.

Trojan horse

A **Trojan Horse** is similar to a *virus* in that it's a program that destroys your data, and a very nasty, evil person has taken the time and trouble and expertise to create it. A Trojan Horse poses as an actual application that ostensibly is useful, such as a word processor or a game. But once you have loaded it on your computer and started to use it, the Trojan Horse lets loose its instructions that destroy information on your disk. Yes, it's hard to imagine who would be mean enough to do this. Unlike viruses, Trojan Horses don't replicate themselves. Please also see *virus* and *Disinfectant.*

troubleshoot

Troubleshoot means to attempt to fix problems as they arise. Of course, the more trouble you have come across in your life, the better you are at troubleshooting. "Experience teaches you to recognize a mistake when you've made it again."

TrueImage

TrueImage is a *page description language* developed by Microsoft for printers, similar to the *PostScript page description language.* A printer that uses TrueImage is considered a *PostScript clone,* and like all other PostScript clones, it can't quite do everything a true PostScript printer can. If you are not heavily PostScript-based, meaning you don't have a big investment in PostScript fonts and you don't create a lot of PostScript graphics, then the less expensive TrueImage printer will be fine for you. Or if your typeface library is TrueType-based, a TrueImage printer would be appropriate.

TrueType

TrueType *fonts* are one of the major types of *scalable* or *outline* fonts, which means they can be printed or displayed on the screen at any size and still always look good. Please see the *font* definition for details and for comparisons with other kinds of scalable fonts.

TrueType also refers to the software that converts the TrueType font outlines into characters you can see on the printed page or the screen. Apple developed the TrueType technology, which is incorporated into System 7 on the Mac and Windows 3.1 and Windows NT on the PC. (You can also use TrueType on the Mac in later versions of System 6 if you install the TrueType *INIT.*)

truncate

Truncate. What a great word. Truncate means to chop off. You may have seen a *dialog box* that warned you that your indents are wider than your page setup margins so part of your document may be truncated when you print (which happens all the time). To fix that, make sure the *indents* in your *ruler* are within the *margins* that you specified in your page setup. That is, if you have specified that you want a one-inch margin on both the left and the right sides of an 8.5 x 11 page, then you only have 6.5 inches left to write on, yes? So if the left indent in your ruler is set at 0, the right indent should be no further over than 6.5 inches (see *indent* for details, if you're not sure about the difference between margin and indent). The page setup margins take precedence over the indents you set in the ruler.

TSR

The acronym **TSR** stands for **t**erminate-and-**s**tay-**r**esident, which refers to a DOS *utility* (little program) stored in memory at all times. Some TSRs, like *disk caching* programs, enhance the operation of your computer without any further intervention on your part. Others are "pop-up" programs, like a calculator or a notepad or a calendar, that you can call on to do special chores while you're using another application. In this case, you can activate the TSR by pressing a certain combination of keys. When you're through, you then close the TSR and it stays accessible in memory should you want it again.

TSRs are great, but it can take a little work to get multiple TSRs to function together harmoniously. You may have to experiment with the order you start them to get things working right, and some combinations just won't fly. The other problem is that in DOS, all your programs have to share a limited amount of memory. The more TSRs you use, the less memory is available for your main applications. However, if you love your TSRs, there are ways to drastically reduce the amount of memory they eat up. You may be able to load them into *high memory* or *expanded* or *extended memory* (see *RAM*), and you can get special utilities that swap them in and out of memory as needed.

t-shirt

Guy defines a **t-shirt** as a person, typically not in management, who believes in egalitarianism, change, and flexibility. Antonym: suit.

tty, TTY

TTY stands for **tele**ty**pe**writer, a low-speed communications device consisting of a printer and a keyboard, sometimes called a "teleprinter." A teletypewriter provides *hard copy* (printed pages). They were often used with early computers as *terminals*. Nowadays, you and I can read the information on our screens instead of having to print it out.

A **t**elecommunications **d**evice for the **d**eaf (TDD) is often called a TTY. A TDD is used by hearing-impaired people to communicate through typing. You need a special *communications program* and a special *modem* to use a TDD.

tune, tuning

You may hear the term **tuning** used to describe the process of adding *hints* to *scalable fonts* to achieve the optimal quality of the characters displayed on screen and when printed. Microsoft was supposed to have spent a great deal of time tuning the TrueType fonts released with Windows 3.1, which is supposed to account for the beauty of these fonts. However, the benefits of all that effort are pretty subtle to the untrained eye. Most people have trouble distinguishing the difference between a tuned font and one that hasn't been tuned.

tweak

Tweak means to change something slightly, to fiddle around with it, often after it appears to be finished. "Yeah, this brochure looks great—let me just tweak the type a little."

Tweak is also used as a noun to describe a person who has just completed something you thought couldn't be done with a computer.

Type 1 font

A **Type 1 font** is the highest-quality kind of *PostScript outline font* (typeface). Type 1 fonts are specially designed to print clearly at small sizes on low-resolution machines. Until just a couple of years ago, only Adobe Systems (the originator of PostScript) knew how to create Type 1 fonts—unless you paid a licensing fee for the Type 1 coding system, you had to create your PostScript fonts as *Type 3 fonts*. But Adobe has since released the Type 1 specifications publicly so now anyone can create Type 1 fonts, including Scarlett (see the colophon at the end of this book). See *font* for details on Type 1 fonts and related information.

Type 2 font

There is no such thing as a **Type 2 font.** It was a technology developed by Adobe Systems that they never released.

Type 3 font

Type 3 fonts are a kind of *PostScript* font that aren't created much anymore. Type 3 was the font technology everyone else had to use while Adobe held onto the secrets to Type 1. Type 3 fonts, though, can have a little more variety within the face, such as gray shades, variable stroke widths, or graduated fills. They cannot be smoothed by *ATM,* but they print easily to *PostScript printers*. See *font* for more details.

typematic rate

If you hold down the "A" key you will see one "a" on the screen at first, but after a short pause you will see "aaaaaaaa." On an IBM-compatible PC, the rate at which each new "a" appears on the screen is called the **typematic rate** (from "type" and "automatic"). The typematic rate was fixed on earlier PCs, and it was too slow, especially when you wanted to move the cursor by holding down one of the arrow keys. Luckily, there were little *utilities* you could get to speed up the typematic rate.

On AT-compatible PCs, you can vary the typematic rate via DOS, but the standard rate is still set the same as on a PC. Even on a very fast PC with, say, an 80486 processor, the cursor will appear to move quite sluggishly through a screen of text. So please speed it up! To change the typematic rate via DOS, use the MODE command (see your DOS manual for details). Since this is a bit tricky and doesn't work with all keyboards, you may still be better off with a utility that does it for you.

typo

Typo is short for "typographical error," those embarrassing and usually stupid mistakes you make while typing. Isn't it amazing how many typos you never see until the job has been compelted? That's why God invented *spelling checkers*.

typography

Typography is the art or process of designing with type and letterforms. It takes into account the choice of a typeface, the size of the text, how closely the letters are spaced, and includes many fine details such as *hanging punctuation,* the proper use of *em* and *en dashes* and oldstyle figures, and on and on.

"Typography" also refers to the general character or appearance of printed matter, rather than the actual process of creating it. You can say, "She's not a great designer, but her typography is always dynamic."

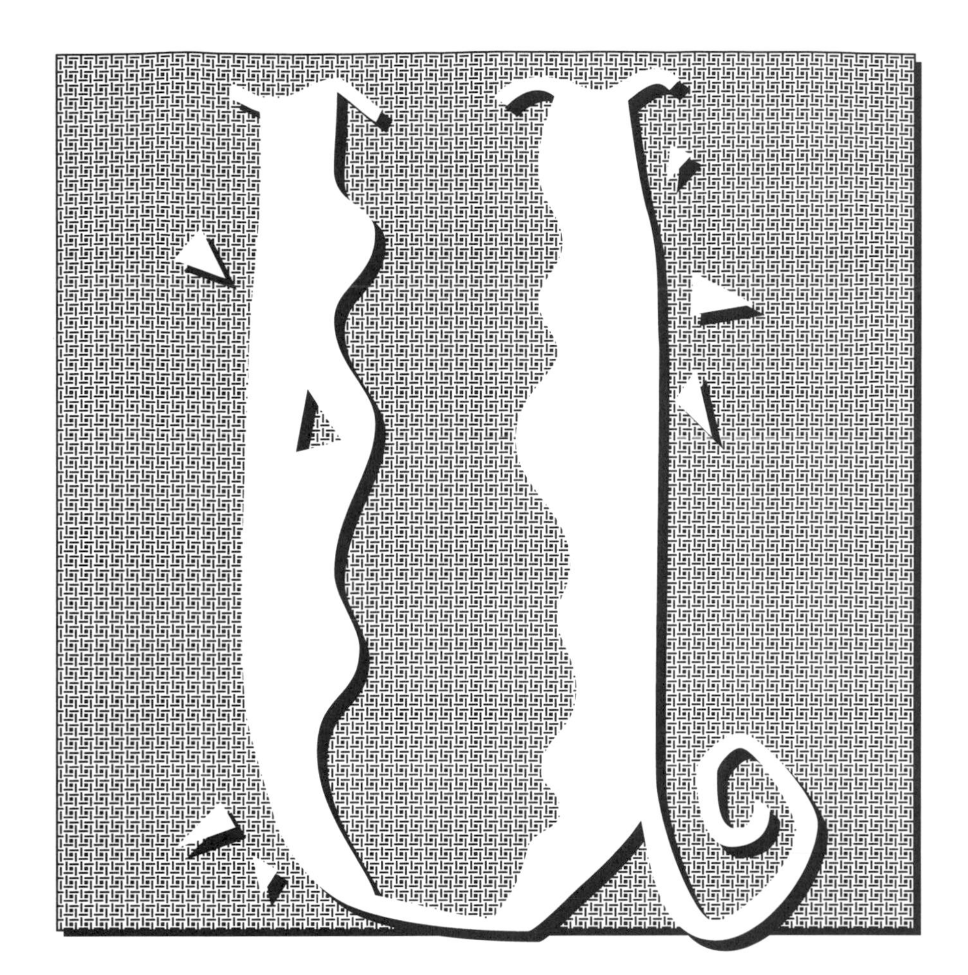

unarchive

To *archive* means to store documents for long-term safekeeping. Often they are *compressed* before they are stored. To **unarchive** means to bring these files back onto your own hard disk, and uncompress them if they were compressed.

un-delete

Un-delete means to retrieve a file that you previously deleted, or threw away. On a Mac, that can sometimes be as easy as grabbing it out of the trash can. Usually, to un-delete a file means using special software that can dig through the mysteries of the disk and pull it back out of the ether. Often it is impossible.

under

See *runs under.*

Undo

Just about every Macintosh and Windows program (and even some DOS programs) has an **Undo** command. The Undo command will undo the very last thing you did. If you erased a paragraph that you didn't mean to, choose Undo. If you selected 46 pages of your word processed document so you could change the font but the cat walked across your keys and those 46 pages disappeared, choose Undo. Undo is always found as the first command in the Edit menu.

The important thing to remember about Undo is that in most applications you can only Undo the very last thing you did. If you lost those 46 pages, then clicked around and selected other text and tried to *paste* something, you can't Undo—those 46 pages are gone. Not even screaming will help. A few applications have multiple levels of Undo, where you can keep undoing different things you did.

On the Mac, the keyboard shortcut for Undo is always Command Z (the Z is the closest possible letter to the Command key).

In Windows, the keyboard shortcut for Undo is Ctrl Z.

Unicode

Unicode is a new system for coding letters and other symbols you see on your screen or in print. Unicode isn't in widespread use yet, but it probably will become a standard over the next few years. Most computers now use a coding sytem called *ASCII* to represent characters on the screen. That is, when you type the letter A, the computer gets a signal made of one *byte*. This byte is eight *bits* of electronic message, something like 01000001, which tells the computer to put the letter A on the screen. With ASCII code, we can create 256 characters by rearranging those eight bits in different ways, characters including caps and lowercase, punctuation, spaces, and things like tabs and line breaks. Well, these 256 characters are not nearly enough to create the characters necessary for non-Roman languages like Japanese or Chinese. So Unicode was invented, which uses two bytes per character. With two bytes, the computer can rearrange the 16 bits (zeros and ones) into over 65,000 different numbers, meaning the computer will be able to handle more than 65,000 characters. Apple's new *QuickDraw GX* fonts take advantage of this technology.

UniFinder

In System 6, **UniFinder** is the unofficial term for when you are not using *MultiFinder*. MultiFinder allows you to open more than one application at a time (please see its definition). Operating without MultiFinder really has no other name, except to say that you are not using MultiFinder. So to abbreviate that, some people use the term UniFinder.

In System 7, there is no MultiFinder or UniFinder—it's just the Finder (although it acts just like MultiFinder, except you can't turn it off).

UNIVAC

UNIVAC is an acronym for **univ**ersal **a**utomatic **c**omputer, the first commercially available and successful computer (over forty systems were sold), introduced by Remington Rand in 1951. It was half the size of a garage. UNIVAC, for a while, was synonymous with "computer."

UNIX

UNIX (pronounced "you nix") is a *multi-user, multi-tasking, multi-platform operating system* widely used by universities and small to mid-sized businesses (an operating system is the software that a computer must have to manage its basic functions and to run other programs). Originally developed at AT&T, UNIX was designed for large *mainframes* and *minicomputers* in the days when one computer was shared by many users (multi-user). With multiple users working simultaneously, UNIX has to do several tasks at the same time (multi-tasking). Multi-platform means UNIX can run on different kinds of computers. The fundamental commands work the same on every computer running UNIX, although the different versions of UNIX differ in some details. UNIX is written in the C *programming language* (also developed at AT&T), which is why it can be transported to so many other platforms.

Some people question the benefits of using UNIX on a single-user personal computer, since it requires large amounts of memory and disk space to accommodate multiple users (the only reason to use UNIX on a single-user computer is if you are already familiar with it and don't want to switch to a new operating system). However, it *is* possible and it sometimes makes sense for more than one person to use a PC or Macintosh at the same time. In this case, UNIX or one of its variants is the operating system of choice (you would also have to equip each user with their own *terminal*). And, since we're rapidly moving into an era when many of us will connect our personal computers to other computers, we'll be encountering UNIX more often.

UNIX is available by that name for PCs, but it's used in several related versions on different platforms (computer systems), including A/UX on the Mac, AIX on IBM workstations, and Mach on the NeXT computer. Some UNIX computers can run the Open Look or Motif graphical user interfaces, which work a lot like the Macintosh or Windows.

unload

If you have to *load* a program into *memory* to use it, you may be able to **unload** (remove) the program to free up the memory it was using. On the PC, *memory resident* utilities (*TSRs,* which work while other programs are running) must be loaded, or placed, into memory. There they remain, taking up memory, whether or not they are doing anything for you, until you unload them. Some TSRs have commands for unloading or "removing" them from memory; if your TSR doesn't have such a command, you have to *restart (reboot)* the computer to unload it.

unprintable characters

Unprintable characters are the characters that may be part of a text file but which cannot be printed because they don't make any mark on the page. These include characters like Tab, Shift, Return, Enter, Alt, Control, Command, Option, or the page break character. Unprintable characters are also called invisible or non-printable characters.

unstuff

The term *stuff* means to *compress* a file so it takes up less disk space. Well, before you can use a stuffed file, you have to **unstuff** it, or *uncompress* it. You usually have to use the same compression utility that stuffed it to unstuff it, but many compression utilities can now create a *self-extracting archive* which you can just double-click to unstuff. Usually these are identified by the extension *.sea* after the file name (such as "Mom's Photo.sea"). We use the term "stuff" because of the very popular compression utility called *Stuffit.*

unzip

To **unzip** means the same as to *unstuff,* as noted directly above. "Zip" only works on PCs, and "stuff" is more popular on Macs. See *zip.* The compression utility PKZip is very popular on PCs.

up

When someone asks you if it's **up,** they are asking if it's functional and ready for use. Unless you are on very intimate terms with this person, they are usually referring to the computer or the printer or the network or some other similar device.

update

When you **update** (as a verb) some files or a program, it means you replace older versions with the current version. For instance, if you keep *backups* of all your files (which of course you do), you may occasionally need to go to those files and update them with the changes that have been made since you last backed them up.

An **update** (as a noun) refers to the new version of a software package that is relatively minor, like from version 4.1 to version 4.2. An *upgrade* would be from version 4.2 to 5.0.

upgrade

When you choose to **upgrade** (as a verb), you are choosing to get a newer, generally more powerful package, either hardware (such as a bigger computer), or software (such as the latest development in your favorite word processing program). "Did you upgrade to a Quadra?"

An **upgrade** (as a noun) refers to the new version of the product which you are hoping to own: "Did you get the upgrade to PageMaker?"

upload

Upload is when you use a *modem* to put one of your files onto a *network* or an *online service;* you load the file up onto the service. You do this so other people can get the file and use it (they need to *download* the file to use it). For instance, I can send my typography column to Desktop Communications magazine in New York by simply sending them an *e-mail* and attaching the word-processed column, as a separate file, to the mail. The file gets uploaded onto the service. The magazine checks their mail, finds the file I sent, and downloads it.

Also, online services are full of all sorts of software that users have uploaded—*utilities, fonts, games,* etc., and which you can download.

uppercase

Capital letters are also known as **uppercase** letters. You can use the word as a noun, as in "Uppercase is harder to read," or as an adjective, as in "Please don't type my name in all uppercase letters."

The word "uppercase" is from the days when all type was created with metal letters. The tiny pieces of metal type were kept in large, flat boxes called cases, with a separate cubby for each letter. There was a case for the capital letters and a separate case for the small letters. In the racks that held the cases, the case for the capital letters was customarily kept above the case for the small letters. Thus the capitals became known as "uppercase" and the small letters as "lowercase." See *font* for some illustrations.

upward compatibility

Software is said to be **upward compatible** if it will work not only on the computer system it was designed for, but also on later or more powerful versions of a system. Upward compatibility is an important feature because you don't want to have to *upgrade* every piece of software you own each time the *operating system* gets upgraded. Also see *backward (downward) compatible.*

user-friendly

User-friendly is a term used to describe a product, machine, or software package that is easy, non-intimidating, and (maybe) actually rather fun to use; that is, it is *friendly* to the *user* (the person who uses it).

An example is the Macintosh computer, which is said to be user-friendly, versus a typical IBM PC equipped with DOS, which is generally not considered to be user-friendly. In DOS, if you want to throw away a file (depending on your particular *system software*) that is called NOBOZOS.PM4, you need to be at the DOS *prompt,* and tell the computer you want to throw a file away, tell it on which disk to find the file, the name of the program the file was created in, the subdirectory where the program is to be found, and the exact name of the file. So you would type something like this (after the C:>): DEL C:\PAGEMAKER\ROBIN\NOBOZOS.PM4. On a Macintosh, you find the picture of the file (or you tell the computer to find something with the letters "robin" in its name). Then you drag the picture, or *icon,* into the icon of the trash can. That's user-friendly.

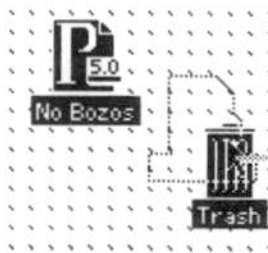

You can see the pointer dragging the icon and putting it in the trash can.

user group

A **user group** is a group of people who get together to give each other help, advice, and support with the particular kind of computer or software they use. Some user groups are small, like five or six people, and some, like the *Boston Computer Society* and *BMUG* have over 10,000 members each, worldwide. User groups usually have a newsletter, a monthly meeting with speakers, libraries of disks to buy, libraries of books to borrow, and an *online bulletin board.* Most larger groups sponsor *SIGs* (special interest groups) where people with similar interests can focus on particular topics each month, like *multimedia* or *desktop publishing* or computing for small businesses. The *MAUG (Micronetworked Apple User Group)* is an electronic group that holds meetings and has a great many services available *online. SeniorNet* is an electronic *network* connecting senior citizens who are computing.

User groups are a particularly integral and vital part of the Macintosh community. Our Mac user group here in Santa Rosa, which is not a very large town, has almost 300 paid members, about 150 of whom show up every month. I run a SIG of this North Coast Mac User Group specifically for new users or people who have been using the Mac for a while but still don't quite know what they're doing. We talk about things in a non-technical way. And we have cookies. I have over 500 people who have requested to be on the mailing list, and about 80 to 120 people show up monthly, depending on the topic.

No matter which computer you use, I strongly recommend you check out the local user groups in your town.

user interface

The **user interface** refers to the way the user and the computer communicate with each other to get the work done. The user interface includes what you see on the screen, the kinds of keys you have to push to get things done, any other devices like a mouse you can use to control the computer, and so on. If you could talk to the computer or it could talk to you, that would be part of the user interface.

Among the different types of user interfaces, the most primitive you'll come across is the *command-line interface.* Here, you're faced with a blank line on which you must type out your commands. The computer responds by carrying out your orders, *if* you typed them correctly— or by rebuking you with terse, barely intelligible error messages if you don't. For instance, in a command-line interface you might type "WORD LETTERMA.DOC" to open a letter you wrote to your mother.

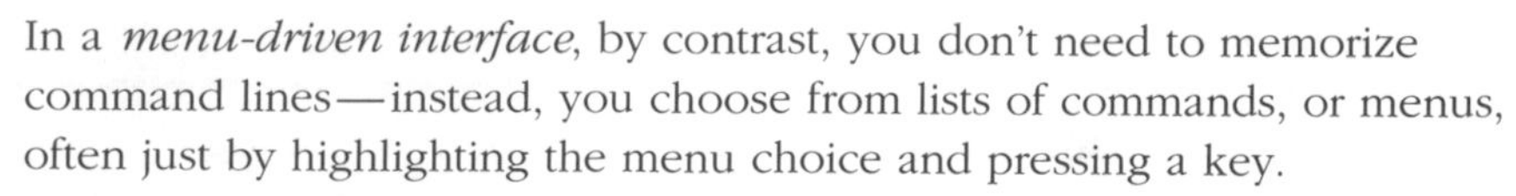

In a *menu-driven interface*, by contrast, you don't need to memorize command lines—instead, you choose from lists of commands, or menus, often just by highlighting the menu choice and pressing a key.

This is a typical icon in a GUI.

Even easier to use, a *graphical user interface* (*GUI,* pronounced "gooey") represents the computer's functions as little pictures, or *icons.* The icons are usually representative of the function; for instance, your letter to mom looks like a little letter named "Letter to Mom," and all you need to do to open it is double-click on the icon with the mouse and pointer. Icons that are representative of their functions are one of the features that makes a computer more *intuitive* to use.

There is also a "pen-based interface" where the user uses a stylus (like a pen, but with no ink) to choose commands and push buttons. And "virtual interfaces" are being developed (see *virtual reality*) where the user would be almost inside the computer and controlling it with her entire body.

The particular user interface you work with depends primarily on the software you're using, less so on the type of computer you have. On a PC, for instance, if you run DOS, you have to work with an exceedingly crude command line interface: all you see is a blinking little light on a bare screen, and you have to type in all the commands yourself. However, every DOS *program* has its own unique user interface; most of them are easier to use than DOS itself, but you have to learn a different interface. And if you run Windows on that same computer, you get a completely different *graphical user interface,* with all those menus and dialog boxes and pretty little icons.

A person tends to bond with one interface or the other, typically with the first one they use. They tend to think it's the best and don't like to switch. Funny how that happens.

USENET

USENET is a worldwide *network* of UNIX-based computers for news distribution, *e-mail,* and special interest *bulletin boards.*

utility

A **utility** is a small program that is not an *application.* That is, you don't usually create something with a utility, but you use it to enhance your work environment. For instance, *AfterDark* is a *screen saver* utility that protects your monitor. *Suitcase* is a utility that lets you manage a large number of *fonts. ATM* is a utility that makes your type display on the screen with smooth edges and curves.

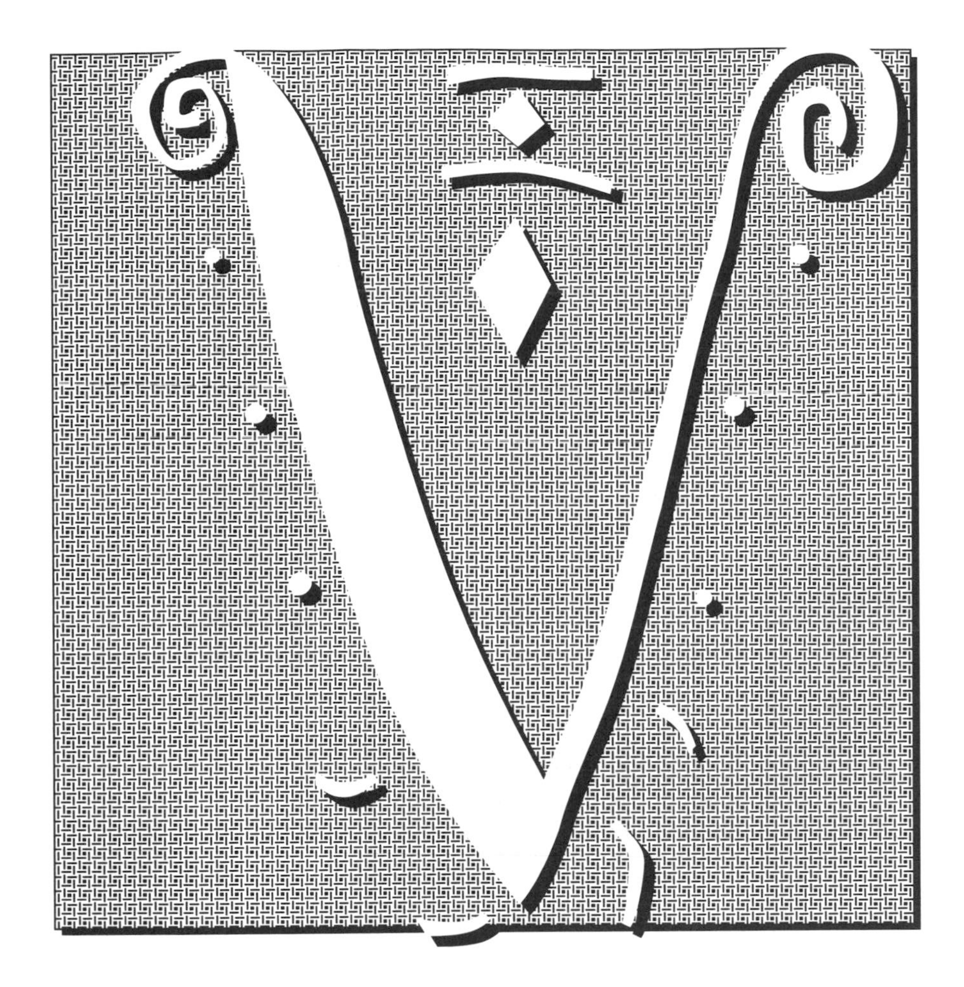

v.32, v.32bis, v.42, v.42bis

These terms (pronounced as "vee dot 32 biss") and similar terms are international standards *(protocols)* set by the CCITT (Consultative Committee for International Telephone and Telegraphy) governing the way modems communicate with each other. A modem protocol specifies the audible signals the modems use to establish connections and the signals they use to actually transfer data. The v.21 was a standard set in 1964. *Bis* means it is the second version.

You don't need to know anything about how the protocols actually work, but you should be sure any modem you buy follows the standards—otherwise, it may not be able to communicate with other modems. It's not enough to know the speed of the modem you're buying; to guarantee success, choose a model that is "compliant," not just "compatible," with the protocol.

Before 1984, the CCITT protocols were used mainly on modems sold outside the United States; 1200 *bps* modems sold in the U.S. still adhere to protocols developed by Bell Labs instead. But beginning with 2400 bps modems, the CCITT protocols have come to be accepted nearly everywhere, including the U.S. The v.32 protocol covers the basic signalling characteristics of 4800 and 9600 bps modems; v.32bis extends up to 14,400 bps, with additional features.

The v.42 protocol encompasses v.32, but adds error correction—a v.42 modem will retransmit as needed to get the data through perfectly. The v.42bis is a standard for data *compression* so files get transmitted faster (but be aware that it won't speed things up on files that have already been compressed). A v.42 or v.42bis modem will work with a v.32 modem, but you won't get the error correction or data compression.

vaccine

Viruses are around that can infect any computer (if you don't know what a virus is, please see that definition). There are *software* applications and *utilities* you can use to protect a computer from viruses, and these are sometimes called **vaccines.** The utility I prefer for the Mac is *Disinfectant,* written by John Norstadt, and available free from many sources (see the definition for *Disinfectant*).

vaporware

Vaporware is a product that the vendor keeps promising is about to arrive any moment *(real soon now)*—but it goes so long past its shipment date that no one believes it will ever really ship. Sometimes it never does. *System 7* was vaporware for a while, since it took two years longer to appear than we were told. Apple's Newton was vaporware for a long while. This book was vaporware for a year (but not because I wasn't trying!).

VAR

VAR (pronounced "var") stands for **v**alue **a**dded **r**eseller. A VAR is a person or company that buys equipment (both software and hardware) directly from the vendors, puts it all together into a package, and sells it to you. This can be great if you need a complex system and you don't have the time to figure out exactly what components you need. A VAR will typically find out your needs, gather up the hardware and software, install it for you, arrange training, and provide support for both the hardware and software. That's why they're called "value added" resellers.

VDT

VDT stands for **v**ideo **d**isplay **t**erminal, which is the same as *monitor* or *screen.*

vector vs. raster graphics

Vector graphics are stored in the computer as a set of mathematical formulas describing the shapes that make up each image. When you display a vector graphic on the screen or print it, these formulas are converted into the patterns of dots you can see. Because the dots are not specified unitl you display or print the graphic, you can change the size of the image without any loss of quality, and the image will always appear at the highest *resolution* of whatever screen or printer you're using. The term vector graphics means exactly the same thing as *object-oriented* (or just object) graphics.

The contrasting term is **raster graphic** (the terms *raster* and *bitmapped* are synonymous). In a raster graphic, the actual dots that make up the image you see are defined when the graphic is created, so the resolution is fixed; changing the size will make the image look coarse or muddy. See *paint program* for an example illustrating the fixed resolution.

Most Macintosh people use the terms *object-oriented* and *bitmapped* rather than *vector* and *raster.* Most PC people use both pairs of terms interchangeably.

version

All software, including *operating systems,* is constantly being updated, debugged, modified, enhanced, and generally improved. When a vendor makes these improvements, they release a new **version** of the product. The new version is indicated by the number after the name of the product; e.g., PageMaker 5.0 (pronounced "five point oh") is a newer version than PageMaker 4.2 (pronounced "four point two").

When the new version is a major *upgrade,* the first number changes; for instance, from 4.2 to 5.0. If it is a minor revision, the second number changes; for instance, from 4.1 to 4.2. If it is a very minor revision, like just a *bug fix,* you'll see a third number or letter; for instance, 4.2a or 6.0.1 (pronounced "six point oh one" or "six point oh point one").

On a Mac, you can determine the current version number by clicking once on the icon representing the program, then choosing "Get Info" from the File menu. Or, if the application is *running* (it's open and you are using it), from the Apple menu choose "About ______."

VESA

VESA stands for **v**ideo **e**lectronics **s**tandards **a**ssociation, a group of PC-industry manufacturers who got together to set standards for *high-resolution* video devices (monitors and video circuits, or *adapters*). If you buy an *add-in video board* or a monitor that complies with the VESA standards, you stand a better chance of getting everything to work together properly and of seeing a quality image on your screen. One caveat: VESA sets minimum standards and recommended standards for some monitors, so choose products that claim to meet or exceed the recommendations, instead of those that only satisfy the minimum requirements. VESA has also come up with a standard called the VL-Bus for *local bus video* circuits.

VGA

VGA, which stands for **v**ideo **g**raphics **a**rray, is currently the most popular standard for PC screen display equipment. Technically, a VGA is a type of *video adapter* (circuitry in the computer that controls the screen). IBM developed the VGA for its PS/2 line of computers (the name "Video Graphics Array" is an IBM trademark), but loads of other manufacturers make VGA *add-in boards* (that plug into a slot in the PC) and VGA *chips* (in some PCs, these VGA chips are built right into the main part of the computer, the *motherboard*). A VGA *monitor* is a monitor that works with a VGA adapter.

A standard VGA system displays up to 640 x 480 pixels (little dots) on the screen, with up to 16 different colors at a time. In lower resolution, 320 x 200 pixels, the screen can show up to 256 colors at once. These specifications are much better than the older video adapter standards, the *CGA* and *EGA*, but they're not good enough for many people. If you're buying a new system or replacing an older video adapter, make sure you get a "SuperVGA" adapter, which can handle higher resolutions (800 x 600 or higher) and many more colors. Remember though, that the higher the resolution and the more colors you have to work with, the slower the display will function, and the more memory you'll need on the card.

Unlike EGA and CGA monitors, VGA monitors are *analog* devices, meaning they can display an infinite range of colors (the number of colors you see is limited by the VGA adapter, not the monitor).

When you're shopping for a VGA monitor, keep several points in mind. First, if you want to use higher resolutions than the VGA standard of 640 x 480, you need a *multiscan* monitor—a plain VGA monitor will *not* work at higher resolutions. Second, some VGA monitors give a sharper image than others. Partly, this depends on the *dot pitch:* a monitor with a smaller dot-pitch (like .28mm) will have better image clarity than one with a larger dot-pitch (like .39mm).

A VGA monitor requires an interface *card* and a cable. You need to know how much *memory* is on the card. You may want to add more memory, especially if you plan to create and use complex graphic or photographic images. The VGA is the current standard right now in monitors, and as such is usually the most readily available.

video accelerator

A **video accelerator** is a *video board* (circuits your computer uses to control the monitor) equipped with specialized chips that speed up the action of your screen and make your programs seem faster in the bargain.

With ordinary video adapters, the computer's main microprocessor is responsible for controlling the color or shade of every dot on the screen individually; every time the screen changes, the microprocessor must re-send the information for all the affected dots, one dot's worth at a time. Not only does that tie up the microprocessor, but a general-purpose microprocessor can't do the job as quickly as a specialized video chip. Even if your computer has a really high-speed processor (like a PC with an *80486* running at 66*MHz*), you'll see noticeable delays every time you scroll the screen, especially with high *resolution* and lots of different colors.

There are two main types of video accelerators. You get the best bang for your buck from the kind that are designed specifically for a particular type of software, like Windows or AutoCAD. Programmable accelerators speed up many different kinds of software, but are more expensive and may not be as fast at any particular task.

video adapter, video board, video card, video display board

Whether you call it a **board,** a **card,** or an **adapter,** it is the same thing: a piece of plastic or fiberglass with electronic *circuits* printed on it, and with *memory* and other *chips* attached. (In the PC world they are usually called adapters; in the Mac world, they are usually called cards or boards.)

In some computers, like most Macintoshes and some PCs, the video adapter is built right into the main computer circuits on the *motherboard.* In others, like most PCs, it comes on a separate *board* or card that plugs into a *slot* in the computer. The video adapter determines the screen's *resolution*—how many colors you can see at one time—whether or not you can see graphics or just plain text, and how fast screen images are displayed. Even if your computer has a built-in adapter, you can probably add a new video board that displays more color (which creates the illusion of higher resolution), or that just runs faster. If you buy a new video card, make sure your computer, your monitor, and the card are all compatible.

Video Overlay Card

The **Video Overlay Card** is a *card* (plastic board with a *printed circuit* and *memory* that you install in your computer) that enhances the video graphics capabilities of Apple IIE and IIGS computers, and some PCs. It includes software, Video Mix, that lets you overlay text, graphics, and animation onto television video.

virtual memory

Virtual memory isn't really *memory* at all. The computer can only work with information that is currently in memory; it cannot work with information sitting on the hard disk (the hard disk is just for storing things, much like your filing cabinet). On some computers and with certain programs, if you are short on real memory you can tell the computer to fake it by **pretending** that any extra hard disk space is memory. This smoke-and-mirrors trick is called *virtual memory.* (Hard disk space is usually used for virtual memory because hard disks are much faster than floppies.)

Let's say you're editing a color photographic image, which requires far more data than will fit into your computer's memory all at once. If you have virtual memory, the computer can work with the entire photo as a single image—it puts the parts of the image that aren't in use at any given moment into virtual memory on the disk, retrieving them automatically when you want to see them. As far as you're concerned, the system works just as if it had more memory than it really does, only a little slower (because the information must be "swapped" back and forth between the disk and real memory). The combination of an adequate amount of real memory plus lots of virtual memory can make your system much more convenient to use, because you can work with bigger documents and run more applications at the same time.

virtual reality

Virtual reality (abbreviated as VR) is fake reality. How's that for an oxymoron? Virtual reality is when something is not real, but it appears and feels so real that it's almost as if it were real. And what is real, after all?

Computers are being used to simulate environments, including sound, touch, and three-dimensional images. The strongest sense of virtual reality is achieved with a three-part system: the stereo headphone, stereoscopic video goggles, and "data gloves." Places like the *M.I.T. Media Lab* have fascinating virtual reality labs set up. The computer image goes in through your eyes and the sounds in through your ears. As you turn your head, the computer images also turn so you can see what is in that direction. The sound moves towards or away from you, depending on how you move. You can reach out and touch things in the computer image; you can create music by banging on instruments; you can feel the pull of electrons as you grab them.

If you've been on the Space Shuttle ride in Disneyland, you have a glimpse into the power of virtual reality. If you've ever fallen in love *online,* you've had a glimpse into the power of virtual reality.

virus

A **virus** is a very interesting thing. It is a software program written by a very intelligent, very skilled, and very evil, sick person—written expressly to do damage to other people's computer systems. Viruses can destroy the data on your computer, corrupt your system software, lock you out of your own machine, eat your programs for lunch; they can even wipe out an entire hard disk.

Viruses travel from computer to computer through floppy disks, networks, and even modems. You don't always know you have a virus; they often have a delayed activation time, so you use the sick program for a while until one day it eats you. (You'd think that a person who can write such a sophisticated program to do so much evil is certainly bright enough to get a real job and direct that energy into making people happy.)

This is a virus scenario: You have a disk that is infected. It doesn't look any different. You put it into your computer. As you use the infected application on that disk, the virus gets into your operating system. You take that disk out and insert another one. The virus jumps from your system onto an application on the floppy disk. You take that disk out, go to another computer in the office and insert it. The virus jumps from that disk onto the other person's System. Ad infinitum. The Macintosh WDEF virus is so contagious that simply inserting an infected disk into a computer infects the computer. It's comparable to somebody with a social disease getting on a bus and everybody on the bus automatically getting the disease. Then everybody on the bus gets off the bus and goes into stores and everybody in the stores gets the disease. Fortunately WDEF is not a terribly devastating virus, just very irritating. (On System 6, just *rebuild your Desktop* to destroy the WDEF; System 7 is immune to WDEF.)

How do you know when you have a virus? Things start acting funny on your computer. Windows may not function properly, printing might not work right, files may be changed, programs may be "damaged." If anything starts acting weird, you can suspect a virus. (Actually, first you should suspect any *TSRs* on your PC, or any *INITs, extensions,* or *cdevs* on your Mac, since they are a much more likely cause of little troubles. Take them all out and put them back in again, one at a time, over a period of a couple of days (or get and use some kind of extension manager software). That way you can pinpoint the offensive creature and remove it. If you think the problem really is a virus, then get virus-protection software; test and "disinfect" your hard disk *and every single disk in your entire office and home.* And never again let anyone put a disk in your computer without checking it first. Safe computing.

Viruses have the potential to cause catastrophic damage; however, most viruses have been held in check by virus-protection software, and with a little knowledge you can easily protect yourself and your loved ones. Be aware of viruses, but not paranoid. Everyone should own and use virus-protection software as a normal part of computer life. Just check any disk someone gives you, and if you have used your disks in the office or at school or at a shop where they rent computer time, be sure to check them again before you insert them into your own computer.

On the Mac, my favorite virus-protection package is called *Disinfectant.* It has a utility that will locate and kill any virus, and it comes with an extension that will quietly check any disk you put in your machine. If an infected disk is inserted, Disinfectant sets off bells and whistles (literally). It is *freeware,* written by a wonderful man named John Norstad at Northwestern University. He spends time to write this program and is constantly updating it to catch new viruses. Can you believe that? He does this out of the kindness of his heart—an intrepid soul battling the forces of evil. Write him a nice thank-you letter and tell him how wonderful he is. See the entry for *Disinfectant* for more details and how to get it.

DOS 6 comes with virus-protection software, and there is a widely used *shareware* utility for PCs called McAfee Anti-Virus.

VisiCalc

VisiCalc was the original personal computer *spreadsheet* application. It fueled the success of the Apple II series during the late 1970s and made its authors a bundle of money. When they failed to catch the IBM PC wave, a look-alike product called *Lotus 1-2-3* was there to fill the void.

VLF

VLF stands for **v**ery **l**ow **f**requency, a type of electrical radiation that is emitted from *CRTs* (cathode ray tubes) like your monitor or television. Please see *radiation.*

VLSI

VLSI stands for **v**ery **l**arge **s**cale **i**ntegration, a term for the technology that enables a great number of transistors and other components to be built into a chip the size of your baby fingernail. (One transistor used to be about as big as a lightbulb.) Four different sources gave me four different figures for the number of components that fit on one chip with VLSI technology, ranging from 5,000 to 1 million. The *68040 chip* holds almost a million components. Incredible.

voice mail

Those automated phone answering systems, the kind that say "If you know your party's extension, enter it now," and "Press 1 for more options" are examples of **voice mail.** You eventually get routed to the person you want to talk to, who of course is rarely there, so you leave a message. Your message is *digitized,* stored on the computer, and when the person returns to the office, they can listen to your voice and the message. Voice mail is controlled by a computer rather than by an electric answering machine like the kind we use at home.

voice recognition

Voice recognition is a computer's ability to understand spoken words and act on them as if you had entered a command through the keyboard or mouse. It's happening right now on computers, and will quickly become more and more sophisticated. Apple is working on a voice recognition technology called "Casper." At the moment you can look at the computer and say, "Casper. Make an appointment with Ms. Pearl," and the computer will find out the days and times you are free, check Ms. Pearl's schedule, arrange a meeting, then report back to you, in a gentle voice, "You are scheduled for a meeting with Ms. Pearl on Wednesday at 10 A.M." It's true.

volatile memory

Volatile memory refers to *memory* that loses its contents when you turn the power off (or when you *crash*). The main memory of a computer, *RAM (random access memory)* is volatile. *Read-only memory (ROM)* is *non-volatile,* meaning it can hold on to the information even after the power is off.

volume

Volume is another name for a disk. Any disk that has an icon or a label that shows up on your screen is considered a volume. This includes one large hard disk that has been *partitioned* into several smaller hard disks—each partition is also a volume.

In DOS, the name of a disk is its "volume label," which can be up to eleven characters long. DOS will tell you the volume label of the current disk if you type the **VOL** command. You can name a disk when you format it, or change the name with the LABEL command by typing something like:

```
LABEL A:DOODADS
```

VR

VR stands for *virtual reality*. Please see that definition.

VRAM

VRAM (pronounced "vee ram") stands for **v**ideo **r**andom **a**ccess ***m**emory,* or video *RAM*. This is a special type of memory used on some *video adapters* to speed up the display of images on the screen. VRAM costs more than regular RAM (DRAM, dynamic RAM), but it does make the screen snappier.

Whichever type of RAM your video adapter uses, the amount of memory determines how many colors you can see on the screen: the more video RAM you have, the more colors your monitor can display and the higher the *resolution* will be.

Analog monitors like the VGA can display an infinite number of colors, they say, because they can vary the intensity of each *color gun* over an infinite range. It's the adapter circuits that determine how many colors are displayed. But anyway, that's not the main point of VRAM.

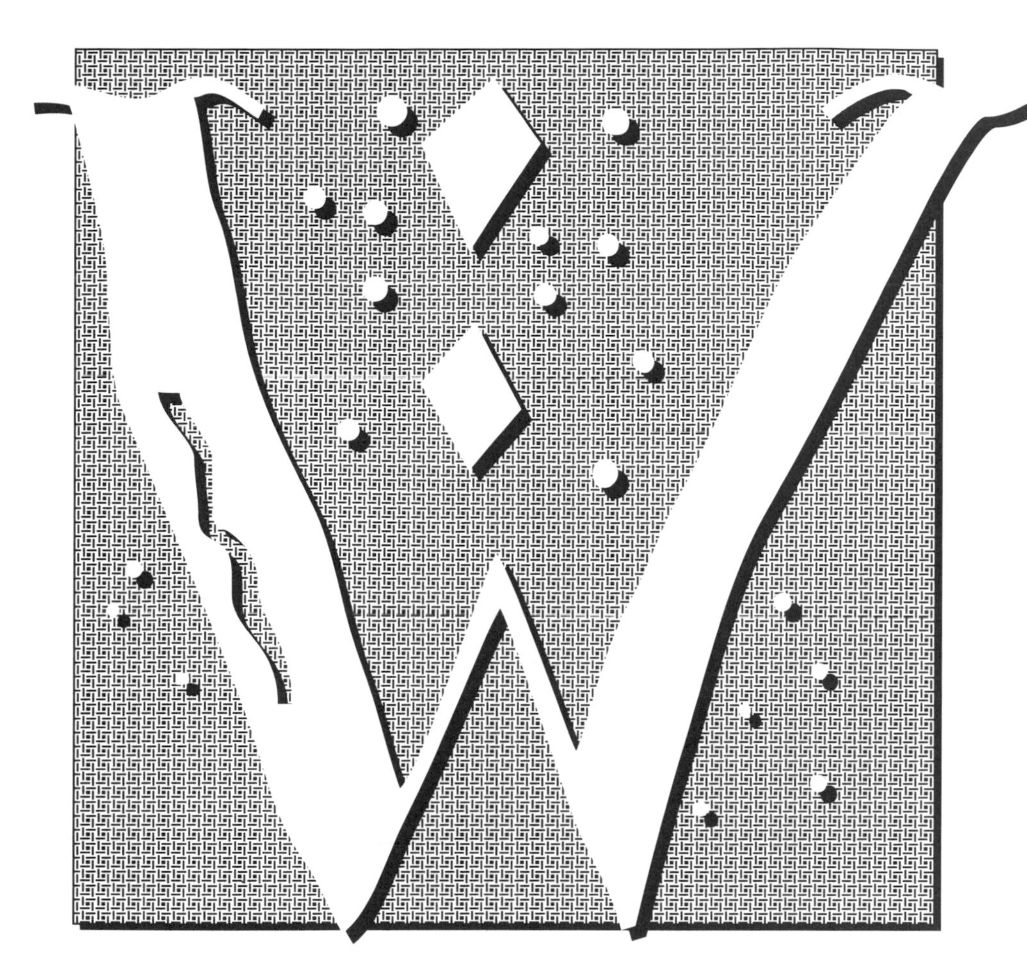

wafer

A **wafer** is a very thin (1/30 of an inch) slice of a salami-shaped *silicon* crystal—silicon being a *semiconductor* material. This thin slice is used in the production of *integrated circuits,* with various etching and layering techniques applied to create the circuits on the wafer. See *chip* for a crude drawing of a wafer.

wallpaper

In Windows, **wallpaper** is a picture or pattern displayed on your *desktop,* which is what you see on the screen "behind" all the windows (you may have to *close* or *minimize* some programs to see the desktop). Wallpaper can be anything from a simple geometric pattern to a full-color photograph of a flower, the StarShip Enterprise, or your sweetheart. Wallpaper images are just ordinary Windows *bitmap* files (in the .BMP format) stored in your WINDOWS directory. You select the file you want using the Desktop icon of the Control Panel program.

The Mac also has wallpaper patterns available for the Desktop. You can even set it up so it changes every five minutes or so. Personally, wallpaper drives me nuts; I can't stand the visual clutter—it's worse than my messy house.

warm boot

If the power to your computer is completely turned off and you start the machine up, that's a *cold boot.* If your computer is already on and you just choose to restart it without turning it off, that's a **warm boot.** Sometimes you don't actually choose to restart, but you have crashed and have no other option but to restart or reset.

WAN

See *wide area network.*

WDEF

WDEF (pronounced "w def" or sometimes just "w d e f") is the name of a particular Macintosh *virus* (programming code that someone has intentionally written that is designed to damage your files). WDEF is particularly infectious because you don't even have to use an application to get infected—you just put an infected disk in your computer and your computer's got the virus. Every disk that you then stick in your computer also picks up the virus and infects every other computer it enters. It's kind of like if someone with a disease got on a bus and everybody on that entire bus instantly caught the disease. Then they all go to work and when they walk in the door everyone in the building gets the disease. And so on. WDEF is transmitted so easily because it infects the "desktop file," an invisible file on every disk, hard or floppy, that keeps track of what's on the disk.

You can eliminate WDEF from any disk by *rebuilding the desktop.* To rebuild the desktop of a floppy disk, hold down the Command and Option keys as you insert the disk. Keep holding them down until you get a message asking if it's okay to rebuild the desktop. Click OK. WDEF will die.

If you are running *System 7,* your hard disk will never get infected by WDEF. If you are on *System 6,* rebuild the hard disk desktop by holding down the Command and Option keys as you turn on your Mac or as you *restart*. Click OK when you get the message. Also see *rebuilding the desktop* for more details on this, and see *virus* and *Disinfectant* for info about viruses and anti-viral software.

wide area network

A **wide area network,** also known as a WAN, is a generic term referring to computers that are *networked* together over a wide geographic area, which requires special software and sometimes hardware. They share information over telephone lines and via radio waves. By contrast, the computers in a LAN (local area network) are connected physically, by cables, and are in close proximity to each other.

widget

A **widget** can be sort of a wildcard term for a real thing for which you don't have an exact name. For instance, a spreadsheet tutorial might ask you, "If the widget broke into forty pieces. . . ."

Some PC users in a *graphical user interface* (using windows and icons) such as *Presentation Manager* or *Windows* call the *scroll box* a **widget.** Please see the definition for *scroll box*.

widow

There is great debate over the exact definition of a widow. I define a **widow** as a very short last line of a paragraph. I usually tell beginners to never leave less than seven characters on the last line, but really the actual number is arbitrary—if it looks too short, it *is* too short. In a wide paragraph, twelve characters may be too short. In a narrow column, six characters may be sufficient. It doesn't matter what you call it or how many letters you define it as, the point is not to do it. Widows are an aesthetically disturbing sight. The worst kind of widow is when there is a hyphenated word and the last part of that word is the last line, as in this very paragraph.

That is really tacky.

The term widow may also refer to the first line of a paragraph that is stranded alone at the bottom of a column, with the rest of the paragraph continued in the next column.

Some people define a widow as the last line of a paragraph that is stranded at the top of the next page or column—but we know that is really an *orphan.* This line is an orphan. Poor thing.

Your software application may give you options to avoid widows and orphans. You would be wise to ascertain exactly what your software considers a widow or an orphan before you use the feature so you are not disappointed or surprised at the results.

wildcard character

A **wildcard character** is a stand-in character when you don't know what the real one is, or you want it to refer to any one of several real characters.

A wildcard character is often used in a *search,* where you are searching for some text or a particular file that either you don't quite remember the spelling of, or you want to find any variations of it. Some applications use a question mark as a wildcard character (?), some use a caret and a question mark (^?), some use an *asterisk* (*).

For instance, if you are looking for the story on Boadicea but you can't remember whether you spelled her name "Boadicea" or "Boadecia," you could search for "Boad?c?a" and the application would find either spelling. Some programs allow you to use an asterisk to mean any number of characters. For instance, you could search for "Boad*" and find any word that started with "Boad."

In DOS, Unix, and other *command-line* operating systems, you can use wildcards to make a single command affect multiple files. In DOS, the asterisk (called a "star") stands for a sequence of characters of any length, while the question mark typically stands for any single character. "Copy file.*" means copy all the files that start with the word "file." and have any *extension* at all (extension being the identifying characters after the dot). For example, if you had three files, "file.bat," "file.exe," and "file.com," the "Copy file.*" command would copy all of them.

Or, if you entered "Copy file00?" and you had three files, "file001," "file002," and "file003," the command would copy all the files that started with "file00" and had 7 characters in the name. But it would not copy "file009x" or "file100.doc," because a ? wildcard only stands in for one character.

Also see *star dot star (*.*).*

WIMP

WIMP is an acronym applied disdainfully by hackers to the graphical user interface: Window, Icon, Menu, Pointing Device (or Pull-down menu). Wimp can also be applied to hackers as an admonition: Don't be one.

Winchester drive

This term now just means a *hard disk*, but the story behind it is endearingly human (as most stories behind-the-scenes are). One of the first hard disks IBM developed had two 30-megabyte disks inside (all hard disks are made up of multiple individual disks, stacked up like coins) and had an *access time* of 30 milliseconds. To the people working on the project, this was reminiscent of a double-barreled Winchester .30-caliber rifle, a "30-30." So internally, the IBM code name for this new hard drive became **Winchester.** This disk drive became the model for most of the hard drives in common use today, and the term was often used to refer to any fixed hard disk, though you don't hear it so much anymore.

windoid

A **windoid** is a little *window* with a difference. Windoids are usually floating *palettes* that stay in front—they never get lost behind other windows. Windoids may have a little close box in the upper left corner, but they usually have no title (no name) in the bar across the top or along the side. They may have a pattern in that title-less bar instead of the horizontal lines of a window. But you can still use that bar to move the windoid around on the screen—just *press-and-drag*. Many windoids are *tear-off palettes* or *menus*.

This is a typical little windoid, or tear-off palette.

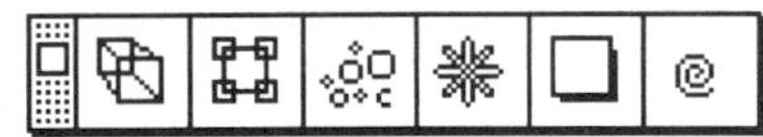

window

A **window** is a rectangular frame on the screen in which you can see and work with a particular application, a particular document, a list of the files on your disk, or some other specific collection of information. This is a visual way to organize the information you're working with into separate sections, making it easier to handle lots of different kinds of information without getting confused.

In *graphical user interfaces* like the Macintosh or Windows, which let you run different applications at the same time, every application has its own window. If the application lets you work with multiple documents at the same time, each document also has its own window.

Windows are typical of *graphical user interfaces* and usually share the features described in the illustration on the next page. When two or more windows are open (on the screen) at once, they can be overlapped; simply click anywhere on the underlying window to bring it to the front.

You may have menu commands in your application that allow you to *tile* the windows, where the windows are all reduced in size and arranged on the screen so you can see them all at once, or to *cascade* them, where they are all overlapped, with the title bar of each window visible.

The illustration below is a Macintosh window; PC windows are similar. If you are unfamiliar with the concept of windows on the Mac, you need a copy of *The Little Mac Book* (by Robin Williams, **not** "Que's Little Mac Book" wannabe).

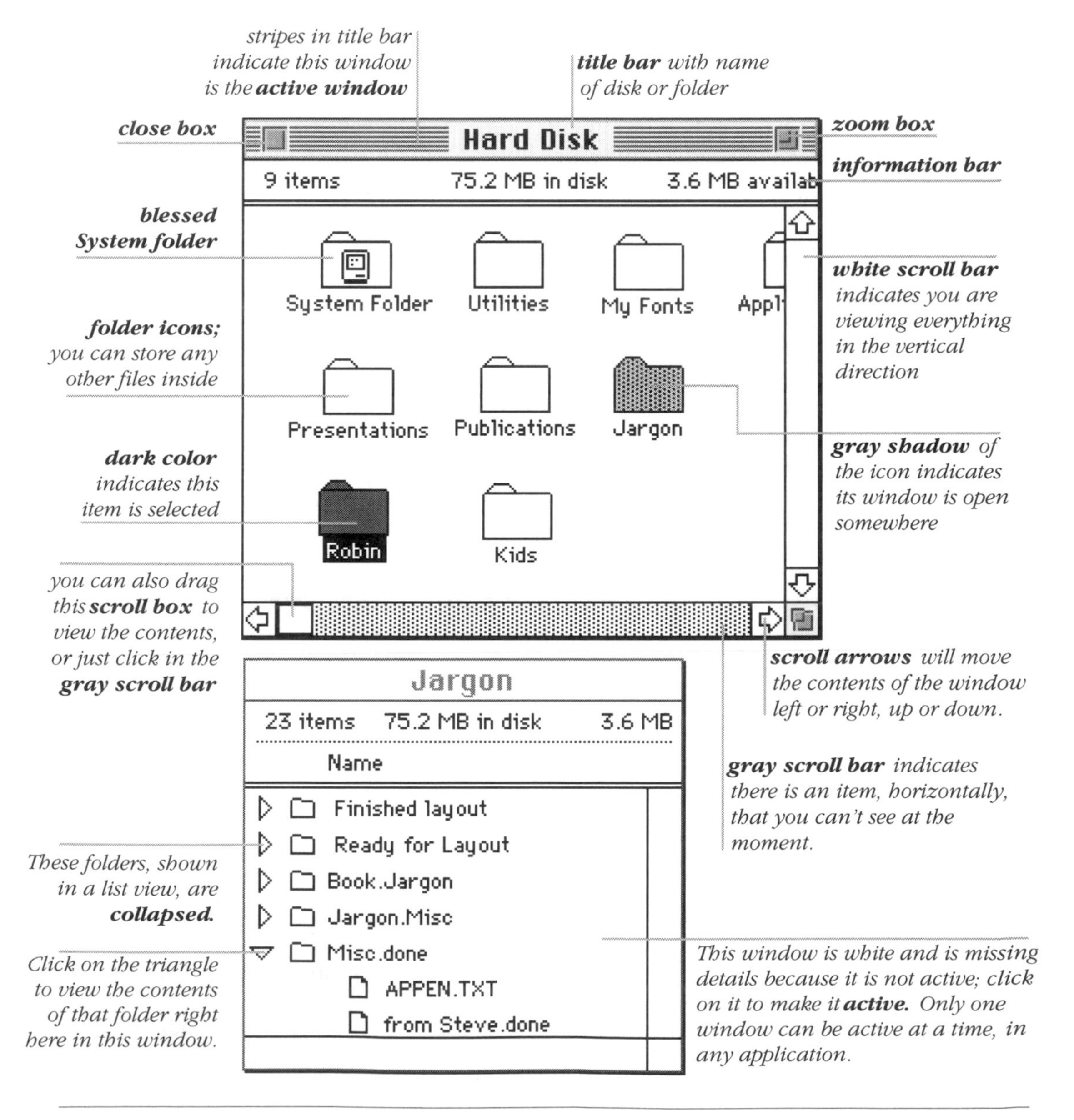

Windows, Microsoft Windows

Windows (with a capital W) refers to **Microsoft Windows,** the software that makes a PC prettier, easier, and more consistent to use—and slower. Windows has become extremely popular, redefining the nature and reputation of the PC.

(Some people say Windows makes a PC into a pale imitation of a Macintosh, but let's put off that squabble for a couple of minutes) *(says Steve; aw, c'mon, says Robin.)*

Windows is usually defined as a *graphical user interface,* or *GUI* (pronounced "gooey"). With a GUI, you interact with the computer by working with symbols on the screen, instead of having to type out commands. As in all other popular GUIs, the Windows screen is organized into rectangular, bordered areas called *windows,* each of which contains a distinct group of information. You can carry out many functions by using a *mouse* to activate little pictures *(icons)* that represent what you want to do. To start a word processing application, for instance, you might activate an icon showing a pen or a printed page.

But behind the scenes, Windows is more than a GUI. For one thing, it creates a consistent system by which your programs communicate with the equipment in your setup—your screen, the printer, the mouse, and anything else you might hook up. With DOS, by contrast, each program is on its own when it comes to controlling these devices, and you have to set up each program separately. In Windows, you have only to set up Windows for your equipment, and **all** your programs can work with that equipment automatically. Windows also changes the way the PC operates, allowing the computer to run more than one program at the same time (multi-task), giving programs access to far more memory than they get with DOS (partly through *virtual memory*), and including rudimentary facilities for networking multiple computers together. By the way, you can still run standard DOS programs while you're running Windows—in fact, you can run more than one at the same time.

Windows relies on DOS, the PC's standard *operating system* (the controlling software) for many important functions. But since Windows modifies and enhances DOS in so many ways, Windows becomes part of the operating system for all practical purposes. The easiest way to describe Windows is to be vague and call it an *environment*—that's what the experts do.

The first version of Windows came out around 1985, but it wasn't until the release of Windows 3.0 in 1990 that Windows really caught on. The upgrade, Windows 3.1, has several advantages over 3.0, including *TrueType,* a *scalable font* technology that lets you print text at any size in any typeface on just about any printer. (Please do upgrade to 3.1 if you have an earlier version.) "Windows for Workgroups" is a version with better *networking* capabilities. *Windows NT* is a complete operating system (see its separate definition, below).

If you have a PC and haven't yet installed Windows, should you? The answer is: maybe. If you're new to computers, or if you still can't figure out how to work your PC, get Windows. Windows is great if you really need its special features, the ones I talked about earlier. But for the jobs people use their computers for the most—like writing or working with spreadhseets or databases—an ordinary DOS program often does the job just as well as Windows software, and the DOS version is usually faster. So if you're already comfortable using DOS, switching to Windows isn't necessarily going to make you more productive, yet you'll have to buy new software and maybe even a more powerful computer. The point is, don't feel you have to have Windows.

OK, now back to the Mac versus Windows issue. Windows is clearly a Macintosh imitator, so the Mac scores for being there first. And the setup process can be a good deal more complicated with Windows than with a Mac, primarily because PCs are much less standardized than Macs. Once you have Windows working, though, the two systems are really pretty similar (Steve says that, cuz he's a PC kinda guy). Purists (that's Robin) passionately dispute the point, but people love a good argument.

Windows NT

The forthcoming **Windows NT** (NT for **n**ew **t**echnology) is a full-fledged operating system in its own right, not just an extension of DOS like the "regular" version of Microsoft Windows. Because NT doesn't depend on DOS, it will be free from DOS's numerous limitations. For instance, NT is free of the infamous 640K limit on *memory* (see *RAM*), and lets you assign long names to your files (you aren't stuck with just 11 characters as in DOS). NT is a 32-bit operating system (meaning it moves data around 32 bits at a time, as opposed to 8 bits in DOS and 16 bits in Windows 3.1). All this should make Windows NT faster and more reliable than ordinary Windows, and make it easier for programmers to develop sophisticated applications. To the user, however, NT will look and work very much like regular Windows. The initial version of NT will run only PCs, but versions for other types of computers are planned as well.

WIN.INI

WIN.INI (short for **Win**dows **ini**tialization) is a file that contains all kinds of settings that determine how *Windows* works. Windows "reads" the WIN.INI file every time it starts up to find out which programs to run automatically, how fast the cursor should blink, what pattern or graphic to display in the background, which printer is the default, which fonts are installed, and the settings for many other options. Some applications use the WIN.INI file to store custom settings for those programs.

WIN.INI is created for you when you first install Windows, and it's automatically modified every time you change Windows' settings. Since WIN.INI is an ordinary text file, you can edit it yourself, if you dare, with just about any text editor or word processor. Just be sure to make a *backup* copy of the file before you change anything, so if Windows doesn't work properly after you make your changes, you can replace your edited version with the backup of the original and all should be as it was before you started monkeying around. See also *SYSTEM.INI*.

wireframe modeling

Wireframe modeling is a way of representing three-dimensional shapes—objects are displayed as if built out of wires. There is no surface, and you can see through to all the wires that create the forms. See *modeling* for a comparison between wireframe modeling, surface modeling, and solids modeling.

wizard

A **wizard** is a computer magician, going beyond the normal run of *power users* and *hackers* and *programmers*. Wizards usually apply their magic to create outstanding and creative programming.

Wizard

The **Wizard,** made by Sharp Instruments, is a small, hand-held *(palmtop)* computer. A Wizard is capable of borrowing information from your personal computer and storing it so you can use the data on the road. You can also put information into your Wizard while on the road, then put that information back into your computer when you get home.

wizziwig

See *WYSIWYG*.

.WKS, .WK1, .WK3 files

These **WK** *extensions* that are tagged on the ends of file names identify worksheets from *Lotus 1-2-3,* a *spreadsheet program*. A WKS file is from Lotus 1-2-3 *version* 1A. A WK1 file is from version 2.0, while a WK3 file comes from version 3 for DOS or version 1 for Windows.

word

In computer jargon, a **word** is the natural data size of a computer, which varies from machine to machine. For instance, in a computer with a *16-bit CPU (central processing unit),* a word is 16 bits; in a computer with an 8-bit CPU, a word is 8 bits. With all else being equal, a computer that can process a 16-bit word is twice as fast as a computer that can process only an 8-bit word.

In *word processing,* a **word** is usually a set of characters between two blank spaces. (In typing classes most of us were taught to add up the total number of characters and divide by five [sometimes six, depending on your teacher] to get the total number of words in a document.) See *word count,* below.

Word

Word, with a capital W, refers to Microsoft Word, a powerful *word processing program*. In fact, it has more power than most people who use a word processor will ever need, comparable to using a helicopter to go across the street. Many have a love/hate relationship with Word. Some people hate it more than they love it.

word count

A **word count** is simply a count of how many words are in a document. Fortunately, any good *word processor* will count the words for you with the click of a button. It can usually count the number of lines, also, or the number of paragraphs or even characters.

On a typewriter, we were taught to count the number of characters and then divide by 5 (or perhaps 6) to find the number of words, assuming there is an average of 5 characters per word. Most word processors, though, do an actual, literal word count. For instance, a word count on the previous paragraph, using the typewriter convention (number of characters divided by 5) would yield 52 words. But my word processor counted each word and told me there are actually 49 words in that paragraph, which there actually are.

word processor, word processing software

A **word processor** is a very wonderful invention. It is a software application very similar to just typing on your old typewriter, but it goes far beyond, even in simple matters like not having to hit carriage returns or not having to press the Tab key for a first-line indent, because those things are automatic. You can remove a paragraph from the first page and stick it in the fourth page and all the rest of the text adjusts. You can copy a section from one piece of writing and paste it into another with the click of a button. You can change the margins, the size of the type, the typeface itself. You can automatically number the pages. You can make columns, tables of contents, outlines, tables, all with the greatest of ease. You can edit cleanly and instantly. To people who are dedicated to the yellow tablet, a No. 2 pencil, and a Pink Pearl, it is definitely a *paradigm switch* to move to the computer. I know, I went through it myself. But the word processor actually inspires better writing because it is so much faster to get the thoughts on paper and it is so easy to edit the writing into a better piece of work. And it's fun.

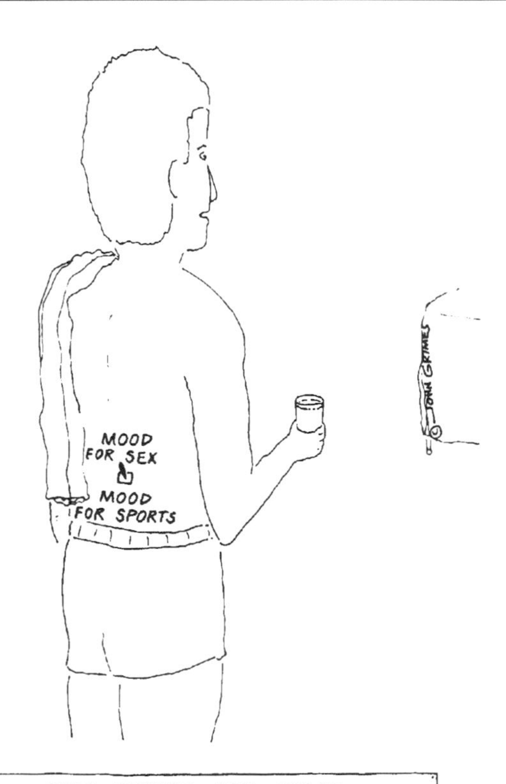

word wrap

On a typewriter (if you are old enough to remember those), when you typed to the end of the line you hit the carriage return and threw the carriage back to the beginning of the next line. Any good word processor on a computer, though, has **word wrap,** which means when the words get to the right margin they just bump themselves down to the next line. You never have to lift a finger from your typing, and in fact you should ***not*** hit the Return or Enter key to force a line ending unless you really always want the line to end at that point, no matter where it appears in a paragraph.

If you let the text word wrap, then change the right margin, the text will automatically flow into the wider or narrower space. If, however, you couldn't resist hitting the Return or Enter key to force a line break, when you change the right margin the lines of text will not adjust to the new dimensions, but will always break where you entered the *hard Return* (where you hit the Return or Enter key).

"Hoe-cake murder," resplendent Ladle Rat Rotten Hut, an tickle ladle basking an stuttered oft. Honor wrote tutor cordage offer groin-murder, Ladle Rat Rotten Hut mitten anomalous woof.

"Wail, wail, wail!" set disk wicket woof. "Evanescent Ladle Rat Rotten Hut! Wares are putty ladle gull goring wizard ladle basking?"

Original text.

"Hoe-cake murder,"
resplendent
Ladle Rat Rotten Hut, an
tickle
ladle basking an stuttered
oft.
Honor wrote tutor cordage
offer
groin-murder, Ladle Rat
Rotten
Hut mitten anomalous
woof.

If I hit Returns at the end of each line, when I change the margins those hard Returns stay there, thinking I really wanted to end the line at that place.

"Hoe-cake murder," resplendent Ladle Rat Rotten Hut, an tickle ladle basking an stuttered oft. Honor wrote tutor cordage offer groin-murder, Ladle Rat Rotten Hut mitten anomalous woof.

If I let the lines word wrap naturally, when I change margins the lines will adjust to the new margins.

WordStar cursor diamond

Once upon a time—in the late 'seventies and early 'eighties, when *CP/M* personal computers held sway—WordStar was by far the most popular word processing program around. One little feature that helped WordStar win so many fans still persists in many DOS programs: the **WordStar cursor diamond.** This refers to the use of certain alphabet keys, in combination with the Control key, to move the cursor around on the screen.

WordStar came out prior to the IBM PC, when many small computers still lacked separate *arrow keys* for cursor control. To overcome this limitation, WordStar let you move the cursor by pressing the Control key along with one of several alphabet keys. These keys, S, D, E, and X, are situated in the shape of a diamond, whose points correspond to the direction of cursor movement (right, left, up, down). Control S moved the cursor to the left, Control D moved it to the right, Control X moved it down a line, and Control E moved it up.

Even if your keyboard has arrow keys, controlling the cursor with the WordStar diamond has an advantage over using arrows: it's faster. Your fingers are already right on top of the alphabet keys, so all you have to do is stick out your pinky to hit the Control key. This is such a neat idea that many programs have adopted the WordStar diamond, at least as an option.

The WordStar diamond works best if the Control key is right beside the A key. These days, unfortunately, most PCs have the Caps Lock key in that position, and put the Control key down by the Spacebar where it's awkward to reach. The solution is to get a utility program that reverses the functions of the Ctrl and Caps Lock keys (see *Control key*).

work disk

A **work disk** is a disk you work on. This disk stores your data files (documents), and is also known as a "data disk." This is opposed to a *master disk* or *original disk*. The term "work disk" also separates that particular sort of disk from a *system disk,* which has a *system* on it to run the computer.

Workplace Shell

In version 2.0 of IBM's OS/2 operating system for PCs, the **Workplace Shell** is the name for the "master" program from which you run your applications and keep track of your files. The Workplace Shell is substantially more sophisticated and more easily customizable than its equivalent in Windows, *Program Manager.*

workstation

A **workstation** is a computer that is bigger and more powerful than the *personal computers (microcomputers)* we have at home, but one that's still used by one person. Actually, the line between personal computers and workstations is fuzzy and getting fuzzier all the time. But the term workstation implies a machine that's used for really intensive jobs like engineering, animation, software development, and professional desktop publishing. A workstation typically has high-quality graphics capabilities and high-capacity *mass storage devices,* and is often *networked.*

The Apple *Quadra,* the top-of-the-line Macintosh, is a personal computer powerful enough to be considered a low-end workstation. A Sun Computer is considered a high-end workstation. Workstations are as powerful as many *minicomputers* combined, but minicomputers are used by more than one person at the same time and don't have sophisticated graphics capabilities.

A workstation can also refer to any computer that is one in a group of *networked* computers, whether it is a personal computer or technically a workstation.

WORM

WORM stands for **w**rite-**o**nce, **r**ead-**m**any. It refers to *optical disk* technology that lets you write (record your own information) onto a *CD*-like disk, and then you can use it (read it) as many times as you like. (A CD, in comparison, cannot be written to at all.)

WORM disks and drives are becoming popular, especially for *archival* (long-term storage) purposes, because they can hold so much data—from about 600MB *(megabytes)* to over 3,000MB—and they store it permanently (write once, remember?) But there is no standard format for WORM disks yet, so if you need to share data with someone else you must both make sure to use the same model. Also see *optical disk* for a comparison of *CD, WORM,* and "erasable optical disk" (floptical disk).

worm

A **worm** is a type of computer *virus,* which is a piece of programming code written with the purpose and intent of causing damage and destruction. A worms usually causes its trouble by sneaking through networks and propagating itself, spawning so many copies in the memory of another computer that the computer crashes. The most infamous worm was the one that invaded the *Internet network,* which links computers throughout the government and in universities all over the country. Unlike most viruses, the Internet worm wasn't developed with evil intent—it was part of a piece of software that was supposed to go into the Internet network and find out exactly how enormous the network was, because nobody knew. But the software had a flaw (a *bug*) in the code that caused it to reproduce itself constantly until it filled up the whole network, and the entire Internet superstructure shuddered to a halt. Ooooh.

wrap

Word wrap refers to text bumping into the right *margin* as you type and wrapping itself back around to the left margin without you having to hit the carriage return. Please see *word wrap.*

Text wrap refers to text wrapping itself around a graphic image. Please see *text wrap.*

wrist rest

A **wrist rest** is a little pad that is placed just below your keyboard, designed for you to rest your wrists upon as you type. You can also get wrist rests that attach to the mouse. This is to help prevent *repetitive strain injuries* such as carpal tunnel syndrome (CTS). Studies seem to show that if you can keep the carpal tunnels open as wide and smooth as possible by not bending the wrists, you will have less risk of getting CTS.

write-protect

If a **disk** is **write protected,** it means it is *locked* so you can't put (write) new information on the disk. You can still retrieve (read) information from the disk—you can open the disk and read any files on it—but you can't change anything. You cannot remove or add files to or from it, you cannot change any of the files, you cannot even change the name of any item, including the disk.

On a 3.5-inch floppy disk there is a little square hole in one corner with a little plastic tab that snaps over to cover the hole. When the hole is open, the disk is write-protected, or *locked.* Just snap the tab over to close the hole to unlock it. (On *high-density* disks you will find two holes, one with a tab and one without a tab. The hole without the tab tells the computer the disk is high-density.)

PC On a 5.25-inch floppy disk, there is a little small, square notch on the side of the disk, the write-protect notch. If you cover that little notch,

on the grid (on the monitor, on the tablet) by the x-y coordinates. The **x-coordinate,** then, is the exact horizontal position of the cursor. Also see *y-coordinate.*

XFCN

XFCN (pronounced "ex function") stands for external function. It's similar to an XCMD as described above, but it is a "function" instead of a "command." A command is like "go to," which makes something happen. A function is like "the time," which gives you back a result, usually a result that the application had to go into the depths of the computer to find.

X (flashing)

When you turn on the Macintosh and you see a flashing **X,** it's a clue to you that the computer cannot find the *System* to start itself up with. This will happen if your kids took the System or Finder icons out of the System Folder before they turned off the machine, or if your hard disk is stuck and can't get spinning (which might even happen if it's too cold). It will happen if your System Folder is on an *external hard disk* that you forgot to turn on. The flashing X is not as serious as the *Sad Mac.* Try inserting a floppy disk that has a System Folder on it to get it up and running. Then call a *power user.*

Actually, you usually see a flashing *question mark* first, which means the Mac is looking for a System Folder. If you try to give it something or if the Mac finds a disk that doesn't have a System Folder, then you get the X.

XGA

XGA is a type of PC *video adapter* (circuitry that controls the image you see on your screen) developed by IBM for its own line of *PS/2* computers. XGA puts out a *resolution* of 1024 x 768 pixels with *16-bit color* (meaning up to 65,000 different colors at a time). Unlike *VGA,* an earlier IBM video adapter, XGA has not become a standard in the PC universe. Many other companies make video *boards* (adapters) with the same 1024 x 768 resolution, but most of them aren't patterned after the XGA. See *VGA* and *SuperVGA.*

x-height

The **x-height** of a typeface is the height of the lowercase letter x. The letter x has been chosen as a standard for judging the visual size of letters because it's the only character that has all four corners touching the outer boundaries—there's no question as to the top or bottom or sides of an **x** as there might be for the letter **c** or **p**.

The relative proportions of x-height to cap height (height of the capital letters) is a determining factor in the *legibility* and *readability* of a typeface. The type designer has complete choice over the size of the x-height in relation to the full body of the characters. And it is the x-height of a typeface that determines how large the typeface appears to be. The following examples are all 24-point type, but they seem to be different sizes because their x-heights are different.

Ryan x   Jimmy x   *Scarlett x*

XMODEM

Xmodem is a *protocol* (set of rules) for transferring information over the phone lines through your *modem*. It's used mostly for transferring information from personal computer *(microcomputer)* to personal computer, as opposed to a protocol like *Kermit* which is used to transfer between two *minicomputers* or *mainframes*. Most *bulletin board services* and *online services* support XMODEM. There are several varieties.

XT

XT is short for e**x**tended **t**echnology. It's the short name people use to refer to the IBM PC/XT and similar computers. IBM'S XT was introduced in 1983 and was the first IBM personal computer to have a hard disk. Like the original IBM PC, it was a predecessor to the AT. It used an *8088 processor*. If someone says "I have an XT," they may well mean they have a non-IBM computer with a hard disk and an 8088 processor.

X-Windows

X-Windows is a windowing system that runs on networks using the Unix operating system. There are also PC versions. Although it looks somewhat similar on the screen, X-Windows is completely different than Microsoft *Windows* for PCs and will not run Microsoft Windows programs. Two competing, much enhanced versions of X-Windows, Motif and Open Look, are being promoted for Unix computers these days.

HONEY, IF I TAKE ONE CLASS THIS WINTER I CAN LIGHTEN MY LOAD THIS SPRING AND SPEND MORE TIME HELPING OUT WITH THE KIDS.
JUST IN TIME, MARGE — THEY'VE BEEN ASKING FOR VEGETABLES AGAIN.
© JOHN GRIMES

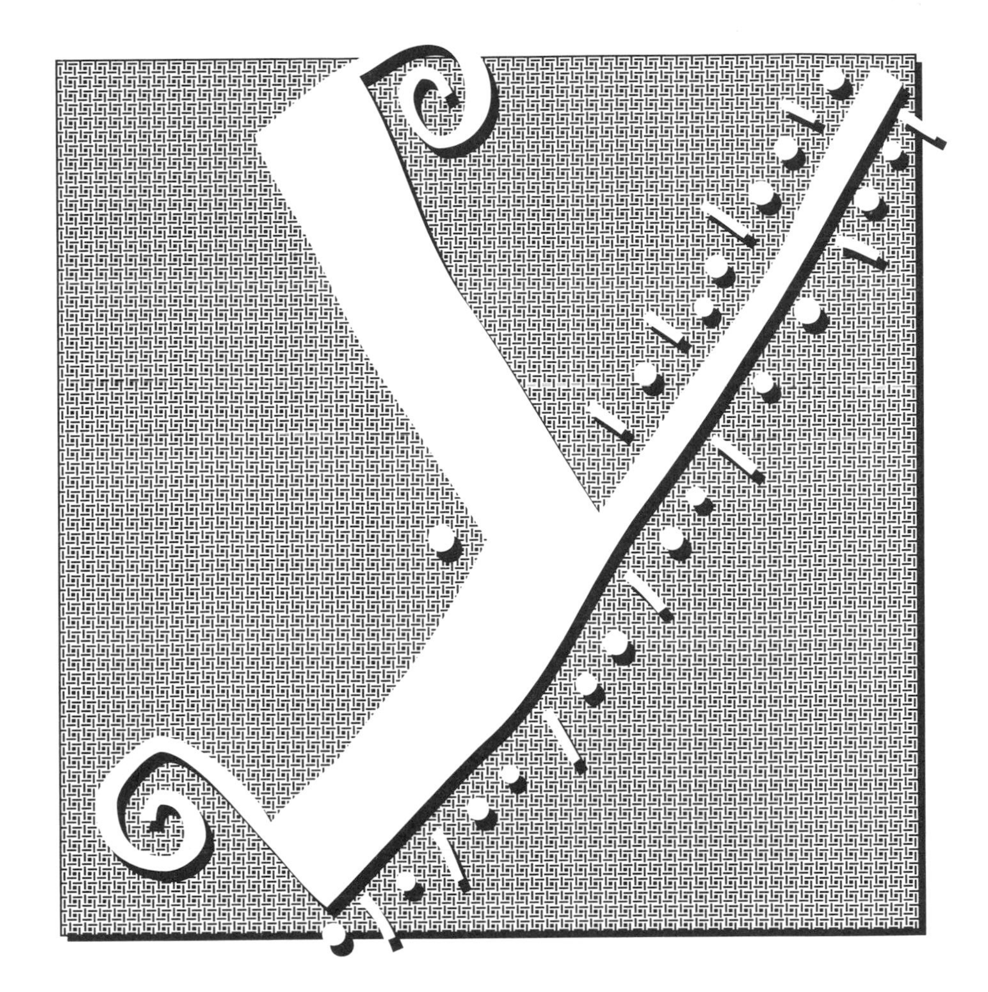

YABA

The acronym **YABA** (pronounced "yah buh") stands for **y**et **a**nother **b**loody **a**cronym. YABAs are often created by *FLAMs*. You can be really hip and use this term to respond disdainfully to questions like, "What is EBCDIC?"

y-coordinate

In a grid, there are rows and columns. The x-axis runs along horizontal rows; the y-axis runs down vertical columns. On display screens and on digitizer tablets, this x-y matrix, or grid structure, is the reference framework. At any moment, the mouse or cursor position can be pinpointed on the grid (on the monitor, on the tablet) by the x-y coordinates. The **y-coordinate,** then, is the exact vertical position of the mouse or stylus. Also see *x-coordinate.*

YMODEM

YMODEM is a *protocol* (set of rules) for transferring information over the phone lines through your *modem*. It's a variation of *XMODEM*. YMODEM's difference is that it can send a batch of files at once. YMODEM-G is a faster variation, but it's fast because it doesn't check for the errors that occur easily on any phone lines that may have a bit of static.

Zapf Dingbats

Zapf Dingbats (pronounce the "p" in Zapf) is the name of a typeface designed by Hermann Zapf (who also designed Zapf Chancery, Palatino, and Optima, among other beautiful typefaces). Zapf Dingbats is a collection of printers' *dingbats* (little decorative characters). This *font* is built into the *ROMs* (read-only memory) of most *PostScript* laser printers, but you can buy it on disk in *Type 1* or *TrueType* format to use with other kinds of printers.

✹④✷ ❁➤♠ ❍➧ ▲⑩◆●➥❁▼♠✌ ✩✣▼❄❒ ✣✼❄✲❍❁■✭✎

ZiffNet

ZiffNet is an *electronic information service* set up by Ziff-Davis, the publishers of many of the best-known computer magazines (among them, PC Magazine, PC/Computing, PC Week, MacUser, and MacWEEK). Ziffnet is a great resource, and it's cheap to use. Instructions for getting started appear in the magazines.

Once you hook up to ZiffNet via a *modem,* you have free run of a large library of *shareware* and *freeware* software—including all the free *utilities* created for PC Magazine and MacUser—that you can *download,* or transfer, to your own computer. You can tap into all kinds of computer-related information, such as product reviews, how-to advice on setting up your system, and tips on using software. And you can communicate directly *(online)* or via *e-mail* messages with other ZiffNet users and with columnists who write in the magazines.

Originally, some of the individual magazines had their own services, such as PC's PC Magnet and ZMac (entry below); people may still refer to PC Magnet and ZMac, but officially, they've been superseded by ZiffNet.

.zip

The *extension* on a file name, **.zip** (pronounced "dot zip") indicates that the file has been *compressed* by a PC shareware compression utility called PKZIP, available on most *bulletin boards* and *onlines services.* Compression makes a file smaller (it takes up less space on a disk). When you use a *bulletin board* or other *online service* to *download* (copy files from the service onto your disk) various programs, fonts, or utilities, they are generally "zipped up" so they take less space on your disk and so they transfer over the modem faster. When you "zip" a file you are making it smaller; when you "unzip" a file, you are returning it to its original size.

ZMac

ZMac (now officially called ZiffNet/Mac) is an *online forum* where you can, among other things, interact with the editors, writers, and columnists for MacUser and MacWEEK magazines. You can read the top stories here in ZMac the week before they appear in MacWEEK. You can *download* all the free stuff like scripts and programming codes and utilities mentioned by the MacUser editors. There are reference databases to help you locate, buy, and sell products.

If you are a *CompuServe* member, to get to the ZMac forum type GO ZMAC at any ! prompt. If you are not a CompuServe member, these are the directions:

- Call (800) 635-6225 (voice) to find your local access number.
- Set up your telecom software with these settings:
  8 bits, **1** stop, and **no** parity.
- Dial the local access number. When connected, press Return.
- At the following prompts, type in the responses:

Host Name:	**CIS**
User ID:	**177000,5200**
Password:	**Z*MAC**
Agreement Number:	**Z12D9014**

ZMODEM

ZMODEM is a *protocol* (set of rules) for transferring information over the phone lines through your *modem*. It's a variation of the *X-* and *YMODEM* family, but if you have a choice, use ZMODEM. ZMODEM's difference is that if your file transfer gets interruped—like maybe you got disconnected for some reason—the transfer will continue the next time you connect. ZMODEM can also send batches of files, like YMODEM, but it does correct for errors, unlike YMODEM.

zone

A **zone** is a subnetwork within a larger network. It's sort of like having zip codes within a city. When you try to send electronic mail, for instance, it's much easier to find a certain address within a particular zone (or zip code) than to find that one spot within the entire city. Different electronic zones on the network can communicate with each other, also.

Each single stop on any network, like a printer or a computer or a file server, is called a *node*.

zoom

Generally, to **zoom** means to make something look larger or smaller on the screen. Many programs have a Zoom command or tool (often a magnifying glass) that lets you display your document or image at differing levels of magnification.

zoom box

On the Macintosh, the **zoom box** is the little box in the upper right corner of a *window.* If you click once in it, the window expands. If you click on it again, the window will reduce, or restore, to the size it was before you enlarged it. This is different from resizing, where you can drag the window to any size by pressing-and-dragging in the *resize box* (in the lower right corner of the window).

In System 7, windows at the *Desktop level* will only zoom open as large as necessary to display all the items. This is a neat little feature. You can hold down the Option key and click the zoom box to enlarge the window to fill the screen, while still allowing room to see the icons on the right side of the screen. Some *applications* also have this feature of only enlarging the document window as large as necessary.

In Windows, the equivalent to the zoom box is the *Maximize* and Restore buttons.

APPENDIX A AND B

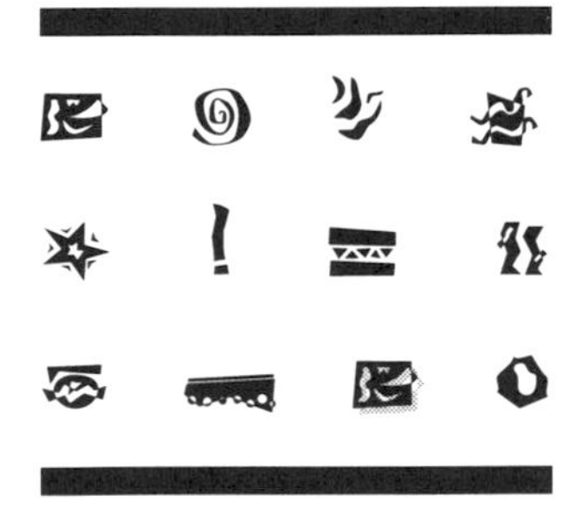

Appendix A

Appendix B

How to Read a Computer Ad

by Pamela Mason and Robin Williams

Computer ads are nothing more than variations on a theme.

Computer ads really *are* written in English—an alien, very technical form of English, but English nonetheless. If you don't know the difference between a *megahertz* and a *megabyte,* don't despair. After you learn a few concepts and become familiar with some common terms, you will be able to break the code and sift through even the most cryptic of ads.

When buying a car, you base your choice on how economical it is, or how many kids and how much stuff it can carry, right? All cars have engines and doors, but all cars certainly aren't the same. When buying a computer, you must make similar types of decisions.

The basic building blocks of any computer are repeated in every ad. Much of what you read is just different names for the same things. The wording in a computer ad, just as with any other form of advertising, will be different from ad to ad. Some ads are easy to decipher, while others take a little more perseverance. This guide will help you become familiar with the basic language and will walk you through the steps to reading an ad. It will help you understand the relationships of the parts to one another. Because there are so many choices in ads, this is not meant to be comprehensive—but it will give you a starting point.

Where to begin?

Are the programs you'll be running text or graphics? Basic word processing, spreadsheet, and database applications usually only need a *monochrome monitor,* while many graphics programs require a *grayscale* or *color monitor.* Ask your local dealer what the minimum requirements are to run the programs you need, or read the software box. This is important because programs *must* have enough hard disk space and memory to run properly. You can then start to look for those specific requirements as you browse the ads.

The basic computer parts

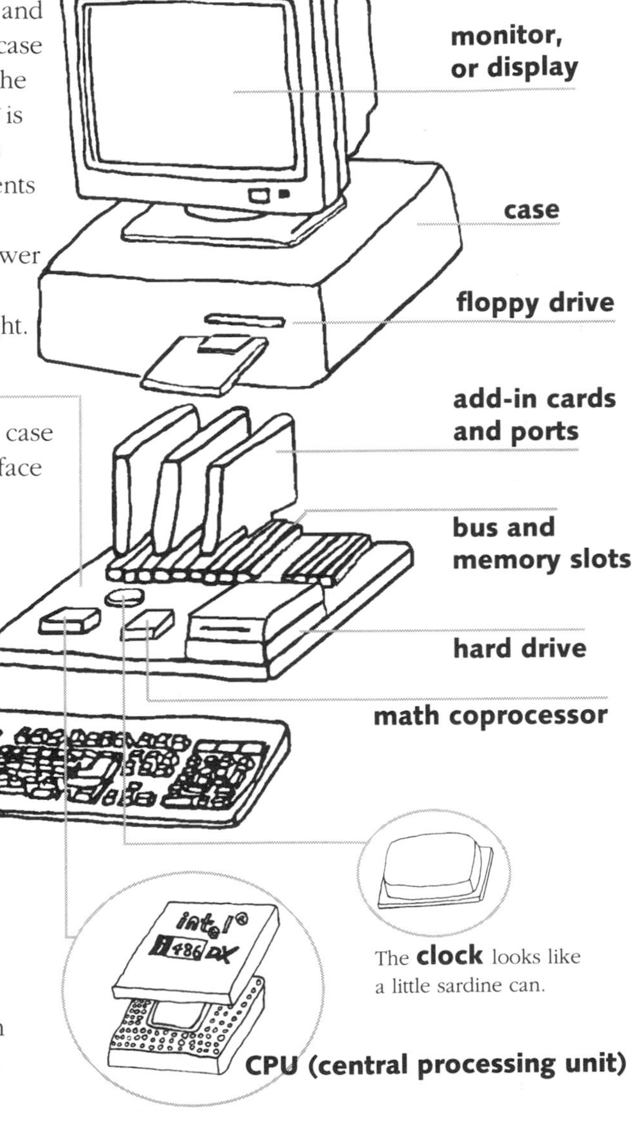

The case

There are desktop cases, mini-, mid- and full-size *tower* cases. The size of the case determines how many *add-in cards* the computer will hold. Because the *CPU* is contained inside the case, sometimes people refer to the case and its contents as the CPU. The one shown here is a desktop model. To visualize what a tower looks like, imagine this desktop case turned on its side and standing upright.

The motherboard

The *motherboard,* if you took off the case and looked inside, is the big, flat surface that everything mounts onto.

The keyboard

Keyboards vary mainly in the amount of total keys, and some of the key positions may vary.

The math coprocessor (FPU, math chip)

The *math coprocessor* looks like the *CPU chip*. It speeds up certain types of programs, CAD programs for example. The computer will either come with the math chip, or have an empty slot where one can be added. Late-model processor chips, like the 80486DX, have a math coprocessor right on the chip.

The hard drive

Hard drives sit in an area called the "bay." You can also store data on optical disks and magnetic tape.

The CPU and the clock

Always consider these two parts together:

(CPU) **(clock)**
BRAIN + SPEED = POWER *(overall performance)*

The ***bigger*** the numbers, the better.

"Also known as . . ."

One thing that makes understanding a computer ad difficult is that there are several terms for the same thing. Actually, **there are technical differences between some of the terms,** but many words are used interchangeably, **as if** they were the same. Below is a chart of "AKAs" ("also known as") to help in your deciphering.

Also Known As . . .			
CPU*	processor	microprocessor	central processing unit
display	monitor	screen	
math coprocessor	coprocessor	math chip	FPU, floating point unit
RAM	memory	SIMMs	random access memory
RAM cache	cache	memory cache	cache memory
drive	disk	disk drive	FD = floppy drive HD = hard drive
high density drive	1.4MB drive	SuperFloppy drive	FDHD (floppy drive, high density)
24-bit color	32-bit color	16.7 million colors	
multisync	multiscan	multimode	multifrequency
card	board	adapter	add-in
CD	CD ROM	CD player	compact disc
MB	megabyte	meg	
motherboard	logic board		
bus	slots	NuBus (Mac only)	
non-interlaced	flicker-free		
Fkeys	function keys		
extended keyboard	expanded keyboard		
80486	486	'486	
80386	386	'386	
80286	286	'286	
68020	020	oh twenty	
68030	030	oh thirty	
68040	040	oh forty	

*sometimes the entire case that contains the actual CPU chip is called the CPU.

Units of measure

When you read a cookbook, you measure with cups and tablespoons; when you follow a blueprint, you measure in feet and inches. You can substitute or vary the quantities as you see fit. Computer parts have units of measure with names like *megahertz* and *kilobytes,* and computer ads offer optional amounts and parts. Some of those strange acronyms and numbers are the units of measure that describe the computer's parts. The following table shows some of the most commonly used acronymns and units of measure. (Some numbers are part and model numbers. For example, the 80486, or just "486" for short, is often referred to as both the model and the type of *CPU* on a PC.)

The acronym:	is short for:	and is a:	which refers to the:
MHz	megahertz	unit of speed	clock or CPU
MB	megabyte	unit of storage size	hard disk, floppy disk, or RAM
K or **KB**	kilobyte	unit of storage size	hard disk, floppy disk, or RAM
ppi	pixels per inch	screen dimension unit	monitor resolution
ns	nanosecond	unit of speed	access time for memory speed
ms	millisecond	unit of speed	access time for hard drives
mm	millimeter	unit of length	dot pitch on a monitor
in	inches	unit of size	floppy disk, floppy drive, or monitor
dpi	dots per inch	size of printed dots	printer resolution

Numbers

Numbers in ads usually represent size or quantity. They can also represent a part or model number. Certain numbers are repeated over and over and are usually associated with a certain part.

Look at the box below labeled PC clock speeds. Notice the numbers are 8, 12, 25, 33, 50, and 66 MHz. Whenever you see one of these, or this particular set of numbers, you will probably be looking at *PC clock speeds.* Start noticing which numbers are commonly found together and with which part they are associated.

Learn to recognize orders of magnitude

Most computers and their parts are bought based on two things: **speed** and **size. Bigger is better** and **faster is better.** Usually, the larger the number, the bigger and/or faster, but there are exceptions to this: you want smaller numbers for access time (ms, milliseconds), for SIMM speeds (ns, nanosecond), and for dot pitch (mm, millimeters).

bigger number means more space to work with		*higher number means more powerful*		*higher number means faster*	
Hard drive storage	**Amt. of RAM**	**(PC) CPU**	**(Mac) CPU**	**(PC) CPU clock speed**	**(Mac) CPU clock speed**
20 MB	256 K	8088	68000	8 MHz	8 MHz
40 MB	1 MB	8086	68020	12 MHz	16 MHz
80 MB	4 MB	80286	68030	25 MHz	20 MHz
120 MB	16 MB	80386	68040	33 MHz	25 MHz
200 MB		80486 *	68050	50 MHz	33 MHz

smaller number means faster			*smaller means better resolution*
CD ROM access time	**Hard disk speed**	**SIMM speed (RAM; memory)**	**dot pitch**
500 ms	65 ms	150 ns	.41 mm
450 ms	42 ms	120 ns	.39 mm
300 ms	28 ms	100 ns	.31 mm
280 ms	17 ms	80 ns	.28 mm
	9 ms	70 ns	.25 mm

*A CPU called the **Pentium** is now shipping. They're not using numbers anymore so the names will get easier to pronounce, but harder to understand.

Model names

Learn to recognize the chronological order of computers and their model names. Below is a list of the Mac and PC model names in the general order in which they were put on the market. Why are there so many? Think of how many makes and models of cars are on the road. For example, there was originally the VW bug, then the VW bus, and now there's Fahrvegnugen.

A computer model may have started out as one name or number when it was first manufactured. Over time, as it improved and evolved technologically, a new name or number was adopted for the next model or "generation." Computer parts evolve in this same way.

Eventually, like a VW bug, the older computer models become obsolete, but that can take many years. Even after a car is discontinued, you can still find it in the want ads and it still works; computers are the same way.

Some PC model names
8088 (IBM XT)
80286 (IBM AT)
DECpc 333sx LP
Gateway 2000 4DX-33
Hundai 466D2
Compudyne 386 SX/20
Compaq ProLinea
80386 SX *
80386 DX *
80486 SX *
80486 DX *
80486 DX/2 *

* You'll often hear PCs referred to by their chip number, rather than their model name.

Some Macintosh model names
Plus, SE, SE/30
Classic, Classic II
LC, LC II, LC III
Mac II, IIx
Mac IIci, IIcx, IIsi, IIfx, IIvx
Performa 200, 400, 600, 600CD
Quadra 700, 900, 950, 800
Centris 610, 650
PowerBooks 100, 140, 145, 160, 170

PCs don't have model names that are worth remembering—that is, there are so many jillions of models (because there are so many different manufacturers, each with many models) you can't keep 'em straight. The closest thing to a model name is the microprocessor type, but even given the same microprocessor, there will be ***major*** *differences in performance, slots, ports, storage, size, everything. So the poor PC buyer gets only general help from knowing what the microprocessor is.*

How to read the computer ad

Computer ads are nothing more than cryptic sentences. Refer to the tables on the previous pages to help decipher numbers and acronyms, to help you see the group relationships, and to identify and match numbers as you go along. Remember, this book is a dictionary, so any term you find unfamiliar is also explained elsewhere in more detail.

First, read an ad one line at a time by following these steps:

1: Find the unit of measure.

Look for the ***unit of measure*** as shown on the ad, such as **MHz** or **Mb** or **inches.** A number (the quantity) will usually be next to the unit of measure. Refer to both the Numbers chart and the Units of Measure chart; sometimes the unit of measure is all the ad gives, and they expect you to know what they are referring to. For instance, MHz is just a speed, and they expect you to know that it is the *clock* speed.

2: Find the feature that the unit of measure describes.

Once you have found both the unit of measure and the **feature** (if there is one), the rest is easy. Often the other words on the line just serve as descriptions. Think of the feature like the subject in a sentence and everything else as the adjectives that describe it.

3: Put these into regular sentence form (English).

This really is the easiest way to understand and de-mystify any ad. Find this line on the PC ad on the opposite page and apply the three rules:

200MB IDE hard disk

MB, or megabytes, is the *unit of measure.* The *feature* is the **hard disk.** Everything else describes the hard disk: it has 200 megabytes of storage space, and it is an *IDE* hard disk, which is a type of standard (see the definition for IDE).

So, translating from cryptic computer tech talk to English, the line now reads:

The PC in this ad has a 200 megabyte hard disk that uses the IDE standard.

Read the ad, line by line, and find these main parts:

Model name or number (it might be just a number, like 386 or 486)

System (CPU) clock speed (it might be just a reference to speed, such as 33 MHz, without telling you it's clock speed)

Hard disk (listed by size in megabytes, MB, and perhaps by speed in milliseconds, ms; it may be abreviated HD, or it may be listed as just a number, such as 4/40, where 4 is the amount of RAM and 40 is the size of the hard disk)

Memory (or RAM, sized in megabytes, MB, see the note in hard disk, above; some computers also have cache memory, sized in kilobytes, K or Kb)

Monitor (see monitor ad)

After finding the main parts, find the type of keyboard and mouse (if included). Then find the number of *ports* and *slots,* and if any software will be included with the system. In this way you can read various ads and compare the features.

The computer ad

1. This ad is for the Macintosh model called the Quadra 800.
2. It has a 68040 *CPU,* with a *math chip.* The CPU is the chip that runs the computer, and the math chip speeds up certain applications, like graphics and engineering.
3. It has a *clock speed* of 33 *megahertz,* which is one of the indicators of speed.
4. It comes with 8 megabytes of RAM, and you can add up to a total of 64 megabytes.
5. Sometimes you will see the memory and hard disk *configuration* stated as in #5: the first number is the amount of RAM, and the second number is the size of the hard disk, both in megabytes. The *CD* means this model also has a CD ROM player.
6. The computer has 5 *NuBus* slots, which are connections inside the computer where you can add *boards* to increase the performance or to connect to other devices.

Apple Macintosh
1 — **Quadra 800**

2 — 68040 with built-in math coprocessor
3 — 33 MHz
4 — 8 MB of RAM, expandable to 64 MB
5 — *(or)* 8/500 CD
6 — 5 NuBus slots
Monitor not included

$5429

1. This ad is for a PC, a '486, which means it has an 80486 *CPU.* The *clock speed* is 50 megahertz, which is the same thing as saying that the CPU is running at 50 MHz, an indication of processing speed.
2. "Display" means "monitor," and the other terms describe what kind of monitor it is.
3. It has 256 kilobytes of *cache memory.*
4. It has 4 megabytes of *RAM (random access memory).*
5. It has an *IDE* (a type of) hard disk that holds 200 megabytes of information.
6. The "1.2MB 5.25" refers to the kind of disks the *floppy drive* will accept; in this case, *high-density* 5.25 floppy disks. High-density drives also accept low-density disks. Some PCs are sold with the newer 3.5-inch floppy drives; some have both types of drives.
7. The kind and number of *ports* (connections on the back of the computer) determine the kinds of devices it can attach to, such as printers, other hard disks, mice, etc.
8. These *operating system* software packages are included with this computer.

1 — **486 50MHz**

2 — 17" non-interlaced mutisync display
3 — 256 K cache
4 — 4 MB RAM
5 — 200 MB IDE hard disk
6 — 1.2MB 5.25 floppy drive
7 — 1 parallel, 2 serial ports
8 — MS DOS 5.0, Windows 3.1

$2950

How to read a monitor ad

Most of this monitor information applies only to PC monitors; Macintosh monitors are much more consistent, and you won't even see or hear of many of these terms when shopping for a Mac monitor.

Although you can buy a monitor separately, understand that the image you see on the screen is produced by a video *card* in the computer. The card is just as important as the monitor in determining the image quality. For example, you may buy a monitor that can display an infinite number of colors, but card may limit you to only 16 or 256 colors. So be sure to choose a matched set.

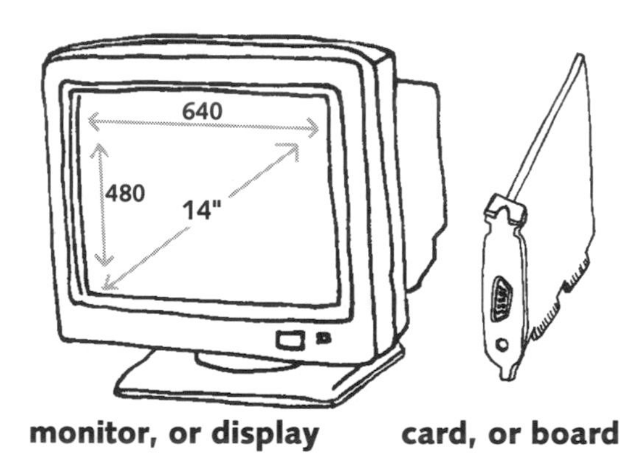

monitor, or display **card, or board**

The main things to look for in a monitor ad:

- Overall **size,** measured in *inches.*
- **Resolution,** measured in *pixels.*
- The **dot-pitch,** measured in *millimeters.*
- **Multisync** or not.
- **Refresh rate** (or rates, if it's multisync).
- **Color** or **monochrome** or **grayscale.** Any new color monitor can display an infinite number of colors—the video card determines how many colors you'll actually be able to see, and at what resolution.
- **Interlaced** or **non-interlaced.**

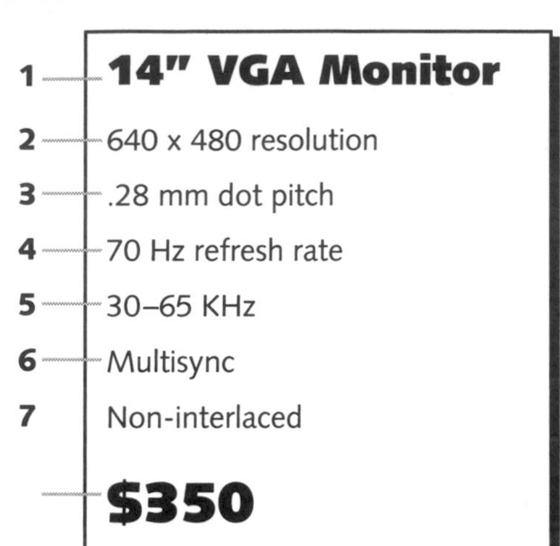

1. This ad is for a 14" Super VGA (PC) monitor.
2. This is the maximum *resolution* (the horizontal-by-vertical pixel count for the whole screen), though the monitor can probably display lower resolutions as well.
3. This is one of those cases where *smaller* is better: a .28 millimeter (mm) dot pitch is better than .31 mm.
4. The ideal *refresh rate* is at least 70 hertz or above. This must match the card's rate.
5. This is the horizontal scan rate. A wide range like this is good.
6. *Multisync* (also known as multifrequency or multiscan) means the monitor can have multiple scan rates; the card you use will give the monitor the ability to toggle between various resolution and color capabilities (the Mac term is "multi-mode").
7. You will often see non-interlaced monitors referred to as "flicker-free." Interlaced monitors are really annoying.

PC monitor options

Monitors are a complicated subject and require much more than an overview to really understand. This is just meant to help guide you through the jungle of terms. Also remember to look words up in the dictionary!

When a monitor is advertised as an individual unit separate from the computer, it usually does not include the *card.* When a monitor is advertised as part of a whole computer system, it usually does include the card.

Resolution

In the PC world, the term *resolution* almost always means the number of *pixels* on the screen, which is one big factor in the clarity of the screen image. (In the Mac world, resolution refers more to the overall image clarity. If you see a reference to the total number of pixels on a Mac screen, that's just another way of stating the monitor *size.*)

320 x 200	**coarse** image on screen
640 x 400	
640 x 480	⇩
800 x 600	
1024 x 768	
1280 x 1024	**crisp** image on screen

When you see an ad that states **1024 x 768 monitor,** the numbers indicate the number of pixels across (1024) by the number of pixels down (768). This is the maximum number of pixels the screen can display, and means the monitor can work with a card that produces an image with a 1024 x 768 pixel resolution. Be sure the monitor you buy can handle all the resolutions the card can produce.

When you see an ad that states: **1024 x 768 1 MB 24-bit card,** it means this *24-bit* card can handle up to this *resolution* (1024 x 768) and the card has 1 megabyte of memory (the amount of memory helps determine the resolution and the number of colors the card can display). Because it is 24-bit color, it can display 16.7 million colors. But you may not be able to get the highest resolution and 16.7 million colors at the same time; often it is a trade-off, one or the other.

Dot pitch

The *dot pitch* refers to the distance between the centers of the pixels. The smaller the space, the clearer the picture.

.41 mm	**lousy**
.39 mm	
.31 mm	⇩
.28 mm	
.25 mm	**excellent**

Color choices

PC **monochrome monitors** may have amber, green, or white dots on a black background (or vice versa). This is the cheapest type of monitor, and it's all right if you will be working primarily with word processing, database, spreadsheet, or simple black-and-white graphics applications (you need a Hercules-compatible card to do graphics on a monochrome monitor).

A **color monitor** is essential if you are doing work with color graphics, but color also really helps for mundane chores like word processing. Get a color screen if you can afford it.

Unless you have an obsolete CGA or EGA screen, the number of colors you see is fixed by the card, not the monitor:

CGA (obsolete) can display **4 colors.**
EGA (obsolete) can display **16 colors.**
VGA can display **256 colors** at low resolution; **16 colors** at high resolution.
Super VGA, XVGA, and other "VGA-and-then-some" adapters vary depending on the manufacturer. Some display only **256 colors,** some are *Hi Color,* with **32,000 colors,** and some are *24-bit color* with **16.7 million colors.**

How to read a memory ad

What memory chips look like

DRAM ("dee ram," dynamic random access memory) and SRAM ("ess ram," static RAM) are little individual chips.

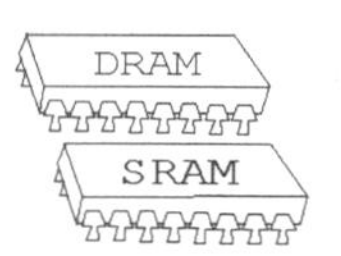

Dynamic RAM is used for main memory.

Static RAM is used for cache memory.

The DRAMs are already put on the SIMMs or SIPs for you when you buy them.

A **SIMM** *has little gold or silver fingers.*

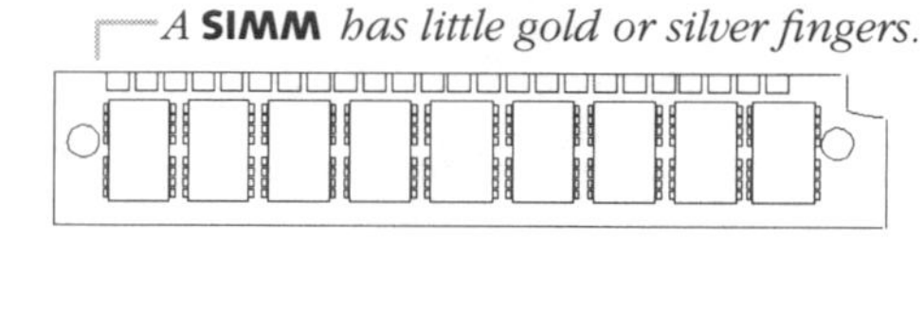

A **SIP** *has little gold or silver legs.*

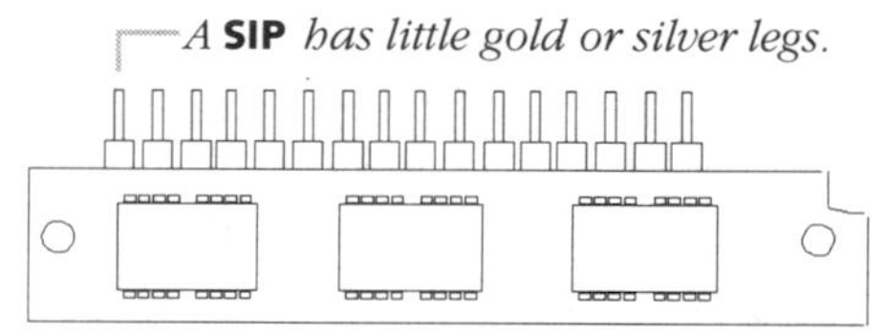

Where memory is added

RAM can be added to:

- the computer
- the video card
- the laser printer

The memory you add to your printer helps your printer handle more fonts and print faster. This can be added via SIMMs or a memory board. You may be able to add memory to the video card to increase the number of colors you can display, or to improve the resolution (some video cards take regular RAM, some take special VRAM). If an ad says that the video card has a *graphics coprocessor* or an *accelerator,* understand that these chips speed up working with graphics, but have nothing to do with memory.

How memory is added

Depending on the type of computer you have, you can add memory in any one or any combination of these three ways. Currently, SIMMs are the most commonly used.

1. The chips can be put on memory boards that plug into the bus slots.
2. The chips can come already mounted on SIMMs or SIPs that plug into the memory slots.
3. The individual chips can be put on the *motherboard.*

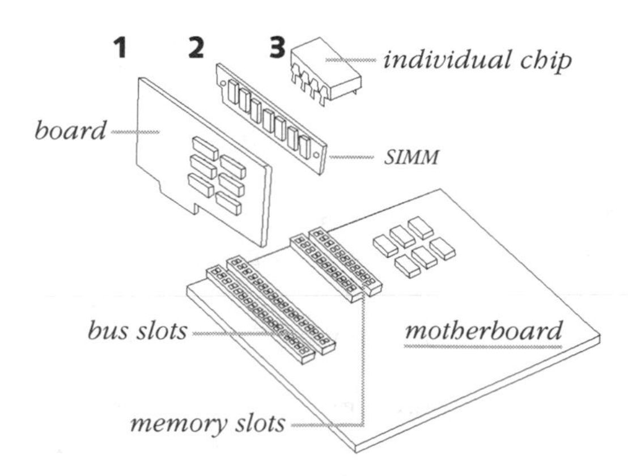

Note: *You can never have too much money, too many fonts, or too much RAM.* Fill up your machine with as much RAM as it will take, especially if you work with graphics. (Then you can just ignore this whole page, and all the other confusing things written about RAM!)

The memory ad

Memory ads, usually selling memory in the form of *SIMMs,* always tell you the two most important things you want to know: **size** (the amount of memory it holds) and **speed** (how fast that memory module can deal with the information that's put into it). The size is measured in *megabytes (MB),* and the speed is measured in *nanoseconds (ns).* The speed of the memory must be able to keep up with the speed of the CPU (the chip that runs the computer).

RAM sizes	RAM speeds
256 K	150 ns
1 MB	120 ns
4 MB	100 ns
16 MB	80 ns
	70 ns

SIMMs are usually added two or four at a time:

If you add four 256 K SIMMs, the total memory you add is 1 MB.

Add four 1 MB SIMMs for a total of 4 megabytes of RAM.

Add four 4 MB SIMMs for a total of 16 megabytes of RAM.

Different models of different computers use different types and amounts of SIMMs. **It is critical that you ask someone knowledgeable to give you specifications on exactly what kind of memory to buy.**

Macintosh SIMM ad

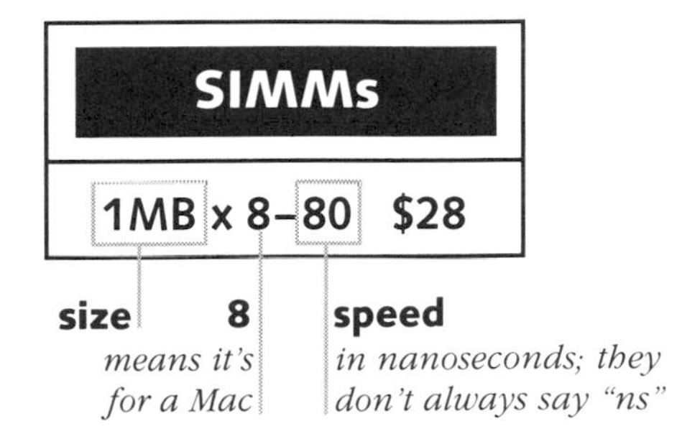

PC SIMM ad

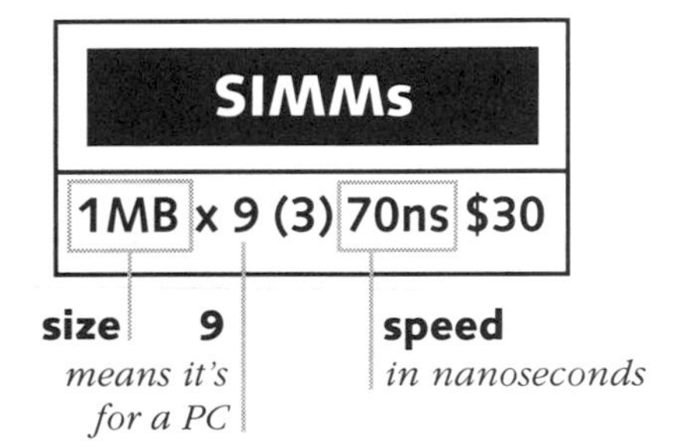

One megabyte "by 8" means there are one million memory locations that can each hold 8 *bits* of information. Think of a million mailboxes, each able to hold eight letters. Mac memory is the "by 8" kind.

One megabyte "by 9" means there is an extra, ninth bit for each of those one million mailboxes. This is the kind of memory in PCs, which use the extra bit for something called "error checking," which Macs don't do.

Bus slots, ports, and add-in cards

PC ads will refer to *8-bit, 16-bit* and 32-bit *cards* and *bus slots*. If you look at the illustrations on the following page, it's easy to see how they are different. The Mac uses *NuBus* cards and bus slots (also illustrated), as well as *processor direct slots (PDS).*

Ports are the types of connectors that you see on the back of the computer. These are attached to the *add-in cards* that snap into the bus slot on the *motherboard.*

When an advertisement says it includes these *ports,* it means the computer includes the card already plugged into the bus (or NuBus) slot. Or it can be built directly into the motherboard in a different way. Either way, it's ready to go and all you have to do is plug the cable from your printer, for instance, into that port. As you can see, the port is built onto the board and when you remove the slot cover and plug it into the bus slot on the motherboard, the connector will poke out of the back of the computer.

Some ports look similar to each other; for example, parallel ports and SCSI ports. *They are not similar in their function*—you must always use the cable that exactly matches the port!

A *local bus* is another type of bus that looks similar to a memory bus. It is a relatively new PC term for a specialized 32-bit bus that allows the computer to process graphics much faster. There is a special video board used with this bus that is different from the ones that go into the ordinary bus slots.

You need to know how many ports (a port includes the board) and slots (slots are empty, without the board) the computer has so you will know how much extra stuff you can add to your computer.

Computer ports

All of the cables plug into matching ports on the back of the computer.

add-in board, or card
to be inserted

port
this will poke through the slot after it is inserted

ADB ports (Mac)

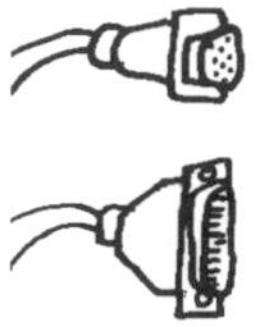

serial and parallel cables

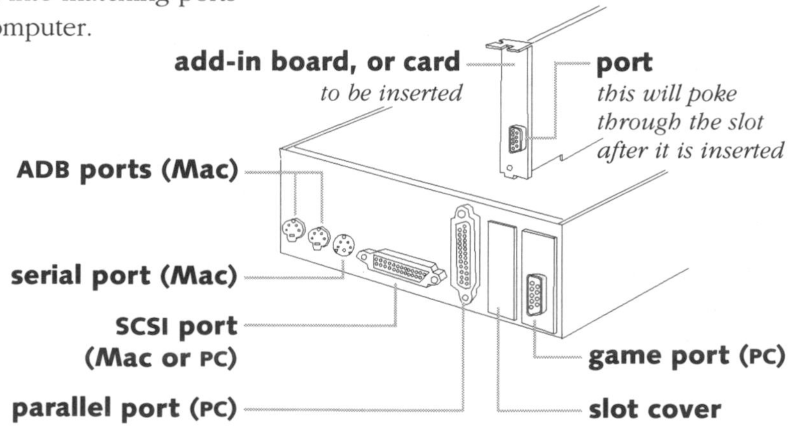

serial port (Mac)

SCSI port (Mac or PC)

parallel port (PC)

game port (PC)

slot cover

PC cards

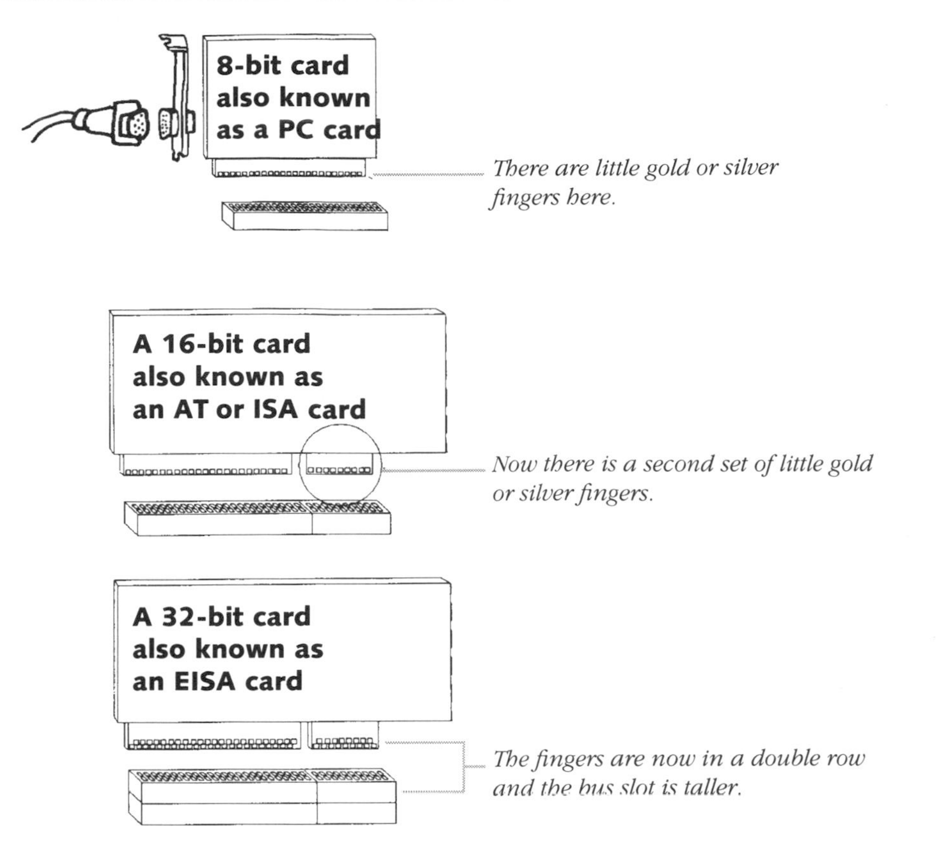

Macintosh NuBus card

A quick overview

Ask your trusted friend

What *operating system(s)* will you use?

Which software programs will you be using?

What are the hard disk requirements for these programs?

What are the memory requirements?

It's wise to buy more storage than you need so you can grow into it, rather than run out because you didn't buy enough.

Learn to recognize these

Model names or numbers

Basic features

Also known as . . .

Units of measure
(use the acronym table)

Orders of magnitude and what they mean *(smaller, larger, etc.)*

Numbers that are often repeated, and the part they go with

Know your minimum requirements

Based on the software you plan to use, figure out the *minimum* requirements you need to have in a computer:

Minimum CPU

Minimum RAM

Total amount of RAM needed

Hard disk size needed *(many applications require a hard disk, but some don't; hard disk space is like money on vacation—you always need twice as much as you thought)*

Be sure to jot down any other special requirements the program calls for, such as the need for a math coprocessor.

How to read any ad

Learn to read one line at a time:

1. Start by finding the unit of measure and its amount.
2. Find the feature, or part that it describes.
3. Translate the cryptic ad line into English.

Repeat this process, line by line, for the rest of the ad.

That's all

That's all there is to it. You have learned to look at computer ads, monitor ads and memory ads—the key to reading other types of ads is just more practice until the language no longer seems foreign.

Good Luck!

CROSS WORD PUZZLE

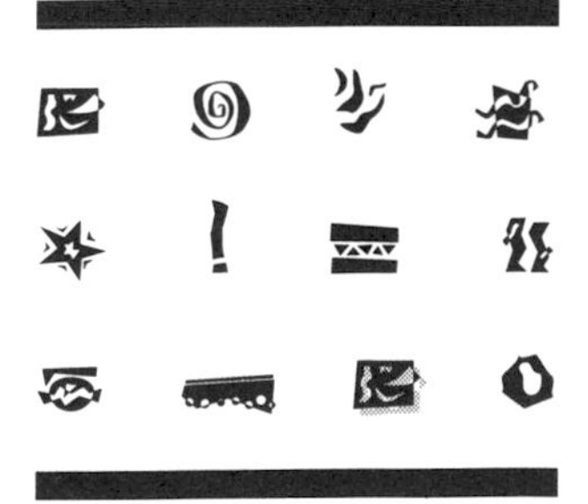

Don't worry—this isn't for techno-nerds! Have fun!

Across

1	first in, first out
4	serial or port
5	you can bump into it
8	found in spreadsheets, tables and jails
11	a means for you to find something
14	allows access to a computer or file
18	press . . .key to continue
19	gofer
20	serious computer enthusiast
22	1024 bytes
25	product sample
27	displayed list of commands
28	put into permanent storage
30	value added reseller
33	information
36	facsimile
37	removable hard disk
38	Rule Number One
39	black printer powder
40	type of macro
41	adjust letter spacing
45	text error
46	Command Z or Control Z
47	combination of sofware, hardware, and electrical wiring, but no wheels
48	connect computers
49	transmit
50	megahertz
51	look for

Down

1	math chip
2	typefaces
3	light amplification from the stimulated emission of radiation
4	not elite
6	resolution
7	volatile memory
9	to keep out
10	newer, more powerful
12	cut, copy, . . .
13	runs your computer
15	special interest group
16	write once, read many
17	component attached to your computer
21	utility from Berkeley Systems
23	portrait or landscape
24	three-letter acronym
26	my marriage
29	America Online
31	one billionth of a second
32	Boston Computer Society
34	make a copy of every important file
35	Apple's first computer
42	binary digit
43	evil program
44	make it bigger

1 2 3
4 5 6 7
8 9 10 11 12 13
14 15 16 17 18
19
20 21
22 23 24
25 26
27 28 29 30
31 32
33 34 35
36 37
38
39
40 41
42 43 44
45 46 47
48
49 50
51

SYMBOLS

A

B

C

D

E

F

G

H

J

K

L

M

N

O

P

S

T

U

V

W

X

Y

Z

How this book was made

Just in case you have any interest in how a book gets created these days, from start to finish, here is a brief rundown. Most authors let other people do the grunt work, but I have this problem with having to be in total control of the entire project, from writing to camera-ready pages. So I grunt myself. Words in italics are defined in the dictionary.

I created this book on my *Macintosh* IIcx (with *dirty ROMs,* eight *megs* of *RAM,* 80-meg *hard disk,* and a 21" *grayscale monitor,* which is a pretty lean machine). I researched through every printed and *electronic* dictionary, resource, glossary, and Macintosh book written. I wrote the text in Microsoft *Word* 5 (which I am begrudgingly starting to like). Steve sent his *PC* contributions, written on his PC, through *CompuServe.* I *downloaded* them and incorporated the text into the Word files. Steve and I still have not met in person.

The Word files, in *hard copy,* went out for *editing,* both grammar edits and tech edits. When those pages started coming in, Shannon and I made the changes in the Word *files.* I designed a *template* in *PageMaker* 4.2 (I do have a developer's version of 5.0, but I made a law that says, "No new technology under deadline"), from which I made one publication for each letter. I *imported* the Word files into the PageMaker publications, where I began to prepare *camera-ready* pages.

I made *screen shots* and illustrations and placed those into position. Steve sent *TIFF* files, screen shots made on his PC, on a PC *disk.* Using the *Apple File Exchange utility,* I opened his PC disks on my Mac, copied the TIFF images over, and imported them straight into PageMaker.

As I prepared final pages, I continued to edit and correct and add, which is why there are sure to be typos here and there. *sigh.*

As the corrected Word files were placed into the PageMaker templates, I gave them to Barbara Sikora and Shannon (my sister), and they indexed each file. As I created final pages, I added to the index, then we edited the index in each separate pub. PageMaker has the greatest indexing feature on the market. After the files had been safely sent to Hassan Herz at TBH TypeCast for *imaging,* I compiled the entire index, creating one huge file from my copies of the 28 separate publications. I edited that monster, poured it onto these pages, and edited some more. And there are probably still typos. *sigh.*

Shannon created the crossword with a neat little program called Crossword Creator from Centron Software Technologies, Inc. (That program is a great way to entertain your kids, also, or have them entertain you.)

The completed pages came back from the *imagesetter* and I sent them off to *Peachpit Press,* who sent it to Hart Graphics, the printer in Austin, Texas, who printed and bound the books. Three weeks later, the books were on my doorstep.

Now, if you are an experienced *desktop publisher,* you know this is a relatively simple, straight-forward, one-color project—just a larger verion of any newsletter. Anybody could do it.

The heads are in Syntax, the body copy is Garamond. The delightful chapter-head font is Scarlett, designed by my daughter, Scarlett. She drew the letterforms on paper and I digitized them in Fontographer, from Altysys, one of the coolest programs around.

MY FONT
I DESIGNED
THIS TYPE
FACE AND
MY MOM
DIGITIZED
IT ON THE
COMPUTER.

A TYPE 1 POSTSCRIPT FONT

You can order this typeface from Scarlett. She charges $5, which includes the disk and mailer. She prefers one dollar bills, but will also accept checks made out to Scarlett Williams.

1275 Fourth Street, Suite 323
Santa Rosa, California 95404

1 2 3 4 5 6 7 8 9 0

!

S C A R L E T T

In case you wanna know

I live in Santa Rosa, California (although I am about to relocate to Santa Fe, New Mexico). I run a monthly Mac user group for new and not-so-new users. I teach here and there and speak to large and small groups as often as possible. I have columns in several magazines on several topics, such as design, typography, and jargon, and write regularly for others. I'm a single mom of three incredible kids. I read a lot, dance a lot, drum a lot, travel a lot.

Steve Cummings is the author or co-author of seven previous books, including such blockbusters as *DeskJet Unlimited* and *Understanding Quicken,* and he has written hundreds of articles for major computer magazines, including PC, PC World, MacWeek, and Macworld.